SOLID STATE
DEVICES AND SYSTEMS
Fourth Edition

atp AMERICAN TECHNICAL PUBLISHERS
ORLAND PARK, ILLINOIS 60467-5756

Gary J. Rockis

Solid State Devices and Systems contains procedures commonly practiced in industry and the trade. Specific procedures vary with each task and must be performed by a qualified person. For maximum safety, always refer to specific manufacturer recommendations, insurance regulations, specific job site and plant procedures, applicable federal, state, and local regulations, and any authority having jurisdiction. The material contained is intended to be an educational resource for the user. American Technical Publishers, Inc. assumes no responsibility or liability in connection with this material or its use by any individual or organization.

American Technical Publishers, Inc., Editorial Staff

Editor in Chief:
    Jonathan F. Gosse
Vice President – Production:
    Peter A. Zurlis
Art Manager:
    James M. Clarke
Multimedia Manager:
    Carl R. Hansen
Technical Editor:
    Scott C. Bloom
Copy Editor:
    Talia J. Lambarki
Cover Design:
    Mark S. Maxwell
Illustration/Layout:
    Nicholas W. Basham
    Melanie G. Doornbos
    Jennifer M. Hines
    Mark S. Maxwell
    Samuel T. Tucker
    Thomas E. Zabinski
CD-ROM Development:
    Robert E. Stickley
    Daniel Kundrat

AEMC is a registered trademark of Chauvin Arnoux, Inc. Chip Quik is a registered trademark of Chip Quik, Inc. IPC is a registered trademark of IPC International, Inc. Mastercool is a registered trademark of Mastercool, Inc. National Electric Code, NEC, and NFPA 70E are registered trademarks of the National Fire Protection Association, Inc. Scopemeter is a registered trademark of John Fluke Mfg. Co., Inc. Underwriters Laboratories Inc. is a registered trademark of Underwriters Laboratories Inc. Velcro is a registed trademark of Velcro Industries.

© 2012 by American Technical Publishers, Inc.
All rights reserved

4 5 6 7 8 9 – 12 – 9 8 7 6 5 4 3 2

Printed in the United States of America

ISBN 978-0-8269-1637-2

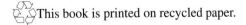This book is printed on recycled paper.

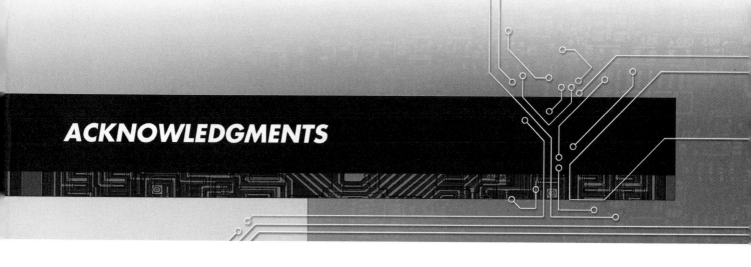

ACKNOWLEDGMENTS

The author and publisher are grateful to the following companies, organizations, and individuals for providing technical information and assistance:

Advanced Assembly Automation, Inc.
AEMC® Instruments
ASCO Valve, Inc.
Baldor Electric Co.
Carrier Corporation
Cooper Industries, Inc.
Cutler-Hammer
David White Instruments
DOE/NREL, NASA/Smithsonian Institution/Lockheed Corp.
DOE/NREL, Robb Williamson
East Penn Manufacturing Co., Inc.
Fluke Corporation
Fluke Networks
GE Lighting
GE Thermometrics
Gordon Brush
GOULD Fiber Optics
Henkel Corporation
H K Wentworth, Ltd.
Honeywell
IBM
Industrial Scientific Corporation
International Rectifier

Kone, Inc.
K-TEK, LLC
Leeson Electric Corporation
Mastercool® Inc.
Micropac Industries, Inc.
Milwaukee Electric Tool Corporation
Mine Safety Appliances Co.
NASA
National Semiconductor
Omni Training
PolyBrite International
Rockwell Automation, Allen-Bradley Company, Inc.
Saftronics, Inc.
SPX Robinair
Square D Company
Stocker Yale, Inc.
Surrette Battery Company
Techspray
TSI Incorporated
U.S. Marine Corps
U.S. Navy
Yellow Jacket Div. Ritchie Engineering Co., Inc.

P. Kevin Gulliver, Jay Buckman, and Dave Carroll
Nida Corporation

CONTENTS

1. Symbols, Circuits, and Safety ... 1
Electrical/Electronic Circuits • Electrostatic Discharge • Circuit Symbols and Terminology • Electrical Safety • NFPA 70E® • Personal Protective Equipment • Lockout/Tagout • Fire Safety

2. Test Instruments ... 23
Types of Test Instruments • Digital Multimeter Operation • Oscilloscope Operation

3. Printed Circuit Board Construction and Troubleshooting ... 51
Printed Circuit Boards • Printed Circuit Board Manufacturing • Mounting Components on a Printed Circuit Board • Printed Circuit Board Identification and Information • Printed Circuit Board Handling • Troubleshooting Printed Circuit Boards • Recycling and Elimination of Hazardous Materials • Conformal Coatings

4. Soldering and Desoldering ... 73
Printed Circuit Board Field Repair • Soldering Aids • The Soldering Process • Soldering and Rework Stations • Technician Effects on Soldering • Soldering Printed Circuit Boards • Desoldering Methods

5. Diode Applications and Troubleshooting ... 107
Diodes • Zener Diodes • Maintaining Semiconductor Devices

6. DC Power Supply Operation and Troubleshooting ... 131
DC Power Supplies • Troubleshooting DC Power Supplies • Power Interruptions

7. Power Sources and Renewable Energy ... 159
Sources of Electrical Energy • Photovoltaic Cells • Wind Turbines • Electrochemical Power Sources • Fuel Cells

8. Transducer Applications and Troubleshooting ... 193
Transducers • Thermistors • Piezoelectric Sensors • Hall Effect Sensors • Ultrasonic Sensors • Radar Systems • Gas Detectors • Radiation Detectors

9 Bipolar Junction Transistors (BJTs) _____217
Bipolar Junction Transistors • Biasing Transistor Junctions • Transistor Operating Characteristic Curves • Transistors as DC Switches • Establishing a Load Line • Biasing Transistors • Power Dissipation • Testing Transistors • Transistor Switching Applications

10 Transistors as Amplifiers _____243
Amplifier Gain • Bandwidth • Decibels • Types of Transistor Amplifiers • Operating Points • Classes of Operation • Input and Output Impedances • Transistor Specification Sheets • Transistor Testers • Transistor Service • Multistage Amplifiers

11 JFETs, MOSFETs, and IGBTs _____265
Field-Effect Transistors (FETs) • Junction Field-Effect Transistors (JFETs) • Metal-Oxide Semiconductor Field-Effect Transistors (MOSFETs) • Insulated Gate Bipolar Transistors (IGBTs) • Multistage Amplifiers

12 Silicon-Controlled Rectifiers (SCRs) _____291
Silicon-Controlled Rectifier Properties • SCR Construction • SCR Troubleshooting • SCR Applications

13 Triacs, Diacs, and Unijunction Transistors _____319
Triacs • Diacs • Unijunction Transistors

14 Operational Amplifiers and 555 Timers _____341
Integrated Circuits • Operational Amplifiers • Integrated Circuit Timers

15 Photonics _____369
Photonics • Light Sources • Light Detection • Fiber Optics

16 Digital Electronics Fundamentals _____411
Digital Electronics • Integrated Circuits • Digital Logic • Digital Pulses • Troubleshooting Digital Electronic Circuits

17 Solid State Relays _____431
Solid State Relay Switching Methods • Solid State Relay Circuits • Solid State Relay Temperature Problems • Solid State Relay Current Problems • Solid State Relay Voltage Problems • Two-Wire Solid State Switches • Three-Wire Solid State Switches • Troubleshooting Solid State Relays • Proximity Sensor Applications • Motor Starter Applications • Soft Start Applications

18 Solid State Technology in Programmable Controllers _____461
Programmable Logic Controllers • PLC Applications • PLC Sections • PLC Interfacing Circuits • Troubleshooting PLC Systems

Appendix _____489

Glossary _____497

Index _____509

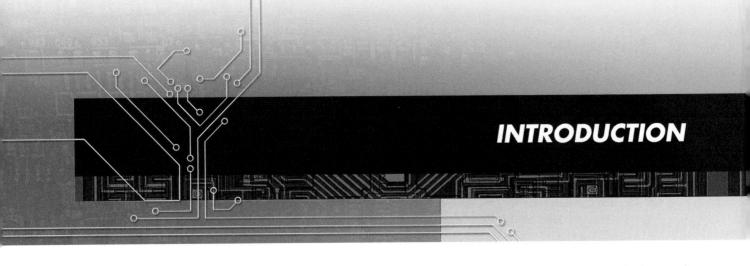

INTRODUCTION

Solid State Devices and Systems, 4th Edition, presents a comprehensive overview of solid state devices and circuitry. This new edition is designed for electricians, students, and technicians who have a basic understanding of electricity. Component and circuit construction, operation, installation, and troubleshooting are emphasized and supported by detailed illustrations. Various practical applications are presented throughout the book as they relate to temperature, light, speed, and pressure control. Electron current flow is used throughout the book. Electron current flow is based on electron flow from negative to positive.

New and expanded topics include test instruments, printed circuit board construction, soldering and desoldering, power sources and renewable energy, photonics, digital electronics, and solid state technology in programmable controllers.

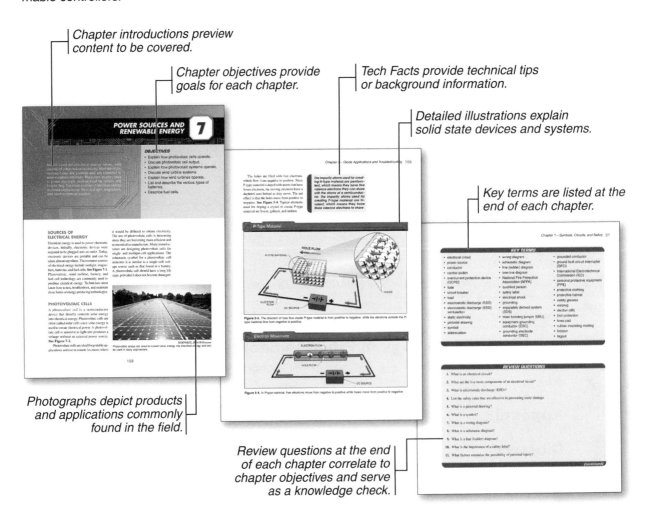

Chapter introductions preview content to be covered.

Chapter objectives provide goals for each chapter.

Tech Facts provide technical tips or background information.

Detailed illustrations explain solid state devices and systems.

Key terms are listed at the end of each chapter.

Photographs depict products and applications commonly found in the field.

Review questions at the end of each chapter correlate to chapter objectives and serve as a knowledge check.

INTERACTIVE CD-ROM FEATURES

Quick Quizzes® reinforce fundamental concepts with 10 interactive questions per chapter.

Illustrated Glossary provides a reference to commonly used terms and includes links to selected illustrations and media clips.

Flash Cards provide a self-study/review tool for matching terms, definitions, and/or images.

Virtual DMM provides an interactive demonstration of digital multimeter functionality.

Interactive Logic Gates provide a learning experience based on common logic gate functions.

Interactive Troubleshooting Simulations provide troubleshooting experiences with electronic components and circuits.

Chapter Reviews provide learners the opportunity to demonstrate comprehension of chapter concepts.

Media Clips consist of animated illustrations or video clips that reinforce and expand upon book content.

ATPeResources.com links to online reference materials that support continued learning.

Information about using the *Solid State Devices and Systems* CD-ROM is included on the last page of this book. To obtain information about related training products, visit the American Technical Publishers website at www.go2atp.com.

The Publisher

SYMBOLS, CIRCUITS, AND SAFETY   1

From computers and video game systems to alarm clocks and stereos, electronics can be found in nearly everything we use. Electronics are composed of individual electronic components, such as resistors, capacitors, inductors, diodes, transistors, and integrated circuits that are connected together to control electron flow. An understanding of electrical safety is important when working with electrical/electronic circuits due to the dangers associated with electrical shock.

OBJECTIVES

- List and describe electrical circuit components.
- Identify types of drawings and diagrams used for control circuits.
- Explain electrical safety.
- Identify personal protective equipment.
- Explain lockout/tagout procedures.
- Explain fire safety.

ELECTRICAL/ELECTRONIC CIRCUITS

An *electrical circuit* is an assembly of conductors and electrical and/or electronic devices through which electrons flow. Electron flow is typically referred to as current flow. When a circuit is complete (closed circuit), current makes a complete trip through the circuit. If the circuit is not complete (open circuit), current does not flow. A broken wire, a loose connection, or a switch in the OFF position stops current from flowing in an electrical circuit.

Note: Electron current flow is based on current flow from negative to positive. Conventional current flow is based on current flow from positive to negative. Electron current flow is used throughout the text.

Electrical Circuit Components

Most electrical circuits include five basic components. The five basic components include a power source, conductors, a control switch, an overcurrent protection device (OCPD), and a load. A power source provides electrical power to a circuit. Conductors connect the individual components. A control switch controls the flow of electrons. An OCPD, such as a fuse or circuit breaker, ensures that a circuit operates safely and within its electrical limits. A load converts electrical energy into another form of usable energy such as motion, light, heat, or sound.

Electrical circuit components may be shown using line diagrams, pictorial drawings, schematic diagrams, and/or wiring diagrams. For example, an automobile interior lighting circuit includes the five components of a typical electrical circuit. **See Figure 1-1.** The power source is a battery, the conductors may be chassis wires or the car frame, the control switch is a plunger-type door switch, the overcurrent protection device is a fuse, and the load is an interior light.

Current Media Clip

> An electrical circuit of a complex machine, such as a computer or automobile, may contain multiple power sources, conductors, control switches, OCPDs, and loads. Therefore, multiple drawings and diagrams may be needed for an individual electrical circuit.

1

Electrical Circuit Components

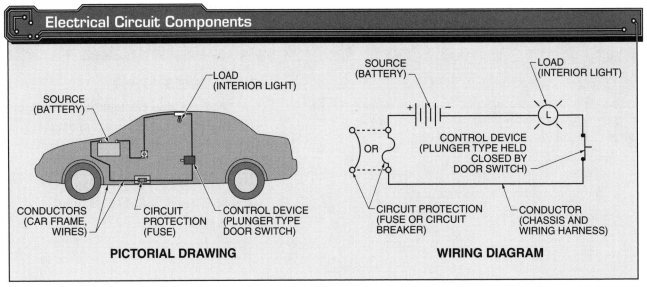

Figure 1-1. Most electrical circuits include a source, conductors, a control switch, an overcurrent protection device (OCPD), and a load.

A *power source* is a device that converts various forms of energy into electricity. The power source in an electrical circuit is normally the point at which to start when reading or troubleshooting a circuit. The components in electrical circuits are connected using conductors. A *conductor* is a material that has very little resistance to current and permits electrons to move through it easily. Copper is the most commonly used conductor material. A *control switch* is a switch that controls the flow of current in a circuit. Switches can be activated manually, mechanically, or automatically.

An *overcurrent protection device (OCPD)* is a disconnect switch with circuit breakers or fuses added to provide overcurrent protection for the switched circuit. A *fuse* is an overcurrent protection device with a fusible link that melts and opens the circuit when an overload condition or short circuit occurs. A *circuit breaker* is an overcurrent protection device with a mechanical or electromechanical mechanism that manually or automatically opens a circuit when an overload condition or short circuit occurs. A *load* is any device that converts electrical energy to motion, heat, light, or sound. Common loads include lights, heating elements, speakers, and motors.

ELECTROSTATIC DISCHARGE

Electrostatic discharge (ESD) is the movement of electrons from a source to an object across a gap. *Static electricity* is an electrical charge at rest. The most common way to build up static electricity is by friction. Friction causes electrons from one source to flow to another material, causing an electron buildup (negative static charge). When a person (negatively charged) contacts a positively charged or a grounded object, all the excess electrons flow (jump) to that object. **See Figure 1-2.**

Electrostatic discharge can be 35,000 V or more. People typically do not feel electrostatic discharges until the discharge reaches 3000 V. Solid state devices and circuits may be damaged or destroyed by a 10 V electrostatic discharge. Electricians and maintenance personnel should wear a wrist grounding strap or other type of grounding device to avoid damage to solid state devices and circuits from ESD.

Safety rules that are effective in preventing static damage include the following:

- Always keep work areas clean and clear of unnecessary materials, especially common plastics.
- Never handle electronic devices by their leads.

- Test grounding devices daily to make sure that they have not become loose or intermittent.
- Never work on ESD-sensitive objects without the proper grounding device.
- Always handle printed circuit (PC) boards by their outside corners.
- Always transport PC boards in antistatic trays or bags.
- Always keep single electronic devices sealed in conductive static shielding when transporting them.

Electrostatic Discharge Workstations

Electrostatic discharge (ESD) workstations should be designed to meet all current standards set in ANSI/ESD S20.20. An ESD workstation should provide a functional, efficient, and safe working environment. **See Figure 1-3.** The workstation should be ergonomically designed for comfort.

To achieve maximum ESD protection, all components that make up the workstation should be grounded. In addition to grounding, protection can be provided by painting the frame with a coat of semielectrically conductive paint.

> *Symbolic representations of electrical circuit components are shown in standards published by both national and international organizations. These representations are provided in ANSI Y32.2, IEEE 315, and IEC 60617.*

The schematic of an ESD workstation shows how the workstation is grounded. **See Figure 1-4.** Circuit cards with ESD-sensitive devices are generally considered safe when placed on an ESD workstation. They may however become susceptible to damage if a module is removed and their connector pins are touched. There must not be any energized equipment placed on the conductive ESD work surface. An ESD workstation is an area for de-energized or "electrically dead" equipment only.

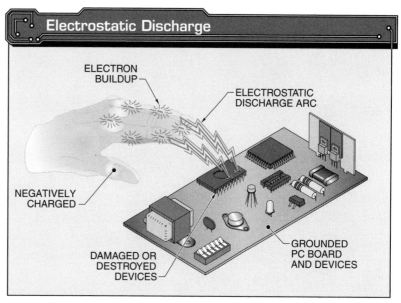

Figure 1-2. *An excess of electrons produces an arc during electrostatic discharge (ESD).*

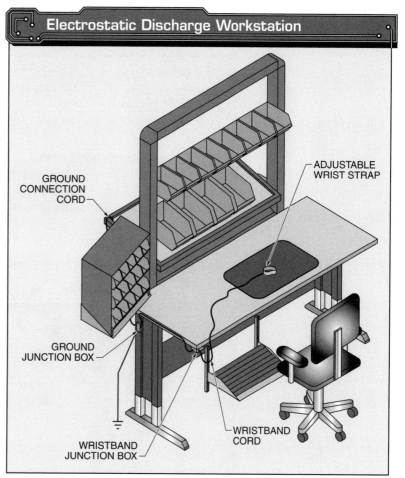

Figure 1-3. *An electrostatic discharge (ESD) workstation should provide a functional, efficient, and safe working environment.*

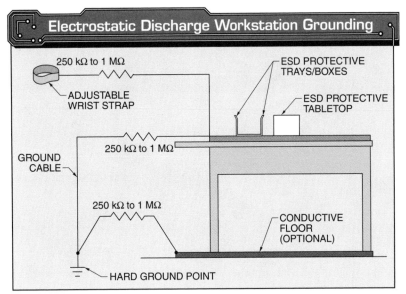

Figure 1-4. *The schematic of an ESD workstation shows how the workstation is grounded.*

If a secure ESD workstation is not available, a field service mat may be used. Once unfolded, a field service mat is assembled and grounded at a common 120 VAC wall outlet. **See Figure 1-5.** If no earth ground is available, an alligator clip is attached to the module or chassis of the equipment being serviced to ensure protection.

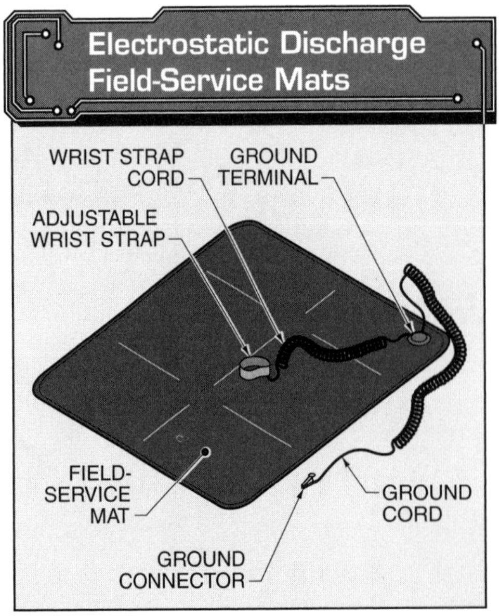

Accessories can be added to an ESD workstation, such as an adjustable chair with an antistatic seat cover or adjustable foot straps. Every source of static electricity should be considered when it comes to ESD protection. **See Figure 1-6.**

Prevention of Electrostatic Discharge Damage. Certain devices are more susceptible to ESD damage than others. A PC board should be checked to see if it has an ESD label. A warning symbol can be used to identify ESD-sensitive devices. **See Figure 1-7.**

When removing or replacing an ESD device or assembly in equipment, the device or assembly should be held with an electrostatic-free wrap, if possible. Otherwise, the device or assembly should be picked up by its body only. Component leads, connector pins, paths, PC boards, and any other electrical connections should not be touched, even though they are covered by a conformal coating.

> *Standard practices for drawing electrical and electronic diagrams are detailed in Electrical and Electronic Diagrams, ANSI Y14.15.*

CIRCUIT SYMBOLS AND TERMINOLOGY

All careers have a certain terminology that must be understood in order to transfer information efficiently. This terminology may include symbols, drawings or diagrams, words, phrases, or abbreviations. Work in the electrical/electronic industry requires an understanding of its terminology, the function of electrical/electronic components, and the relationship between each component in a circuit. With this understanding, a technician is able to read drawings and diagrams, understand circuit operation, and troubleshoot problems. Drawings and diagrams used to convey electrical information include pictorial drawings, electrical symbols and abbreviations, wiring diagrams, schematic diagrams, and line diagrams.

Figure 1-5. *An ESD field-service mat can be used to work on equipment if an ESD workstation is not available or the equipment cannot be moved.*

Chapter 1—Symbols, Circuits, and Safety 5

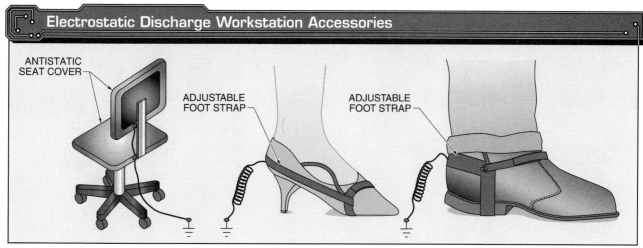

Figure 1-6. *Accessories can be added to an ESD workstation, such as a chair with an antistatic seat cover or adjustable foot straps.*

Figure 1-7. *All ESD-sensitive components should be labeled so that they are handled properly.*

Pictorial Drawings

A *pictorial drawing* is a drawing that shows the length, height, and depth of an object in one view. Pictorial drawings show physical details of an object as seen by the eye. **See Figure 1-8.**

Electrical Symbols and Abbreviations

A *symbol* is a graphic element that represents a quantity or unit. Symbols are used to represent electrical components on electrical and electronic diagrams. An *abbreviation* is a letter or combination of letters that represents a word that is used for a device, connection, or process. **See Figure 1-9.**

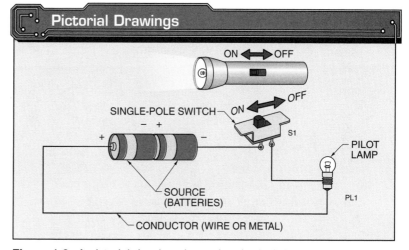

Figure 1-8. *A pictorial drawing shows the physical details of components as seen by the eye.*

6 SOLID STATE DEVICES AND SYSTEMS

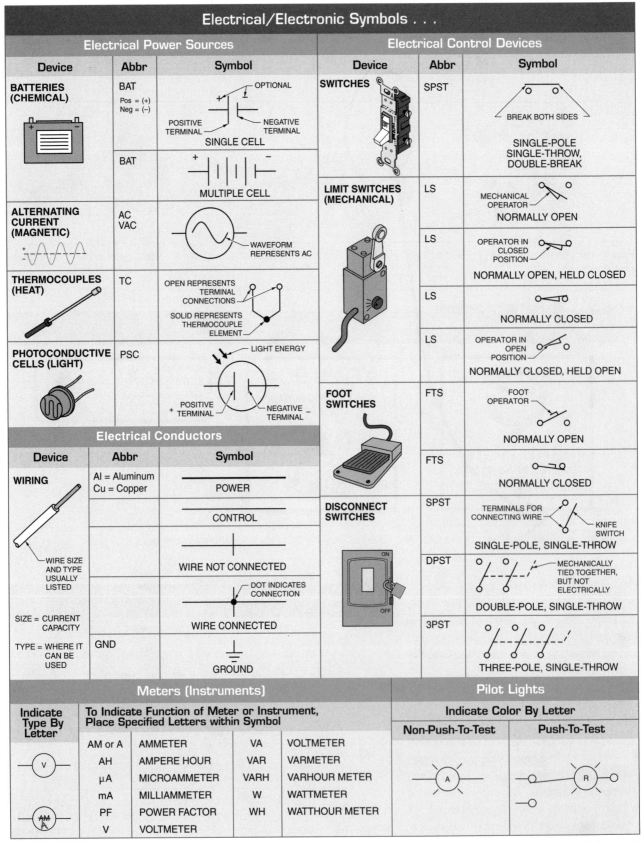

Figure 1-9. *Symbols are used to conveniently represent electrical components in diagrams of most electrical and electronic circuits.*

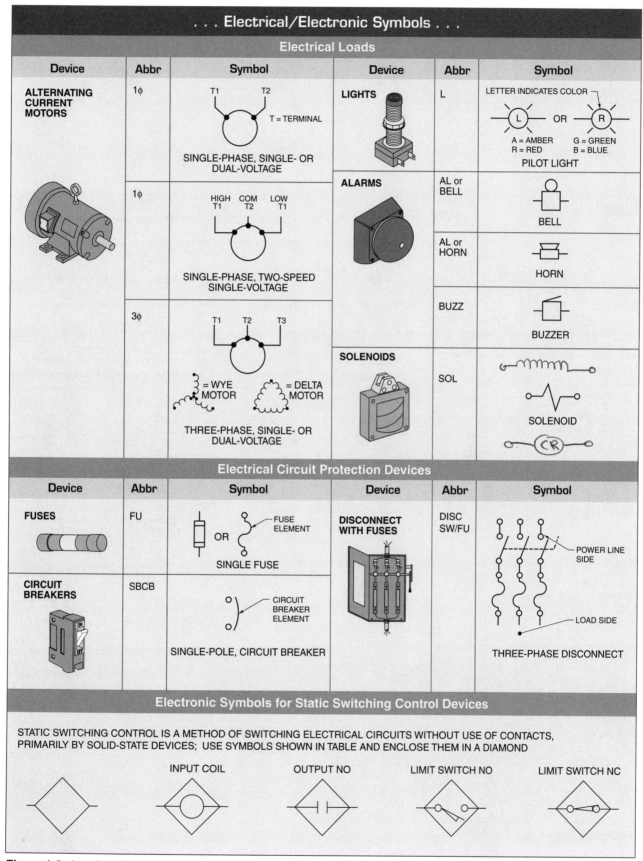

Figure 1-9. *(continued)*

8 SOLID STATE DEVICES AND SYSTEMS

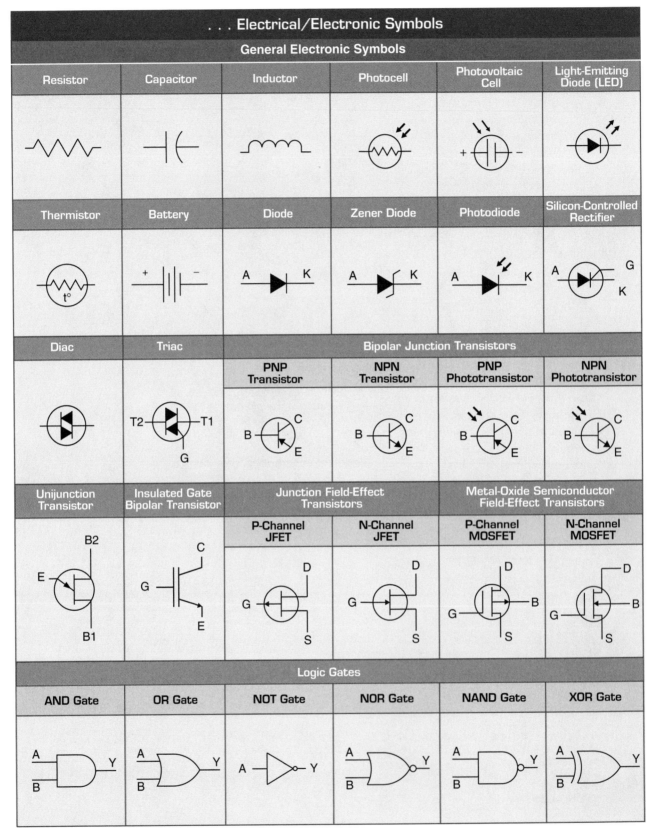

Figure 1-9. *(continued)*

Wiring Diagrams

A *wiring diagram* is a diagram that shows the electrical connections of all components in a piece of equipment. Wiring diagrams show, as closely as possible, the actual location of each component in a circuit. Wiring diagrams often include details of the type of wire and the kind of hardware by which wires are fastened to terminals. **See Figure 1-10.**

A wiring diagram is similar to a pictorial drawing except that the components are shown as rectangles or circles. The location or layout of the parts is accurate for the particular equipment. All connecting wires are shown connected from one component to another. Wiring diagrams are used widely by electricians when installing electronic equipment and by technicians when maintaining such equipment.

Schematic Diagrams

A *schematic diagram* is a diagram that shows the electrical connections and functions of a specific circuit arrangement with graphic symbols. Schematic diagrams do not show the physical relationship of the components in a circuit. The term schematic diagram is normally associated with electronic circuits.

Schematic diagrams are intended to show the circuitry that is necessary for the basic operation of a device. Schematic diagrams are not intended to show the physical size or appearance of a device. In troubleshooting, schematic diagrams are essential because they enable an individual to trace a circuit and its functions without regard to the actual size, shape, or location of the component, device, or part. **See Figure 1-11.**

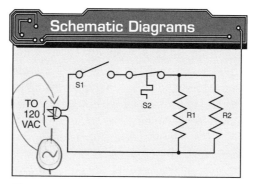

Figure 1-11. *In a schematic diagram, the actual position of components is not shown. Rather, components are laid out, so the circuitry is easily read.*

Electrical schematics were originally drafted by hand using standardized templates or preprinted adhesive symbols. Today, electrical computer-aided drafting (E-CAD) software is often used.

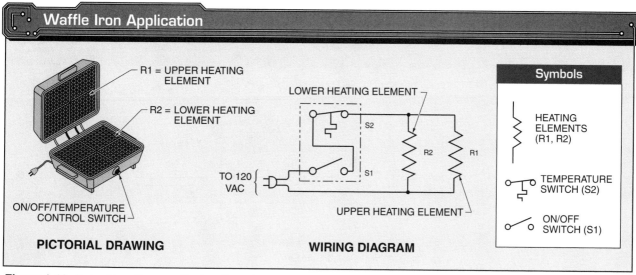

Figure 1-10. *In a wiring diagram, the location of components is generally shown as close to the actual circuit configuration as possible.*

Line (Ladder) Diagrams

A *line (ladder) diagram* is a diagram that shows the logic of an electrical circuit or system using standard industry symbols. A line diagram is used to show the relationship between circuits and their components but not the actual location of the components. Line diagrams provide a fast, easy understanding of the connections and use of components. The term line diagram is typically associated with industrial electrical/electronic circuits. Line diagrams are also referred to as ladder diagrams. **See Figure 1-12.**

The arrangement of a line diagram should promote clarity. Graphic symbols, abbreviations, and device designations are drawn per standards. The circuit should be shown in the most direct path and logical sequence. Lines between symbols can be horizontal or vertical but should be drawn to minimize line crossing.

Line diagrams are often incorrectly referred to as one-line diagrams. A *one-line diagram* is a diagram that uses single lines and graphic symbols to indicate the path and components of an electrical circuit. One-line diagrams have only one line between individual components. **See Figure 1-13.** A line diagram, on the other hand, often shows multiple lines leading to or from a component (parallel connections).

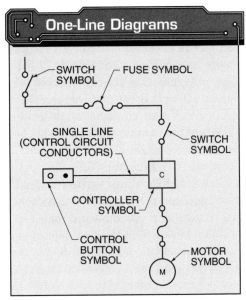

Figure 1-13. *A one-line diagram is a diagram that uses single lines and graphic symbols to indicate the path and components of an electrical circuit.*

Care must be taken when using electrical symbols to design or communicate electrical circuit operation. Electrical circuit operation may be changed by using incorrect electrical symbols. This may create a hazardous situation. One problem that occurs is that limit switch operation is commonly misinterpreted when using electrical circuit diagrams.

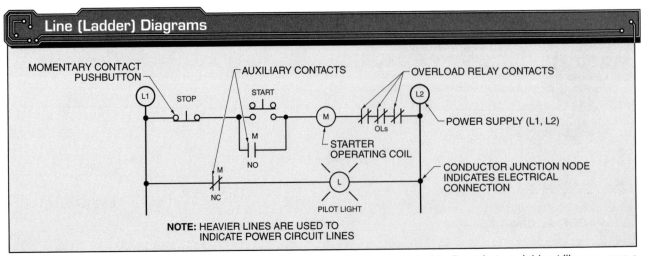

Figure 1-12. *A line (ladder) diagram consists of a series of symbols interconnected by lines that are laid out like rungs on a ladder to indicate the flow of electrons through the various components of a circuit.*

For example, a circuit contains four limit switches that are used to control four lamps. **See Figure 1-14.** Lamp 1 (L1) is controlled by limit switch 1 (LS1). Limit switch 1 includes a normally open (NO) contact. Lamp 1 (L1) is not energized (not turned ON) until an object activates the limit switch operator and closes the NO limit switch contacts.

Lamp 2 (L2) is controlled by limit switch 2 (LS2). Limit switch 2 (LS2) also includes a NO contact. However, the NO contacts are shown in their held closed position. This is often done when a switch would normally be found in the held position, such as a limit switch that detects when a door is closed. Anytime the door is closed, the limit switch NO contacts are held closed and L2 is energized. For any limit switch symbol, the limit switch contacts are always NO when the moving part of the symbol is drawn below the terminal connections.

Lamp 3 (L3) is controlled by limit switch 3 (LS3). Limit switch 3 (LS3) includes a normally closed (NC) contact. Lamp 3 (L3) is energized (turned ON) before an object activates the limit switch operator.

Lamp 4 (L4) is controlled by limit switch 4 (LS4). Limit switch 4 (LS4) also includes a NC contact. However, the NC contacts are shown in their held open position. This is often done when a switch would normally be found in the held position. For any limit switch symbol, the limit switch contacts are always NC when the moving part of the symbol is drawn above the terminal connections. Line diagrams are designed to show circuit operation and include switches in their normal position and their held (actuated) position.

ELECTRICAL SAFETY

A technician must be knowledgeable of the dangers associated with electrical power and the potential hazards on the job. Safe work habits, proper personal protective equipment (PPE), and proper procedures minimize the possibility of personal injury.

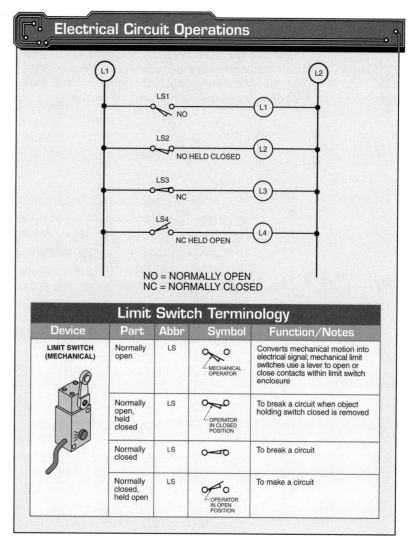

Figure 1-14. *Care should be taken when using electrical symbols to design or communicate electrical circuit operations because these operations may be changed.*

Technicians must work safely at all times. Electrical and/or electronic safety rules must be followed when working with equipment to help prevent injuries from electrical energy sources. Electrical safety has been advanced by the efforts of the National Fire Protection Association (NFPA), Occupational Safety and Health Administration (OSHA), and state safety laws. The *National Fire Protection Association (NFPA)* is a national organization that provides guidance in assessing the hazards of the products of combustion. The NFPA sponsors the development of the National Electrical Code® (NEC®).

> *The National Electrical Code® is adopted for use by local (village, town, or city), county or parish, and state authorities. Typically, these authorities are represented by building officials who issue permits for jobs and make periodic inspections of work in progress. Additionally, they certify that completed buildings meet all applicable codes before occupancy occurs.*

National Electrical Code®

The National Electrical Code® is one of the most widely used and recognized consensus standards in the world. The purpose of the NEC® is to protect people and property from hazards that arise from the use of electricity. Improper procedures when working with electricity can cause permanent injury or death. Many city, county, state, and federal agencies use the NEC® to set requirements for electrical installations. The NEC® is updated every three years. Electrical safety rules include the following:

- Always comply with the NEC® and state and local codes.
- Use equipment and components approved by Underwriters Laboratories, Inc. (UL).
- Before removing any fuse from a circuit, be sure the switch for the circuit is open or disconnected. When removing fuses, use an approved fuse puller and break contact on the line side of the circuit first. When installing fuses, install the fuse first into the load side of the fuse clip, then into the line side.
- Inspect and test grounding systems for proper operation. Ground any conductive component or element that is not energized.
- Turn off, lock out, and tag out any circuit that is not required to be energized when maintenance is being performed.
- Always use PPE and safety equipment.
- Perform the appropriate task required during an emergency situation.
- Use only a Class C rated fire extinguisher on electrical equipment. A Class C fire extinguisher is identified by the color blue inside a circle.
- Always work with another individual when working in a dangerous area, on dangerous equipment, or with high voltages.
- Do not work when tired or taking medication that causes drowsiness unless specifically authorized by a physician.
- Do not work in poorly lighted areas.
- Ensure there are no atmospheric hazards such as flammable dust or vapor in the area.
- Use one hand when working on a live circuit to reduce the chance of an electrical shock passing through the heart and lungs.
- Never bypass fuses, circuit breakers, or any other safety device.

Qualified Persons

To prevent an accident, electrical shock, or damage to equipment, all electrical work must be performed by qualified persons. A *qualified person* is a person who has special knowledge of the construction and operation of electrical equipment or a specific task and is trained to recognize and avoid electrical hazards that might be present. NFPA 70E® Part II, *Safety-Related Work Practices,* Chapter 1, *General,* Section 1-5.4.1, *Qualified Persons,* provides additional information regarding the definition of a qualified person. A qualified person does the following:

- determines the voltage of energized electrical parts
- determines the degree and extent of hazards for use of the PPE
- plans jobs to perform work safely on electrical equipment by following all NFPA, OSHA, manufacturer, state, and company safety procedures and practices
- performs the appropriate task required during an accident or emergency situation
- understands electrical principles and follows all manufacturer procedures and approach distances specified by the NFPA
- understands the operation of test equipment and follows all manufacturer procedures

- informs other technicians and operators of tasks being performed and maintains all required records

Safety Labels

A *safety label* is a label that indicates areas or tasks that can pose a hazard to personnel and/or equipment. Safety labels appear in several ways on equipment and in equipment manuals. Safety labels use signal words to communicate the severity of a potential problem. The three most common signal words are danger, warning, and caution. **See Figure 1-15.**

Electrical Shock

An *electrical shock* is a shock that results any time a body becomes part of an electrical circuit. Electrical shock effects vary from a mild sensation, to paralysis, to death. Also, severe burns may occur where electricity enters and exits the body. The severity of an electrical shock depends on the amount of current in milliamperes (mA), the length of time the body is exposed to the current, the path electricity takes through the body, and the physical size and condition of the body through which the current passes. **See Figure 1-16.**

Safety Label	Box Color	Symbol	Significance
DANGER — HAZARDOUS VOLTAGE • Ground equipment using screw provided. • Do not use metallic conduits as a ground conductor.	red	 ⚠	**DANGER** – Indicates an imminently hazardous situation which, if not avoided, will result in death or serious injury
WARNING — MEASUREMENT HAZARD When taking measurements inside the electric panel, make sure that only the test lead tips touch internal metal parts.	orange	 ⚠	**WARNING** – Indicates a potentially hazardous situation which, if not avoided, could result in death or serious injury
CAUTION — MOTOR OVERHEATING Use of a thermal sensor in the motor may be required for protection at all speeds and loading conditions. Consult motor manufacturer for thermal capability of motor when operated over desired speed range.	yellow	 ⚠	**CAUTION** – Indicates a potentially hazardous situation which, if not avoided, may result in minor or moderate injury, or damage to equipment; May also be used to alert against unsafe work practices
 **WARNING** Disconnect electrical supply before working on this equipment.	orange	 ⚡	**ELECTRICAL WARNING** – Indicates high-voltage location and conditions that could result in death or serious injury from an electrical shock
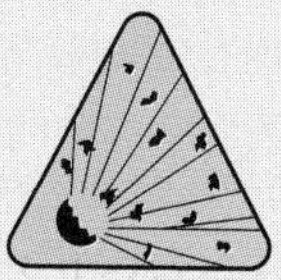 **WARNING** Do not operate the meter around explosive gas, vapor, or dust.	orange	 💥	**EXPLOSION WARNING** – Indicates location and conditions where exploding electrical parts may cause death or serious injury

Figure 1-15. *Safety labels are used to indicate a situation with different degrees of likelihood of death or injury to personnel.*

14 SOLID STATE DEVICES AND SYSTEMS

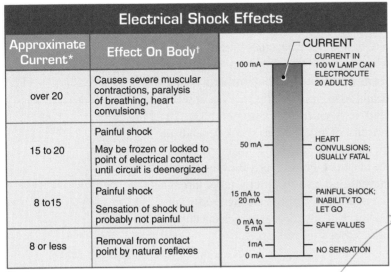

Figure 1-16. *Electrical shock is a condition that results any time a body becomes part of an electrical circuit.*

[Create a safety path]

Electrical shock can be prevented. Anyone working on electrical equipment should have respect for all voltages, have knowledge of the principles of electricity, and follow safe work procedures. All technicians should be encouraged to take a basic course in cardiopulmonary resuscitation (CPR) so that they can aid a coworker in the event of an emergency.

During an electrical shock, the body of a person becomes part of an electrical circuit. The resistance the body of a person offers to the flow of electricity varies. Sweaty hands have less resistance than dry hands. A wet floor has less resistance than a dry floor. The lower the resistance, the higher the current. As current increases, the severity of the electrical shock increases.

If a person is receiving an electrical shock, power should be removed as quickly as possible. If power cannot be removed quickly, the victim must be removed from contact with live parts. Action must be taken quickly and cautiously. Delay may be fatal. Individuals must keep themselves from also becoming a casualty while attempting to rescue another person. If the equipment circuit disconnect switch is nearby and can be operated safely, the power should be shut off. Excessive time should not be spent searching for the circuit disconnect. In order to remove the energized part, insulated protective equipment, such as a hot stick, rubber gloves, blankets, wood poles, and plastic pipes, can be used if accessible.

After the victim is free from the electrical hazard, help is called and first aid begun as needed. The injured individual should not be transported unless there is no other option and the injuries require immediate professional attention.

Grounding. *Grounding* is the connection of all exposed non-current-carrying metal parts to the earth. Grounding provides a direct path to the earth for unwanted fault current without causing harm to persons or equipment. Electrical circuits are grounded to safeguard equipment and personnel against the hazards of electrical shock. Proper grounding of electrical tools, motors, equipment, enclosures, and other control circuitry helps prevent hazardous conditions. However, improper electrical wiring or misuse of electricity causes destruction of equipment and fire damage to property as well as personal injury.

Grounding is accomplished by connecting the circuit to a metal underground water pipe, the metal frame of a building, a concrete-encased electrode, or a ground ring in accordance with the NEC®. To prevent problems, a grounding path must be as short as possible and of sufficient size as recommended by the manufacturer (minimum 14 AWG copper), never be fused or switched, be a permanent part of the electrical circuit, and be continuous and uninterrupted from the electrical circuit to the ground.

A ground is provided at the main service equipment or at the source of a separately derived system (SDS). A *separately derived system (SDS)* is a system that supplies electrical power derived (taken) from transformers, storage batteries, photovoltaic (solar) systems, and generators. The majority of separately derived systems are produced by the secondary of the distribution transformer.

The neutral ground connection must be made at the transformer or at the main service panel only. The neutral ground connection is made by connecting the neutral bus to the ground bus with a main bonding jumper. A *main bonding jumper (MBJ)* is a connection at the service equipment that connects the equipment grounding conductor, grounding electrode conductor, and grounded conductor (neutral conductor). **See Figure 1-17.**

An *equipment grounding conductor (EGC)* is an electrical conductor that provides a low-impedance ground path between electrical equipment and enclosures within the distribution system. A *grounding electrode conductor (GEC)* is a conductor that connects grounded parts of a power distribution system (equipment grounding conductors, grounded conductors, and all metal parts) to the NEC®-approved earth grounding system. A *grounded conductor* is a conductor that has been intentionally grounded.

Ground Fault Circuit Interrupters. A *ground fault circuit interrupter (GFCI)* is a device that protects against electrical shock by detecting an imbalance of current in the normal conductor pathways and opening the circuit. When current in the two conductors of an electrical circuit varies by more than 5 mA, a GFCI opens the circuit. A GFCI is rated to trip quickly enough (1/40 of a second) to prevent electrocution. **See Figure 1-18.**

A potentially dangerous ground fault is any amount of current that may deliver a dangerous shock. Any current over 8 mA is considered potentially dangerous depending on the path the current takes, the physical condition of the person receiving the shock, and the amount of time the person is exposed to the shock. Therefore, GFCIs are required in such places as dwellings, hotels, motels, construction sites, marinas, receptacles near swimming pools and hot tubs, underwater lighting, fountains, and other areas in which a person may experience a ground fault.

A GFCI compares the amount of current in the ungrounded (hot) conductor with the amount of current in the neutral conductor. If the current in the neutral conductor becomes less than the current in the hot conductor, a ground fault condition exists. The amount of current that is missing is returned to the source by some path other than the intended path (fault current).

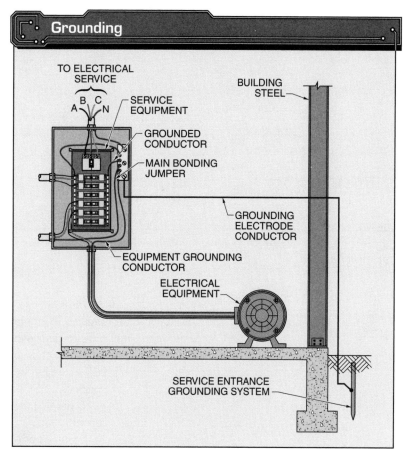

Figure 1-17. *Grounding provides a direct path for unwanted (fault) current to the earth without causing harm to individuals or equipment.*

GFCI protection may be installed at different locations within a circuit. Direct-wired GFCI receptacles provide ground fault protection at the point of installation. GFCI receptacles may also be connected to provide protection at all other receptacles installed downstream on the same circuit. GFCI circuit breakers, when installed in a load center or panelboard, provide GFCI protection and conventional circuit overcurrent protection for all branch-circuit components connected to the circuit breaker.

Plug-in GFCIs provide ground fault protection for devices plugged into them. These plug-in devices are often used by personnel working with power tools in an area that does not include GFCI receptacles.

16 SOLID STATE DEVICES AND SYSTEMS

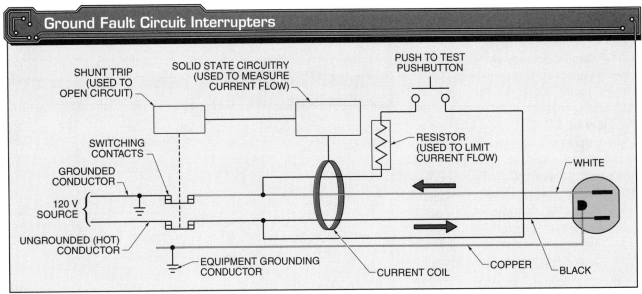

Figure 1-18. *A GFCI compares the amount of current in the ungrounded (hot) conductor with the amount of current in the neutral conductor.*

Portable GFCIs can be moved easily from one location to another. Portable GFCIs commonly contain more than one receptacle outlet protected by an electronic circuit module. Portable GFCIs should be inspected and tested before each use. GFCIs have a built-in test circuit to ensure that the ground fault protection is operational.

A GFCI protects against the most common form of electrical shock hazard, the ground fault. A GFCI does not protect against line-to-line contact hazards, such as a technician holding two hot wires or a hot and a neutral wire in each hand. GFCI protection is required in addition to NFPA grounding requirements.

International Electrotechnical Commission (IEC) 1010 Safety Standard

The *International Electrotechnical Commission (IEC)* is an organization that develops international safety standards for electrical equipment. IEC standards reduce safety hazards that can occur from unpredictable circumstances when using electrical test equipment such as DMMs. For example, voltage surges on a power distribution system can cause a safety hazard.

NFPA 70E®

The National Fire Protection Association standards in NFPA 70E®, *Standard for Electrical Safety in the Workplace,* addresses "electrical safety requirements for employee workplaces that are necessary for the safeguarding of employees in their pursuit of gainful employment." Per NFPA 70E, "Only qualified persons shall perform testing work on or near live parts operating at 50 V or more." NFPA 70E was written at the request of OSHA and has become the standard for electrical safety in the electrical industry. Its methods for protection are more detailed than OSHA requirements.

NFPA 70E is the basis for the OSHA Subpart S, *Electrical Regulations*. For technicians, NFPA 70E addresses requirements such as PPE and safe approach distance requirements that could be encountered in jobs such as installing temporary power.

PERSONAL PROTECTIVE EQUIPMENT

Personal protective equipment (PPE) is clothing and/or equipment worn by a

Personal Protective Equipment
Media Clip

technician to reduce the possibility of injury in the work area. The use of PPE is required whenever work may occur on or near energized, exposed electrical circuits. For maximum safety, PPE must be used as specified per OSHA 29 Code of Federal Regulations (CFR) Part 1910, Subpart 1 – *Personal Protective Equipment* (1910.132 through 1910.138) and other applicable safety mandates.

All PPE and tools are selected for at least the operating voltage of the equipment or circuits to be worked on or near. Equipment, device, tool, or test equipment must be suited for the work to be performed. PPE includes protective clothing, head protection, eye protection, ear protection, hand protection, foot protection, back protection, knee protection, and rubber insulating matting. **See Figure 1-19.**

Protective Clothing

Protective clothing is clothing that provides protection from contact with sharp objects, hot equipment, and harmful materials. Protective clothing made of durable material, such as denim, should be snug yet allow for ample movement. Clothing should fit snugly to avoid danger of becoming entangled in moving machinery. Pockets should allow convenient access but should not snag on tools or equipment. Soiled protective clothing should be washed to reduce the flammability hazard.

Head Protection

Head protection requires the use of a protective helmet. A *protective helmet* is a hard hat that is used in the workplace to prevent injury from the impact of falling and flying objects and from electrical shock. Protective helmets resist penetration and absorb impact force. Protective helmet shells are made of durable, lightweight materials. A shock-absorbing lining keeps the shell away from the head to provide ventilation and protection against impact. Protective helmets are identified by class of protection against specific hazardous conditions.

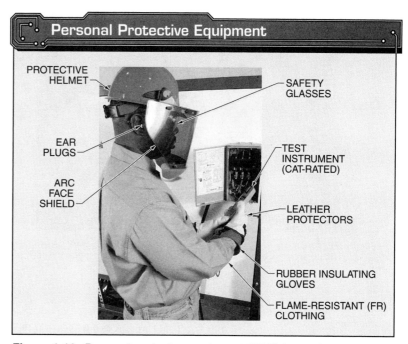

Figure 1-19. *Personal protective equipment (PPE) is used to reduce the possibility of an injury.*

Eye Protection

Eye protection must be worn to prevent eye or face injuries caused by flying particles, contact arcing, and radiant energy. Eye protection must comply with OSHA 29 CFR 1910.133, *Eye and Face Protection*. Eye protection standards are specified in ANSI Z87.1, *Occupational and Educational Eye and Face Protection*. Eye protection includes safety glasses, face shields, goggles, and arc blast hoods. **See Figure 1-20.**

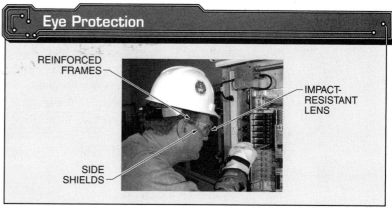

Figure 1-20. *Eye protection must be worn to prevent eye injuries caused by flying particles.*

18 SOLID STATE DEVICES AND SYSTEMS

Fluke Corporation
Proper test and personal safety equipment must be used when testing any electrical circuit.

Safety glasses are an eye protection device with special impact-resistant glass or plastic lenses, reinforced frames, and side shields. Plastic frames are designed to keep the lenses secured in the frame if an impact occurs and minimize the shock hazard when working with electrical equipment. Side shields provide additional protection from flying objects. Tinted-lens safety glasses protect against low-voltage arc hazards.

Ear Protection

Ear protection is any device worn to limit the noise entering the ear and includes earplugs and earmuffs. An *earplug* is an ear protection device made of moldable rubber, foam, or plastic and is inserted into the ear canal.

Power tools and equipment can produce excessive noise levels. Technicians subjected to excessive noise levels may develop hearing loss over a period of time. The severity of hearing loss depends on the intensity and duration of exposure. Noise intensity is expressed in decibels. A *decibel (dB)* is a unit of measure used to express the relative intensity of sound.

Hand Protection

Hand protection includes gloves worn to prevent injuries to hands caused by cuts or electrical shock. The appropriate hand protection required is determined by the duration, frequency, and degree of the hazard to the hands.

Foot Protection

Foot protection is protection consisting of safety shoes worn to prevent foot injuries that are typically caused by objects falling less than 4′ and having an average weight of less than 65 lb. Safety shoes with reinforced steel toes protect against injuries caused by compression and impact. Insulated rubber-soled shoes are commonly worn during electrical work to prevent electrical shock. Foot protection must comply with ANSI Z41, *Personal Protection—Protective Footwear*.

Back Protection

A back injury is one of the most common injuries resulting in lost time in the workplace. Back injuries are the result of improper lifting procedures. Back injuries are prevented through proper planning and work procedures. Assistance should be sought when moving heavy objects. When lifting objects from the ground, ensure the path is clear of obstacles and free of hazards. When lifting objects, the knees are bent and the object is grasped firmly. The object is lifted by straightening the legs while keeping the back as straight as possible. The object is kept close to the body and steady.

Knee Protection

A *knee pad* is a rubber, leather, or plastic pad strapped onto the knee for protection. Knee pads are worn by technicians who spend considerable time working on their knees or who work in close areas and must kneel for proper access to equipment. Knee pads are secured by buckle straps or Velcro® closures.

Rubber Insulating Matting

Rubber insulating matting is a floor covering that provides technicians protection from electrical shock when working on live electrical circuits. Dielectric, black fluted rubber matting is specifically designed for use in front of open cabinets or high-voltage equipment. Matting is used to protect technicians when voltages are over 50 V.

LOCKOUT/TAGOUT

Lockout is the process of removing the source of electrical power, removing all sources of hazardous energy, and installing a lock that prevents the power from being turned ON. To ensure the safety of personnel working with equipment, all electrical, pneumatic, and hydraulic power is removed and the equipment must be locked out and tagged out. *Tagout* is the process of placing a danger tag on the source of electrical power and sources of hazardous energy, which indicates that the equipment may not be operated until the danger tag is removed. Per OSHA standards, equipment is locked out and tagged out before any installation or preventive maintenance is performed. **See Figure 1-21.**

A danger tag has the same importance and purpose as a lock and is used alone only when a lock does not fit the disconnect device. A danger tag must be attached at the disconnect device with a tag tie or equivalent and must have space for the technician's name, craft, and other company-required information. A danger tag must withstand the elements and expected atmosphere for the maximum period of time that exposure is expected.

Lockout/tagout is used in the following situations:
- power is not required to be ON for a piece of equipment to perform a task
- when removing or bypassing machine guards or other safety devices
- the possibility exists of being injured or caught in moving machinery
- when jammed equipment is being cleared
- the danger exists of being injured if equipment power is turned ON

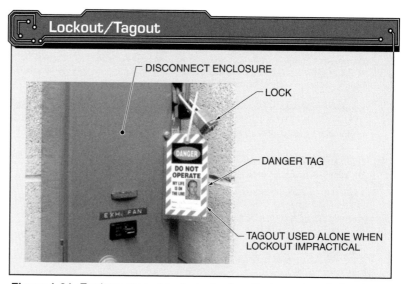

Figure 1-21. *Equipment must be locked out and/or tagged out before installation, preventive maintenance, or servicing is performed.*

FIRE SAFETY

Fire safety requires established procedures to reduce or eliminate conditions that could cause a fire. Guidelines in assessing hazards of the products of combustion are provided by the NFPA. Prevention is the best strategy to ward off potential fire hazards. Technicians must take responsibility in preventing conditions that could result in a fire. This includes proper use and storage of lubricants, oily rags, and solvents, and immediate cleanup of combustible spills. The chance of fire is greatly reduced by good housekeeping.

Hazardous Locations

The use of electrical equipment in areas where explosion hazards are present can lead to an explosion and fire. This danger exists in the form of escaped flammable gases such as naphtha, benzene, and propane. Coal, grain, and other dust suspended in the air can also cause an explosion. Article 500 of the NEC® and Article 235 of the NFPA 70E cover hazardous locations. **See Figure 1-22.** Any hazardous location requires the maximum in safety and adherence to local, state, and federal guidelines and laws, as well as in-plant safety rules. Hazardous locations are indicated by class, division, and group.

Refer to Chapter 1 Quick Quiz® on CD-ROM

Hazardous Locations—Article 500

Hazardous Location – A location where there is an increased risk of fire or explosion due to the presence of flammable gases, vapors, liquids, combustible dusts, or easily-ignitable fibers or flyings.

Location – A position or site.

Flammable – Capable of being easily ignited and of burning quickly.

Gas – A fluid (such as air) that has no independent shape or volume but tends to expand indefinitely.

Vapor – A substance in the gaseous state as distinguished from the solid or liquid state.

Liquid – A fluid (such as water) that has no independent shape but has a definite volume. A liquid does not expand indefinitely and is only slightly compressible.

Combustible – Capable of burning.

Ignitable – Capable of being set on fire.

Fiber – A thread or piece of material.

Flyings – Small particles of material.

Dust – Fine particles of matter.

Classes	Likelihood That a Flammable or Combustible Concentration is Present
I	Sufficient quantities of flammable gases and vapors present in air to cause an explosion or ignite hazardous materials
II	Sufficient quantities of combustible dust are present in air to cause an explosion or ignite hazardous materials
III	Easily ignitable fibers or flyings are present in air, but not in a sufficient quantity to cause an explosion or ignite hazardous materials

Divisions	Location Containing Hazardous Substances
1	Hazardous location in which hazardous substance is normally present in air in sufficient quantities to cause an explosion or ignite hazardous materials
2	Hazardous location in which hazardous substance is not normally present in air in sufficient quantities to cause an explosion or ignite hazardous materials

Class I Division I:
- Spray booth interiors
- Areas adjacent to spraying or painting operations using volatile flammable solvents
- Open tanks or vats of volatile flammable liquids
- Drying or evaporation rooms for flammable vents
- Areas where fats and oils extraction equipment using flammable solvents is operated
- Cleaning and dyeing plant rooms that use flammable liquids that do not contain adequate ventilation
- Refrigeration or freezer interiors that store flammable materials
- All other locations where sufficient ignitable quantities of flammable gases or vapors are likely to occur during routine operations

Class II Division I:
- Grain and grain products
- Pulverized sugar and cocoa
- Dried egg and milk powders
- Pulverized spices
- Starch and pastes
- Potato and wood flour
- Oil meal from beans and seeds
- Dried hay
- Any other organic materials that may produce combustible dusts during their use or handling

Class III Division I:
- Portions of rayon, cotton, or other textile mills
- Manufacturing and processing plants for combustible fibers, cotton gins, and cotton seed mills
- Flax processing plants
- Clothing manufacturing plants
- Woodworking plants
- Other establishments involving similar hazardous processes or conditions

Hazardous Locations		
Class	Group	Material
I	A	Acetylene
I	B	Hydrogen, butadiene, ethylene oxide, propylene oxide
I	C	Carbon monoxide, ether, ethylene, hydrogen sulfide, morpholine, cyclopropane
I	D	Gasoline, benzene, butane, propane, alcohol, acetone, ammonia, vinyl chloride
II	E	Metal dusts
II	F	Carbon black, coke dust, coal
II	G	Grain dust, flour, starch, sugar, plastics
III	No groups	Wood chips, cotton, flax, nylon

Figure 1-22. *Article 500 of the NEC® covers hazardous locations.*

KEY TERMS

- electrical circuit
- power source
- conductor
- control switch
- overcurrent protection device (OCPD)
- fuse
- circuit breaker
- load
- electrostatic discharge (ESD)
- electrostatic discharge (ESD) workstation
- static electricity
- pictorial drawing
- symbol
- abbreviation
- wiring diagram
- schematic diagram
- line (ladder) diagram
- one-line diagram
- National Fire Protection Association (NFPA)
- qualified person
- safety label
- electrical shock
- grounding
- separately derived system (SDS)
- main bonding jumper (MBJ)
- equipment grounding conductor (EGC)
- grounding electrode conductor (GEC)
- grounded conductor
- ground fault circuit interrupter (GFCI)
- International Electrotechnical Commission (IEC)
- personal protective equipment (PPE)
- protective clothing
- protective helmet
- safety glasses
- earplug
- decibel (dB)
- foot protection
- knee pad
- rubber insulating matting
- lockout
- tagout

REVIEW QUESTIONS

1. What is an electrical circuit?
2. What are the five basic components of an electrical circuit?
3. What is electrostatic discharge (ESD)?
4. List the safety rules that are effective in preventing static damage.
5. What is a pictorial drawing?
6. What is a symbol?
7. What is a wiring diagram?
8. What is a schematic diagram?
9. What is a line (ladder) diagram?
10. What is the importance of a safety label?
11. What factors minimize the possibility of personal injury?

(continued)

REVIEW QUESTIONS (continued)

12. List at least five electrical safety rules.

13. What is personal protective equipment (PPE)?

14. At what amperage does electricity cause paralysis of breathing?

15. What two overcurrent protection devices (OCPDs) are used to automatically stop the flow of electricity in a circuit that has an overcurrent condition?

16. What is lockout?

17. What is tagout?

18. What practice will reduce the chance of fire?

19. Besides flammable gases, what else can cause an explosion or fire in a hazardous location?

20. What is the lowest voltage that can harm or destroy solid state devices?

TEST INSTRUMENTS 2

The proper test instrument must be selected for each job. Test instruments must be used for the tasks for which they were designed. Before any test instrument is used, the user manual should be consulted for correct operation.

OBJECTIVES
- List and describe common test instruments.
- Explain digital multimeter operation.
- Explain oscilloscope operation.

TYPES OF TEST INSTRUMENTS

Electronic test instruments are used by technicians to aid in the taking of various electrical measurements. Care must be taken when using electronic test instruments because damage to the instrument or personal injury may result from improper or unsafe usage. The owner's manual must be consulted and all functions of a test instrument fully understood before using it.

Many types of test instruments are used by technicians. Most test instruments are available in either analog or digital versions and are designed to perform specific functions. Electrical/electronic test instruments include continuity testers, receptacle testers, voltage testers, multimeters, oscilloscopes, and digital logic probes.

Processes require the use of instrumentation to measure variables such as temperature, humidity, pressure, vacuum, flow, conductivity, voltage, current, resistance, speed, time, and other process operating conditions. Without the ability to accurately measure process variables, standards, quality control, and documentation would be limited.

Instruments installed as part of making a product are permanent instruments used to monitor that piece of equipment. Test instruments used as part of a troubleshooting system or when performing preventative maintenance tasks are of the portable type.

Continuity Testers

A *continuity tester* is a test instrument that tests for a complete path for current to flow. A continuity tester is used to test switches, fuses, and grounds. It is also used for identifying individual conductors in multiwire cable. Continuity testers give an audible indication when there is a complete path. Most multimeters have a built-in continuity test mode. **See Figure 2-1.**

The continuity test mode is commonly used to test components, such as switches, fuses, electrical connections, and individual conductors. In continuity test mode, a multimeter emits an audible response (beep) when there is a complete path. The indication of a complete path can be used to determine the condition of a component as open or closed. For example, a good fuse has continuity, whereas a bad fuse does not have continuity. The main advantage of using the continuity test mode of a multimeter is that listening for an audible response is often more desirable than reading a resistance measurement. An audible response allows an electrician to concentrate on the testing procedures without looking at the display.

WARNING: A continuity tester must only be used on de-energized circuits or components. Voltage applied to a continuity tester can cause damage to the test instrument and/or harm to the electrician. Always check a circuit for voltage before conducting a continuity test.

Receptacle Testers

A *receptacle tester* is a device that is plugged into a standard receptacle to determine if the receptacle is properly wired and energized. **See Figure 2-2.** Some receptacle testers include a ground fault circuit interrupter (GFCI) test button that allows the receptacle tester to be used on GFCI receptacles.

When testing a receptacle, the indicator light code indicates whether the receptacle is wired correctly. The situation of having the hot and neutral wires reversed is a safety hazard and must be corrected. Improper grounds are also a safety hazard and must be corrected.

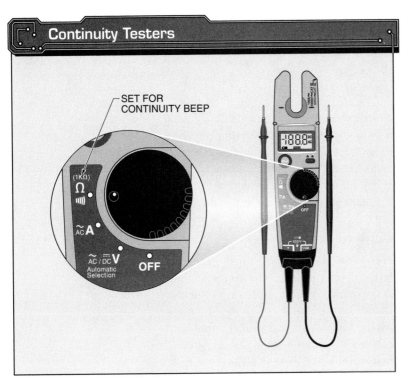

Figure 2-1. *A continuity tester tests for a complete path for current to flow.*

Continuity testers only indicate continuity by sound on a circuit or component. They only operate with circuits that have very low resistance (typically 40 Ω or lower) and will not indicate the actual resistance measurement of the circuit or component being tested.

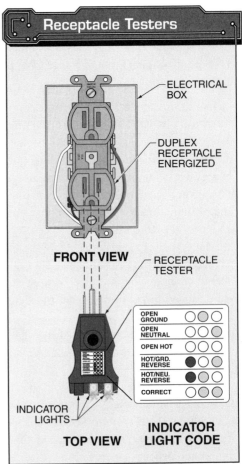

Figure 2-2. *Receptacle testers are plugged into standard receptacles to determine if the receptacle is properly wired and energized.*

Branch Circuit Identifiers

A *branch circuit identifier* is a two-piece test instrument that includes a transmitter that is plugged into a receptacle and a receiver that provides an audible beep when located near the circuit to which the transmitter is connected. Branch circuit identifiers are used to identify a particular circuit breaker. Before working on any electrical circuit that does not require power, the circuit must be de-energized. Normally, branch circuits are de-energized by turning off the circuit breakers of the circuit and applying a lockout/tagout. Often, the circuit breaker in the circuit to be de-energized is not clearly marked or identifiable. Turning off an incorrect circuit breaker may unnecessarily require loads to be reset. **See Figure 2-3.**

Voltage Indicators

A *voltage indicator* is a test instrument that indicates the presence of voltage when the test tip touches, or is near, an energized hot conductor or energized metal part. The tip glows and/or the device emits a beep when voltage is present at the test point. Voltage indicators are used to test receptacles, fuses/circuit breakers, breaks in cables, and other applications in which the presence of voltage must be detected. **See Figure 2-4.**

Voltage indicators are available in various voltage ranges (a few volts to hundreds of volts) and in the different voltage types (AC, DC, or AC/DC) for testing various types of circuits. Voltage indicators rated for 90 VAC to 600 VAC are used to test power-supply circuits. Voltage indicators rated for 24 VAC to 90 VAC are used to test low-voltage control circuits. Some models of voltage indicators are available to test for magnetic fields (AC, DC, or permanent magnets) or are used to test solenoids, transformers, and other types of coils for voltage.

> *Test instruments use standardized symbols and abbreviations to denote electrical functions. Abbreviations may vary internationally, but symbols are independent of language for quick recognition.*

AEMC® Instruments

Figure 2-3. *A branch circuit identifier provides an audible beep to identify the circuit breaker or fuse for a particular receptacle without disconnecting the power.*

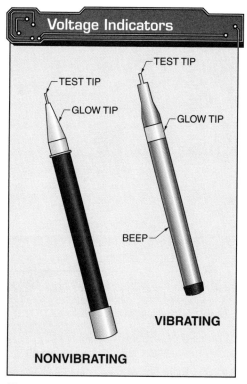

Figure 2-4. *A voltage indicator glows to indicate the presence of voltage when the indicator touches, or is near, an energized hot conductor or energized metal part.*

Voltage indicators are inexpensive, small enough to carry in a pocket, easy to use, and nonconductive. They also indicate a voltage without touching any live parts of a circuit, even through conductor insulation. Voltage indicators only indicate that voltage is present but do not indicate the actual voltage amount. They may not provide an indication that voltage is present, even when voltage is present, when the wire being tested is shielded.

Voltage Testers

A *voltage tester* is a device that indicates approximate voltage level and type (AC or DC) by the movement of a pointer on a scale (and vibration on some models). Voltage testers contain a scale marked 120 VAC, 240 VAC, 480 VAC, 600 VAC, 120 VDC, 240 VDC, and 600 VDC. **See Figure 2-5.**

is energized (hot), if a system is grounded, if fuses or circuit breakers are good, or if a circuit is a 115 VAC, 230 VAC, or 460 VAC circuit. Before using a voltage tester or any voltage-measuring instrument, the voltage tester should be checked on a known energized circuit that is within the voltage rating of the voltage tester to verify proper operation.

WARNING: If a voltage tester does not indicate a voltage, a voltage that can cause an electrical shock may still be present. The voltage tester may have been damaged during the test from too high a voltage. Always retest a voltage tester on a known energized circuit after testing a suspect circuit to ensure the voltage tester is still operating correctly.

An advantage of a voltage tester is that electricians can concentrate on the placement of test leads instead of reading the tester. Also, the tester has lower impedance (resistance) than a voltage indicator or digital multimeter. GFCIs are designed to trip at approximately 6 mA (0.006 A). The lower impedance of voltage testers allows them to be used for testing GFCIs.

Multimeters

A *multimeter* is a meter that is capable of measuring two or more electrical quantities. Multimeters can be used to measure electrical functions, such as voltage, current, continuity, resistance, capacitance, frequency, and duty cycle. Multimeters may be analog or digital. **See Figure 2-6.** Standard multimeters (analog or digital) can be equipped with attachments (adapters) that allow the multimeters to measure many types of variables.

Analog Multimeters. An *analog multimeter* is a meter that can measure two or more electrical properties and displays the measured properties along calibrated scales using a pointer. Analog multimeters use electromechanical components to display measured values. Most analog multimeters have several calibrated scales, which correspond to the different selector switch settings (AC, DC, and R) and placement of the test leads (mA jack and 10 A jack).

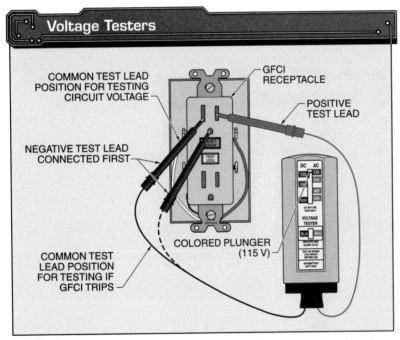

Figure 2-5. *A voltage tester is a device that indicates approximate voltage level and type by the movement and vibration of a pointer on a scale. Voltage testers can be used to check ground fault receptacles.*

Multimeters Media Clip

Voltage testers are used to take voltage measurements anytime voltage of a circuit to be tested is within the rating of the tester and an exact voltage measurement is not required. Exact voltage measurements are not required when testing to determine if a receptacle

Chapter 2—Test Instruments 27

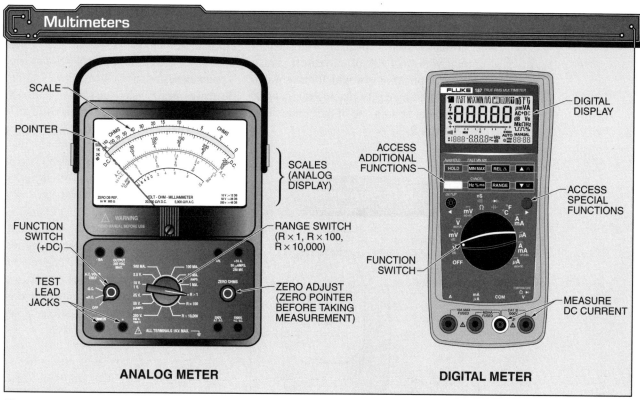

Figure 2-6. *Multimeters are portable test instruments that measure two or more electrical properties and display the measured properties as numerical values.*

When reading a measurement on an analog multimeter, the correct scale must be used. The most common measurements made with analog multimeters are voltage, resistance, and current. Analog multimeters may also include scales for measuring decibels (dB) and checking batteries. Analog test instruments and meters are less susceptible to noise (magnetic field coupling) and do not display ghost readings as easily as digital meters.

Digital Multimeters. A *digital multimeter (DMM)* is a meter that can measure two or more electrical properties and displays the measured properties as numerical values. Basic digital multimeters can measure voltage, current, and resistance. Advanced digital multimeters may include functions that measure capacitance and/or temperature. The main advantages of a digital multimeter over an analog multimeter are the ability to record measurements and ease in reading the displayed values.

Digital multimeters display measurements as numerical numbers, not a scale position. They are not likely to be misread unless the prefixes and symbols accompanying numerical values are misapplied. Most digital multimeters are autoranging. Once a measuring function is selected, such as VAC, the meter will automatically select the best meter range for taking the measurement (400 mV, 4 V, 40 V, 400 V, or 4000 V). Digital multimeters are more accurate than analog multimeters (typical voltage accuracy specifications are between 0.01% and 1.5%). Digital multimeters have high input impedance (resistance), which will not load down sensitive circuits when taking a measurement.

Clamp-On Ammeters

A *clamp-on ammeter* is a meter that determines the current in a circuit by measuring the strength of the magnetic field around a single conductor. Clamp-on ammeters

Ghost Voltage Media Clip

measure currents from 0.01 A or less to 1000 A or more. A clamp-on ammeter is normally used to measure current in a circuit with over 1 A of current. It is also used in applications in which current can be measured by easily placing the jaws of the ammeter around one of the conductors. **See Figure 2-7.**

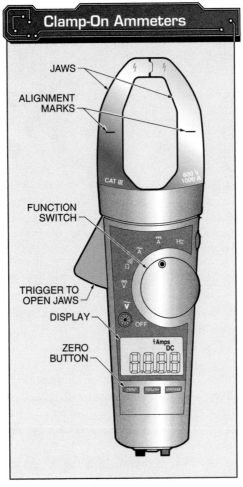

Figure 2-7. *A clamp-on ammeter measures the current in a circuit by measuring the strength of the magnetic field around a single conductor.*

Current is a common troubleshooting measurement because only a current measurement can be used by an electrician to determine how much a circuit is loaded or if a circuit is operating correctly. Current measurements vary because current can vary at different points in parallel or series/parallel circuits. Current in series circuits is constant throughout the circuit. The largest amount of current in a parallel circuit is at a point closest to the power source. Current decreases as the system distributes it to each of the parallel loads. Any variation that is excessively high must be investigated. An excessively high current measurement may indicate that a partial short exists on one of the lines and a small amount of current is flowing to ground.

Technicians must ensure that clamp-on ammeters do not pick up stray magnetic fields by separating conductors being tested as far as possible from other conductors during testing. If stray magnetic fields are possibly affecting a measurement, several measurements at different locations along the same conductor must be taken.

Visual Inspection Test Instruments

Technicians are required to work on systems and in areas where everything cannot easily be seen. Problems that cannot be seen, such as blocked pipes and ducts, leaks behind walls, and obstacles that may be in the way when trying to install devices in panels, can be accounted for with a visual inspection.

Visual inspection test instruments are used to look at areas that are typically out of sight. These instruments can be as simple as a light and flexible fiber cable. Visual inspection test instruments can include cameras, display screens, and extension cables that allow viewing objects 100′ or more away. **See Figure 2-8.** With this type of instrument, a technician can record and/or take pictures of any problems or obstacles that are viewed.

> *To take accurate measurements and readings, a test instrument must be properly handled. Dirt and moisture must be kept out of the jacks to avoid false readings. Abrasives, cleaners, or other solvents must not be used for cleaning.*

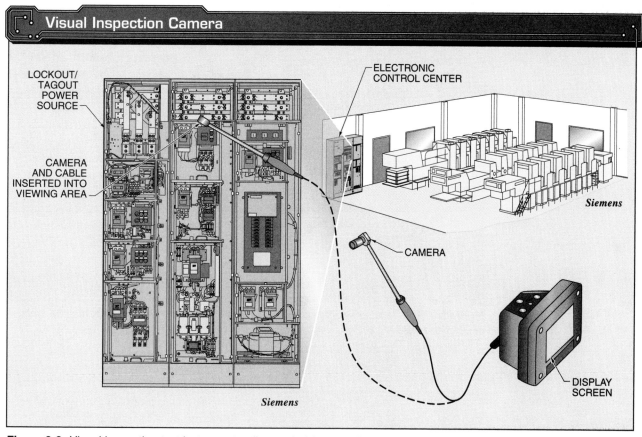

Figure 2-8. Visual inspection test instruments allow technicians to view areas behind panels and inside ducts that typically are out of normal sight.

Temperature Probes

A *temperature probe* is the part of a temperature test instrument that measures the temperature of liquids, gases, surfaces, and pipes. The temperature probe required depends on the material being measured, temperature measurement range, and accuracy required. Temperature is measured during troubleshooting because the resistance of most materials changes as the temperature of the materials change. An increase in temperature decreases the performance of electrical equipment and destroys insulation.

Contact Temperature Probes. A *contact temperature probe* is a device used for taking temperature measurements at a single point by direct contact with the area being measured. Contact temperature probes are used to measure the temperature of various solids, liquids, and gases, depending on the type of probe used. Contact temperature probes are connected directly to a temperature measuring test instrument, or the probes are connected to a temperature module between the probe and instrument. **See Figure 2-9.**

Noncontact Infrared Temperature Probes. A *noncontact temperature probe* is a device used for taking temperature measurements on energized circuits or on moving parts. An *infrared temperature meter* is a noncontact temperature meter that senses the infrared energy emitted by a material. All materials emit infrared energy in proportion to the temperature at the surface of the material. Infrared temperature meters are commonly used to take temperature measurements of electrical distribution systems, motors, bearings, switching circuits, and other equipment where electrical heat buildup is critical. **See Figure 2-10.**

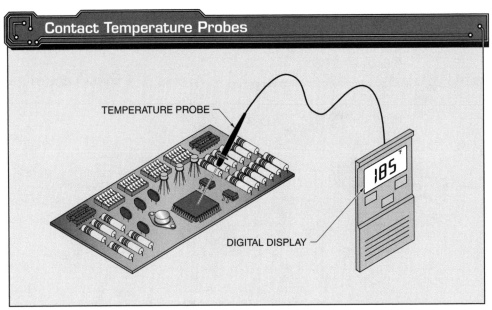

Figure 2-9. *A contact temperature probe takes temperature measurements by directly contacting components.*

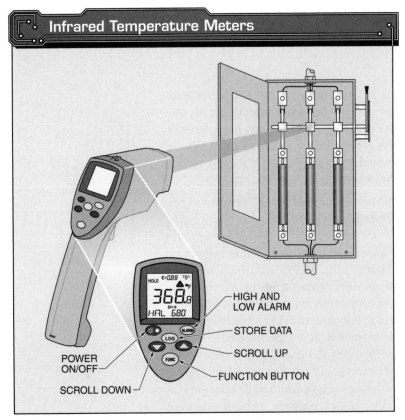

Figure 2-10. *An infrared temperature meter takes temperature measurements by measuring the infrared energy emitted by components.*

Frequency Meters

A *frequency meter (frequency counter)* is a test instrument that is used to measure the frequency of an AC signal. A frequency meter can be a stand-alone meter whose primary function is to measure frequency or part of a digital multimeter as a frequency-measuring function. Frequency meters designed primarily for measuring the frequency of a signal have a wide frequency measuring range—from a few hertz to a gigahertz (GHz) or more. Digital multimeters that include a frequency measuring function can measure a wide range of frequencies but have a much lower measuring range, typically 5 Hz to 1 MHz. **See Figure 2-11.**

Oscilloscopes

Most electronic circuits are designed to take a signal (power, voice, data, etc.) and change it by amplifying, filtering, storing, displaying, or converting the signal from analog to digital or digital to analog. The testing of electronic circuits and equipment often requires measuring or observing electrical signals. Electrical signals are measured by using meters (voltmeters and ammeters) or are observed by using oscilloscopes. An *oscilloscope* is a test instrument that provides a visual display of voltages. An oscilloscope displays the waveform for the voltage in a circuit and allows the voltage level, frequency, and phase to be measured. The two types of oscilloscopes are bench and handheld. Both types of oscilloscopes include the same basic features. **See Figure 2-12.**

A *bench oscilloscope* is a test instrument that displays the shape of a voltage waveform and is used mostly for bench testing electrical and electronic circuits. Bench testing is testing performed when equipment being tested is brought to a designated service area. Bench-type oscilloscopes are used to troubleshoot digital circuit boards, communication circuits, TVs, DVD players, computers, and other types of electronic circuits and equipment.

A *handheld oscilloscope* is a test instrument that displays the shape of a voltage waveform and is typically used for field testing. Field testing is testing performed when the test instrument is taken to the location of the equipment to be tested. Most handheld oscilloscopes are a combination oscilloscope and digital multimeter. Handheld oscilloscopes can also be referred to as portable oscilloscopes, Scopemeters®, PowerPads™, power quality meters, power analyzers, or power meters.

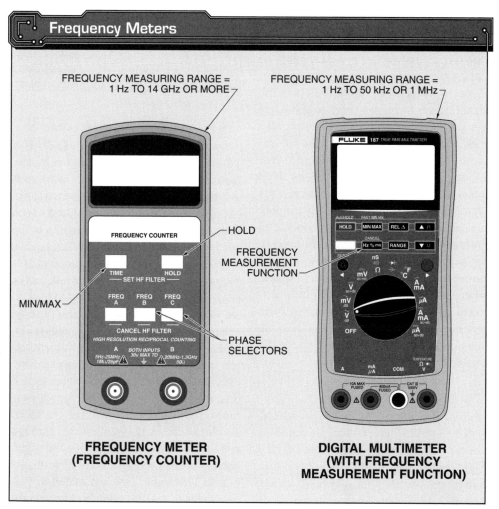

Figure 2-11. *Frequency is measured with stand-alone frequency meters or digital multimeters with a frequency measurement function.*

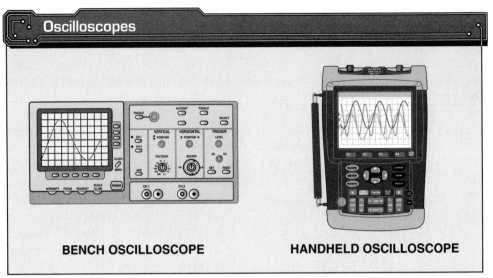

Figure 2-12. *Bench oscilloscopes and handheld oscilloscopes provide a visual display of voltages.*

Signal (Function) Generators

A *signal (function) generator* is a test instrument that produces test signals or pulses for testing purposes. The injected signal is measured at various points as the signal travels through circuits to test when the signal disappears. The signal is also measured when it is overly attenuated (reduced in energy), distorted, or clipped. **See Figure 2-13.** Much of the work involved when troubleshooting electronics involves signal tracing. Based on test instrument measurements, circuits may be operating correctly, in need of repair, or in need of replacement. Signal tracing follows a signal through a circuit, device, or system to find where the signal disappears or becomes distorted. Signal tracing is accomplished using test instruments to measure an electronic signal either already within or intentionally injected into a system or circuit.

Digital Logic Probes

A *digital logic probe* is a special DC voltmeter that detects the presence or absence of a signal. Displays on a digital logic probe include logic high, logic low, pulse light, memory, and TTL/CMOS. **See Figure 2-14.** The high light-emitting diode (LED) lights when the logic probe detects a high logic level (1). The low LED lights when the logic probe detects a low logic level (0). The pulse LED flashes relatively slowly when the probe detects logic activity present in a circuit. Logic activity indicates that the circuit is changing between logic levels.

Time-Domain Reflectometer

A *time-domain reflectometer (TDR)* is a cable tester used primarily for detecting faults in electrical conducting cables such as telephone cables. The TDR emits a series of pulses through the cable and looks for a fault, which will reflect voltage back to the test instrument.

When a pulse encounters a problem area such as a frayed wire or open circuit, part or all of the pulsed energy is reflected back to the TDR. The instrument measures the time in nanoseconds (ns) or microseconds (μs) it takes for the pulse to reach the problem and reflect back. A calculation is made using the distance, velocity of propagation, and time (distance = velocity of propagation ÷ time). Velocity of propagation is the speed at which electrical signals travel through conductors relative to the speed of light. Factors affecting the velocity of propagation include cable size and material. It is important to follow manufacturer recommendations when using TDRs since there are many variations of cables.

Chapter 2—Test Instruments 33

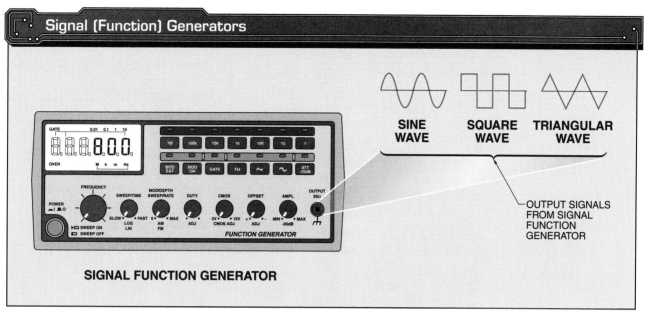

Figure 2-13. Signal (function) generators are test instruments that provide a known input signal to components, circuits, or systems for testing purposes.

Figure 2-14. A digital logic probe is a special DC voltmeter that detects the presence or absence of a signal.

Optical Time Domain Reflectometers

An *optical time domain reflectometer (OTDR)* is a test instrument that is used to measure cable attenuation. An OTDR uses a laser fiber optic light source that sends short pulses into a fiber and analyzes the light scattered back. The light source decays with fiber attenuation. Attenuation is produced by reflections from splices, connectors, and any areas in the cable that cause problems. Based on the amount of signal reflected back, the type and location of a fault is displayed. **See Figure 2-15.**

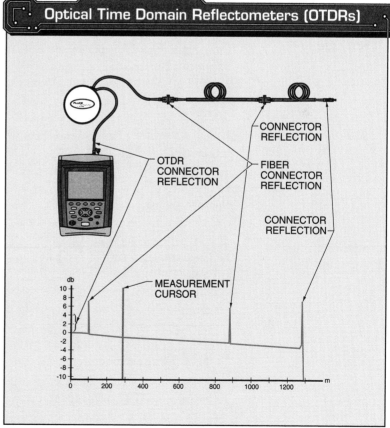

Figure 2-15. An optical time domain reflectometer (OTDR) is used to measure fiber optic signal loss (attenuation), length, and fiber endface integrity.

Test Instrument Safety

Proper equipment is required when taking measurements with test instruments. To take and interpret a measurement, the test instrument must be properly set to the correct measuring position and properly connected into the circuit to be tested. There is always the possibility that the test instrument or meter will not be properly set to the correct function or to the correct range and/or will be misread. Test instruments not properly connected to a circuit increase the likelihood of an improper measurement or meter reading, or creating an unsafe condition.

> Even if a test light does not appear to be illuminated, voltage can still be present and dangerous. For example, the LED indicators in the lights may be difficult to see in direct sunlight even though they are illuminated.

Electrical Measurement Safety. Electrical measurements are taken with test instruments designed for specific tasks. Test instruments such as voltage testers, analog and digital multimeters, and clamp-on ammeters measure electrical quantities, such as voltage, current, power, and frequency. Each meter has specific features and limits. User manuals detail specifications and features, proper operating procedures, safety precautions, warnings, and allowed applications.

Conditions can change quickly as voltage and current levels vary in individual circuits. General safety precautions for using test instruments include the following:

- When a circuit does not have to be energized (as when taking a resistance measurement or checking diodes and capacitors in a circuit), lockout and tagout all equipment and circuits to be tested.
- Never assume a test instrument is operating correctly. Check the test instrument that will be measuring voltage on a known (energized) voltage source before taking a measurement on an unknown voltage source. After taking a measurement on an unknown voltage source, retest the test instrument on a known source to verify the meter still operates properly.
- Always assume that equipment and circuits are energized until positively identified as de-energized by taking proper measurements.
- Never work alone when working on or near exposed energized circuits that may cause an electrical shock.
- Always wear personal protective equipment appropriate for the test area.
- Ensure that the function switch of a test instrument or meter is set to the proper range and function before applying test leads to the circuit. A test instrument set to the wrong function can be damaged. For example, damage can occur when test leads contact an AC power source while the meter is set to measure resistance or continuity.
- Ensure that the test leads of the test instrument or meter are connected properly. Test leads that are not connected to the correct jacks can be dangerous. For example, attempting to measure voltage while the test leads are in the amperage jack produces a short circuit. For additional safety, use a meter that is self-protected with a high-energy fuse. Follow the operating instructions in the user manual or the directions on the graphic display for proper test lead connections and operating information.
- When using a test instrument or meter, ensure that the function switch is set to the proper range and function before connecting leads to a circuit. When using a graphic display meter, verify that the proper screen (voltage or current measurement) is selected.
- Start with the highest range when measuring unknown values. Using a range that is too low can damage the meter. Attempting a voltage or current measurement above the rated limit can be dangerous.

- Connect the ground test lead (black) first and the voltage test lead (red) next. Disconnect the voltage test lead (red) first and the ground test lead (black) next, after taking measurements.
- Whenever possible, connect test leads to the output side (load side) of a circuit breaker or fuse to provide better short circuit protection.
- Never assume that a circuit is de-energized or equipment is fully discharged. Capacitors can hold a charge for a long time—several minutes or more. Always check for the presence of voltage before taking any other measurements.
- Check test leads for frayed or broken insulation. Electrical shock can occur from accidental contact with live components. Electrical test equipment should have double-insulated test leads, recessed input jacks on the meter, shrouds on the test lead plugs, and finger guards on test probes.
- Use meters that conform to the IEC 1010 category in which they will be used. For example, to measure 480 V in an electrical distribution feeder panel, a meter rated at CAT III–600 V or CAT III–1000 V is used.
- Avoid taking measurements in humid or damp environments.
- Ensure that no atmospheric hazard, such as flammable dust or vapor, is present in the area.
- Use one hand when working on a live circuit to reduce the chance of an electrical shock passing through the heart and lungs.

DIGITAL MULTIMETER OPERATION

A digital multimeter (DMM) is commonly used for measuring electrical quantities, such as voltage, amperage, or resistance, in electrical and electronic circuits. Although DMM features and operations may vary, most abbreviations, symbols, and terminology are standardized in the industry.

Digital Multimeter Abbreviations and Symbols

An abbreviation is a letter or combination of letters that represent a word. Standard abbreviations are used for DMMs to represent a quantity or term for quick recognition. For example, quantities such as voltage and amperes are identified with abbreviations. **See Figure 2-16.** A symbol is a graphic element that represents a quantity, unit, or device. Symbols can be quickly recognized and interpreted regardless of the language.

Meter Abbreviation Symbols
Media Clip

Digital Displays

A *digital display* is an electronic device that displays readings as seven-segment numerical values. There is a greater possibility of incorrectly reading an analog multimeter than a digital multimeter. Errors can occur when reading a digital display if the prefixes, symbols, and/or decimal point displayed are not properly interpreted. **See Figure 2-17.**

Mine Safety Appliances Co.
Digital test instruments can provide greater accuracy than analog test instruments.

36 SOLID STATE DEVICES AND SYSTEMS

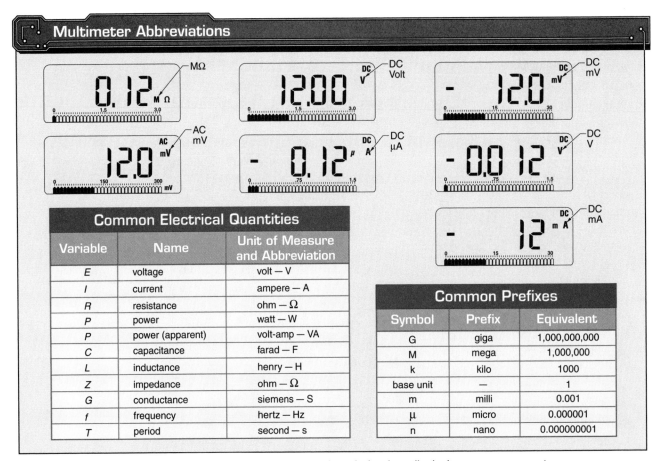

Figure 2-16. *Test instruments use multiple abbreviations and symbols when displaying a measurement.*

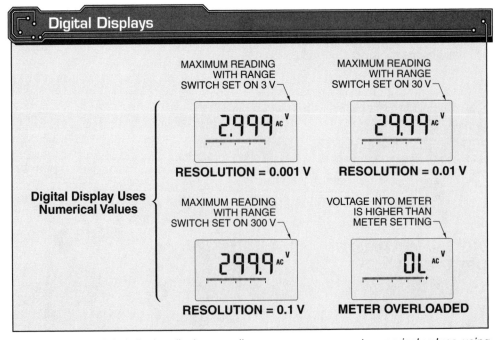

Figure 2-17. *A digital display displays readings as seven-segment numerical values using either a light-emitting diode (LED) display or a liquid crystal display (LCD).*

AC Voltage Measurements

The procedures for AC voltage measurement may vary slightly with different DMMs. Caution should always be exercised when measuring AC voltages over 24 V with any DMM. **See Figure 2-18. WARNING:** Ensure that no body part contacts any part of the circuit, including the metal contact points at the tip of the test leads.

To measure AC voltage with a DMM, use the following procedure:

1. Set the function switch to AC voltage. Set the range to the highest voltage setting, if voltage in the circuit is unknown.
2. Plug the black test lead into the common jack.
3. Plug the red test lead into the voltage jack.
4. Connect the DMM test leads to the circuit. Common industrial practice is to connect the black test lead to the grounded (neutral) side of the AC voltage.
5. Read the voltage measurement displayed on the DMM.

DC Voltage Measurements

The DC voltage measurement procedure is similar to the AC voltage measurement procedure. Measurements of DC voltages exceeding 60 V and DC measurements taken near any battery require extra caution. **See Figure 2-19. WARNING:** A battery can be shorted by contact with a metal object between the positive and negative battery terminals.

To measure DC voltage with a DMM, use the following procedure:

1. Set the function switch to DC voltage. If the DMM includes more than one DC setting and the circuit voltage is unknown, select the highest setting.
2. Plug the black test lead into the common jack.
3. Plug the red test lead into the voltage jack.
4. Connect the DMM test leads to the circuit. The black test lead is connected to the negative polarity test point (circuit ground) and the red test lead to the positive polarity test point. Reverse the test leads if a negative sign (–) appears to the left of the measurement displayed.
5. Read the voltage measurement displayed on the DMM.

Measuring AC Voltage Media Clip

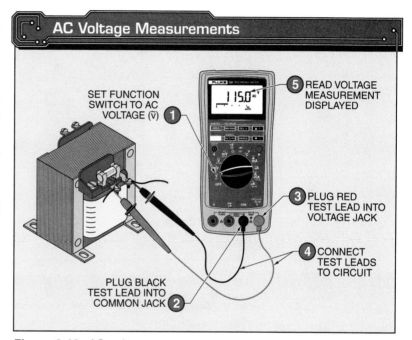

Figure 2-18. *AC voltage measurements may vary slightly with different DMMs. Manufacturer procedural manuals should always be consulted for proper operating procedures.*

Resistance Measurements

A DMM measures the resistance of a circuit or the resistance of a component removed from a circuit. **See Figure 2-20. WARNING:** Voltage applied to a DMM that is set to measure resistance will damage the DMM, even if the DMM has internal protection.

Resistance measurements indicate the condition of a circuit or component. Resistance and current are inversely related. For example, the higher the resistance in a circuit, the lower the current flow through the circuit.

38 SOLID STATE DEVICES AND SYSTEMS

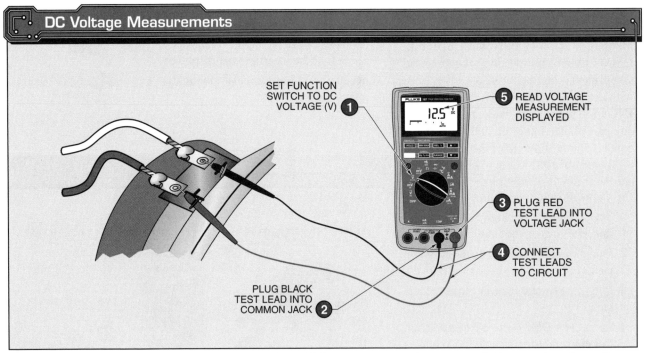

Figure 2-19. *DC voltage measurements exceeding 60 V and DC measurements taken near any battery require extra caution.*

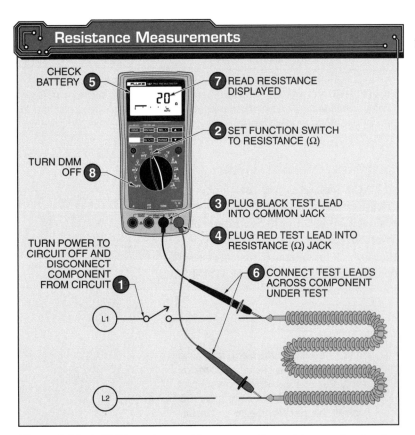

Figure 2-20. *A DMM can be used to measure the resistance of a circuit or component removed from a circuit.*

To measure resistance with a DMM, use the following procedure:

1. Check that all power is OFF to the circuit under test. Always remove any device under test from the circuit.
2. Set the function switch to resistance mode on the DMM. The DMM should display "OL" and the "Ω" symbol when the DMM is in the resistance mode.
3. Plug the black test lead into the common jack.
4. Plug the red test lead into the resistance jack.
5. Ensure that the DMM batteries are in good condition. The battery symbol is displayed when batteries are low.
6. Connect the leads across the device under test. Ensure that there is good contact between the test leads and the component leads.
7. Read the resistance measurement displayed on the DMM.
8. After completing all resistance measurements, turn the DMM OFF to prevent battery drain.

In-Line Current Measurements

Care is required to protect the ammeter, circuit, and the user when measuring AC or DC current with an in-line ammeter. In-line current measurements are not recommended if the DMM is not fused. **See Figure 2-21. WARNING:** Always ensure that the function switch position corresponds to the connection of the test leads. The DMM can be damaged if the test leads are connected to measure current, but the function switch is set for a different measurement, such as voltage or resistance.

To measure in-line current with a DMM, use the following procedure:

1. Set the function switch to the proper position for measuring the AC or DC current level (A, mA, or µA).
2. Plug the black test lead into the common jack.
3. Plug the red test lead into the current jack. The current jack may be marked A, mA, or µA.
4. Turn the power OFF to the circuit or device under test and discharge all capacitors, if possible.

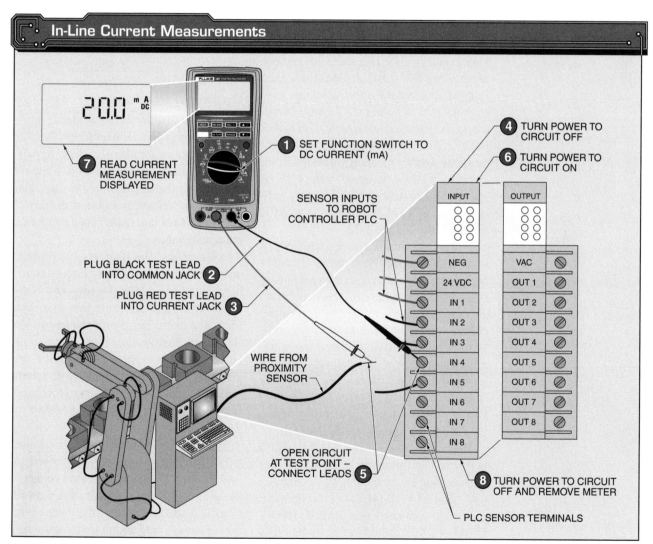

Figure 2-21. *Care is required to protect the ammeter, circuit, and user when measuring AC or DC current with an in-line ammeter.*

5. Open the circuit at the test point and connect the test leads to each side of the opening.
6. Turn the power ON to the circuit under test.
7. Read the current measurement displayed on the DMM.
8. Turn the power OFF and remove the ammeter from the circuit.

Proper safety precautions and PPE must be used to ensure that no body part contacts a live circuit during in-line current measurements. Clamp-on ammeters and accessories may be used for additional safety.

Digital Multimeter Voltage Protection

To protect against transient voltages, protection must be built into the test equipment used. In the past, the industry standard followed was IEC 348. This standard has been replaced by IEC 1010. A DMM designed to the IEC standard offers a higher level of protection. A higher CAT number indicates an electrical environment with higher power available, larger short-circuit current available, and higher energy transients. For example, a DMM designed to the CAT III standard is resistant to higher energy transients than a DMM designed to the CAT II standard. **See Figure 2-22.**

Power distribution systems are divided into categories because a dangerous high-energy transient voltage, such as a lightning strike, is attenuated or dampened as it travels through the impedance (AC resistance) of the system and the system grounds. Within an IEC 1010 standard category, a higher voltage rating denotes a higher transient voltage withstanding rating. For example, a CAT III–1000 V (steady-state) DMM provides better protection than a CAT III–600 V (steady-state) DMM.

Between categories, a DMM with a higher voltage rating might not have higher transient voltage protection. For example, a CAT III–600 V DMM has better transient protection compared to a CAT II–1000 V DMM. A DMM should be chosen based on the IEC overvoltage installation category first and the voltage second.

OSCILLOSCOPE OPERATION

An oscilloscope is an instrument that graphically displays an instantaneous voltage. The oscilloscope is one of the most useful measuring instruments used in electronic circuit diagnostics and bench testing. **See Figure 2-23.** Bench testing is testing performed when devices or circuits are brought to a designated service area. Oscilloscopes are used to troubleshoot digital circuits, communication circuits, factory process instrumentation, machine control circuitry, and computers. Besides showing a voltage waveform in a circuit, oscilloscopes also allow the voltage level, frequency, and phase to be measured. Oscilloscopes are available in basic and specialized types that can display different waveforms simultaneously.

Graphical Displays

A graphical display shows electrical information on a graph. When circuits include rapidly fluctuating signals, stray signals, and phase shifts, problems are easily detected on the display. In some troubleshooting situations, a graphical display of circuit voltage has more meaning than a numerical value.

An oscilloscope displays the voltage under test on a graph. The graph contains horizontal and vertical axes. The horizontal (x) axis represents the time. The vertical axis (y) represents the amplitude (level) of the voltage waveform. The horizontal and vertical lines divide the graph into equal divisions. The divisions are used to measure the voltage level and frequency of the displayed waveforms. **See Figure 2-24.**

An oscilloscope is connected in parallel with a circuit or component under test. A probe at the end of each test lead connects the oscilloscope to a circuit. A 1X probe (1 to 1) is used to connect the input of the oscilloscope to the circuit under test when the test voltage is within the voltage range of the scope.

IEC 1010 Overvoltage Installation Categories

Category	In Brief	Examples
CAT I	Electronic	• Protected electronic equipment • Equipment connected to (source) circuits in which measures are taken to limit transient overvoltage to an appropriately low level • Any high-voltage, low-energy source derived from a high-winding-resistance transformer, such as the high-voltage section of a copier
CAT II	1φ receptacle-connected loads	• Appliances, portable tools, and other household and similar loads • Outlets and long branch circuits • Outlets at more than 30′ (10 m) from CAT III source • Outlets at more than 60′ (20 m) from CAT IV source
CAT III	3φ distribution, including 1φ commercial lighting	• Equipment in fixed installations, such as switchgear and polyphase motors • Buses and feeders in industrial plants • Feeders and short branch circuits and distribution panel devices • Lighting systems in larger buildings • Appliance outlets with short connections to service entrance
CAT IV	3φ at utility connection, any outdoor conductors	• Refers to the origin of installation, where low-voltage connection is made to utility power • Electric meters, primary overcurrent protection equipment • Outside and service entrance, service drop from pole to building, run between meter and panel • Overhead line to detached building

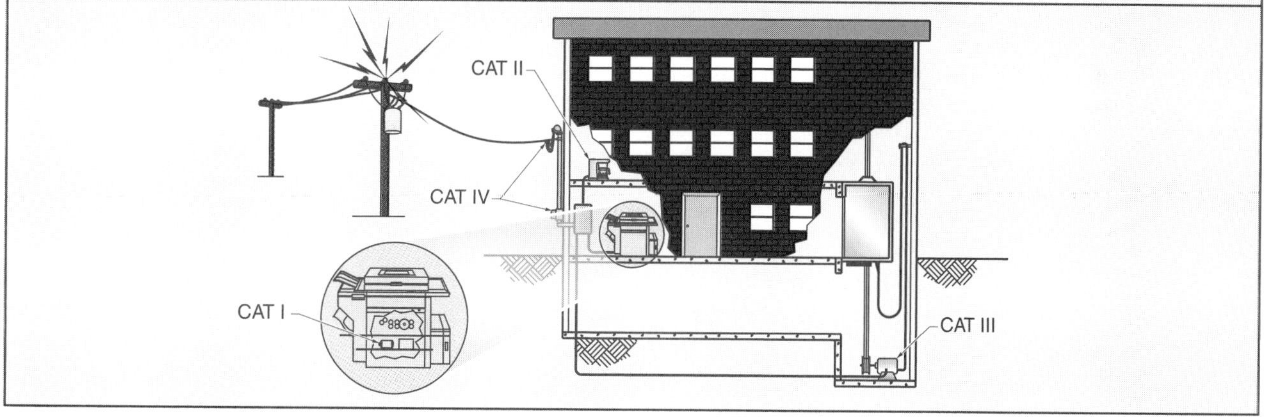

Figure 2-22. *The applications in which a DMM may be used are classified by the IEC 1010 standard into four overvoltage installation categories.*

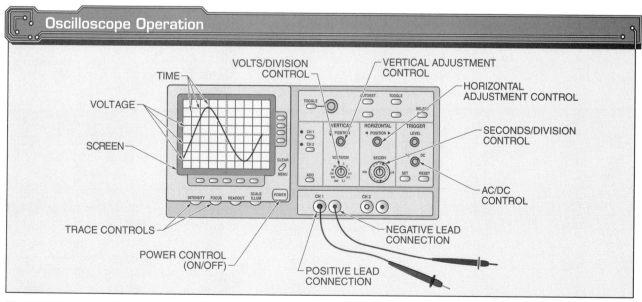

Figure 2-23. *An oscilloscope is a test instrument that displays an instantaneous voltage image on a graph.*

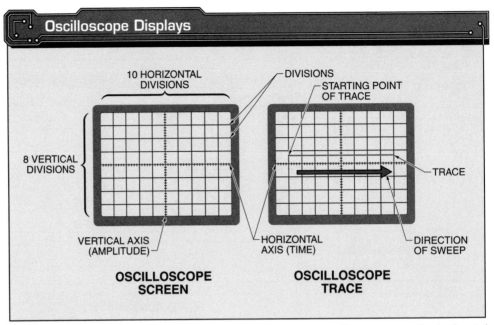

Figure 2-24. *A trace that is established on the screen of an oscilloscope uses the horizontal (x) axis to represent time and the vertical (y) axis to represent the amplitude of the voltage waveform.*

A 10X probe (10 to 1) is used to divide the input voltage by 10. The oscilloscope graphic voltage reading equals 10 times the normal rated voltage when a 10X probe is used. The amount of measured voltage displayed on the screen must be multiplied by 10 to obtain the actual circuit voltage when using a 10X probe. When the measured scope voltage is 25 V when using a 10X probe, the actual circuit voltage is 250 V (25 V × 10 = 250 V).

Oscilloscope Trace. An oscilloscope trace is established on the screen before the circuit to be tested is connected. An *oscilloscope trace* is a reference point/line that is visually displayed on the face of the oscilloscope screen. It is normally positioned over the horizontal centerline on the screen.

The starting point of the oscilloscope trace is located near the left side of the screen. The movement of the displayed trace across the oscilloscope screen is known as a sweep. The sweep of an oscilloscope trace is from left to right on the screen.

A *dual-trace oscilloscope* is an oscilloscope that displays two traces simultaneously. Dual-trace oscilloscopes include two separate input channels (channel 1 and channel 2) through which the two signals are sent into the oscilloscope for viewing and comparison. Dual-trace oscilloscopes are used for such applications as monitoring the input signal and output signal of a circuit simultaneously. By means of the dual trace, any signal gain, loss, distortion, or other changes to the signal are seen. **See Figure 2-25.**

Manually Operated Controls. Manually operated controls are used to adjust a waveform for viewing. Typical manually operated oscilloscope adjustment controls include intensity, focus, horizontal positioning, vertical positioning, volt/division, and time/division.

Intensity is the brightness level of a displayed object. The intensity control sets the level of brightness on the displayed voltage trace. The intensity level is kept as low as possible, consistent with keeping the trace in focus. *Focus* is the sharpness of a displayed object. The focus control adjusts the clarity of the displayed voltage trace.

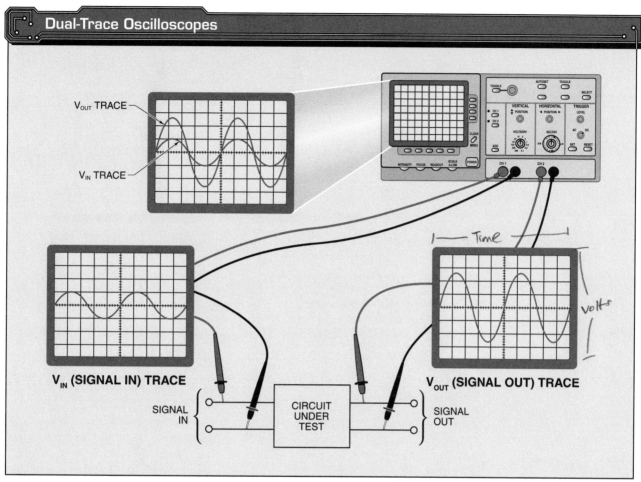

Figure 2-25. A dual-trace oscilloscope is used to view two signals, such as an input and an output, simultaneously.

Horizontal positioning is the left or right shifting of a displayed voltage trace. The horizontal control shifts the starting point of the trace so that the displayed waveform begins on the vertical axis. *Vertical positioning* is the up and down shifting of a displayed voltage trace. The vertical control shifts the up and down position of the displayed waveform so that the center of the waveform is on the horizontal axis. **See Figure 2-26.**

Volt/division (volts per division) is the number that each horizontal division on a screen represents. The volt/division control selects the height of the displayed waveform. When a waveform occupies four horizontal divisions and the volt/division control knob is set at 20, the peak-to-peak voltage ($V_{p\text{-}p}$) equals 80 V (4 × 20 = 80 V). **See Figure 2-27.**

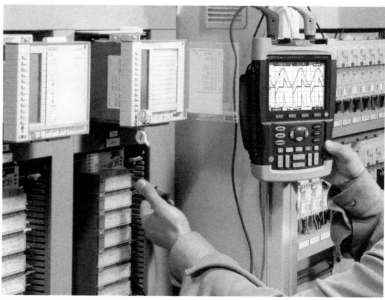

Fluke Corporation
Handheld oscilloscopes are used when signal measurements must be made in the field.

44 SOLID STATE DEVICES AND SYSTEMS

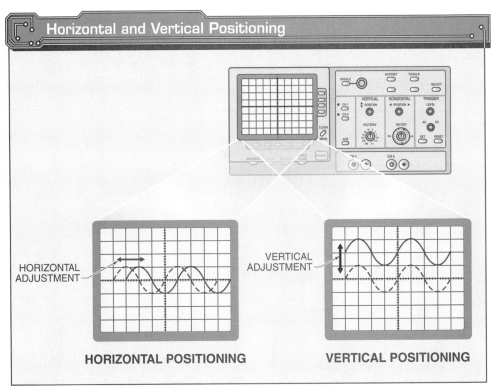

Figure 2-26. *Manual controls on an oscilloscope, such as intensity, focus, horizontal positioning, vertical positioning, volts per division, and time per division, must be adjusted to view waveforms for usable information.*

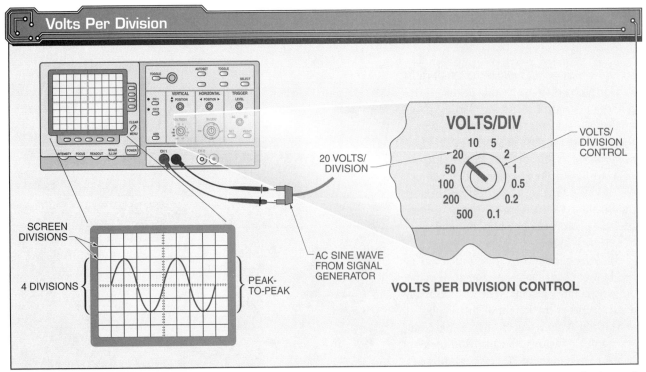

Figure 2-27. *The volts per division control selects the height of the displayed waveform by setting the number of volts each horizontal screen division represents.*

Time/division (time per division) is the length of time a cycle takes to sweep across a screen. The time/division control selects the width of the displayed waveform. When the time/division control is set at 10, each vertical screen division equals 10 ms (milliseconds). If one cycle of the waveform equals four divisions, the displayed time equals 40 ms (4 × 10 = 40 ms). **See Figure 2-28.**

AC Voltage Measurements

To measure AC voltage using an oscilloscope, the test probes are connected to the points in the circuit where the AC voltage is to be measured. The negative lead of the scope is connected to the ground side of the circuit. **See Figure 2-29.**

To measure AC voltage with an oscilloscope, use the following procedure:

1. Turn the power switch ON and adjust the trace intensity on the screen.
2. Set the AC/DC control switch to AC.
3. Set the volt/division control knob to display the voltage level under test. Set the control to the highest value if the voltage is unknown.
4. Connect the oscilloscope probes to the AC voltage under test.
5. Adjust the volt/division control knob to display the full waveform of the voltage under test.
6. Set the time/division control knob to display several cycles of the voltage under test.
7. Adjust the waveform by using the vertical control to set the lower edge of the waveform on one of the lower lines.
8. Measure the vertical amplitude of the waveform by counting the number of divisions displayed ($V_{p\text{-}p}$).

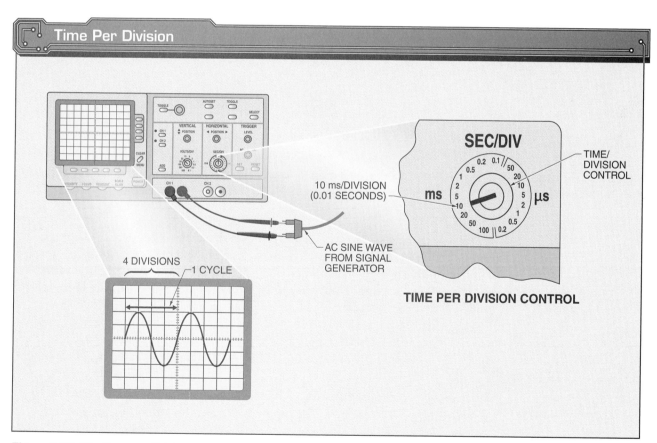

Figure 2-28. *The time per division control selects the width of the displayed waveform by setting the length of time (per division) each cycle takes to move across the screen.*

Oscilloscope AC Voltage Measurements

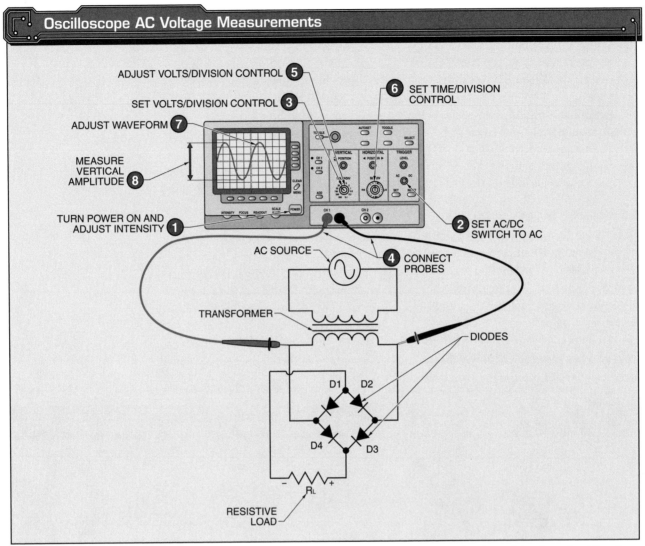

Figure 2-29. To measure AC voltage using an oscilloscope, the test probes are connected to the points where the AC voltage is to be measured. In this case, the probes are located across the secondary of the step-down transformer.

To calculate V_{rms}, first calculate $V_{p\text{-}p}$. $V_{p\text{-}p}$ is calculated by multiplying the number of divisions by the volt/division setting. When a waveform occupies four divisions and the volt/division setting is 10, $V_{p\text{-}p}$ equals 40 V (4 × 10 = 40). V_{max} equals 20 V (40 ÷ 2 = 20). V_{rms} equals 14.14 V (20 × 0.707 = 14.14).

DC Voltage Measurements

To measure DC voltage using an oscilloscope, test probes are connected to the points in the circuit where the DC voltage is to be measured. The negative lead of the oscilloscope is connected to the ground side of the circuit. The voltage is positive if the trace moves above the centerline. The voltage is negative if the trace moves below the centerline. **See Figure 2-30.**

To measure DC voltage with an oscilloscope, use the following procedure:

1. Turn the power switch ON and adjust the trace intensity on the screen.
2. Set the AC/DC control switch to DC.
3. Set the volt/division control knob to display the voltage under test. Set the control to the highest value possible if the voltage level under test is unknown.

4. Connect the oscilloscope probes to the ground point of the circuit under test.
5. Set the vertical control knob so that the displayed line is in the center of the screen. The displayed line represents 0 VDC.
6. Remove the oscilloscope probe from the ground point and connect it to the DC voltage under test. The displayed voltage moves above or below the oscilloscope centerline depending on the polarity of the DC voltage under test.
7. Measure the vertical amplitude of the voltage from the centerline by counting the number of divisions from the centerline.

The number of displayed divisions is multiplied by the volt/division setting to determine the DC voltage under test. When a waveform is located at three divisions above the centerline and the volt/division control knob is set at 5 V, the voltage equals 15 VDC (3 × 5 = 15 V).

Frequency Measurements

In AC applications, it may be necessary to measure the frequency (hertz) of the circuit. Where high accuracy is required, a frequency meter may be used, but an oscilloscope gives an adequate reading for most frequency measurement applications. An oscilloscope also shows any distortion present in the signal under test. To measure frequency, the oscilloscope probes are connected in parallel with the circuit or component under test. **See Figure 2-31.**

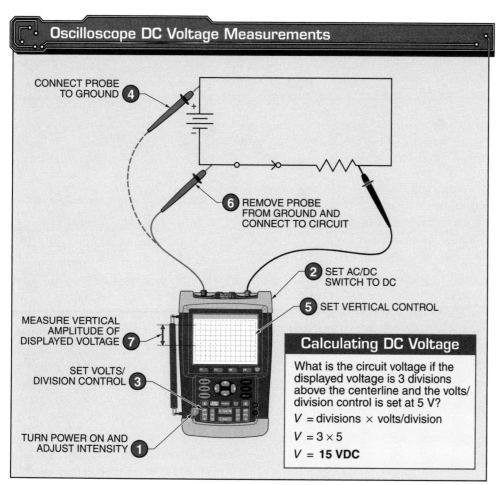

Figure 2-30. *To measure DC voltage using an oscilloscope, the test probes are connected to the points in the circuit where the DC voltage is to be measured.*

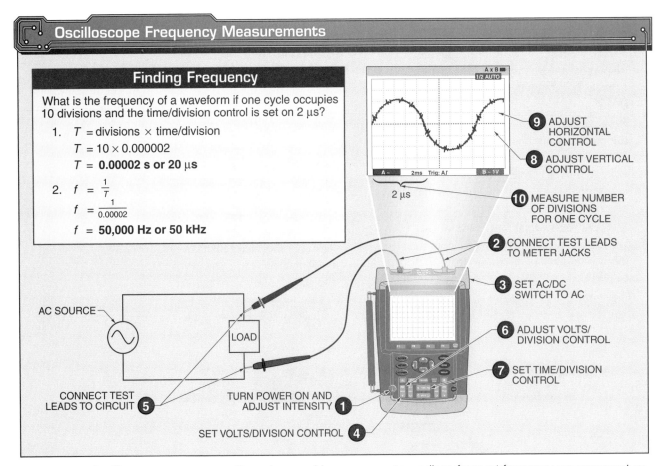

Figure 2-31. Oscilloscopes or power quality meters provide very accurate readings for most frequency measurement applications by displaying any distortion present in the circuit voltage being tested.

Refer to Chapter 2 Quick Quiz® on CD-ROM

To measure frequency with an oscilloscope, use the following procedure:

1. Turn the power switch ON and adjust the trace intensity.
2. Connect the test leads to the meter jacks.
3. Set the AC/DC control switch to AC.
4. Set the volt/division control knob to display the voltage level under test. Set the control to the highest value if the voltage level is unknown.
5. Connect the oscilloscope probes to the AC voltage under test.
6. Adjust the volt/division control knob to display the vertical amplitude of the waveform under test.
7. Set the time/division control knob to display approximately two cycles of the waveform under test.
8. Set the vertical control knob so that the center of the waveform is on the horizontal centerline of the oscilloscope screen.
9. Set the horizontal control so that the start of one cycle of the waveform begins at the vertical centerline on the oscilloscope screen.
10. Measure the number of divisions between the start point and the end point of one cycle.

To calculate frequency, the number of divisions is multiplied by the time/division setting. The number of divisions is 10 with the time/division control knob set at 2 μs. The time period equals 20 μs (10 × 0.000002 = 0.00002). To determine the frequency of the waveform, 1 is divided by the time period. The frequency equals 50 kHz (1 ÷ 0.00002 = 50,000).

KEY TERMS

- continuity tester
- receptacle tester
- branch circuit identifier
- voltage indicator
- voltage tester
- multimeter
- analog multimeter
- digital multimeter (DMM)
- clamp-on ammeter
- temperature probe
- contact temperature probe
- noncontact temperature probe
- infrared temperature meter
- frequency meter (frequency counter)
- oscilloscope
- bench oscilloscope
- handheld oscilloscope
- signal (function) generator
- digital logic probe
- optical time domain reflectometer (OTDR)
- dual-trace oscilloscope
- intensity
- horizontal positioning
- vertical positioning
- volts/division (volts per division)
- time/division (time per division)

REVIEW QUESTIONS

1. What is a test instrument?
2. What is a digital multimeter (DMM)?
3. What is an oscilloscope?
4. What is a digital logic probe?
5. Explain test instrument safety.
6. Discuss electrical measurement safety.
7. Describe how AC voltage measurements are taken with a DMM.
8. Describe how DC voltage measurements are taken with a DMM.
9. Explain measuring resistance with a DMM.
10. Explain measuring in-line current with a DMM.
11. Discuss DMM voltage protection.
12. Describe how AC voltage measurements are taken with an oscilloscope.
13. Describe how DC voltage measurements are taken with an oscilloscope.
14. Explain measuring frequency with an oscilloscope.

PRINTED CIRCUIT BOARD CONSTRUCTION AND TROUBLESHOOTING 3

The purpose of a printed circuit board is to provide electrical connections between the components mounted on it. There are several types of printed circuit boards and manufacturing processes. Today, complex software is used to design and lay out the circuitry found within printed circuit boards. An understanding of how printed circuit boards are designed and built helps technicians to troubleshoot them and their components.

OBJECTIVES

- Describe printed circuit boards.
- List and describe the types of printed circuit boards.
- Describe printed circuit board manufacturing.
- Discuss how components are mounted on a printed circuit board.
- Explain printed circuit board identification.
- Discuss concerns of handling printed circuit boards.
- Explain printed circuit board troubleshooting.
- Discuss recycling and the elimination of hazardous materials.

PRINTED CIRCUIT BOARDS

A *printed circuit (PC) board* is a board of insulating material with electrical interconnections attached to its surface to create a circuit for components mounted on it. **See Figure 3-1.** The purpose of PC boards is to provide electrical paths to connect components to a circuit. The conducting paths are placed on one or both sides of the insulating material. Complex circuits require multilayer boards constructed with conducting paths on both sides of the insulating material and between layers.

Traces are used to connect two or more solder pads. A *trace* is a conductive path found on various layers of a PC board. A trace may also be referred to as a foil. Plated through holes connect traces that are not located on the same layer of a PC board. A *solder pad* is a small, flat conductor to which component leads are soldered. A *bus* is a large trace or foil that extends around the edge of a PC board to provide a conducting path from several sources. A *solder mask* is a permanent protective coating that protects circuitry from any unwanted solder.

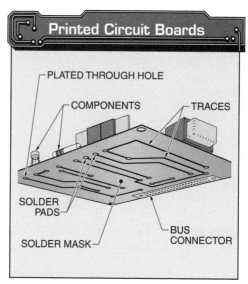

Figure 3-1. *A PC board is made of an insulating material with conducting paths that connect components to form a circuit.*

PC boards may include edge card connectors, which are multiple terminals on one edge of a PC board. Edge card connectors allow PC boards to be connected to a system with the least amount of hardware.

Printed Circuit Board Material

PC boards are typically made of preimpregnated (prepreg) material upon which copper traces are added and components may be mounted. **See Figure 3-2.** Several types of prepreg materials can be used to make PC boards, such as epoxy, phenolic, bismaleimide, and polyimide. Each of these materials has various benefits, such as good mechanical strength, fire resistance, and the ability to be used for applications with high operating temperatures.

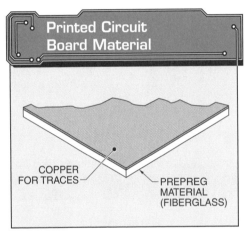

Figure 3-2. *PC boards are made of prepreg material upon which copper traces are added.*

The most commonly used base material for PC boards is FR-4 PC board laminate. The acronym FR stands for flame retardant, and the number 4 indicates the type of resin, which is woven glass epoxy resin. FR-4 PC board laminate is strong enough to protect circuitry placed on it and is common in consumer, industrial, and military applications. This laminate is also commonly referred to as fiberglass.

The National Electrical Manufacturers Association (NEMA) standard L1 1-1998, Industrial Laminating Thermosetting Products, provides information on FR-4 material. The grade designation FR-4 was established for consistency in the PC board manufacturing industry.

Printed Circuit Board Types

The three main types of PC boards are single-sided, double-sided, and multilayered. **See Figure 3-3.** The various types of PC boards are distinguished by their layers of traces. In addition to the main types of PC boards, specialty PC boards, such as flexible PC boards and hybrid integrated circuits, may be used for special applications.

Single-Sided Printed Circuit Boards. A *single-sided PC board* is a PC board that has traces on only one side of an insulating material. A single-sided PC board is the simplest type of PC board. When there are too many traces for a single-sided PC board, a double-sided PC board may be used.

Double-Sided Printed Circuit Boards. A *double-sided PC board* is a PC board that has a layer of traces on each side of an insulating material. Plated holes are used to interconnect the two layers of traces. A *plated hole* is a hole drilled through a double-sided PC board and plated with a conductive material that interconnects two layers of traces. When there are too many traces for a double-sided PC board, a multilayered PC board may be used.

Multilayered Printed Circuit Boards. A *multilayered PC board* is a PC board that has several layers of traces separated by layers of an insulating material. Multilayered PC boards have been rapidly replacing single- and double-sided PC boards. The purpose of a multilayered PC board is to provide a greater range of routing paths and higher density component location for complex electronic circuits.

Several layers of circuits are stacked on top of one another and then interconnected. The stack of layers is called a layup. In a multiple stack layup, the top and bottom layers are positioned with the aid of tooling pins. The layup is placed on a laminator to press the individual boards into one multilayer PC board. Multilayered PC boards are commonly used in automotive, communication, industrial, computer gaming, and consumer products.

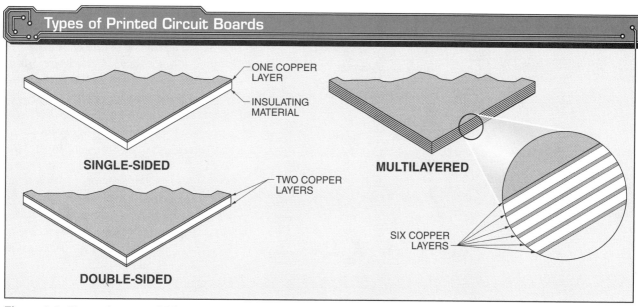

Figure 3-3. *The main types of PC boards are single-sided, double-sided, and multilayered.*

The various layers of traces in multilayered PC boards are interconnected with vias. A *via* is a plated hole drilled through one or more layers of traces and insulating material that provides an electrical connection between the layers. Vias are based on the plated-hole method of interconnection used for double-sided PC boards. The main types of vias are through-hole, blind, and buried. **See Figure 3-4.** A *through-hole via* is a via that passes completely through a PC board and connects all layers of conductors. A *blind via* is a via that is exposed only on one outer layer of a PC board. A *buried via* is a via that connects only conductors on the interior layers of a PC board and cannot be seen. Buried vias are practically impossible to repair.

The layer build-up method may be used to construct multilayered PC boards. With the layer build-up method of PC board construction, conductors and insulation layers are alternately deposited on a PC board. **See Figure 3-5.** This method produces copper interconnections between layers and minimizes heat expansion. Reworking the internal connections in built-up layers is difficult, if not impossible. Internal damage may require that the board be discarded.

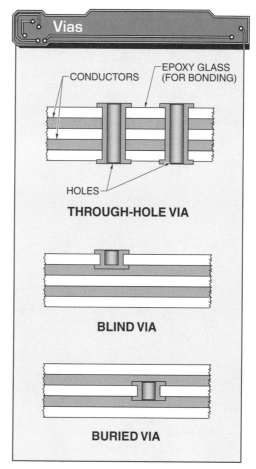

Figure 3-4. *The main types of vias are through-hole, blind, and buried.*

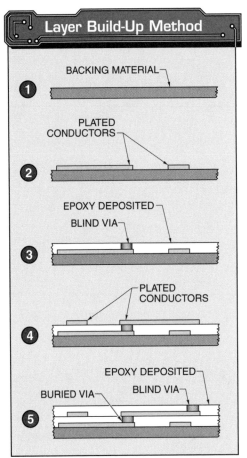

Figure 3-5. With the layer build-up method of PC board construction, conductors and insulation layers are alternately deposited on the board.

Specialty Printed Circuit Boards. In addition to the major types of PC boards, flexible PC boards and hybrid integrated circuits may be used for electronics. **See Figure 3-6.** A *flexible PC board* is a PC board that bends or flexes. It is used in applications where a PC board may need to conform to the design of a product. Flexible PC boards may be found in cameras, cell phones, and membrane keyboards. A *hybrid integrated circuit* is a combination of two or more electronic components mounted and interconnected through a common PC board. Hybrid integrated circuits serve as a customized electronic function or subsystem and are packaged as single devices and connected directly to a PC board.

PRINTED CIRCUIT BOARD MANUFACTURING

The manufacturing of PC boards includes design, building, and testing. The PC board circuit is designed and laid out using computer software and then the PC board is physically built. The more technicians know about PC boards, the more likely they are to discover manufacturing defects. Heat, vibration, open circuits, and short circuits affect the life and repair of a PC board.

Designing Printed Circuit Boards

The circuitry of a PC board is designed and entered into electronic design automation software. **See Figure 3-7.** A PC board design file is then sent to the manufacturer. The patterns for the circuitry, solder masks, and other various functions, such as drilling, are all specified in the design file. Computer-aided design (CAD) tools drive the computer-aided manufacturing (CAM) process and computer numerical control (CNC) machines that drill the holes and cut the boards.

Electronic design automation software lays out the electronic components and connections in the most readable manner. The software typically includes a comprehensive library of electronic symbols, manufacturing specification options, and simulation and specialized layout tools.

National Semiconductor
Some electronic components may require manufacturing in a clean environment.

Specialty Printed Circuit Boards

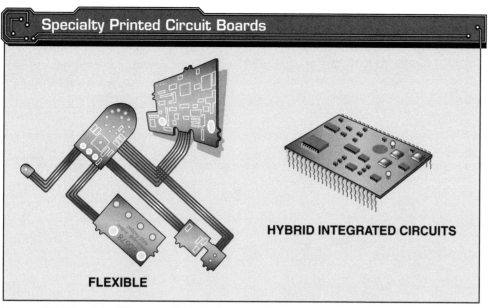

Figure 3-6. *Specialty PC boards include flexible PC boards and hybrid integrated circuits.*

Printed Circuit Board Design

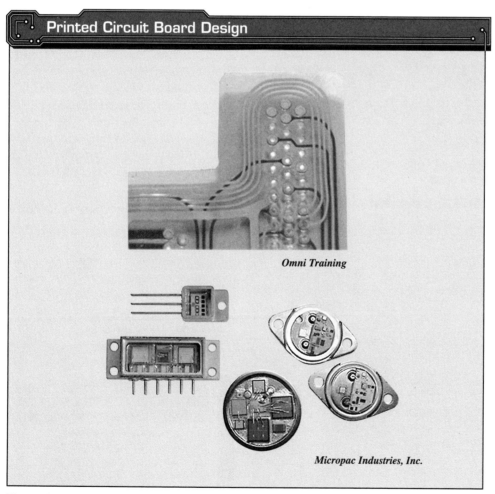

Figure 3-7. *Electronic design automation software is used to create patterns for electronic circuits, solder masks, and various other functions, such as drilling.*

The circuit design often starts as a sketch or drawing on paper before it is finished in CAD format. PC boards can be laid out manually with CAD software or automatically with an autorouter. However, the PC board designer, rather than software, often has a better idea of how to arrange circuitry. Typically, the best results are achieved by using both the manual and automatic layout methods.

Electronic circuits contain components such as transistors, diodes, and resistors on a single circuit board.

Building Printed Circuit Boards

PC boards are designed and manufactured to meet customer specifications. The PC boards must go through several steps to achieve the correct size, circuitry, and protective coating before they can be tested and assembled. **See Figure 3-8.** The following procedure is typically used to manufacture a double-sided PC board:

1. Prepare the material necessary to create the PC board, based on the information provided by the customer.
2. Drill holes and place the PC board in a plating rack, and immerse in a copper plating solution.
3. Transfer a negative image of the circuitry to the PC board surface, and cover the PC board with plating resist, except where copper circuitry is to remain.
4. Clean and place the PC board into an additional copper plating solution. Plate the copper to customer specifications.
5. Apply tin plating as etch resist to protect the circuitry during plating resist removal.
6. Chemically remove the plating resist. The tin plating protects the copper circuitry. This allows the excess copper to be exposed so that it can be removed later.
7. Remove all excess copper. The copper etching process removes all copper not protected by the tin plating. With all excess copper removed the desired circuitry remains.
8. Once the appropriate amount of copper is removed, remove the protective tin/etch resist using a chemical process.
9. Apply a solder mask to the surface of the PC board to protect the circuitry from receiving any unwanted solder during fabrication and assembly.
10. Apply a solder coat to the surface of the exposed copper where connections should be made to assist in the assembly of PC boards and their components.

Creating Circuitry. PC board manufacturers typically use two methods for creating circuitry on PC boards. The two methods are the subtractive and additive methods.

The subtractive (etch-down) method is the process of removing conductive metal from a PC board to leave only traces for PC board circuitry. The method starts with a board or substrate covered with a layer of copper over the entire PC board. Copper not needed for the circuitry is removed from the PC board, leaving only traces for PC board circuitry.

Two subtractive methods of creating circuitry on PC boards are chemical etching and photoengraving. Chemical etching can be performed using etching solutions, such as ferric chloride or ammonium persulfate. Photoengraving can be performed by using ultraviolet light and photoresistant materials.

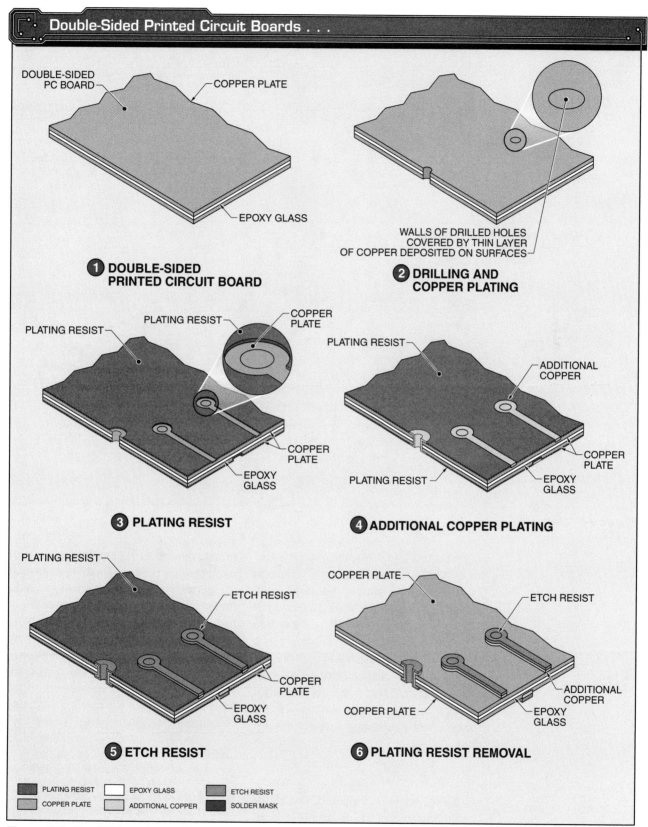

Figure 3-8. The PC board manufacturing process includes preparing materials, drilling, and chemically removing excess copper.

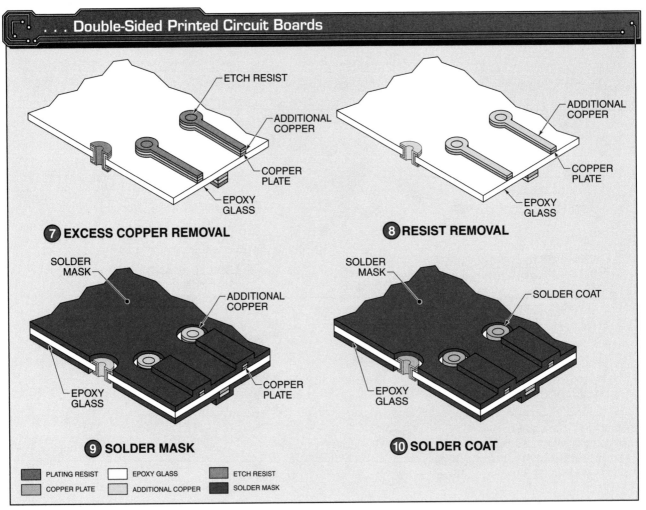

Figure 3-8. *(continued)*

The additive (plate-up) method involves the direct deposition of conductive metal on a PC board to create circuitry. **See Figure 3-9.** Typically, copper is deposited electrolytically in the additive method. First, the surface of a PC board is etched. Etching provides a reactive surface for a catalyst to be applied. A catalyst is used to make electroplating more effective. The additive process is used when conductive layers must be thinner than the traces currently being manufactured through other methods.

Testing Printed Circuit Boards

Once a PC board is completed, and before components are added to its surface, it is tested for quality assurance. Each circuit connection must be correct on the finished PC board. In large manufacturing facilities, one type of testing used is the bed of nails test. The bed of nails test is used to make contact with the copper traces, solder pads, and vias on one or both sides of the PC board. **See Figure 3-10.** Computers are used in conjunction with the bed of nails test to send current through each contact point. The computer program controls the amount of current sent to the test points, analyzes the data received, and determines if the circuit passes the test.

The additive method of creating circuitry on PC boards may be used to avoid some of the disadvantages of the subtractive methods, such as the unnecessary waste of copper from etching.

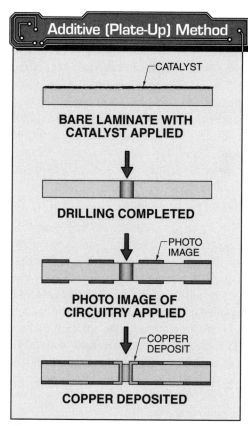

Figure 3-9. The additive (plate-up) method involves the direct deposition of conductive metal on a PC board to create circuitry.

The purpose of the test is to check the electrical integrity of the PC board. It is essential to make sure that all circuits are complete and that the appropriate circuits correlate to the original design. Using a bed of nails test ensures that every circuit is tested. Each bed is custom designed to make sure all pins are aligned with the configuration of the PC board. Current is passed between one test point to another and the signal generated is measured at the second test point. If a signal is detected, the circuit is okay. If a signal is not detected, the PC board is pulled out of production. Every circuit must be tested on the basic PC board before any components are added.

Once the PC board has been populated with components, further testing is required to determine if the complete PC board is functional. During a functional test, a computer analyzes the performance of all the functional parameters of the PC board circuits. **See Figure 3-11.**

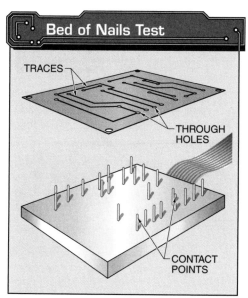

Figure 3-10. A bed of nails test is used to test a PC board by sending current through each contact point.

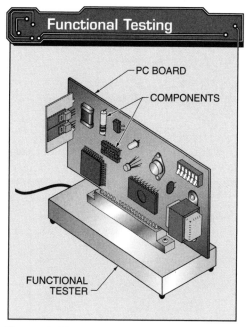

Figure 3-11. During a functional test of a populated PC board, the performance of all the functional parameters of the total circuitry is analyzed.

MOUNTING COMPONENTS ON A PRINTED CIRCUIT BOARD

Components are mounted on a PC board to complete an electrical circuit. Components are electrically connected with through-hole technology and surface mount technology. Typically, like components can be mounted with either through-hole technology or surface mount technology based on component packaging.

Through-Hole Technology

Through-hole technology is the technology and methods used to mount components on a PC board that involves placing the leads (thin wires) of the components through small, conductive through-holes in the PC board and soldering the leads to the through-holes. In mass production of assemblies using through-hole technology, a wave solder machine is used to solder the leads. **See Figure 3-12.**

Wave soldering involves a tank of molten solder configured to produce waves across the PC board surface, which allows some of the solder to reach the exposed side of the PC board. The solder reacts to the exposed metallic areas, which are not protected by a solder mask to create a reliable mechanical and electrical connection. Since multiple contacts can be soldered at one time, the process is much faster than soldering by hand.

Surface Mount Technology

Surface mount technology (SMT) is the technology used to directly mount components on the solder pads of a PC board. **See Figure 3-13.** Using SMT requires that a solder paste consisting of glue, flux, and solder be applied at the point of contact to hold components in place until the solder is melted. The components that are used with SMT are typically referred to as surface mount devices (SMDs). Once the components are in place, the PC board is reflowed in a reflow oven to make the final connection.

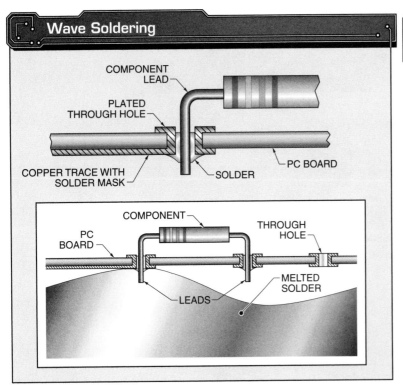

Figure 3-12. *In mass production assemblies using through-hole technology, a wave solder machine is used for soldering with surface mount devices (SMDs) once the components are in place.*

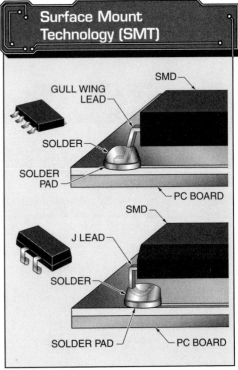

Figure 3-13. *Surface mount technology (SMT) uses a reflow oven to melt the solder once the surface mount devices are in place.*

Steps in the reflow process include preheating, flux activation, reflow, and cool down. During preheating, the temperature of the PC board is raised to a point where the solvent in the paste begins to evaporate. During flux activation, flux cleans and removes oxides from the component leads and solder pads. During reflow, the two surfaces are soldered. Finally, in the cool down step, the solder cools and bonds the components to the PC board. Although SMT requires much care in the placement of components, it eliminates the time-consuming drilling process and necessary space-consuming connection pads required for through-hole technology.

Surface Mount Devices. A *surface mount device (SMD)* is an electronic component that is soldered directly to the traces of a PC board. **See Figure 3-14.** There is a wide variety of SMD packages used for semiconductors that includes diodes, transistors, and integrated circuits. There is such a wide variety of SMD packages because of the various applications for the devices in various packages. SMDs tend to be much smaller than conventional components. SMDs are used more often in automated assembly lines than conventional components.

Small outline transistors (SOTs) are SMD packages with two or three terminals for diodes or transistors. The passive, rectangular SMD components are mainly resistors and capacitors. For certain applications, tantalum capacitors are also manufactured in this format.

Henkel Corporation
Surface mount devices (SMDs) are soldered directly to the traces of a PC board.

Surface Mount Devices (SMDs)	
Resistor	
Capacitors	
Diode	
Transistors	
Small outline integrated circuit (SOIC)	

Figure 3-14. *There is a wide variety of SMD packages used for semiconductors including diodes, transistors, and integrated circuits.*

PRINTED CIRCUIT BOARD IDENTIFICATION AND INFORMATION

A *silk-screen layer* is a layer of text that is normally applied to the outer surface of a PC board. Silk-screen layers hold information about PC boards, such as part numbers and instructions. **See Figure 3-15.** This information helps technicians to assemble, troubleshoot, and use the PC boards. The component identification information is typically silk-screened in white. When there is enough space, a great deal of information can be added to a PC board on a silk-screen layer. When space is limited, standard bar codes and new technology, such as 2D matrix laser technology, must be used.

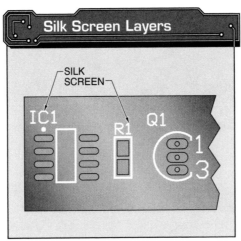

Figure 3-15. *The silk-screen layer of a PC board holds information about the PC board.*

Bar Codes

A *bar code* is a series of vertical lines and spaces of varying widths that are used to represent data. **See Figure 3-16.** Bar codes may be used on PC boards to store coded information. Information is retrieved from the bar code sequence using an optical scanning bar code reader. The bar code reader may be stationary or handheld, such as those used in grocery and retail stores, for a wide range of applications.

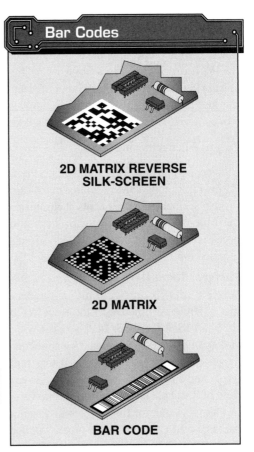

Figure 3-16. *Bar codes contain information used for PC board troubleshooting and replacement.*

2D Matrix Bar Codes. A *2D matrix bar code* is a two-dimensional bar code that can represent more data than a standard, one-dimensional bar code. It is a permanent, machine-readable code that is applied to a PC board. A 2D matrix bar code must be durable enough to withstand the manufacturing process, especially during PC board cleaning. It must also be able to store as much information needed in the allocated space without affecting circuit performance.

A 2D matrix bar code is a method of storing a large amount of data in a very small space. Laser marking technology is used to permanently apply a 2D matrix bar code to most PC board surfaces. Some of the coded data may be necessary to replace or successfully troubleshoot a PC board. Therefore, technicians should be aware of the data stored in this format and know how to retrieve the information.

PRINTED CIRCUIT BOARD HANDLING

A PC board must be handled carefully. PC boards should be handled by their edges so that oil and dirt from hands are not deposited onto the board. A PC board should be installed gently and fit snugly into its edge card connector. When removing a PC board from an edge card connection, the PC board should be gently rocked. **See Figure 3-17.** In many cases, manufacturers will design an extraction system to make PC board removal from an edge card connection safe and easy. Electrostatic discharge from improperly handling a PC board can also damage the board.

and therefore, would not be identified as the cause of further problems. The damage caused by ESD may not be visible because it occurs at the microscopic level. A PC board must be stored in a specially manufactured antistatic bag.

To prevent ESD from damaging a PC board and electronic components, any static charge buildup should be discharged by touching a conductive surface, such as grounded conduit, before touching the PC board and electronic components. Manufacturers recommend wearing a grounding wrist strap when working with sensitive electronic components to prevent ESD. When placing a PC board inside electronic equipment, the PC board should be handled by its insulated edge. Also, equipment being tested or serviced should be grounded.

TROUBLESHOOTING PRINTED CIRCUIT BOARDS

Troubleshooting is the systematic elimination of the various parts of a system, circuit, or process to locate a defective part. Operational failure of a PC board is typically caused by excessive heat, warping, scratches, or tears. **See Figure 3-18.** Technicians use troubleshooting skills on a regular basis to solve problems.

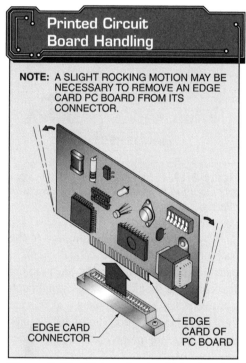

Figure 3-17. PC boards should be handled carefully during removal and installation.

Printed Circuit Board Electrostatic Discharge

A PC board can be damaged if static electricity is allowed to transfer from the hands or clothing to the PC board. Electrostatic discharge (ESD) can destroy a PC board immediately or cause future problems. Uncontrolled ESD from PC board handling may not be considered during troubleshooting

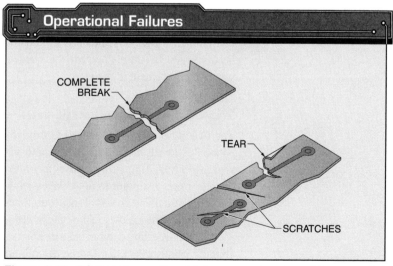

Figure 3-18. PC board operational failures may be caused by excessive heat, warping, scratches, or tears.

Effective troubleshooting involves quickly diagnosing and fixing a problem. Technicians must recognize symptoms and use them as clues to determine the cause of a problem. Therefore, it is important to know how a system and its components work. In order to become skilled at troubleshooting, a technician must have the ability to do the following:
- comprehend the information provided in technical manuals
- establish and follow a logical troubleshooting sequence
- use test equipment
- analyze and interpret test results

It is also important for a technician to use schematics, block diagrams, wiring diagrams, and flow charts to understand circuit functions A technician should always use the correct repair procedures and use quality workmanship when making repairs. If a component must be substituted temporarily, good workmanship should be used for the temporary repair. Bare wires should never be left hanging where they may touch each other. Components should never be left dangling by their leads. If component leads are supposed to hold the component in place, the leads should be secure so that the component does not vibrate or move.

Components should be properly mounted on a PC board to avoid problems with circuits. **See Figure 3-19.** Small components may be mounted by their leads. Large components may have brackets or clips to hold them in place. Brackets and clips should not be discarded because they will be needed when a replacement is made.

Printed Circuit Board Inspection

The two common methods used to troubleshoot PC boards are signal tracing and visual inspection. Signal tracing uses a current source at one test point to detect a signal at the other test point. A visual inspection can uncover problems quickly, if they can be seen.

Before energizing equipment, a technician should visually check for shorts; discolored or leaking components; loose, broken, or corroded connections; and damaged resistors or capacitors. After energizing equipment, a technician should visually check for smoking components, sparking, and open traces.

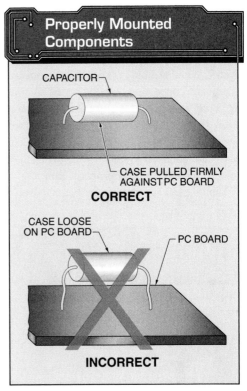

Figure 3-19. *Components should be properly mounted in order to avoid problems with circuits.*

When locating and repairing any of the problems listed under the visual check, a note of what was found and how the problem was corrected should be made. Even though a problem may have been fixed, a technician must prove beyond speculation that the equipment is operating properly and that no other problems exist.

Shorts. Any terminal or connection that is close to the conductive surface of a module, chassis, or any other terminal should be examined for the possibility of a short. A short in any part of the power supply can cause considerable damage. Therefore, any problem that may cause a short circuit should be corrected. For example, an enclosure may be shaken to listen for any rattles, which may indicate stray drops of solder, bits of wire, nuts, or screws to be removed. Failure to remove these materials could result in a short circuit.

Discolored or Leaking Components. A discolored or leaking component indicates that there is a short that should be located. If the equipment has a fuse, it should be determined why the fuse did not blow. In many cases, fuses that are too large may have been installed, or there may be a short across the fuse holder.

Loose, Broken, or Corroded Connections. Any connection that is not in good condition, such as one that is loose, broken, or corroded, may be a source of problems. If the troubled spot of a connection is not causing a current problem, it may cause a problem in the future. The connection should be repaired once the troubled spot is identified.

Damaged Resistors and Capacitors. A resistor that is discolored or charred indicates that it has been subjected to an overload. An electrolytic capacitor damaged from overloading will have a whitish deposit at the seal around the terminals. Whenever resistors and capacitors are discovered to be damaged, they should be checked for shorts. The part should be replaced after signal tracing and other diagnostics have been conducted and the problem has been identified.

Smoking Components. If any component smokes, or if boiling or sputtering sounds are heard, power should be removed immediately. A smoking component indicates that a short circuit was missed in the first inspection. A DMM should be used to check the component again, starting in the area where there is smoking.

Sparking. If sparking can be seen after tapping or heard after shaking a module or chassis, the location of the loose connection should be identified. The connection should then be repaired.

The visual inspection of a PC board can also be performed using an automated process, which is useful in high-production facilities. With automated visual inspection, software is used to inspect a digitized image of the PC board.

Open Traces. If a PC board is suspected of having an open trace, an ohmmeter can be used to check each section of the trace for continuity. **See Figure 3-20.** If the resistance increases dramatically on a trace, an open circuit is likely. As the PC board is inspected, a certain amount of flexing may indicate a hairline crack that would not have been noticed at first.

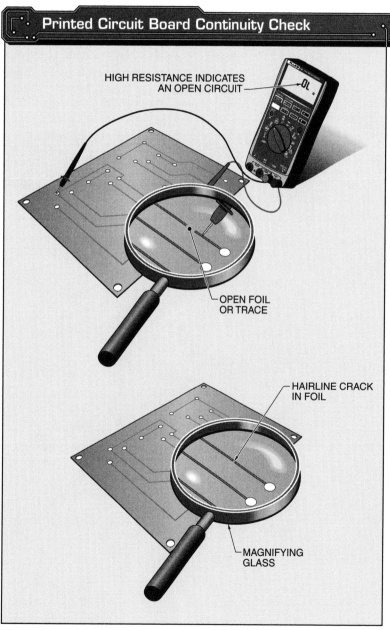

Figure 3-20. *An ohmmeter can be used to check for open circuits on a PC board.*

Printed Circuit Board Swapping

PC board swapping is the removal and replacement of a defective PC board with a properly working PC board. **See Figure 3-21.** Although swapping can help to troubleshoot PC boards, it is not a flawless technique. One disadvantage of PC board swapping is that keeping a set of spare PC boards can be very expensive. All spare PC boards on hand should be problem-free and include the most recent modifications.

Removing a PC board may break loose the corrosion on connectors. The connectors and PC board may work after reinstalling the PC board if the problem was indeed a corroded connection. Moving or twisting a PC board or connection may temporarily cure a problem in the system but not in the PC board.

When troubleshooting by PC board swapping, a technician should try to pinpoint the problem on a particular PC board by substituting one PC board at a time. Once the problem is identified, the PC board can be exchanged or repaired. A defective PC board should be labeled as defective so that it will not be used again. If a unit still does not work after all PC boards have been swapped, the original PC boards should be reinstalled, and the technician should look for other problems.

Another disadvantage is that a spare PC board may develop the same problem as the original. In complex digital systems, it is not unusual for an installation to develop problems under daily operating conditions. Problems may develop that the original designer was unable to anticipate.

Component Substitution Troubleshooting

Substituting components that appear to be defective for properly working components is a common method of troubleshooting. **See Figure 3-22.** If a table lamp does not light, for example, the existing bulb is substituted for a new bulb to determine if the bulb is defective. If the new bulb does not work, the power source should be checked by connecting a known working lamp or other appliance to the same electrical outlet. Also, the reverse substitution method could be used by placing the existing bulb in a lamp that is known to work. Both methods represent substitution troubleshooting.

A typical PC board may have dozens of passive and active components, all working together to perform one or more functions in a system. Checking these parts individually can be difficult and time-consuming. If available, a manufacturer checklist or procedure should be used to troubleshoot components. As a general rule, substitution is preferred whenever it will take significantly less time than any other solution.

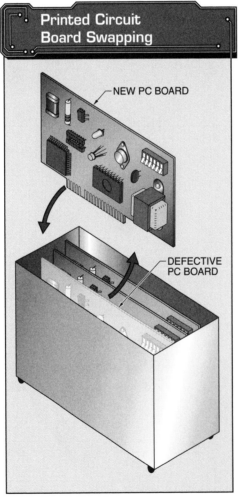

Figure 3-21. *Defective PC boards may be swapped with properly working PC boards.*

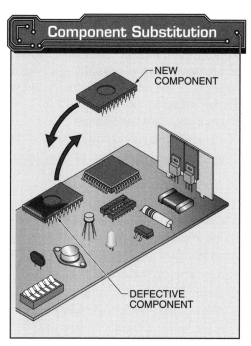

Figure 3-22. *Component substitution troubleshooting may be used to find problems by replacing components suspected of being defective with properly working components.*

Of course, there must be a good reason to believe that substitution troubleshooting may identify or fix a problem. When working with equipment, a technician develops an intuitive sense for symptom diagnosis that may help in knowing when to try substitution. An available substitute also dictates whether a component should be substituted for another. If a substitute is not available, a technician must consider how much it will cost, or how long it will take, to obtain one. Downtime for equipment or impact on an assembly line should be considered as a cost.

RECYCLING AND ELIMINATION OF HAZARDOUS MATERIALS

Most electronic devices should not be discarded into a landfill. Although many devices do not contain hazardous waste products, they may contain many resources that can be recycled and reused. These devices can be dismantled and separated into plastic, glass, and various metals, which can become useful resources. **See Figure 3-23.** However, PC boards must be removed from their devices because some of the materials in PC boards may be hazardous. The most common hazardous material found in PC boards is the lead in the solder. It is critical that technicians know how to properly dispose of PC boards and other electronics that must be discarded.

Instead of off-site disposal, several companies offer recycling services to reclaim resources contained in a PC board and its components. The amount of money paid as salvage for a PC board is determined by the volume and type of its salvageable items and the presence of any hazardous materials. PC boards containing hazardous materials, such as mercury, may have no recyclable value. In fact, there may be a fee for their proper disposal.

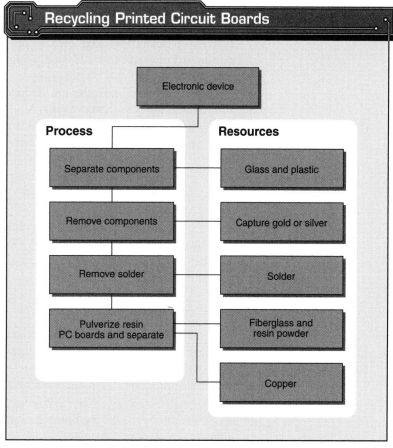

Figure 3-23. *Most materials from electronic devices can be recycled and reused.*

Directives for electronic devices are designed to reduce the impact of electrical equipment on the environment during disposal. The two main directives are the Restriction of Hazardous Substances (RoHS) Directive and the Waste Electrical and Electronic Equipment (WEEE) Directive.

The intent of the RoHS and WEEE directives is to make electrical and electronic equipment manufacturers environmentally responsible for their products. There are financial incentives for manufacturers to design products with resources that can be recovered, recycled, and reused.

RoHS Directive

The Restriction of Hazardous Substances (RoHS) Directive is a European Union directive that prohibits the import of consumer electronics containing lead, mercury, chromium, and other chemicals considered hazardous to the environment. The RoHS Directive is enforced at the homogeneous material level. A *homogeneous material* is a material that cannot be broken down by common mechanical processes. Homogeneous materials include alloys, plastic, glass, ceramics, certain liquids, and other low-density materials that are consistent in composition.

The RoHS Directive is enforced at the homogeneous level, but actually, heterogeneous (nonhomogeneous) materials are more common than homogeneous materials. A *heterogeneous material* is a material that can be broken down by common mechanical processes. Examples of heterogeneous materials include PC boards, cable assemblies, and individual electronic components.

WEEE Directive

The objective of the Waste Electrical and Electronic Equipment (WEEE) Directive is a European Union directive to reduce electrical and electronic equipment waste; increase the recovery, recycling, and reuse of electronic equipment; and improve the disposal processes of nonrecyclable materials. The directive targets original equipment manufacturers (OEM).

However, technicians should be aware of the impact this directive has on their working environment. The ten categories of electrical and electronic equipment targeted by the WEEE Directive include the following:

- large household appliances
- small household appliances
- IT and telecommunications equipment
- consumer equipment
- lighting equipment
- electrical and electronic tools (excluding large-scale stationary industrial tools)
- toys, leisure, and sports equipment
- medical devices (excluding all implanted and infected products)
- monitoring and control equipment
- automatic dispensers

> Both the RoHS Directive and the WEEE Directive have become law in the European Union. Although the United States has yet to adopt these directives as federal law, the state of California has passed a law called the Electronic Waste Recycling Act of 2003 (EWRA), which is similar to the directives.

CONFORMAL COATINGS

Manufacturers are constantly looking for ways to improve the reliability of their products. One way to improve the reliability of an assembly is to use a conformal coating. A conformal coating is a thin layer of material applied to the surface of a PC board and its components to provide protection.

Conformal coatings are applied to electronic assemblies to prevent damage from corrosion, dust, moisture, and stress. They also extend the life of a PC board by reducing damage from handling during construction, installation, and usage. Various types of conformal coatings are available, each with different characteristics that improve reliability. **See Figure 3-24.** The types of coatings include acrylic, silicone, urethane, epoxy, and paraxylene.

Conformal Coating Types

Type of Coating	Ease of Use	Characteristics
Acrylic	Easy to apply and easy to remove overall	Tough and transparent; does not show great resistance to abrasion or chemicals
Silicone	Easy to apply and easy to remove in small sections	Tough abrasion resistance; good moisture and humidity resistance; useful over a wide range of temperatures; low toxicity
Urethane	Difficult to apply and very difficult to remove	Tough and exhibits great resistance to solvents; excellent abrasion resistance; low moisture permeability; low temperature flexibility
Epoxy	Difficult to apply and very difficult to remove	Coating is very hard, usually opaque, and resists the effects of moisture; excellent chemical and abrasion resistance; may require extra precaution in safe handling
Paraxylene	Applied by vacuum deposition, requires the preprocessing of boards, and uses very specialized equipment	Provides uniform coating; sensitive to contaminants

Figure 3-24. *Various types of conformal coatings are available, each with different characteristics.*

Even though most coatings are liquid, the coating must be thick enough to remain on the PC board surface as a uniform film. It cannot be too thick or the surface tensions can cause voids. It is extremely important that surface contaminants be removed before a conformal coating is applied. Furthermore, because certain fluxes have a corrosive nature, they must be removed thoroughly. If flux becomes trapped under the conformal coating, a potential failure could occur in the future.

Applying Conformal Coatings

Conformal coatings are supplied as aerosol sprays or liquids. **See Figure 3-25.** In a production environment, conformal coatings can be applied using the spray, brush, dip, or flow method. The spray and brush methods are likely to be the most convenient for most repair applications.

Applying conformal coating using a brush is fairly simple. However, a technician must remember that many of the coatings that can be applied by brush are cured by moisture in the air. Therefore, these coatings will begin to cure as soon as their container is opened. Applying moderate heat using a convection oven or infrared lamp may accelerate curing. Heat curing will improve the physical strength of the coating.

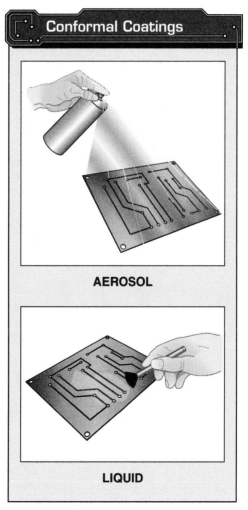

Figure 3-25. *Conformal coatings can be found in aerosol and liquid form.*

Applying conformal coating using the spray method is more difficult than using the brush method. Electrical conductors, such as connectors, wiring terminals, and jumpers, may need to be masked prior to coating to preserve their conductivity.

Safety is always an issue when using conformal coatings. There should be adequate ventilation when applying any type of conformal coating. Manufacturers usually provide a material safety data sheet (MSDS) that explains handling and exposure guidelines.

Removing Conformal Coatings

Because of the characteristics of conformal coatings, they must be removed before any work can be done on PC boards. The coating must be removed from the lead area of the component and the widest point of the component body.

Conformal coatings can be removed by chemical, thermal, or mechanical means. The best method of removal depends on the type of coating used. Standards J-STD-001 and IPC-A-610 established by the IPC® should be followed when removing conformal coatings.

Chemical Removal of Coatings. A liquid conformal coating solvent can quickly and precisely remove most conformal coatings, including acrylics and silicones. Many of these solvents are available in tubes. **See Figure 3-26.**

Thermal Removal of Coatings. The thermal removal of conformal coatings involves using thermal parting tools to soften and scrape away the coatings. Soldering irons should never be used for coating removal because the high temperatures will cause the coatings to char and possibly damage PC board materials.

A thermal parting tool may have a regulated power supply. The thin, blade-like instruments act as heat generators and will maintain the heat levels necessary to remove conformal coatings. Thermal parting tips can be changed easily to suit the amount of coating to be removed. **See Figure 3-27.**

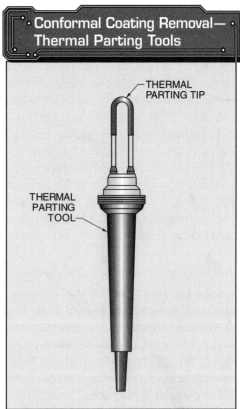

Figure 3-27. *Thermal parting tips may be used to remove conformal coatings by heating them and then scraping them away.*

Hot-Air Jet Removal of Coatings. A hot-air jet uses controlled temperature-regulated air to soften or break down coatings. By controlling the temperature, flow rate, and shape of the jet, coatings may be removed

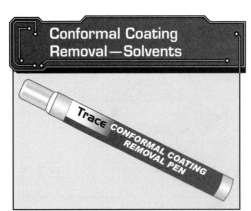

Figure 3-26. *A liquid conformal coating solvent delivered in a tube can quickly and precisely remove most conformal coatings, including acrylics and silicones.*

from almost any workpiece configuration without causing damage. When using a hot air jet, it should not be allowed to physically contact the workpiece surface. Delicate work handled in this manner allows observation of the removal process. **See Figure 3-28.**

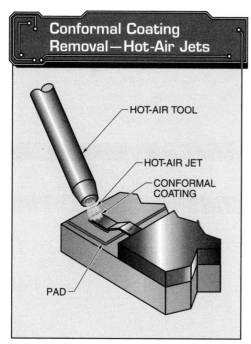

Figure 3-28. *A hot-air jet with temperature, flow rate, and shape control may be used to remove conformal coatings without causing damage.*

Power Tool Removal of Coatings. The power tool method involves a grinding or cutting action to mechanically remove coatings. **See Figure 3-29.** Abrasive grinding/rubbing methods are used for thin conformal coatings, while abrasive cutting methods are used for thick coatings. This method allows the consistent and precise removal of coatings without the damage caused from heating the electric components. A variable-speed motor permits fingertip control and proper speed for ease of handling.

A variety of rotary abrasive material and cutting tools are available. These specially designed tools include ball mills, burrs, and small rotary grinders.

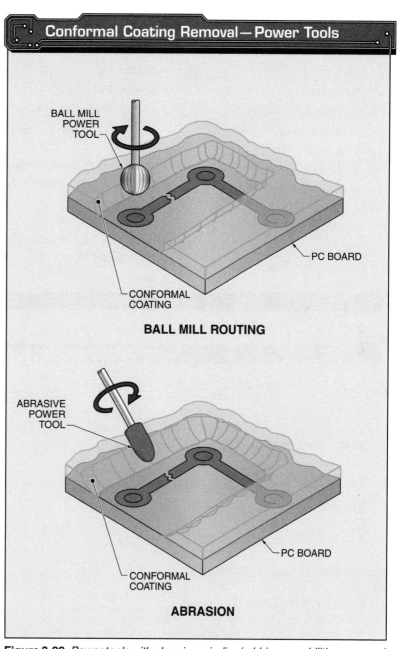

Figure 3-29. *Power tools with abrasive grinding/rubbing capabilities are used to remove conformal coatings from PC boards.*

The ball mill is the most efficient cutting tool. Different mill sizes are used to enter small areas where thick coatings need removal. Rubberized abrasives of the proper grade and grit are used to remove coatings from surfaces. Rotary bristle brushes work better than rubberized abrasives on contoured or soldered connections because the bristles conform to surface irregularities.

Refer to Chapter 3 Quick Quiz® on CD-ROM

KEY TERMS

- printed circuit (PC) board
- trace
- solder pad
- solder mask
- single-sided PC board
- double-sided PC board
- plated hole
- multilayered PC board
- preimpregnated (prepreg) material
- via
- through-hole via
- blind via
- buried via
- flexible PC board
- hybrid integrated circuit
- electronic design automation software
- through-hole technology
- surface mount technology (SMT)
- surface mount device (SMD)
- small outline transistor (SOT)
- silk-screen layer
- bar code
- 2D matrix bar code
- electrostatic discharge (ESD)
- PC board swapping
- component substitution troubleshooting
- homogeneous material
- heterogeneous material
- Restriction of Hazardous Substances (RoHS) Directive
- Waste Electrical and Electronic Equipment (WEEE) Directive
- conformal coating

REVIEW QUESTIONS

1. What is a printed circuit (PC) board?
2. What is a trace?
3. What is a solder mask?
4. What is preimpregnated (prepreg) material?
5. What are the three main types of PC boards?
6. List and describe the different types of vias.
7. What is a flexible PC board?
8. Define hybrid integrated circuits.
9. Describe the process of manufacturing double-sided PC boards.
10. Describe through-hole technology.
11. Explain surface mount technology (SMT).
12. What is a surface mount device (SMD)?
13. Explain silk-screen layers.
14. Define bar code.
15. Define 2D matrix bar code.

SOLDERING AND DESOLDERING

Soldering is the process of making an electrical and mechanical joint between certain metals by joining them with solder. Without the process of soldering, printed circuit boards would be of little use because components could not be electrically and mechanically joined to them. Desoldering is just as important as soldering. Desoldering allows the bad components of a printed circuit board to be replaced by new components without having to discard the printed circuit board.

OBJECTIVES

- Describe how printed circuit (PC) boards are inspected.
- Explain PC board repair.
- Describe how soldering aids are used for PC board repair.
- Describe the soldering process.
- Explain soldering and rework stations.
- Describe the methods of desoldering.
- Discuss electrostatic discharge workstations.

PRINTED CIRCUIT BOARD FIELD REPAIR

A PC board may often need to be repaired in the field. Repairs to PC boards in the field include repairing breaks in foil, PC board corners and pads, or lifted foil. In addition, defective resistors or capacitors may be replaced and edge card connectors may be cleaned in the field.

For complete specifications and guidelines for the rework and repair of PC boards, refer to IPC-7711/7721. Formerly the Institute for Printed Circuits, IPC® is an association that creates standards for PC board design and manufacturing and also electronics assembly for the electronics interconnection industry.

Repairing Printed Circuit Board Foil

If a break is located, any loose or separated foil should be cut away. The removal of the loose foil to a secure point enables a repair to be made. Once the foil has been cut back to a secure point, an appropriate solvent should be used to remove any protective coatings from the foil section to be repaired.

See Figure 4-1. CAUTION: Care must be exercised when cutting so that hairline cracks, caused by excessive flexing, are not created in the PC board.

When the protective coating is removed, it is helpful to clean the surface of the foil. In the past, either fine steel wool or a light abrasive cloth was used to remove oxidation before soldering. However, these materials often proved harmful to a PC board with smaller traces or higher density of components. Manufacturer recommended solvents and cleaners are now used to avoid damaging PC boards.

U.S. Navy
Soldering can be used to connect a replacement component to a PC board.

74 SOLID STATE DEVICES AND SYSTEMS

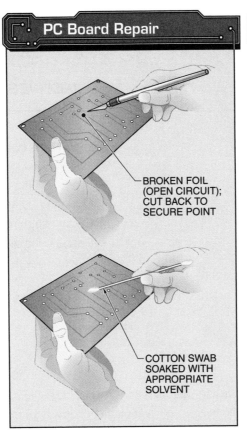

Figure 4-1. *Broken foil should be cut back to a secure point. The foil section to be repaired should be cleaned with an appropriate solvent to remove any protective coatings.*

Repairing Small Breaks in Foil. If there is a small break in foil, a short piece of bare jumper wire soldered across the open foil is sufficient to bridge the gap for emergency or field repairs. **See Figure 4-2.** To repair a small break in foil, apply the following procedure:

1. Tin a short piece of jumper wire by melting a small amount of solder to the wire.
2. Place the jumper wire across the break.
3. Secure the jumper wire.
4. Solder and allow the solder to flow along the length of the break so that the tinned jumper wire becomes part of the circuit.

Note: Only enough heat to melt and flow the solder should be applied. If components are in close proximity, the jumper wire may be wrapped around the leads to secure it before soldering.

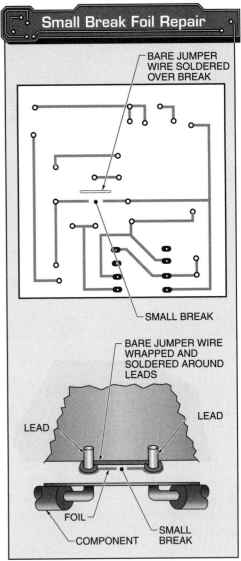

Figure 4-2. *For small breaks in foil, a short piece of bare jumper wire can be soldered across the open foil.*

A three-step repair procedure using pen technology also can be used to repair small cracks or breaks on a PC board. This procedure requires the use of a flux remover pen, a conductive pen, and an overcoat pen. The procedure allows technicians to quickly perform quality PC board repairs without the use of expensive equipment. **See Figure 4-3.** The three-step repair procedure for repairing small cracks or breaks in foil is as follows:

1. Rub a flux remover pen tip over the area to be cleaned until fluxes and contaminants are removed from the surface of the affected area.
2. Use a conductive pen to fill the gap on the damaged foil.
3. Once the trace is repaired, use an overcoat pen to apply a coating to protect the repaired trace.

Repairing Large Breaks in Foil. When breaks in foil are over 1″ in length, it is often best to use a piece of insulated jumper wire to bridge the circuit gap. In many cases, the insulated jumper wire eliminates the need to follow the continual path of the foil. It also prevents contact with the bare conductor. The insulated jumper wire may be on the front side or back side of the PC board. Once the repair is complete, a conformal coating should be applied to the PC board by using a brush or spraying with an aerosol. **See Figure 4-4.**

Repairing Printed Circuit Board Corners

Poor handling procedures may damage the corners or edges of PC boards. Most damaged corners or edges of PC boards can be repaired using commercially available PC board repair kits. These repair kits offer a means of saving high-value PC boards by providing the materials to replace damaged corners, edges, and holes. Repair kits include all the tools and materials needed to successfully and safely repair a PC board.

Repairing Printed Circuit Board Pads

Pad repair kits include the tools and materials needed to replace lifted or damaged surface mount pads and ball grid array (BGA) pads on PC boards. Through-hole repair kits are also available and widely used. Pad and through-hole repair kits include several different tin-plated circuit frames, allowing for a large quantity of usable replacement land patterns to be available. The tin-plated circuit frames can be used on both lead-free and tin-lead finish PC boards.

> *Even if special no-clean fluxes have been used, it is recommended that a flux remover pen be used to ensure that no residue or contaminants remain on a PC board.*

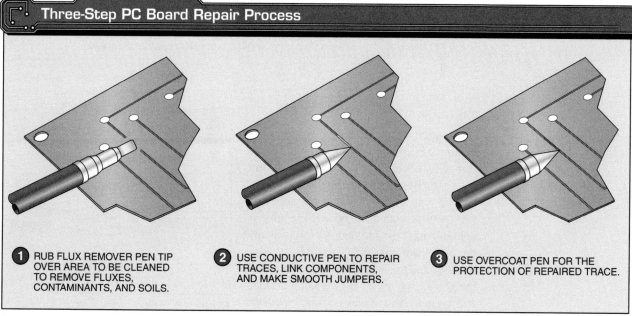

Figure 4-3. *A three-step PC board repair procedure requires the use of a flux remover pen, a conductive pen, and an overcoat pen.*

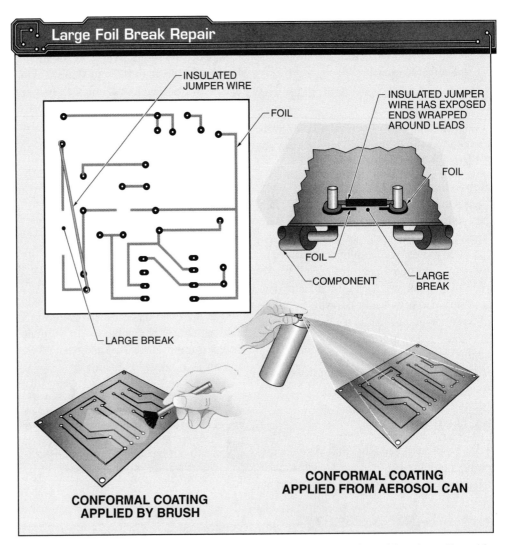

Figure 4-4. *For large breaks, a piece of insulated jumper wire can be soldered on either side of the PC board and a conformal coating applied once the repair is complete.*

Repairing Lifted Foil

A common problem caused by excessive heat on a PC board is foil separation. If the foil is lifted and not damaged, it should be reattached to the PC board with glue. Epoxy-type glues are used for adhering foil to an insulated PC board. The glue is applied under the foil and the foil is firmly pressed down onto the PC board. **See Figure 4-5.** The foil is held firmly in place until the glue sets. If necessary, the foil is cleaned and any protective coating that may have been removed is reapplied. Dry film adhesive kits are also widely used to repair lifted foil.

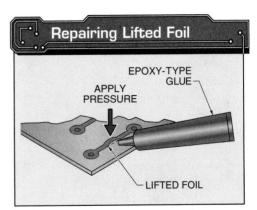

Figure 4-5. *Epoxy-type glues and a reasonable amount of pressure are used to reattach lifted foil without further repair.*

Replacing Defective Resistors or Capacitors

It may be possible to quickly replace a defective resistor or capacitor without removing a PC board from the chassis by using the crush-and-cut procedure. **See Figure 4-6.** If the leads extending from the defective component are long enough for a replacement part to be soldered to them, the component leads may be cut where they enter the defective component. The new component may be soldered to the component leads. If the defective component leads are short, the component is cut in half. Excess material is removed by crushing the defective component to take advantage of the lead length within the component. The defective component leads are looped, and the replacement components are soldered to the defective component leads.

Cleaning Edge Card Connectors

Edge card connectors can be cleaned using a contact protector/enhancer, which is available in aerosol or pen dispenser form. **See Figure 4-7.** A contact protector/enhancer is used to clean and protect contacts, lubricate contacts for the easy insertion of a PC board, and prevent oxidation. Contact protector/enhancers are noncorrosive and safe for use on gold and platinum contacts.

SOLDERING AIDS

Soldering aids are instruments designed specifically for PC board repair. They are made of nonmagnetic steel, so solder will not adhere to them. A soldering aid can have a fork, hook, reamer, knife, or brush end. **See Figure 4-8.** The fork end of a soldering aid straddles wire for looping and guiding. The hook end is used to hook wires. The reamer end cleans lug holes and is used for deburring. The knife end removes surplus solder. The brush end cleans solder connections. There are various types of brushes available. Soldering aids also include heat sinks and thermal shunts.

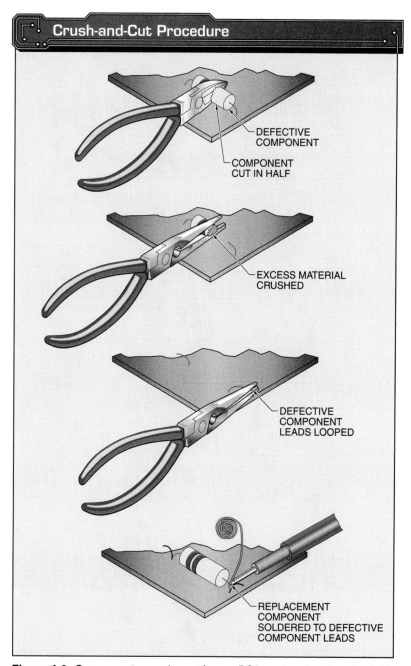

Figure 4-6. *Components may be cut from a PC board and replaced quickly using the crush-and-cut procedure.*

It is standard practice to use soldering aids and other handheld instruments for PC board soldering and repair. Handheld instruments allow a technician to solder and closely inspect and repair detailed portions of a PC board. However, it is more common to solder mass-produced PC boards using wave solder machines.

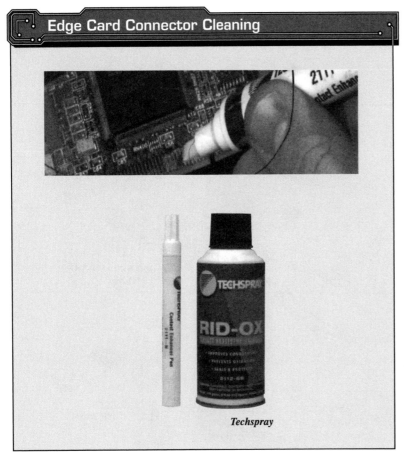

Figure 4-7. *Cleaning edge card connectors using a contact protector/enhancer is a maintenance procedure used to protect contacts, lubricate PC boards, and prevent oxidation on gold and platinum contacts.*

Brushes

Brushes are used in a variety of soldering applications. Brushes can be used for PC board preparation, solder ball removal, edge connector cleaning, and other aspects of PC board repair. **See Figure 4-9.**

Electrostatic discharge (ESD) must be considered when using brushes. Static electricity can be generated by a variety of sources. The most common source of static electricity is friction caused by rubbing two materials together. This process is called the "triboelectric effect." A material that inhibits the generation of static charges from the triboelectric effect is classified as an antistatic material. Antistatic materials used to make brushes include wood, hog bristle, horse hair, and goat hair. These materials can be used in ESD-sensitive areas.

The most critical characteristics of a brush that determine its electrostatic properties are the fiber, handle material, and type of grounding. If the mounting location is not grounded, then a grounding strap or a 1 MΩ resistor should be used between the brush and an electrical ground. The various brushes available include conductive, dissipative, antistatic, and insulative brushes.

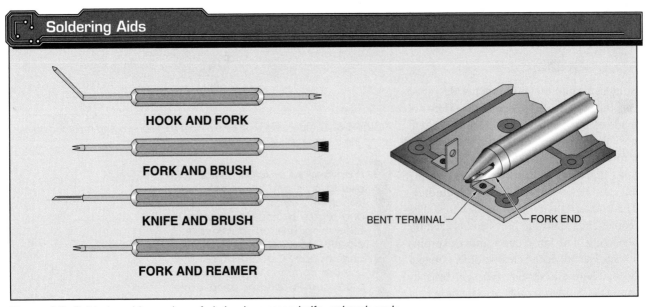

Figure 4-8. *Soldering aids can have fork, hook, reamer, knife, or brush ends.*

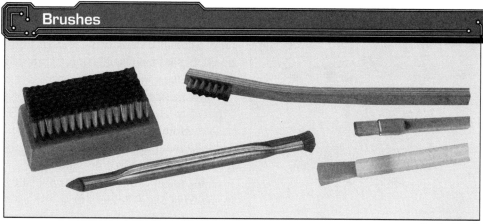

Gordon Brush

Figure 4-9. *Brushes can be used for PC board preparation, solder ball removal, edge connector cleaning, and other aspects of PC board repair.*

Conductive Brushes. Conductive brushes have a low surface resistivity and are used when it is necessary to remove static buildup from the work surface. Conductive brushes provide a path between the charged surface and ground. A technician must select the best size and style of brush for an application.

Dissipative Brushes. Dissipative brushes have a surface resistivity that is higher than a conductive brush but still conductive enough to avoid static buildup in the brush. These brushes are used to remove any static charge that could attract particles, such as dust, to the PC board work surface. Dissipative brushes should be used with a wrist strap or other grounding system.

Antistatic Brushes. Antistatic brushes are made from low-charging materials that are nearly neutral with the triboelectric effect. This means that they neither give nor take on electrons and essentially remain electrically neutral. In highly voltage-sensitive applications, antistatic brushes should be used only with an antistatic liquid or immersed in antistatic liquid before use. A popular antistatic liquid is isopropyl alcohol.

Insulative Brushes. Insulative brushes have a high surface resistivity. They are an inexpensive, water-resistant alternative to antistatic brushes. Insulative brushes are best used in an environment where static is not critical to the immediate task or where the brush is submerged in liquid during use.

It has been found that when soldering in space without gravity, gas bubbles can form pores in a solder joint and reduce the strength of the joint. On Earth, solder joint strength depends on gravity and convection.

Heat Sinks and Thermal Shunts

Heat sinks and thermal shunts are soldering aids used to draw heat away from components and protect them from heat damage. The terms "heat sink" and "thermal shunt" are often used interchangeably. Many types, shapes, and sizes of heat sinks and thermal shunts are available. When space is critical, or when a small component is involved, a special heat sink called a "seizer" is used. **See Figure 4-10.**

Before heat is applied to solder a joint, a heat sink is attached to the leads of sensitive components, such as diodes, transistors, and microchips. Because of its high heat capacity and high thermal conductivity, copper is usually used for heat sinks. Aluminum can also be used for heat sinks, but it has lower heat conductivity.

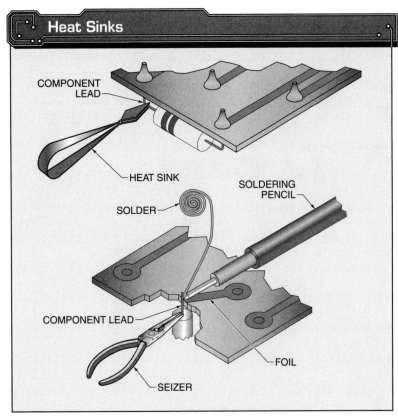

Figure 4-10. *A heat sink applied to a component lead protects the component from heat damage. A seizer is used where space is critical and a small heat sink is necessary.*

THE SOLDERING PROCESS

Soldering is the process of making a sound electrical and mechanical joint between certain metals by joining them with solder. A good solder joint gives a metallurgical bond to metals. *Solder* is an alloy consisting of specific percentages of two or more metals. In electrical work, an alloy usually consists of tin (Sn) and lead (Pb). In electronics work, especially with PC boards, lead-based solders are being replaced by lead-free (Pb-free) solders.

The soldering iron tip transfers thermal energy from the solder station through its tip to the solder connection. The conductivity of a soldering iron tip determines how fast the thermal energy can be transferred from the heater to the solder joint. The shape and size of the soldering iron tip also affect its performance. Temperature is important, but it is not the major factor in determining the supply of heat. The length and size of the soldering iron tip determine heat flow capability. The actual shape of the soldering iron tip establishes how well heat is transferred from it to the connection.

Most metals react with oxygen when exposed to air, especially if the metal is heated. Both oxygen exposure and heat are present during soldering. Solder flux is formulated to remove a thin film of oxide from the metal and keep further oxidation from taking place. Solder flux is also used to make the solder and metal surfaces of the component leads and PC board pads bond more easily with each other. During the soldering process, heat from the moving soldering iron tip liquefies the solder and simultaneously vaporizes the flux to remove oxide film. **See Figure 4-11.**

Soldering Iron Tips

Good soldering skills are essential for repairing PC boards. Careless work creates unnecessary damage. When soldering, the soldering iron tip must heat the metal to solder-melting temperature before actual soldering can take place. The tip should be held directly against one side of the component lead. **See Figure 4-12.** Solder should be applied from the opposite side. The solder-melting temperature is reached in a matter of 1 sec to 5 sec. Therefore, the soldering iron tip and the solder should be applied simultaneously. **CAUTION:** Be sure to apply solder to the opposite side of the component lead, not directly to the soldering iron tip. If the soldering procedure is done correctly, the solder will flow across the joint, firmly bonding the component lead to the PC board.

When soldering, exposing PC board components to heat beyond their thermal tolerance may damage the components beyond repair. Unnecessary replacements can be costly, which is why the proper heat sinks and thermal shunts should be used.

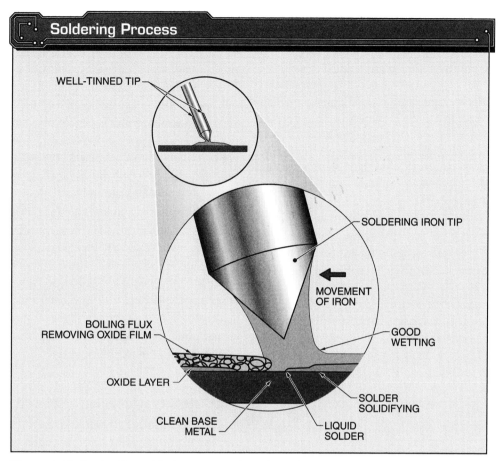

Figure 4-11. *During the soldering process, heat from the moving soldering iron tip liquefies solder and simultaneously vaporizes flux to remove oxide film.*

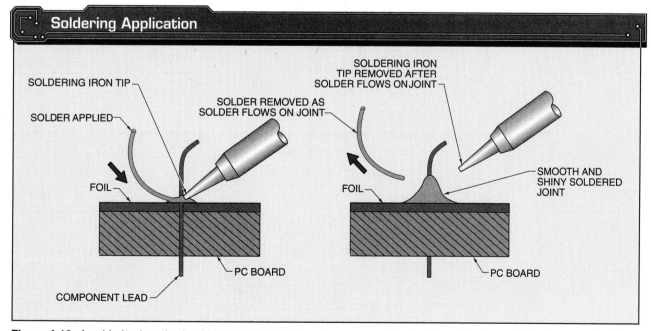

Figure 4-12. *A soldering iron tip should be held to one side of a component lead while solder is applied to the opposite side.*

Selecting Soldering Iron Tips. Selecting the proper tip will improve heat transfer to the soldering application. The correct tip should have similar shape and dimension of the object being soldered. Flat tips produce a larger contact area with a connection than do conical tips. Flat tips also tend to transfer heat more efficiently. The tip should match the width of the object being soldered. A wide range of tip sizes and shapes are available for use with multiple applications. **See Figure 4-13.**

The amount of heat and how it is controlled are critical factors in the soldering process. The tip of the soldering iron transfers heat from the soldering station to the work. The size and shape of the tip are mainly determined by the type of work to be performed. The tip size and the wattage of the heating element must be capable of rapidly heating to the melting temperature of solder.

After the proper soldering iron tip is selected and attached to the soldering iron, the technician may control the heat of the soldering iron by using a regulated soldering station. Ideally, the solder joint should be brought to the proper temperature rapidly and held there for a short period of time. In most cases, soldering should be completed within 1 sec to 5 sec. When soldering a smaller connection, heat may be controlled by using a smaller soldering tip.

When selecting a soldering tip, more emphasis should be placed on heat transfer efficiency rather than absolute tip temperature. Factors such as tip shape, tip condition, power output of the soldering iron, and time on the joint will impact heat transfer efficiency. These factors should be taken into account when monitoring, controlling, and defining the soldering process.

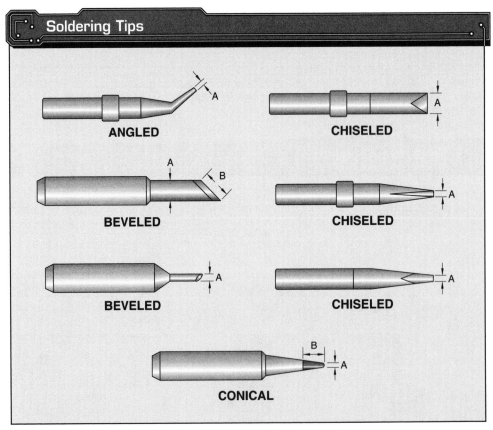

Figure 4-13. *A wide range of tip sizes and shapes are available for use with multiple applications. The tip of the soldering iron transfers heat from the soldering station to the work. The shape and size of the tip are mainly determined by the type of work to be performed.*

Care of Soldering Iron Tips. One of the most common causes of tip failure is the loss of the protective layer of solder, which causes the tip working surface to become oxidized. This is commonly referred to as a de-tinned tip. A de-tinned tip will minimize the ability of the tip to accept solder. Without a properly tinned tip, it is virtually impossible to transfer sufficient heat to metals to be joined. **See Figure 4-14.**

a sponge is to expose the entire sponge to the water source and then squeeze to remove excess water. It should be damp but not dripping wet. Synthetic sponges are not recommended for soldering tips because they may include elements or contaminants, which could actually reduce the life of the tip.

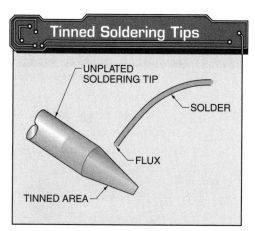

Figure 4-14. *Without a properly tinned soldering tip, it is virtually impossible to transfer sufficient heat to metals to be joined.*

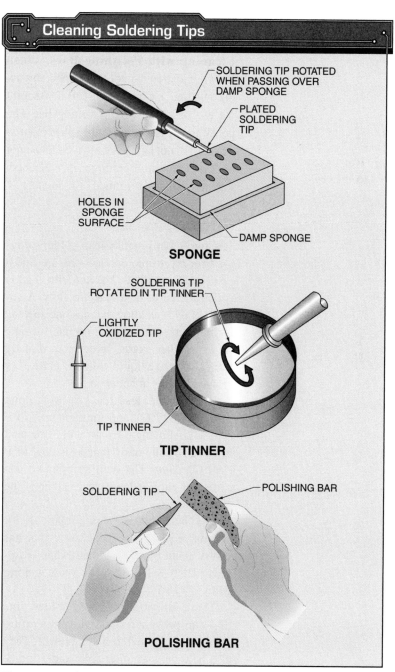

Figure 4-15. *Soldering tips may be cleaned with a sponge, tip tinner, or polishing bar.*

An oxidized tip will also lose its wettability and cannot supply heat to some of the part being soldered. Wetting occurs when solder covers the tip and acts as a heat transfer medium. Wetting means the molten solder leaves a continuous permanent film on the metal surface. Wetting can only be done properly on the clean surface of a tip. If the tip does not have good wettability, it will only contact the part over a tiny area and will not be able to transfer heat efficiently.

All dirt and grease must be removed from the metal surface of the tip. Using light abrasives and/or fluxes to remove these contaminants produces highly solderable and wettable surfaces. Soldering tips may be cleaned with a sponge, tip tinner, or polishing bar. **See Figure 4-15.**

Cleaning with Sponges. Sponges with holes can provide more surface area and allow contaminants to fall to the bottom of a sponge tray. The best way to dampen

Cleaning with Tip Tinners. Tip tinner is used to remove light oxidation from soldering tips. Cleaning with a tip tinner is necessary when a sponge does not work to remove oxidation. A tip is rotated in the tip tinner until a bright tinning appears on the surface of the tip. Solder should be applied immediately to re-tin the surface of the tip. Tip tinner should not be overused because it will reduce the useful life of the tip.

Cleaning with Polishing Bars. When heavy oxidation cannot be removed by using a tip tinner, cleaning the tip may require the use of an abrasive polishing bar. The polishing bar is applied to the soldering iron when cool, not hot. Care must be taken not to remove the iron plating. Once cleaned, the tip should be re-tinned immediately.

Lead Solders

The two most popular lead solder alloys used for electronic connections are Sn63/Pb37 and Sn60/Pb40. The Sn63/Pb37 alloy is considered to be a eutectic alloy. A *eutectic alloy* is an alloy that has one specific melting temperature with no intermediate stage. In other words, solder goes directly from a solid to a liquid with no "plastic" or "elastic" phase in between.

For example, lead has a melting point of 621°F (327°C), and tin melts at 450°F (232°C). Alloying 63% tin with 37% lead forms a eutectic alloy that melts at 374°F (190°C). *Note:* The melting point of this particular alloy is lower than either of the parent metals.

The Sn60/Pb40 alloy is a relatively inexpensive general-purpose alloy that has a very small plastic regional range of approximately 8°F. It also has a low melting point of 374°F (190°C).

These alloys allow rapid solder time that can prevent excessive temperature from being applied to a component. They also help prevent unreliable solder joints. Soldering must be focused on the initial solder connection.

A solder connection is not made from a shiny hot melt glue. A solder connection consists of an alloy that combines metallurgically with the surface of another metal to form an intermetallic layer. The intermetallic layer forms the electrical and mechanical connections. Any joint that exhibits a high resistance will adversely affect the operation of a circuit. A good lead solder joint is smooth and shiny. **See Figure 4-16.**

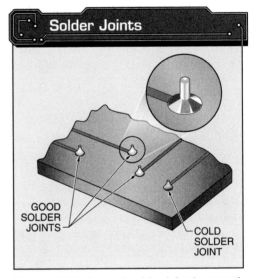

Figure 4-16. *A good solder joint is smooth and shiny.*

Solder joints that are overheated generally have a dull, bumpy exterior. Overheated joints are often mistakenly called "cold solder joints." A *cold solder joint* is a defective solder joint that results when the parts being joined do not exceed the liquidus temperature of the solder. A cold solder joint can be identified as jagged shapes on the surface of a joint. Overheated and cold solder joints indicate a lack of proper wetting techniques when solder was applied. It should be obvious from the jagged shapes of the solder joint that the solder did not flow properly within the joint.

Eutectic tin-lead solders Sn63/Pb37 and Sn60/Pb40 have been in use for a long time. Tin-lead solder has many advantages including a melting point of 374°F (190°C). This temperature is high enough to remain solid in normal applications but still low enough to not seriously damage a PC board and its components. Tin-lead solder also has a fairly long shelf life when properly stored. Even though there are advantages to using lead-based solder, modern concerns over lead exposure in the environment have forced the industry to seek lead-free alternatives.

There is currently no time limit on the availability of spare parts containing lead solder placed on the market before the restriction of hazardous substances (RoHS) directive deadline. Tin-lead solder continues to be available, as there are many types of products that are not covered by the RoHS directive.

Solder Fluxes

Reliable solder joints can only be achieved when the soldering surfaces are clean and free of oxide. Surfaces are usually cleaned with solvents and abrasives. However, almost immediately after cleaning, oxidation of the surface begins again. Most metals react with oxygen when exposed to air. This is particularly true if the metal is heated. Both oxygen exposure and heat are present in the soldering process. Solder flux is formulated to remove a thin film of oxide from metal and keep further oxidation from taking place. Flux is also used to make the solder and metal surfaces of the component leads and PC board pads more likely to bond with each other. Therefore, flux application is very important in the entire soldering process.

Technicians must be careful in the selection of fluxes and use good cleaning techniques to remove hazardous residues. Cleaners and defluxers are available in spray and aerosol form. Cleaners and defluxers remove the most stubborn residues without requiring brushing. Aggressive brushing can damage leads and does not necessarily clean under components, such as large-scale integrated circuits. The strong stream from an aerosol dispenser will blast residues off the surface, while the solvent flows and cleans under components. **See Figure 4-17.**

Figure 4-17. *Cleaners and defluxers can remove stubborn residues from PC boards.*

Temperature Considerations for Fluxes. The application of thermal energy will affect flux performance. The typical ingredients of electronic grade flux may include acid, alcohol, and water.

The boiling point of alcohols and some acids are well below standard soldering temperatures. With low boiling points, it is important not to apply too much heat too quickly during the soldering process. In such a situation, flux will evaporate before it has time to activate and promote wetting of the solder. With the higher process temperature, higher oxidation rate, and weaker wetting force of lead-free alloys, a stronger flux may be needed. An increase in the volume of flux in solder wire, from about the 1.0% typically used to amounts in excess of 2%, is also likely.

No-Clean Fluxes. Strong, aggressive flux may require some form of cleaning of the PC board after soldering. This requires an additional process, which increases costs. Most electronics manufacturers have gone to no-clean flux because cleaning materials can pose their own environmental problems. The PC board industry is trying to speed up the production process of PC boards and address environmental problems by switching to no-clean flux. The purpose of no-clean flux is to eliminate the cleaning of PC board residues after soldering.

Flux Residues. The residues left on a PC board from flux have led to problems, such as dendrite formation and PC board corrosion. *Dendrite formation* is the formation of crystals as a molten metal solidifies that results in a brittle solder joint. After soldering, the flux residue contains ingredients that tend to make it sticky and pick up contaminants from the air. This combination of flux ingredients and contaminants can result in the formation of conductive compounds. Therefore, corrosive damage to the PC board and short circuits become possibilities.

Most corrosion on a PC board is typically easy to see. However, the formation of dendrites (conductive paths) is much less obvious and takes longer to create damage. The term "dendrite" comes from the Greek word "dendron" meaning tree. Dendrites grow branches until a bridge is formed between two foils. This conductive bridge can result in a short circuit. **See Figure 4-18.**

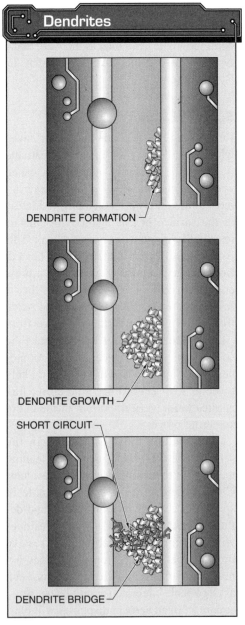

Figure 4-18. *Dendrites grow branches until a bridge is formed between two foils. This conductive bridge can result in a short circuit.*

H K Wentworth, Ltd.
Flux pens are used to easily apply flux to electronic components prior to soldering.

Flux-Core Solders. *Flux-core solder* is solder that contains flux in its core. Flux-core solder is typically found in wire form. It may be available in large rolls or small tubes. During manufacturing, the center of the wire is filled with the proper amount of flux required for soldering. **See Figure 4-19.** Generally, no additional flux is required when soldering with flux-core solders. The surfaces to be soldered should not be visibly oxidized when soldering with this type of flux.

Lead-Free Solders

With support from the major electronics manufacturers, lead-free solder alloys have been developed in many forms. The most promising lead-free solder alloys exhibit acceptable structural and conductive properties. In certain cases, they are eutectic solders. The trade-off, however, is that they require higher working temperatures and additional steps. This means that extensive investments in new equipment and process designs are needed to implement lead-free soldering.

Many new challenges have arisen around the use of lead-free solder. For instance, lead-free solder has a significantly higher melting point than lead-based solder. The higher melting point of the solder presents the challenge of limited use with heat-sensitive electric components in plastic packaging. To overcome this problem, solder alloys with a high silver content and no lead have been developed with a melting point slightly lower than traditional lead-free solders. However, the melting point for this type of lead-free solder is much higher than it is for traditional tin-lead solder. The new lead-free solder is typically called "SAC alloy solder." This term along with a numerical designation indicates the percentage of metals that make up the alloy.

SAC Alloy Solders. An SAC alloy solder is solder containing tin (Sn), silver (Ag), and copper (Cu). The abbreviation "SAC" comes from the element symbols for tin, silver, and copper. Element symbols can be found in the periodic table of elements. SAC alloy solder is typically composed of a percentage of each of these metals. For example, SAC alloy 305, or SAC305, is a mixture of 96.5% tin, 3% silver, and 0.5% copper, while SAC387 contains 95.5% tin, 3.8% silver, and 0.7% copper.

Lead-Free Soldering versus Lead Soldering. One of the major differences between lead-free and lead solder is the temperature at which the solder melts. Lead solder melts at 361°F (183°C). Lead-free solder typically melts at three different temperatures 423°F (217°C), 430°F (221°C), and 441°F (227°C). Conventional tin-lead solder melts at temperatures 62°F (34°C) to 80°F (44°C) lower than lead-free solder.

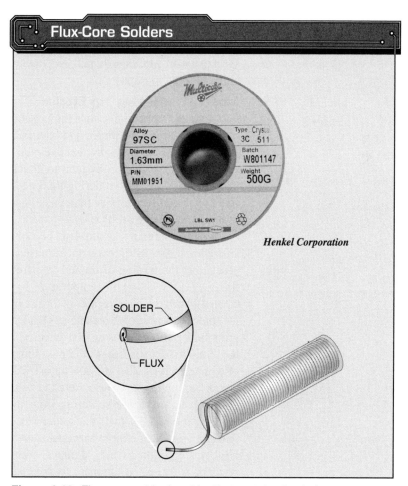

Figure 4-19. *Flux-core solder is solder that contains flux in its core.*

Henkel Corporation
Surface mount devices may be very small and require precision when soldering them to printed circuit boards.

The wetting (spread) characteristics of lead-free solder are very poor in comparison to tin-lead solder. Lead-free solder wets slower than tin-lead solder. Poor wetting is characteristic of lead-free solder because the absence of lead reduces the flowability of the solder. Poor wetting occurs with respect to not only PC boards and components but also soldering tips.

Lead-Free Soldering Tip Erosion. The two main materials that cause the erosion of soldering iron tips are tin and lead-free (Pb-free) flux. Tin is a more reactive metal than iron and will naturally tend to attack the iron plating of the soldering tip. Lead-free flux is more aggressive than the flux of ordinary tin-lead (Sn-Pb) alloys.

High temperatures could also be considered a factor in tip erosion because it accelerates the erosion of tin and lead-free fluxes. The elevation in temperature has created some confusion regarding the use of the soldering iron. Some technicians tend to press harder on the soldering iron to increase heat transfer. This increases the possibility of damaging the soldering iron tip and PC board. Technicians may be tempted to raise the temperature of the soldering iron tip, thinking it will help melt the solder more quickly. This may do more harm than good. Higher temperatures may damage components or cause the laminate to separate from the PC board foil. The increased tip temperature also accelerates erosion of the tip. The increased tin in the solder and the new types of fluxes associated with lead-free soldering also cause tips to erode.

Guidelines for Lead-Free Soldering. Excessive process temperatures can lead to thermal damage of PC boards. During the hand soldering process, the following guidelines should be taken into consideration to avoid thermal damage:

- Using a tip of the correct size maximizes the contact area the tip makes with the joint and improves heat transfer efficiency.
- Existing tip temperatures can be used in most applications, provided the correct tip shape is implemented. Higher tip temperatures may be needed for thermally demanding applications.
- The ability of a soldering iron to input the correct rate of temperature rise to a solder joint is important with respect to both flux activation and the final solder joint temperature.
- Flux content, by volume, is likely to increase, which will help heat transfer but may have postsolder cleaning implications. Activating the flux at the correct rate is vital.
- Dwell time, or the length of time a tip is applied for a given temperature, is important. Dwell time must be long enough for proper wetting.

The implementation of lead-free alloys will place more emphasis on control of the process. The hand soldering process will need to be more fully defined and the tip shape, power output, thermal transfer efficiency, and absolute tip temperature should be specified.

As part of the RoHS Directive, most countries require newly manufactured PC boards to be free of six hazardous substances: lead, mercury, cadmium, hexavalent chromium, polybrominated biphenyl, and polybrominated diphenyl ether.

Tip Maintenance for Lead-Free Soldering. Efforts should be made to extend the life of a soldering iron tip. To extend tip life when using lead-free solder, the following recommendations should be considered:

- A soldering tip should be kept clean. Properly cleaned tips are bright and shiny. Keeping the tip clean ensures that the maximum heat will be at the tip surface.
- A soldering tip should be wiped on a damp natural cellulose sponge. If the tip is rotated while wiping, flux and contaminants will not build up on the tip. In addition to sponges, some soldering stands have cups with brass shavings in them. When the soldering tip passes through the shavings, solder is removed.
- A soldering iron should always be placed in a stable stand regardless of whether it is being used.
- The working end of a soldering tip should be covered with solder during idle periods.
- A soldering iron should not be allowed to sit idle at operating temperatures for extended periods. Operating at a temperature higher than necessary speeds oxidation.
- A temperature of 800°F (426.7°C) or less should be maintained whenever possible. Operating at a higher temperature dramatically increases the formation of iron oxide.
- A soldering tip should never be wiped on rough materials, such as dry sponges, rags, or paper towels. Sulfur-free, pure cellulose sponges that are wet to the touch should be used.
- The best way to minimize soldering tip maintenance is to find a quality solder. Solder that has high metal purity and a high tin content should be used.

CAUTION: Excessive wiping causes the temperature of the soldering tip to drastically rise and fall. The temperature fluctuation on different metal layers in the tip causes the tip to repeatedly expand and contract. This cycling can lead to metal fatigue and failure. Avoid the practice of dipping the tip into flux in order to clean it. Flux is corrosive and will erode the tip.

Soldering tip temperature regulation is critical, both during soldering and when idle, to prevent the tip from attaining excessive temperatures. Improved temperature regulation should be combined with an optimized tip profile that is designed to maximize the thermal energy transfer from the heater to the soldering site. Nominal tip temperature is 720°F (382°C), which is the same temperature recommended for hand soldering with tin-lead alloys. A lead-free solder joint is formed at 494.6°F (257°C).

Tip temperature of around 716°F (380°C) is sufficient to enable satisfactory lead-free alloy and flux soldering. This results in good wetting speed and high-quality joint formation. The ability to preset the temperature with no overshoot greatly retards erosion of the tip. It also reduces flux charring, leading to a corresponding reduction in the soldering force applied by operators and less frequent tip cleaning.

Idle Temperature of Soldering Tips. Measures taken to reduce tip temperature when a soldering iron is idle also play a large role in slowing the tip erosion caused by the use of lead-free solder and flux. A soldering iron spends a significant portion of its useful life idle. By significantly reducing the tip temperature during this time, mechanisms that reduce tip life can be further slowed.

Technicians soldering with lead-free materials should pay close attention when matching a tip to the geometrics of a solder joint. This practice optimizes wetting speed without increasing the tip temperature set point. Using the correct size tip maximizes the contact area between the tip and joint, improving heat transfer efficiency. Ideally, tip and target dimensions should be the same.

Avoiding Bad Solder Joints

Bad solder joints result in improper connections of components and can lead to damaged components and circuits. Therefore, efforts should be made to prevent bad solder joints. To achieve a proper solder joint, the following conditions should be avoided:
- applying too little solder to a joint produces a weak joint and causes a component to work loose over time
- applying too much solder to a connection could bridge component leads or PC board traces, causing unwanted conducting paths
- soldering with too little heat produces a cold solder joint and results in a poor connection
- soldering with too much heat may damage a component or cause the foil on a PC board to lift from the surface
- soldering with no flux or too little flux would not remove all contaminants, which would result in a poor connection

Soldering Safety

Tests have shown that soldering with lead and lead-free alloys creates a level of breathable airborne pollutants that exceeds the safety threshold set by OSHA. In particular, lead soldering may generate airborne lead oxide and fumes from colophony. Lead oxide is produced from lead-based solder, and colophony is produced from solder flux.

Because exposure to these fumes may be a health hazard, suitable measures to protect health must be carried out. When reasonable, exposure should be avoided. If that is not practical, then the process should be adequately controlled using fume extractors and proper ventilation techniques.

As a technician, it is necessary to follow instructions on safe working practices given by the employer including the correct use and adjustment of control measures, such as local extraction ventilation. When required, protective equipment should be worn, such as respirators. Suitable gloves, protective clothing, and eye protection may also be appropriate for certain work. It is important to become familiar with the material safety data sheets (MSDSs) provided by manufacturers and always follow safety rules. For safe soldering, it is important to observe the following rules:
- A soldering iron should always be returned to its stand when not in use and should never be set down on a workbench.
- Soldering should be done in a well-ventilated area. The smoke formed while soldering is mostly created from the flux and can be toxic or irritating. Keeping the head to the side of the work will prevent smoke inhalation.
- Hands should be washed after handling solder. Solder may contain lead, which is a poisonous metal.
- Safety glasses should be worn for protection from splatter and lead clippings.

SOLDERING AND REWORK STATIONS

A *soldering station* is a workstation with all of the tools required for soldering. Soldering stations are single purpose in that they are primarily designed only to solder components. However, they can be used to remove components when necessary.

A *rework station* is a workstation with all of the tools required for soldering and desoldering plus many of the accessories required for rework. Rework stations are dual purpose. They are intended to remove a variety of components safely and then resolder them into a workable unit. The primary considerations when looking at soldering stations and rework stations are their wattage and temperature range. Soldering stations and rework stations may have unregulated or regulated temperature. **See Figure 4-20.**

Rework Stations

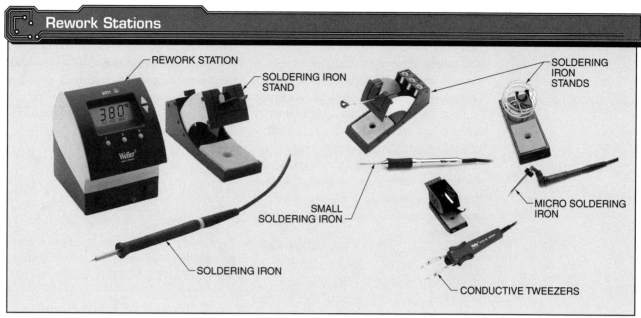

Cooper Industries, Inc.

Figure 4-20. A rework station is a workstation with all the tools required for soldering and desoldering plus many of the accessories required for rework.

Wattage

Because a soldering station has a higher wattage does not necessarily mean that it has a hotter soldering iron. Soldering stations with higher wattage can deliver more power to a solder joint, allowing for a larger surface area or bigger connection to be soldered. A lower wattage soldering station may not be able to keep its temperature in a large surface area or with a large connection, resulting in a cold solder joint.

Temperature

A soldering temperature of 700°F (371°C) to 800°F (427°C) is typically recommended for single-layer and double-sided PC boards. It may seem unnecessary to heat a soldering iron tip from 700°F to 800°F when solders typically have a melting point below 441°F (227°C). This is useful because higher temperatures store heat in the tip, which speeds up the melting process. Technicians must be aware of this "dwell time" or time spent on the joint. If the dwell time is too long, the excessive heat created by elevated temperatures will transfer to the PC board and damage it or the components.

A typical dwell time may vary from 1 sec to 5 sec depending upon the size of the joint. Soldering and developing the proper dwell time requires practice. If done properly, the technician can complete the solder joint without applying too much pressure and form an intermetallic connection, which is critical for a reliable mechanical and electrical solder joint.

Unregulated Soldering Stations

An unregulated soldering station typically runs at the level of power consumption it was designed to produce. With a soldering station of unregulated temperature, wattage is a poor indicator of its overall temperature performance. Wattage is an indicator of how quickly the station will heat up the soldering iron to its maximum temperature. It is also an indicator of how fast the iron will create new heat for transfer into the solder joint. It does not necessarily indicate what temperature is present at a given time.

As long as power is supplied to the tip, the tip will continue to get hotter until it reaches equilibrium. In other words, the heat will be conducted away from the tip to the surrounding air as quickly as it is generated. The temperature of an unregulated soldering iron can reach 1100°F (593.3°C), which is much hotter than necessary for soldering.

Heat transfers out of the tip more quickly when it is in contact with the work surface of a PC board. The rate at which heat transfer occurs depends on the size and shape of the tip and the amount of contact surface the tip has with the solder connection. The wattage or applied power sustains the working temperature.

When the soldering iron is not being used for more than 5 min to 10 min, it should be turned OFF. If the soldering iron is left ON, the tip will need to be refinished or replaced fairly often. If the soldering iron is left ON overnight, the tip will most likely be ruined, requiring replacement the next day. This is typically known as burnout.

To prevent burnout, the soldering tip should be kept tinned with a thin coat of solder during soldering and when the soldering iron is sitting idle in the holder. Often, the soldering tip is wiped clean before the soldering iron is returned to the holder and not re-tinned until the next soldering task. This causes rapid burnout. New soldering tips should be tinned as soon as they are heated to the solder melting point. Otherwise, burnout can occur within minutes.

Regulated Soldering Stations

A regulated (temperature-controlled) soldering station has a thermostatic control to ensure that the temperature of the soldering iron tip is maintained at a fixed level within certain limits. This guarantees that the temperature does not overshoot certain limits and that the output will be relatively consistent.

Regulated soldering stations consist of benchtop units, which have special low-voltage soldering irons plugged into them. Some versions may have built-in digital temperature readouts. The station will have a coded knob that enables the technician to vary the setting. In many units a thermocouple is built into the tip shaft to monitor the temperature. Regulated soldering stations allow for greater soldering consistency compared to unregulated soldering irons. Overall, this means fewer cold joints, heat-damaged components, and heat-damaged pads and traces. The quality of work will also be much better than soldering with unregulated soldering irons.

PC Board Repair Stations

A PC board repair station is used to repair solder connections on PC boards. A repair station power unit provides power for all devices connected to the unit. The unit will provide low voltage and standard voltage. **See Figure 4-21.** Control units may include the following:

- digital display and pushbutton selectors
- regulated or "spike-free" power switching currents for attached electrical hand tools to eliminate damage to a component from electrostatic discharge
- milling, drilling, grinding, and cutting tools that use a flexible shaft, rotary-drive machine
- heat gun for the thermal removal of conformal coatings
- resistive and conductive tweezers for soldering applications
- thermal wire stripper for removing polyvinyl chloride (PVC) and other synthetic wire coverings

PC Board Repair Station Accessories

Repair stations with certain microelectronic circuits and components require special accessories. A basic understanding of these tools will uncover why further training may be necessary for their application. The equipment is essential for making repairs to current miniature and microminiature PC boards.

Basic repair/rework station accessories include a magnifying lens, PC board holder, high-intensity light, and a soldering iron holder. Repair stations for making microminiature repairs may also include a stereoscopic-zoom microscope. **See Figure 4-22.**

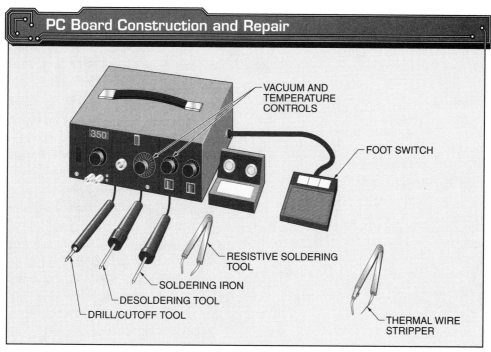

Figure 4-21. *A repair station contains the tools required for the extensive repair of PC boards.*

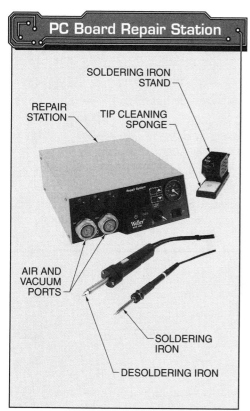

Cooper Industries, Inc.

Figure 4-22. *Repair stations typically include air and vacuum ports to help with desoldering.*

Electrostatic-Discharge-Safe Soldering Stations

Electrostatic discharge (ESD) safety ensures that static energy from an individual will be grounded via a plastic soldering iron handle or solder station plastic housing. Otherwise, the static charge would transfer to the soldering tip and destroy the component being soldered. The solder station essentially discharges the static charge of the body to ground every time the soldering iron is picked up. ESD-safe soldering stations should be used with ESD workstations whenever possible.

TECHNICIAN EFFECTS ON SOLDERING

A technician controls many of the soldering parameters that determine how much heat ends up at the solder joint. In addition, the technician selects the soldering station and the shape of the soldering tip. The technician can also vary the position of the soldering iron tip, the time on contact, and the pressure of the tip against the solder joint.

The reliability of a solder joint depends on the selected equipment and the ability of the technician to apply the proper soldering techniques. Choosing a specifically designed soldering iron tip should result in efficiency and/or a reduction in the risk of PC board damage.

SOLDERING PRINTED CIRCUIT BOARDS

The heat produced by a soldering iron can ruin a PC board. The heat intensity of a soldering iron is determined by the wattage of the device. To minimize the chance of overheating, it is best to use a soldering iron with a rating of no more than 40 W.

Soldering Procedure for Printed Circuit Boards

Properly soldering components to a PC board ensures that the components are secured and that the components and solder complete the circuit for the intended operation. To solder a PC board, apply the following procedure:

1. Inspect the PC board and the component leads to make sure they are clean. Older components may be oxidized and require cleaning. The surface of the traces should be shiny and free of smudges, fingerprints, dirt, and grease.
2. Secure the work.
3. Select a tip with the correct shape to maximize the contact area with the solder joint.
4. Tin the iron with a small amount of solder before cleaning, and clean the tip of the hot soldering iron on a damp sponge.
5. Check and recheck the value, orientation, and location of components before soldering them to the PC board.
6. Inspect the desoldered connections. Determine where the solder should go. Be sure that no unwanted solder is left behind that will create a solder bridge and short the circuit.
7. Heat all parts of the joint with the soldering iron, and apply sufficient solder to form an adequate joint. The formation of the joint should take from 1 sec to 5 sec to complete. During this time, the flux will start to activate.
8. Do not move any parts until the solder has cooled. Once the heated tip is removed, the solder joint will begin to solidify.
9. Inspect the soldered connection. A magnifying lens may show details or problems not seen with regular vision.
10. Trim component leads as required. Remove unwanted flux left on the surface of the PC board. It is highly recommended that leads be trimmed before soldering. However, standards state that if leads are trimmed after soldering, the solder joint should be either reflowed or inspected for damage, such as fracturing.

Requirements for Soldering Surface Mount Devices

A *surface mount device (SMD)* is an electronic component that is soldered directly to the traces of a PC board. Products using SMDs include cell phones, global positioning system (GPS) equipment, laptop computers, microwave ovens, calculators, and many industrial control applications.

Technician requirements for soldering surface mount devices include good vision and a steady hand. A technician should also have the following materials available:

- solder
- solder wick
- solder flux in a pen or to be applied by a brush
- solder flux remover
- regulated soldering iron with a small, clean tip
- heat gun for desoldering large devices
- metal shield to deflect heat away from other components when using heat gun
- magnification device
- fine tip tweezers for placing parts

- dental pick or utility knife for lifting devices
- swabs for cleaning

It is possible to bend a lead on an SMD when handling it. For example, leads can be damaged when picking up an SMD. Furthermore, fingers should never be used to manipulate an SMD. Tweezers or a vacuum pencil should be used. When using tweezers, the component should be gripped around its body. *Note:* It may be necessary to clamp the PC board in a movable vise or stick it to a small block of wood with double-sided tape to steady it when soldering.

Soldering Surface Mount Devices

Soldering SMDs can be difficult because no holes are drilled through the PC board to mount components. When mounting chip components, care must be taken not to overheat them. **See Figure 4-23.** To solder an SMD, apply the following procedure:

1. Tin the PC board pad with a low-temperature soldering iron.
2. If chip semiconductors are being soldered, use a properly grounded soldering iron, and apply all ESD safety procedures.
3. Allow the tinned pads to cool.
4. Place the SMD in position using tools such as tweezers or a vacuum pencil. When using tweezers, the component should be gripped across the body. Never use fingers to manipulate an SMD.
5. Place the soldering iron tip on the PC pad near the component. The soldering iron tip should not come in contact with the component.
6. As the pad heats up, apply solder of a small diameter and a liquid flux or solder paste, if available.
7. After one side of the device has been soldered, move to the other side, and solder the opposite side in the same manner.
8. Hold the device in place by applying light pressure with tweezers, if needed.
9. Remove excess solder by placing a soldering iron on a solder wick.
10. Perform a final check using magnification to make sure the solder is smooth and fully bonded to the PC board on each side of the chip.

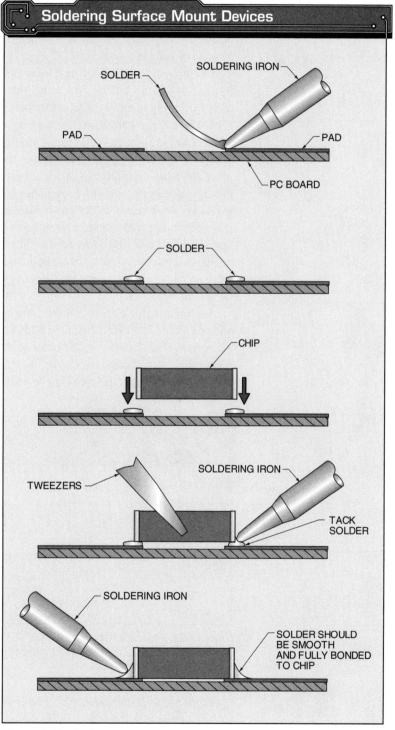

Figure 4-23. *Surface mount devices (SMDs) are electric components that are soldered directly to the traces of a PC board.*

To solder a 14 pin SMD, fine tweezers should be used to set the SMD on top of a pad. The pad should be heated with solder and the SMD adjusted so that it lines up with the pad. The rest of the pins should be soldered one at a time. *Note:* Do not solder adjacent pins. Start with the pin at the opposite corner to the pin already soldered, and solder each pin.

The procedure for soldering SMDs can be performed using solder paste rather than directly tinning the pads. The solder paste should be applied carefully with a syringe or screened into place with a paste mask. Although the solder paste method can be more difficult than simply using solder, it may yield better results. Care must be exercised when using solder paste. Solder paste contains microscopic, easily ingested lead particles, which are hazardous. When using solder paste, rubber gloves should be worn and directions from the manufacturer should be followed.

An SMD solder joint should form a concave shape that blends with the body of a component and solder pad. Solder joint appearance may change depending on the component and lead type. **See Figure 4-24.**

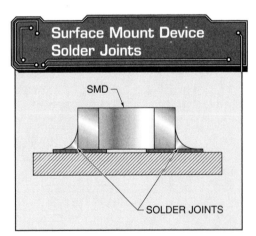

Figure 4-24. *The solder joints for SMDs should be smooth, concave, and blend with the body of the SMD and solder pad.*

When a technician ventures into the area of surface mount technology (SMT) soldering, it is a good idea to begin looking into training based on IPC standards. Surface mount technology (SMT) makes minimal use of plated through-hole technology and maximum use of multilayer PC boards. Component leads are not placed through the PC board. Rather, components and their interconnecting leads are placed on the PC board surface. Soldering SMDs is a skill that is learned by understanding the soldering process and is developed with practice.

Soldering Multipin Surface Mount Devices

Soldering a multipin SMD requires proper placement of the SMD and precise soldering to avoid bad connections. **See Figure 4-25.** To solder a multipin SMD, apply the following procedure:

1. Brush on a thin coat of paste flux to the SMD pads on the PC board. Paste flux will help hold the part in place when soldering.

2. Carefully place the SMD on the pads using a set of tweezers. Be sure the part is centered on the pads. Make sure pin 1 on the SMD aligns with pin 1 on the PC board.

3. Select the proper tip and clean off any extra solder with a wet sponge as the tip begins to heat. Add a small drop of solder to the tip of the soldering iron.

4. Tack one pin into place, and check the alignment of the SMD on the pads. Tack another pin into place on the opposite corner of the part to secure the SMD in place. Recheck the alignment of the SMD in relation to the PC board pads.

5. With all the pins aligned, heat each pin using the solder to reflow on the PC board and secure the pins. *Note:* As a precaution, the pins can be reheated one more time by moving the soldering iron tip across each pin for 1 sec to 2 sec to make sure the proper amount of heat has been applied.

6. Closely inspect the work using magnification to make sure there is no excess solder left or bridging between components. Remove excess solder with a solder wick.

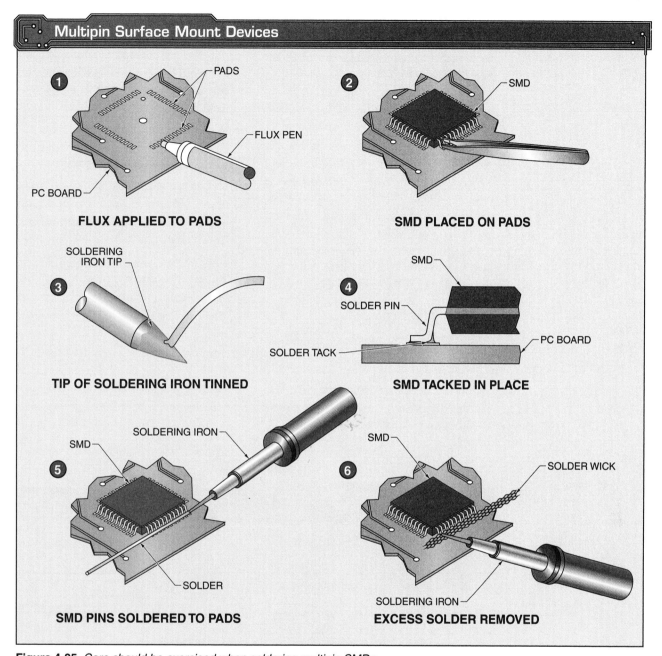

Figure 4-25. *Care should be exercised when soldering multipin SMDs.*

DESOLDERING METHODS

There are several methods currently being employed to remove solder from a joint. The type of removal system to be used will depend on the location of the repair. For example, a solder wick (desoldering braid) and a handheld desoldering pump may be used in the field. If the PC board can be taken off-site for repair, a vacuum desoldering pump on a soldering workstation may be used. The type of components being removed will also determine the tool to be used. The removal of SMDs requires more sophisticated equipment than the removal of through-hole components.

A technician desoldering a PC board must understand that the PC board is more important than the component being replaced. It is typically much easier and less expensive to replace a component than to replace a PC board.

Solder Wicks

A *solder wick* is a fine copper braid impregnated with flux that is used to remove solder. **See Figure 4-26.** When used properly, a solder wick can remove almost all of the solder from a connection and component. The performance of a solder wick can be enhanced by lightly applying flux to the wick prior to use. One problem with a solder wick is that it oxidizes over time. The drying process will cause the flux to fall out of the mesh, therefore making it unusable. If an expiration date is not on the solder wick, a magic marker can be used to record the purchase date on it. By looking at the date and condition of the solder wick, a technician can determine its usefulness.

A solder wick along with a soldering iron can be used to remove solder. To remove solder using a solder wick, apply the following procedure:

1. Ensure that the soldering iron is hot and tinned. Desoldering usually requires more heat than soldering. Extra flux is helpful.
2. Lay the end of the solder wick on top of the connection that is to be desoldered, and press the soldering iron firmly into the solder wick. Watch for solder to appear in the wick while holding it in place.
3. When solder has been drawn about ½″ from the end of the wick, remove the soldering iron and solder wick.
4. Cut off the used end of the solder wick about ¼″ above the point at which solder is visible. Solder has not been drawn up that far, but the flux may have boiled out.
5. Repeat the previous steps until the component can be removed.
6. Remove the solder wick and soldering iron at the same time. Two or more applications may be required for each lead. Keep in mind that where PC board holes are plated through, there will be more solder to remove than there is for a single-sided PC board.

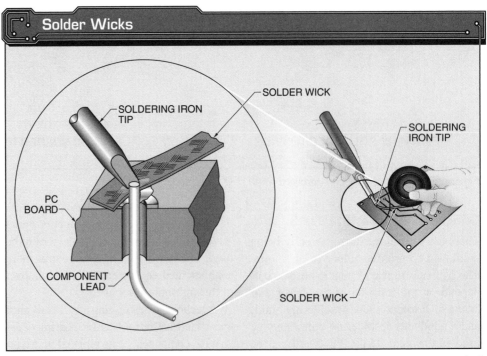

Figure 4-26. *Solder can be removed by placing a solder wick (a specially woven strand of copper wire) on a solder joint prior to applying a hot soldering iron.*

Desoldering Pumps

A vacuum desoldering pump may be used to remove molten solder. The vacuum action is created by a spring-loaded plunger. Once the spring-loaded plunger is released by the push of a button, the molten solder is drawn up into the pump. It may take one or two attempts to remove all of the solder. **See Figure 4-27.**

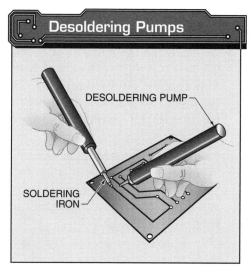

Figure 4-27. *Solder can be removed by using a desoldering vacuum pump and a soldering iron.*

When a lot of desoldering needs to be completed, an electric continuous desoldering pump may be used. An electric continuous desoldering pump is a combination of a soldering iron with a hollow tip and a vacuum pump. This unit is typically found in a production environment where rework is done.

Desoldering Through-Hole Components

When removing a through-hole component, the PC board is generally positioned so that the component leads to be desoldered are facing down. After flux is applied, a hot soldering iron tip is held under and against a component lead. **See Figure 4-28.** Solder flows to the soldering iron tip but may be removed with a wet sponge. This procedure is repeated until sufficient solder is removed from the component lead to free that end of the component. When both leads are loose, the component is lifted from the PC board, and the board is cleaned thoroughly.

CAUTION: The component to be removed should never be pried or forced loose. Any attempt to force a component loose may result in broken or separated foil. If any solder is left in the terminal hole after the component leads have been removed, lightly apply the soldering iron tip to the hole to melt the solder. Clear the melted solder from the hole using a solder wick.

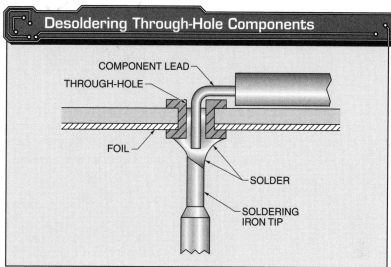

Figure 4-28. *A soldering iron can be used to desolder through-hole components.*

Desoldering Surface Mount Devices

Surface mount devices can be desoldered in a manner similar to soldering them. Typically, heat is applied to the solder using conductive tweezers, soldering irons with fitted tips or hot air tools.

When leaded components are removed from a PC board, the process typically requires removing excess solder before removing the components. With SMDs, another technique must be used. The SMD requires that the technician must first reflow all of the solder connection, remove the component, and then remove the excess solder.

Desoldering with Conductive Tweezers. An SMD can be removed from a PC board using conductive tweezers. *Conductive tweezers* are tweezers that are heated in a manner similar to a soldering iron and used to solder and desolder small SMDs. All joints must have flux applied prior to component removal. **See Figure 4-29.**

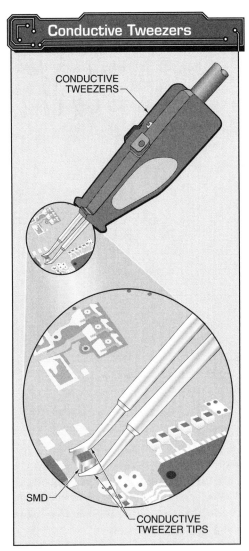

Figure 4-29. *Conductive tweezers are used to solder and desolder small SMDs.*

Flux helps solder to reflow on the solder pad while heat is transferred and distributed. This reduces spikes on the surface of the pads when the component is lifted from the surface of the PC board. To remove an SMD using conductive tweezers, apply the following procedure:

1. Select the proper tip size for removal of the component.
2. Clean and re-tin the tips to provide better contact with the terminations and speed up both the heat transfer and rework process.

   Note: Some tips cannot be tinned. Contact tools are the fastest method of removing components with lead-free alloy but require more skill and experience than other methods.

3. Gently grasp the component terminations, ensuring good contact until solder reflow is observed. This can take from 1 sec to 5 sec, depending on the component type and the PC board construction.
4. Remove the component using tweezers, being careful not to lift or twist the component until the solder reflows. Lifting and twisting too early could result in the pads being lifted.

   Note: If the component is held in place using adhesive from wave soldering, a slight twisting action may be required to break the adhesive bond. The twisting action should be applied after reflow of the solder.

5. Inspect the pads on the surface of the PC board for any damage or lifting, and clean the rework area if necessary.

Desoldering with Fitted Tips. The easiest way to remove multipin SMDs may be to use a soldering iron with a fitted tip. The tip should be just wide enough to contact all ends of the component to be removed. **See Figure 4-30.**

The soldering iron should be positioned so that the tip makes good contact with both ends of the component. Preheating the PC board can reduce the possibility of creating blisters or thermal shock. When the solder melts, the component can be removed. After the PC board has cooled for a few minutes, the excess solder can be removed from the area where the component was removed. A solder wick works well at removing solder.

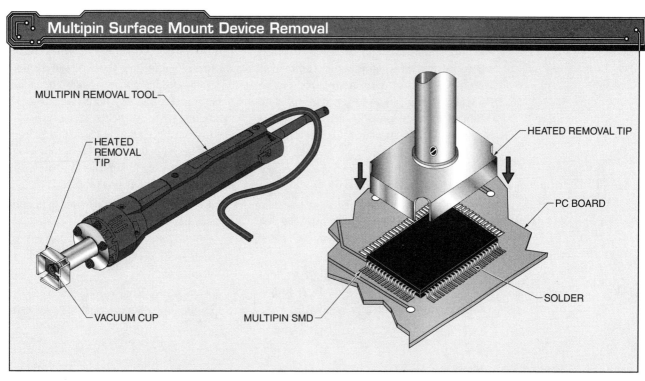

Figure 4-30. *The easiest way to remove a multipin SMD may be to use a soldering iron with a fitted tip wide enough to contact all ends of the SMD to be removed.*

Desoldering with Hot Air. Hot air from a heat gun can be used to remove SMDs from PC boards. The careful selection of a good heat gun is important. Many heat guns on the market are designed for heat shrinking or paint removal. These guns usually produce too much heat, too much airflow, or too wide an airflow. A heat gun that is well-suited for SMD rework and offers a thin stream of low-velocity air should be selected. A wide stream of air would heat too wide an area of the PC board. A high-velocity stream of air could blow parts off the PC board. **See Figure 4-31.**

In the 1960s, the space industry began to make the first serious use of photovoltaic technology to provide power aboard spacecraft. Through the space programs, photovoltaic technology advanced, its reliability was established, and the cost began to decline.

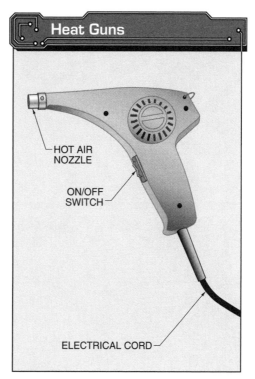

Figure 4-31. *Heat guns can be used to remove an SMD from a PC board.*

Most SMD vendors, especially the ceramic-chip capacitor manufacturers, recommend that hot air be used for rework because hot air simplifies the removal of high pin-count components. To remove SMDs using hot air and a heat gun, apply the following procedure:

1. Apply gentle heat and flux with the heat gun about 2″ to 3″ from the PC board. Heat from directly above to avoid blowing parts off the PC board.
2. Wait about 40 sec to 60 sec for the solder to liquefy. Do not bring the heat gun in too close because the PC board will heat unevenly.
3. Lift the component with tweezers or a dental pick. Do not pry. Wait for the solder to completely liquefy.

A professional hot air tool can be used when a large amount of rework is needed on a variety of PC boards. It precisely controls heat to easily remove SMDs without damage to surrounding components. **See Figure 4-32.**

Hot air pencils may be used to remove SMDs. They may be better suited than heat guns to remove SMDs from PC boards. Advantages of hot air pencils include the following:

- no opportunity to bend or deform SMD leads or damage PC board pads
- little opportunity to introduce contaminants into a solder joint
- may be used with either lead-free or lead solder alloys because they never touch the solder joint
- not likely to damage ceramic chip capacitors due to no direct contact with the leads

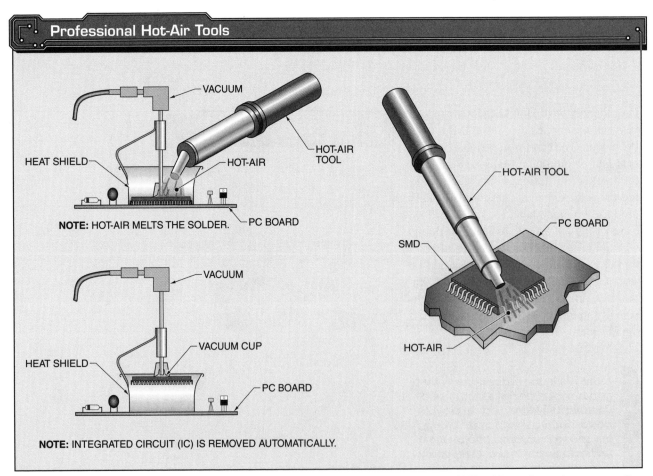

Figure 4-32. *A professional hot air tool precisely controls heat to easily remove an SMD without damaging surrounding components. A hot air pencil can also be used to remove an SMD.*

Desoldering with Chemicals. Chemicals, such as Chip Quik®, can be used to remove SMDs. Chip Quik® is used to remove SMDs that are melted onto pins. It can be used to repair a poor solder connection or remove components from a PC board when making a field repair. **See Figure 4-33.** To remove SMDs using Chip Quik®, apply the following procedure:

1. Apply flux to all leads with a syringe or applicator.
2. Uniformly melt the Chip Quik® on all pins. Maintain the alloy in molten state long enough to release the chip.
3. Lift the chip from the PC board with tweezers or a dental pick.
4. Thoroughly clean the area with a cotton swab dipped in flux while applying heat. Once the area is at room temperature, clean it thoroughly with an alcohol pad.

5S ORGANIZATION SYSTEMS

The 5S organization system is a workplace organizational system that stands for sort, straighten, sweep or shine, standardize, and sustain. The system is used to reduce waste and optimize productivity through maintaining an orderly workplace. Visual cues are used to achieve more consistent operational results. Routines and systems, such as the 5S organization system, that maintain organization and orderliness are essential for maintaining a smooth and efficient flow of activities when performing work.

> *In 1835, the inception of digital electronics started when the electronic circuit or gate was invented by Joseph Henry. In the next century, the vacuum tube was invented along with numerous devices.*

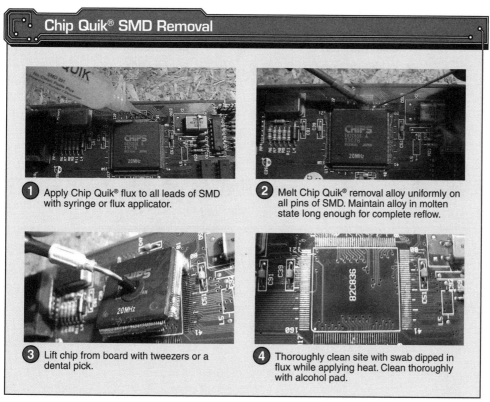

Figure 4-33. *Chip Quik® may be used to repair a solder connection or remove components from a PC board when making a field repair.*

104 SOLID STATE DEVICES AND SYSTEMS

For example, a basic soldering rework station has numerous parts, pieces, tools, and supplies, and it is easy to lose track of their location. A 5S soldering workstation organizer can be used to organize the soldering rework station, and therefore improve work efficiency by saving time. **See Figure 4-34.**

The main goals of the 5S organization system are improved workplace morale, safety, and efficiency. The system is based on the concept that by assigning a location for each item needed, time is not wasted looking for needed items. In addition, it is easy to identify when an important or expensive tool is missing from its designated location by using the system.

Refer to Chapter 4 Quick Quiz® on CD-ROM

Omni Training

Figure 4-34. *The 5S organization system can be used to organize soldering workstations.*

KEY TERMS

- printed circuit (PC) board
- surface mount technology (SMT)
- inspection mirror
- foil
- ball grid array (BGA)
- heat sink
- thermal shunt
- brush
- conductive brush
- dissipative brush
- antistatic brush
- insulative brush
- soldering
- solder
- soldering iron
- soldering iron tip
- eutectic alloy
- cold solder joint
- solder flux
- dendrite formation
- flux core solder
- soldering station
- rework station
- surface mount device (SMD)
- desoldering
- solder wick
- conductive tweezers
- heat gun
- hot air pencil

REVIEW QUESTIONS

1. What is soldering?
2. What is desoldering?
3. What are soldering aids?
4. What is solder?
5. Describe how to clean soldering iron tips.
6. What is a eutectic alloy?
7. What is a cold solder joint?
8. What is solder flux?
9. What is the purpose of no-clean solder flux?
10. Define dendrite formation.
11. What is a flux core solder?
12. Describe SAC alloy solders.
13. What is a soldering station?
14. What is a rework station?
15. Define surface mount device (SMD).
16. What is a solder wick?
17. Explain desoldering pumps.
18. What are conductive tweezers?

DIODE APPLICATIONS AND TROUBLESHOOTING 5

Diodes are semiconductor devices that allow electrons to flow through them in only one direction. Diodes are typically made from silicon and germanium. They were first made of a metal plate coated with the semiconductor material selenium. Semiconductor diodes are the most widely used diodes because of their small size and weight. The unique properties of diodes come from the semiconductor materials from which they are made. Diodes may be used in rectifiers, clipping circuits, and clamping circuits.

OBJECTIVES

- Understand diode operation.
- List diode materials.
- Explain forward bias.
- Explain reverse bias.
- Understand diode testing.
- List diode applications.
- Explain zener diode operation.
- Understand zener diode testing.
- List zener diode applications.
- Explain how to maintain semiconductor devices.

DIODES

A *diode* is an electronic component that allows electrons to pass through it in only one direction. When the diode was first developed, it was made of a metal plate coated with a selenium material. Semiconductor diodes are the most widely used diodes because of their small size and weight. Diodes have unique properties because of the semiconductor material from which they are made. The schematic symbol representing a diode consists of a triangle that represents the anode and a straight line that represents the cathode. Electrons flow from the cathode to the anode, or against the triangle. **See Figure 5-1.**

Diode Materials

Diodes are typically made from semiconductors. A *semiconductor* is material that has the electrical conductivity characteristics between that of a conductor (high conductivity) and an insulator (low conductivity). The two most common semiconductor materials used to make diodes are silicon and germanium. In their natural state, silicon and germanium are pure crystals. Pure crystals do not have enough free electrons to support electron flow.

A *free electron (carrier)* is an electron in a material that allows the conduction of electricity. By adding certain impurities through a process called doping, the materials become very conductive when a voltage of the proper polarity is applied.

Diodes Media Clip

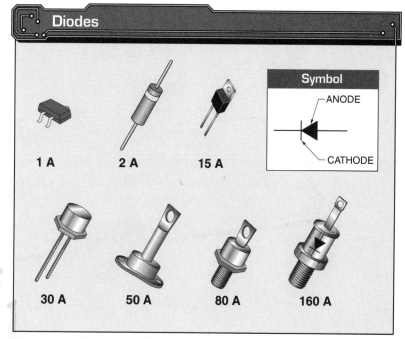

Figure 5-1. *Diodes have the unique property of allowing electrons to pass through them in only one direction.*

Doping. Pure silicon and germanium crystals are prepared for use in diodes by doping. *Doping* is the addition of impurities to the crystal structure of a semiconductor to allow electron flow. In doping, some of the atoms in the crystal are replaced with atoms from other elements to create N-type material and P-type material.

N-type material is semiconductor material created by doping a region of a crystal with atoms of an element that has more electrons in its outer shell. Adding these atoms to the crystal results in more free electrons (carriers) that support the flow of electrons. Electrons flow from negative to positive through the crystal when voltage is applied to N-type material. The semiconductor material is referred to as N-type material because its electrons have a negative charge. **See Figure 5-2.**

Some elements commonly used for creating N-type material are phosphorous, arsenic, bismuth, and antimony. The amount of doping material used generally ranges from a few parts per billion to a few parts per million. By controlling the amount of doping material in a crystal, the manufacturer can control the operating characteristics of the semiconductor.

P-type material is semiconductor material created by doping a region of a crystal with atoms of an element that has fewer electrons in its outer shell, which creates empty spaces, or holes. *Holes* are the missing electrons in a crystal structure. These holes are represented as positive charges. To create P-type material, a crystal is doped with atoms of an element that has fewer electrons in its outer shell than the crystal. The direction of hole flow inside P-type material moves from positive to negative. The direction of electron flow outside the P-type material moves from negative to positive. **See Figure 5-3.**

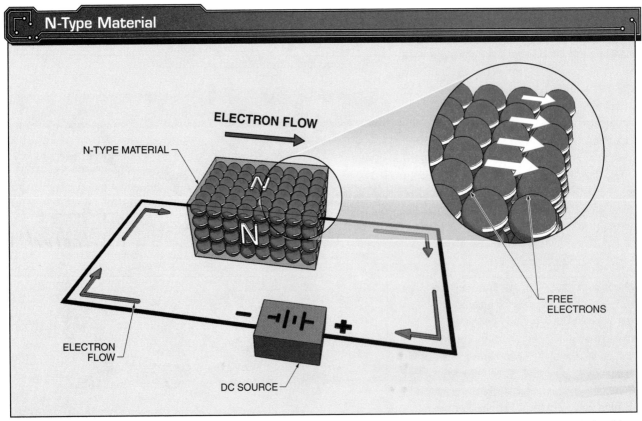

Figure 5-2. *When a voltage is applied to an N-type material, electrons flow from negative to positive through the help of free electrons called carriers.*

The holes are filled with free electrons, which flow from negative to positive. Since P-type material is doped with atoms that have fewer electrons, the moving electrons leave a depleted area behind as they move. The net effect is that the holes move from positive to negative. **See Figure 5-4.** Typical elements used for doping a crystal to create P-type material are boron, gallium, and indium.

> *The impurity atoms used for creating N-type material are pentavalent, which means they have five valence electrons they can share with the atoms of a semiconductor. The impurity atoms used for creating P-type material are trivalent, which means they have three valence electrons to share.*

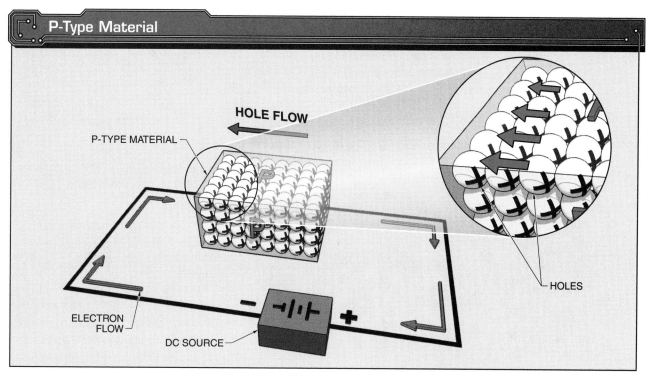

Figure 5-3. *The direction of hole flow inside P-type material is from positive to negative, while the electrons outside the P-type material flow from negative to positive.*

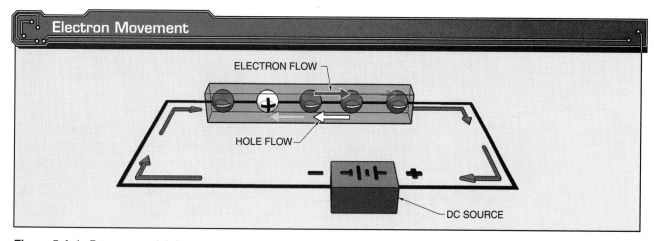

Figure 5-4. *In P-type material, free electrons move from negative to positive while holes move from positive to negative.*

Depletion Region. Diodes have the ability to block electron flow in one direction and pass electrons in the opposite direction. This is made possible by the doping process, which creates N-type material and P-type material. The junction of N-type material and P-type material in a semiconductor forms the depletion region. A *depletion region* is a thin, neutral area created at the junction of P-type and N-type materials, where the two materials exchange carriers. **See Figure 5-5.** Because the depletion region is very thin, it responds rapidly to voltage changes. When voltage is applied to the diode, the depletion region either blocks electron flow or allows electrons to pass.

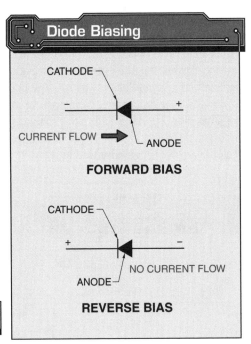

Figure 5-6. *In forward bias, electrons flow from cathode to anode. In reverse bias, electrons do not flow.*

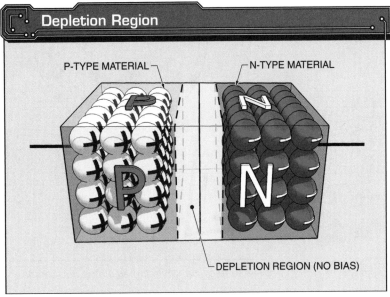

Figure 5-5. *A depletion region is formed at the junction of the P-type and N-type materials, where the materials exchange carriers.*

Diode Operation Principles

A diode has a relatively low resistance in the forward-bias direction, and a high resistance in the reverse bias direction. The anode (triangle) represents the P-type material, and the cathode (straight line) represents the N-type material. Electrons flow from the cathode to the anode, or against the triangle, when forward-biased. When a negative polarity is applied to the anode and a positive polarity is applied to the cathode, the diode is in reverse bias and there is no electron flow. **See Figure 5-6.**

Manufacturers mark diodes in different ways to indicate the cathode and the anode. **See Figure 5-7.** Diodes may be marked with the schematic symbol, or there may be a band at one end to indicate the cathode. Some manufacturers use the shape of the diode package to indicate the cathode end. Typically the cathode end is beveled or enlarged to ensure proper connection. When unsure, the user can check the polarity of a diode with a digital multimeter (DMM) set to diode test mode.

Forward Bias. *Forward-bias voltage* is voltage applied with polarity that allows a diode to act as a conductor. Forward-bias voltage occurs when the anode of a diode is positive and the cathode is negative. When a diode is forward-biased, the polarity of the voltage source causes the free electrons and holes to move toward the depletion region. **See Figure 5-8.** The carriers bridge the depletion region and cause it to close. In this condition, carriers are available from one end of the diode to the other, allowing electrons to flow with little resistance.

Chapter 5—Diode Applications and Troubleshooting 111

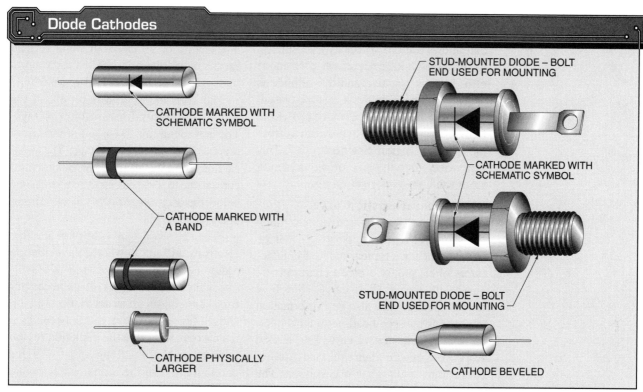

Figure 5-7. *Manufacturers use a variety of methods to indicate the cathode end of a diode. With stud-mounted diodes, the threaded end may be an anode or cathode. Check manufacturer specifications for proper installation.*

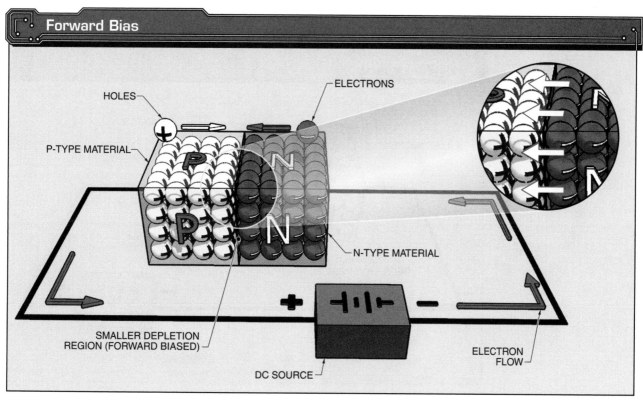

Figure 5-8. *When a diode is forward-biased, the polarity of the voltage source moves the free electrons and holes toward the depletion region, causing it to close and allow electrons to flow with little resistance.*

Reverse Bias. *Reverse-bias voltage* is voltage applied with opposite polarity that allows a diode to act as an insulator. Reverse-bias voltage occurs when the anode of a diode is negative and the cathode is positive. When a diode is reverse-biased, the polarity of the voltage source causes the free electrons and holes to move away from the depletion region. **See Figure 5-9.** This action increases the size of the depletion region and prevents electron flow.

Diode Characteristic Curve. A *diode characteristic curve* is a curve on a graph that shows the relationship between voltage and current for a typical diode. Manufacturers often supply a diode characteristic curve with diodes. These curves show how diodes operate. They also give information concerning the specifications of the diodes.

A diode characteristic curve indicates the response of a diode when subjected to different forward- and reverse-bias voltages. The axes of the graph provide specific information. The horizontal axis of the characteristic graph represents the voltage applied to a diode. The vertical axis represents the electron flow (current) through the diode. The intersection of the horizontal and vertical axes is called the origin. The origin represents zero on both axes. **See Figure 5-10.**

The horizontal line to the right of the origin represents forward-bias voltage. The horizontal line to the left of the origin represents reverse-bias voltage. The vertical line above the origin represents forward current due to forward bias. The vertical line below the origin represents reverse current due to reverse bias.

When forward-bias voltage is applied, electrons will not flow until the depletion region is closed. *Forward breakover voltage* is the forward-bias voltage necessary for a semiconductor to act as a conductor. Only a few tenths of a volt is needed for the carriers to bridge the depletion region (about 0.6 V for silicon and 0.3 V for germanium). On the diode characteristic curve, the depletion region is indicated by the short flat line that runs parallel to the horizontal axis.

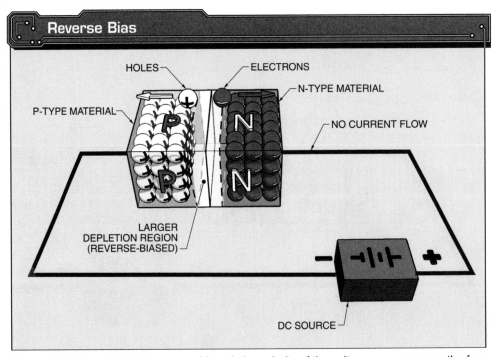

Figure 5-9. *When a diode is reverse-biased, the polarity of the voltage source moves the free electrons and holes away from the depletion region, making the region larger where electrons do not flow because of the high resistance.*

Once the depletion region is closed, resistance across the diode is very low and current increases rapidly. If the forward-bias voltage remains, the diode continues to allow electrons to flow freely. *Forward operating current* is the current range in which a semiconductor can safely operate once it reaches the forward breakover voltage. When the forward operating current is high, a heat sink may be needed to dissipate the heat generated.

Depending on the physical size and rating of a diode, typical forward current ranges from a few milliamperes (mA) to several hundred amperes (A). The value of the current depends on the type of diode. The current rating indicates the maximum continuous current level a diode can safely handle.

CAUTION: A replacement diode should have at least the same forward current rating as the original diode.

Diodes are also rated for the maximum reverse-bias voltage they can withstand. *Peak inverse voltage (PIV)* is the maximum reverse-bias voltage that a diode can withstand without breaking down and allowing electron flow. The PIV ratings for most diodes used in industry range from a few volts to several thousand volts. Reverse bias applied to a diode that exceeds its PIV rating breaks down the diode and allows electrons to pass freely. *Avalanche current* is the flow of electrons when a diode breaks down and allows electrons to pass freely, which can damage the diode. To avoid avalanche current, diodes with the correct rating must be used.

CAUTION: A replacement diode must have an equal or greater PIV rating than the original diode.

Diode Testing

In most cases, diodes are tested with a digital multimeter (DMM). The polarity of the diode can be determined with a DMM, and the diode can also be checked for opens and shorts. Diode testing includes determining polarity and checking for opens and shorts.

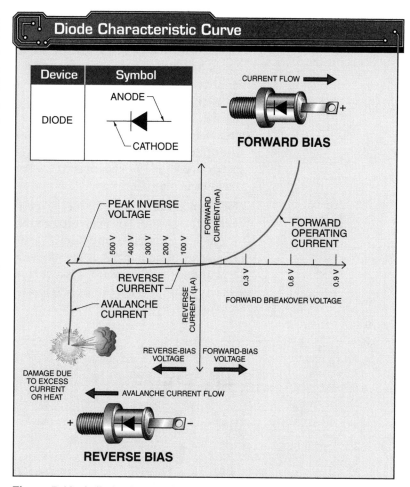

Figure 5-10. *A diode characteristic curve indicates the response of a diode when subjected to different forward- and reverse-bias voltages.*

CAUTION: Because DMMs vary somewhat from manufacturer to manufacturer, consult the operator's manual before conducting any diode tests with them.

Polarity. A DMM set to measure resistance can be used to determine which end of a diode is the cathode and which end is the anode. This is possible because the DMM is a voltage source with a definite polarity.

Externally, the polarity of the DMM may be marked positive (+) and negative (−). It may also be identified by a color-coding system—usually red for positive and black for negative. Internally, however, the voltage source, or battery, actually determines the external polarity.

Testing Diodes Media Clip

The forward and reverse bias of an unknown diode can be determined if the diode is placed between known polarities one way, and then placed in the opposite direction. The diode indicates a low resistance in forward bias and a high resistance in reverse bias. **See Figure 5-11.** Since the polarity of the source is known, the end connected to the negative lead during forward bias must be the cathode, and the end connected to the positive lead must be the anode.

Opens and Shorts. The best way to test a diode is to measure the voltage drop across the diode when it is forward-biased. Testing a diode using a DMM set to measure resistance may not indicate whether a diode is good or bad. Testing a diode that is connected in a circuit with a DMM may give false readings because other components may be connected in parallel with the diode being tested.

A good diode has a voltage drop across it when it is forward-biased and acts as a conductor. The voltage drop is between 0.5 V and 0.8 V for the most commonly used silicon diodes. Some diodes are made of germanium and have a voltage drop between 0.2 V and 0.3 V.

A voltage drop is the amount of reduced voltage that occurs across a component when moving from the positive side to the negative side of the power supply.

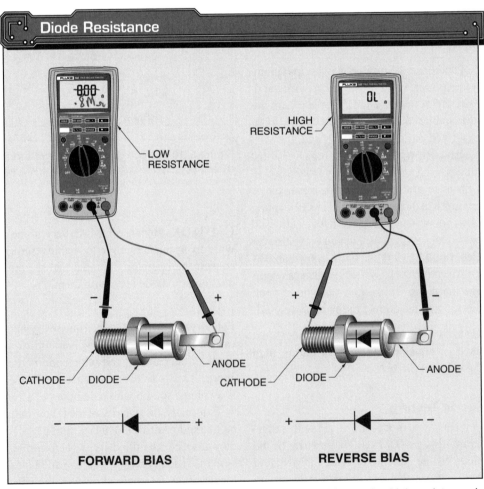

Figure 5-11. *A diode indicates a low resistance in forward bias and a high resistance in reverse bias.*

A DMM in the diode test mode is used to test the voltage drop across a diode. In this position, the DMM produces a small voltage between the test leads. The DMM displays the voltage drop when the leads are connected across a diode. **See Figure 5-12.** To test a diode using the diode test mode on a DMM, apply the following procedure:

1. Ensure that all power in the circuit is OFF. Test for voltage using a DMM set to measure voltage to ensure power is OFF.
2. Set the DMM to the diode test mode.
3. Connect the DMM leads to the diode. Record the reading.
4. Reverse the DMM leads. Record the reading.

The DMM displays a voltage drop between 0.5 V and 0.8 V (for a silicon diode) or 0.2 V and 0.3 V (for a germanium diode) when a good diode is forward-biased. The DMM displays an OL reading when a good diode is reverse-biased. The OL reading indicates that the diode is acting like an open switch. An open (bad) diode does not allow electrons to flow through it in either direction. The DMM displays an OL reading in both directions when the diode is open. A shorted diode gives the same voltage drop reading in both directions. This reading is typically 0.4 V. If a diode is open or shorted, it must be replaced.

Henkel Corporation
Diodes may be found in various electrical equipment, such as flat panel TVs.

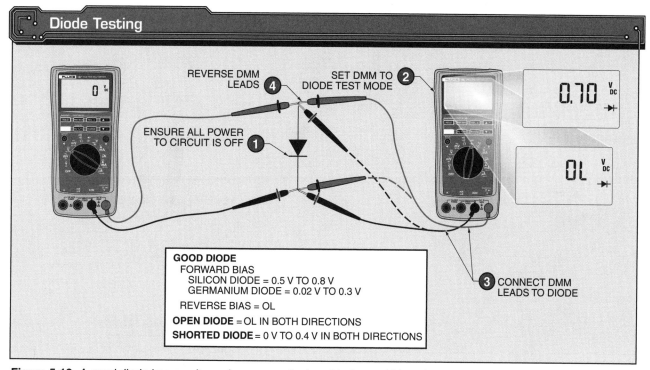

Figure 5-12. *A good diode has a voltage drop across it when it is forward-biased and conducting electricity.*

Diode Installation and Service

To ensure that diodes operate properly, the following points must be observed when installing or servicing a diode:

- Observe voltage specifications. Check the power line voltage to make sure the diode is properly rated.
- Do not overload diodes, even momentarily. Double check circuits, polarities, component sizes, and wiring before installation.
- Stud-mounted diodes must be fastened properly and securely to their heat sinks to ensure good heat flow. **See Figure 5-13.** Do not overtighten the stud or it will stretch or strip. The best way to install diodes in heat sinks is to use a torque wrench. Manufacturer specifications should be followed.
- Carefully observe the manufacturer's recommended use of heat sinks. **See Figure 5-14.** If heat cannot escape, the diode may be damaged. Be sure air can circulate around the device. Also, watch for excessively high temperatures caused by other components such as nearby lamps, motors, or heaters.
- When diodes are mounted on heat sinks, silicon grease is often applied to the surfaces of the metal where contact is made. Silicon grease allows heat to transfer more rapidly.

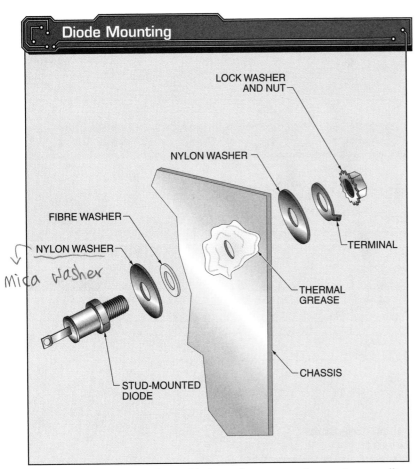

Figure 5-13. A chassis along with thermal grease can be used as an effective heat sink to conduct excess heat away from a diode.

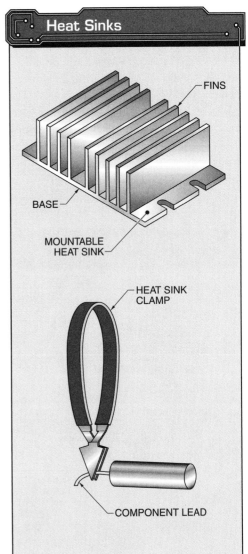

Figure 5-14. Heat sinks are designed to draw heat away from the body of a semiconductor device.

- The amount of heat developed by soldering irons and pencils can damage semiconductor devices. Care must be exercised when semiconductor devices, such as diodes, are removed or replaced during use of soldering equipment. In most cases, heat-sensitive components can be protected through the use of a heat sink. A small metallic clamp is designed to draw heat away from the body of a semiconductor device. When a heat sink is not available, an alligator clip or a pair of long-nose pliers may be substituted. Pliers may require a rubber band or wire wrapped around the handles to maintain good thermal contact. **See Figure 5-15.**

CAUTION: Heat sinks should not be removed immediately. In most cases, allow 30 sec to 45 sec before removing a heat sink.

Diode Power Capacity and Derating

The maximum current at which a diode may operate in a normal environment is limited primarily by the temperature rise at the PN junction. It is also limited by the size of the heat sink provided to dissipate the heat. The power capacity of most diodes is rated at an operating temperature of 25°C, or approximately room temperature. As the ambient temperature increases beyond this point, the diode can no longer dissipate as much heat and must be derated. *Derating* is the act of operating a diode at less than maximum operating current. Derating tables are available to determine this value.

When it is necessary to operate a diode at or close to the maximum current rating, heat sinks must be used. Heat sink tables are available to determine the proper size for a particular job and should be consulted.

Diode Applications

Diode applications vary. By connecting a diode in series, it can stop electron flow from entering another component. By connecting a diode in parallel, it can bypass potentially damaging currents. By using series-parallel arrangements, diodes can direct electron flow as required. Diode applications include rectifiers, clipping (limiting) circuits, and clamping circuits.

Rectifiers. Many electronic devices require direct current (DC) power for operation. However, alternating current (AC) power is generally the most readily available power. Therefore, AC power must be converted to DC power before it can be used to operate electronic devices. *Rectification* is the changing of AC to DC. A *rectifier* is a device consisting of diodes that convert AC power to DC power by allowing electrons to flow in only one direction.

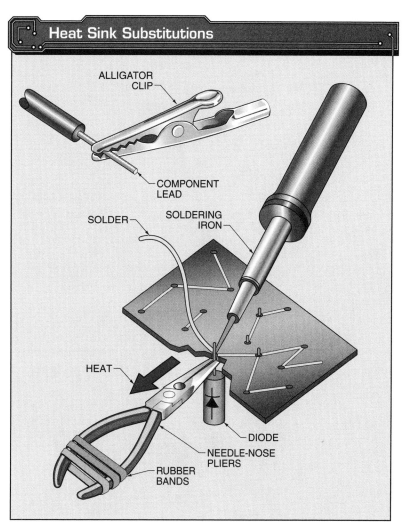

Figure 5-15. *Alligator clips and needle-nose pliers may be used as substitutes for heat sinks.*

The simplest form of rectifier is the half-wave rectifier. **See Figure 5-16.** A *half-wave rectifier* is a circuit containing an AC source, a load resistor (R_L), and a diode that permits only the positive half cycles of the AC sine wave to pass, creating pulsating DC. Half-wave rectification is accomplished because electrons are allowed to flow only when the anode is positive and the cathode is negative. Electrons are not allowed to flow through the half-wave rectifier when the cathode is positive and the anode is negative.

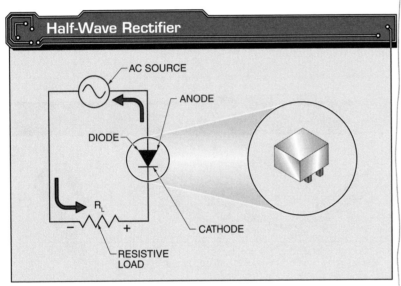

Figure 5-16. *A half-wave rectifier is the simplest form of a rectifier circuit, consisting of a load resistor and a diode across an AC source.*

Various solid state components can be found in mobile devices.

A half-wave rectifier can affect the AC voltage by rectifying or cutting out the negative output voltage. **See Figure 5-17.** Since the signal now travels in only one direction 50% of the time, it is classified as pulsating DC. A half-wave rectifier is often mounted on a PC board. The half-wave rectifier is only one type of power supply circuit in which diodes are used.

Diode Clipping (Limiting) Circuits. It is necessary to limit the amount of voltage present in certain electronic circuits, especially where large voltage changes are expected. A diode clipping (limiting) circuit is also referred to as a voltage limiter. A *voltage limiter* is an electronic circuit that consists of diodes used to control voltage where large voltage changes are expected. A simple voltage limiter can be constructed from a resistor and a reverse-biased diode. The bias voltage and the voltage required to forward bias the diode determine the maximum positive peak output of the voltage limiter.

A bias voltage of 3 V is connected to the positive terminal of the bias voltage source. The bias voltage source is connected to the cathode of the diode. This polarity causes the diode to be reverse-biased and no electrons flow through the circuit. For electrons to flow, a positive signal of 3.6 V or greater must be applied to the circuit. A value of 3.0 V is necessary to neutralize the reverse bias, and a value of 0.6 V is needed to make the diode conduct. For any value less than 3.6 V, the diode acts as an open switch. **See Figure 5-18.**

However, when a signal of 3.6 V or greater is applied, the diode conducts and effectively shorts out the output. The positive output of the half cycle is limited to the bias voltage of 3.6 V. The entire negative half of the cycle appears across the output since the diode is an open circuit.

Derating tables, usually available from electronics manufacturers, account for several factors, including ambient temperature, the temperature of the whole system, and heat dissipation.

Chapter 5—Diode Applications and Troubleshooting **119**

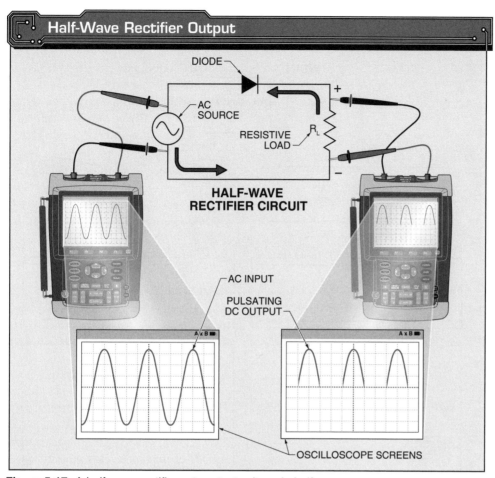

Figure 5-17. *A half-wave rectifier cuts output voltage in half.*

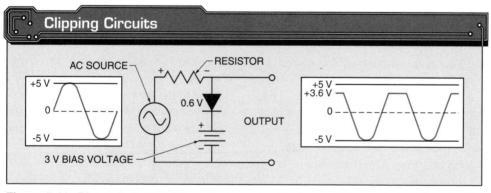

Figure 5-18. *Bias voltage plus the voltage required to forward bias a diode determine the maximum positive peak output of a voltage limiter.*

Clamping Circuits. A *clamping circuit* is a circuit that holds voltage or current to a specific level. The component commonly used as a clamper (clamping device) is the diode. During the clamping process, the diode should not significantly change the shape of a waveform. **See Figure 5-19.**

An oscilloscope can be used to identify the difference between a diode clipping circuit, which distorts a voltage waveform by cutting off the voltage at a certain level, and a clamping circuit.

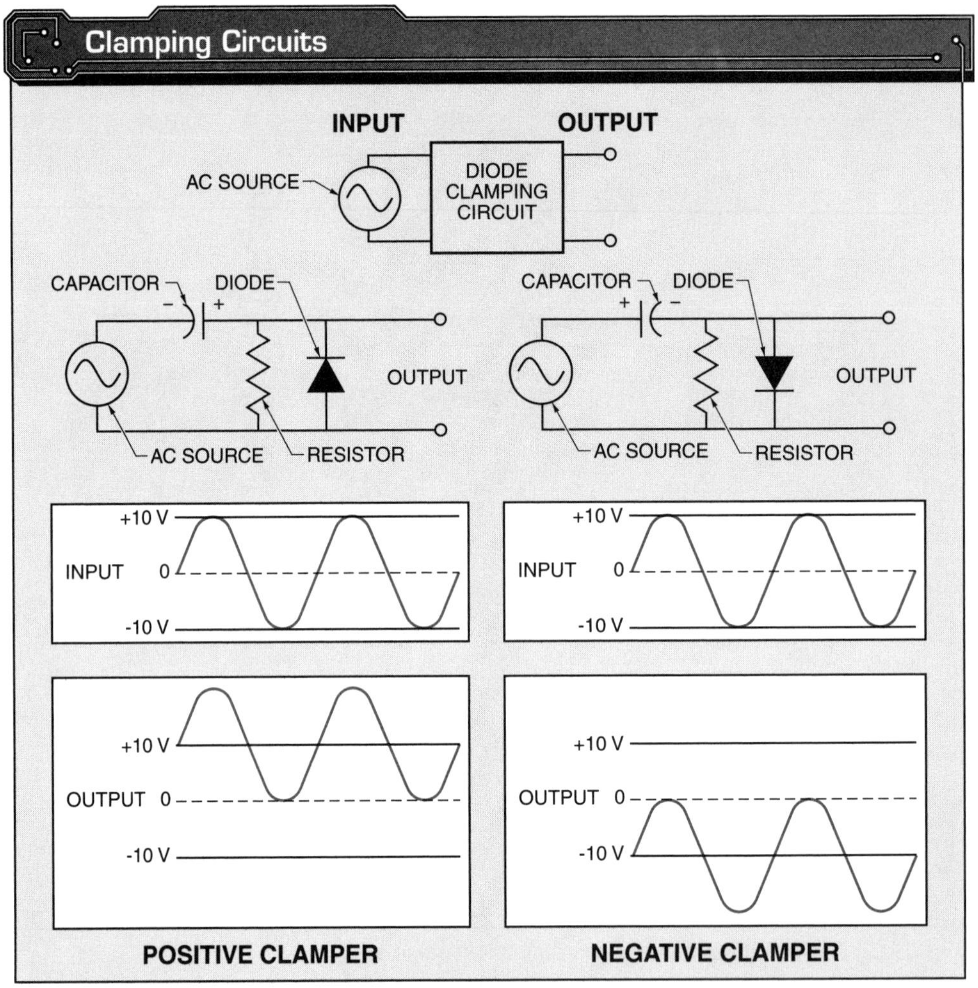

120 SOLID STATE DEVICES AND SYSTEMS

Clamping Circuits

Figure 5-19. A diode clamping circuit effectively raises or lowers the positive or negative peaks of a signal to zero or a desired reference level.

The main function of a clamper is to provide a DC reference level for a signal voltage. The clamper effectively raises or lowers the positive or negative peaks of a signal to zero or some other desired reference level. **See Figure 5-20.** In a diode clamping circuit schematic, inputs A through E represent the first AC cycle appearing at the input terminals. Outputs A through E represent the corresponding output voltages. The solid-line portion of each segment shows only the part of the cycle that is affected at that point in time.

The portion of the cycle shown in input A charges capacitor C1 through forward-biased diode D1. The electron flow (current) when capacitor C1 is charging is indicated by the direction of the arrow. With the polarity shown, capacitor C1 charges to the maximum positive value of the AC cycle. As the capacitor charges, the capacitor voltage is always equal to and opposite of the AC input signal voltage. The opposite voltages cancel each other, producing an output of 0 V.

As the positive portion of the cycle decreases, the diode becomes reverse-biased. The capacitor tries to discharge some of its stored potential through resistor R1. However, the RC time constant is always long compared to the time duration of the input voltage. With resistor R1 large enough, the capacitor loses only a small amount of its charge over a complete AC cycle.

Clamping Circuit Operation

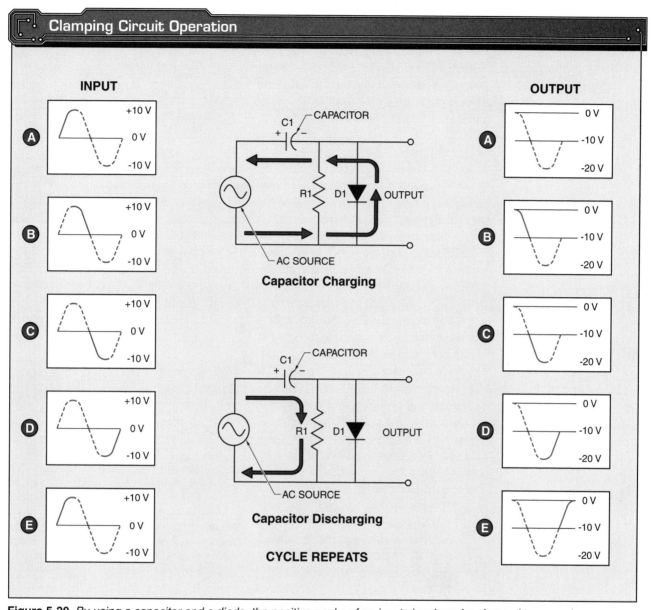

Figure 5-20. *By using a capacitor and a diode, the positive peaks of an input signal can be clamped to ground.*

As the AC input signal voltage continues to decrease (input B), the capacitor acts like a battery and opposes the signal voltage. The negative voltage at the output (output B) represents the difference between the charge on the capacitor and the decreasing AC voltage. Eventually, the point is reached in input B where the signal returns to zero. The corresponding voltage output B only represents the charge on the capacitor and ranges from 0 V to –10 V.

In inputs C and D, the AC input signal voltage is shown in its negative polarity region. All of the circuit conditions described for input B still apply. However, instead of opposing the negative-going signal voltage, the capacitor voltage now aids it. When the full negative value of the signal is reached, the corresponding output is shown in outputs C and D. *Note:* The output voltage is twice the value of either the positive or negative alternation at the input.

In input D, the signal returns to zero. The corresponding output voltage is the charge on the capacitor, which ranges from –20 V to –10 V. The cycle repeats as shown in input A. However, the output waveform in output A occurs only on the first input cycle because the capacitor is charging for the first time. The corresponding output waveform for all later cycles is output B. *Note:* The diode clamping circuit clamps the positive peaks of the input signal to ground. A circuit that clamps the negative peaks to ground is made by reversing the diode.

A diode clamping circuit can be used when a reference level other than zero is needed. A clamper can provide an off-ground reference level. A *biased clamping circuit* is a diode clamping circuit that uses a battery to reverse-bias the diode. It is similar to the diode clamping circuit. The difference is that the biased clamping circuit produces an output reference level equal to the battery voltage. **See Figure 5-21.**

The battery applies 3 V of forward bias to the diode. When the diode conducts, capacitor C charges to approximately 3 V. The first negative alternation at the input also causes the capacitor to charge. However, on the negative alternations, the capacitor voltage opposes the input voltage. Therefore, the output voltage cannot be less than +3 V.

On the positive alternations, the capacitor voltage aids the input voltage. Therefore, the maximum positive voltage is equal to the peak-to-peak input voltage plus the bias voltage. In this case, the maximum positive voltage is 23 V.

ZENER DIODES

A *zener diode* is a semiconductor device used as a voltage regulator. The zener diode looks similar to other silicon diodes. **See Figure 5-22.** Its purpose is to act as a voltage regulator either by itself or in conjunction with other semiconductor devices. A voltage regulator is an electronic circuit that maintains a relatively constant value of output voltage over a wide range of operating situations. In a schematic, the zener diode symbol differs from a standard diode symbol in that the normally vertical cathode line is bent slightly at each end.

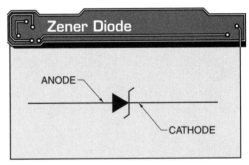

Figure 5-22. *A zener diode symbol differs from a standard diode symbol in that the typically vertical cathode line is bent slightly at each end.*

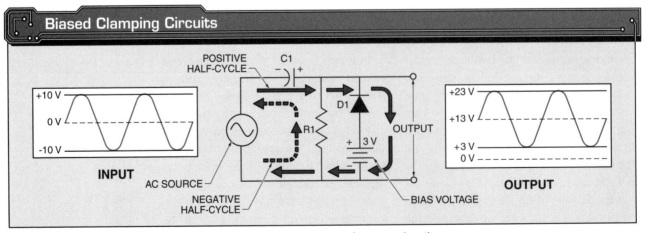

Figure 5-21. *A biased clamping circuit can be used for a voltage reference other than zero.*

The zener diode is unique because it is most often used to conduct electricity under reverse-bias conditions. Standard diodes usually conduct in the forward-bias condition and can be destroyed if the reverse voltage or bias is exceeded. Because the zener diode typically operates in reverse breakdown, it is often called an avalanche diode.

Zener Diode Operation

The operation of a zener diode is best understood through the use of a characteristic curve. **See Figure 5-23.** The overall forward and reverse characteristics of a zener diode are similar to that of the standard diode. When a source voltage is applied to the zener diode in the forward direction, there is a breakover voltage and forward current. When a source voltage is applied to the zener diode in the reverse direction, the current remains very low until the reverse voltage reaches reverse (zener) breakdown. The zener diode then conducts heavily, or avalanches. Reverse current in a zener diode must be limited by a resistor or other device to prevent diode destruction. The maximum amount of current a zener diode can handle is determined by diode size. Like the forward voltage drop of a rectifier diode, the reverse voltage drop (zener voltage) of a zener diode remains constant despite large current fluctuations.

A zener diode is capable of being a constant voltage source because of the resistance changes that take place within the PN junction. The constant voltage can be seen by studying the reverse characteristic curve. When a source voltage is applied to a zener diode in the reverse direction, the resistance of the PN junction remains high and should produce only leakage currents in the microampere range. However, as the reverse voltage is increased, the PN junction reaches a critical voltage and the zener diode avalanches. As the avalanche voltage is reached, the normally high resistance of the PN junction drops to a low value and the current increases rapidly. The current is limited generally by a circuit resistor or resistance R_L. **See Figure 5-24.** The breakdown current is not destructive to a zener diode. The current becomes destructive if it is excessive or the heat-dissipating capabilities of the zener diode are exceeded.

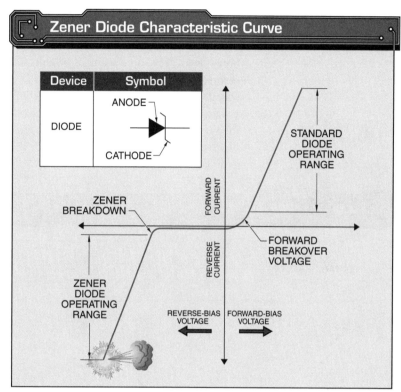

Figure 5-23. *A zener diode typically operates in reverse breakdown and is often referred to as an avalanche diode.*

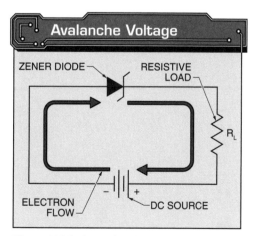

Figure 5-24. *As avalanche voltage is reached, the typically high resistance of the PN junction drops to a low value and the junction current increases rapidly.*

Zener Diode Voltage Ratings. The zener diode is manufactured through a highly controlled doping process. The doping process allows manufacturers to predetermine the operating range of a zener diode. Typical values of zener voltages may range from a few volts to several hundred volts.

When a zener diode is rated at a specific voltage, it is not necessarily the same value of voltage that begins to cause the diode to break down. The rated value represents the reverse voltage across the zener diode at a specific zener test current. This value falls within the initial breakdown range and at the maximum reverse current that a zener diode can safely handle. **See Figure 5-25.**

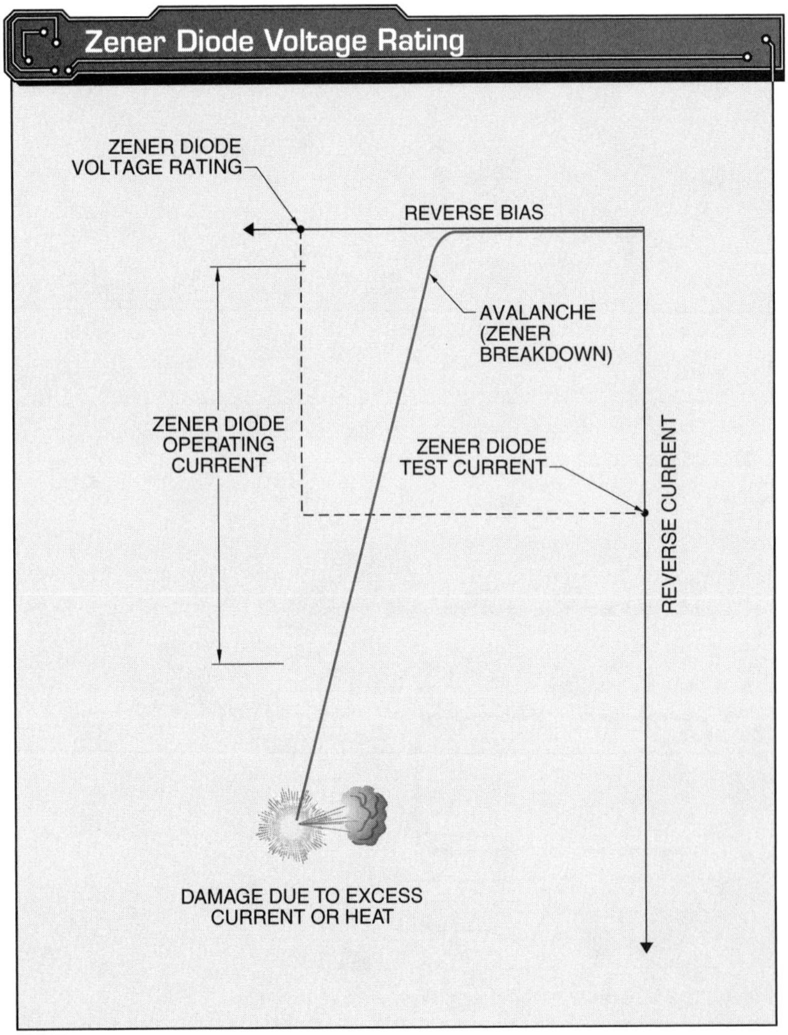

Zener Diode Tolerances. A zener diode is designed to have a specific breakdown voltage rating. Typically the voltage is a close approximation of the voltages necessary for circuit control, for example, 4.7 V, 5.1 V, 6.2 V, and 9.1 V. Like resistors, zener diodes are manufactured according to certain tolerances. The standard zener breakdown voltage tolerances are ±20%, ±10%, and ±5%. When precision is required, zener diodes are also available in tolerances of ±1%.

Obtaining Other Voltages. When a voltage that is not normally obtainable from standard zener diodes must be regulated, several zener diodes can be connected in series to achieve proper voltage. **See Figure 5-26.** Zener diodes do not necessarily have equal breakdown voltages because the series arrangement is self-equalizing. However, their power ratings and current ranges should be similar and the loads should be matched to avoid damaging any of the zener diodes.

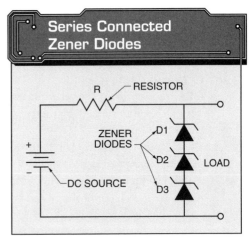

Figure 5-26. *When it is necessary to regulate a voltage not typically obtainable from standard zener diodes, several zener diodes can be connected in series to achieve the proper voltage.*

Zener Diode Power Rating. Zener diodes range from one millionth of a watt up to 50 W. The most popular zener diodes are those rated at 1 W and smaller. The power dissipation rating of a zener diode, like other

Figure 5-25. *The rated voltage value of a zener diode is based on a specific zener test current.*

diodes, is given for a specific operating temperature. The power rating is usually based on an ambient temperature of 25°C.

Temperature Considerations. The zener breakdown voltage varies considerably with changes in ambient temperature. For this reason, manufacturers frequently list the zener voltage temperature coefficient. *Temperature coefficient* is the percentage of change in voltage per degree change in temperature. Typically, the zener voltage change is about 0.1% per degree centigrade.

Zener Diode Testing

A zener diode provides voltage regulation. If it fails, the zener diode must be replaced in order to return the circuit to proper operation. Occasionally, the zener diode may appear to fail only in certain situations. An *intermittent* is an electronic device that works only part of the time. To check for intermittents, the zener diode must be tested while in operation. An oscilloscope is used for testing the characteristics of a zener diode in an operating situation. **See Figure 5-27.**

An oscilloscope is a measuring device used to display voltage waveforms. An oscilloscope displays the dynamic operating characteristics of the zener diode. If the zener diode is good, the appropriate test display will be shown. The horizontal axis of the oscilloscope represents the voltage across the zener diode while the vertical axis represents the current. This test is often used for production testing and sorting diodes.

Zener Diode Applications

The zener diode is used in a variety of circuit applications. Because of its unique operating characteristics, it may be applied to very simple as well as very complex circuitry. However, the basic operating principle of the zener diode never changes. Zener diode applications include voltage regulator, clipping and limiting, and oscilloscope calibration.

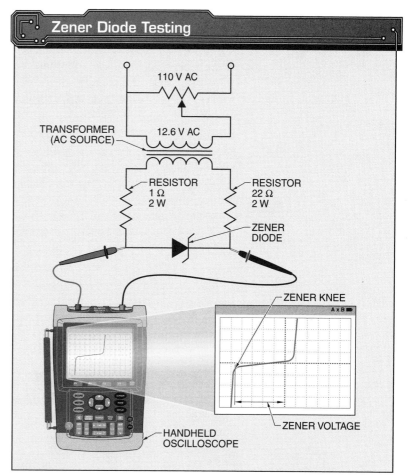

Figure 5-27. *A zener diode that fails intermittently must be tested with an oscilloscope while it is operating in a circuit.*

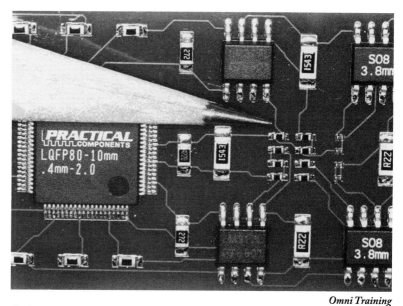

Omni Training

Surface mount diodes are very small and may be used to create full-wave bridge rectifiers.

Voltage Regulators. Zener diodes can be used as voltage regulators. A voltage regulator is an electronic circuit that maintains a relatively constant value of output voltage over a wide range of operating situations.

Voltage regulators include series regulators and shunt regulators. A series regulator is a circuit with a voltage regulator in series with a load. The series regulator allows a preset voltage to the load, regardless of the regulator input voltage or the current demand of the load. A shunt regulator is a circuit with a voltage regulator at the output of a power supply and in parallel with a load. The shunt regulator absorbs or shunts excess voltage at the load. **See Figure 5-28.** The total current through resistor R1 is the sum of the zener diode current and the load current. Resistor R1 is typically referred to as a current-limiting resistor.

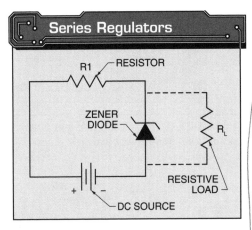

Figure 5-28. *A shunt regulator includes a zener diode used as a voltage regulator installed across the load. A current limiting resistor is required to prevent destruction of the diode, power supply, and load.*

The zener diode, operating as a voltage regulator, maintains a specific voltage across the load. If the DC source voltage increases, the zener diode absorbs the voltage by increasing the current flow through the diode. This causes current flow through current-limiting resistor R1 to increase and the voltage drop across resistor R1 to increase. The voltage across the load remains the same. If the DC source voltage decreases, the current through the zener diode decreases, causing the voltage at the load to increase. The current through resistor R1 drops, and the voltage drop across the resistor decreases.

If the load decreases, the zener diode acts the same as when the DC source voltage increases. Internal current increases through the zener diode, current increases through resistor R1, and the voltage drop across the resistor increases to compensate for the decreased load. If the load increases, the zener diode decreases internal current flow and the voltage drop across the resistor decreases to compensate for the increased load.

Shunt regulators operate at a given voltage, as long as the DC source voltage and load stay within limits. If the DC source voltage or load causes the voltage across the shunt regulator to decrease below the regulated voltage, all internal regulation stops until either the DC source voltage is increased or the load is decreased.

Clipping (Limiting) Circuits. A zener diode that is supplied with alternating current and connected as a shunt regulator is capable of limiting either the positive or negative parts of an AC cycle. **See Figure 5-29.** On the positive segment of the signal, the zener diode conducts almost immediately after the signal passes through zero, leaving no output signal across it. On the negative segment of the cycle, the zener diode does not conduct until the applied voltage reaches V_z. At V_z the diode conducts, clipping the remainder of the signal (zener clipping). If the zener diode is reversed in the circuit, the positive signal will be clipped.

If both the positive and negative portions of the signal are to be clipped, then two zener diodes must be wired back-to-back. **See Figure 5-30.** Two zener diodes wired back-to-back may be used to produce a relatively simple square wave signal. The two zener diodes may also be used as a protection circuit. For example, the zener diodes could be wired across a speaker. In a protection circuit, the zener diodes limit or clip any high-voltage spikes that could damage the speaker.

Chapter 5—Diode Applications and Troubleshooting 127

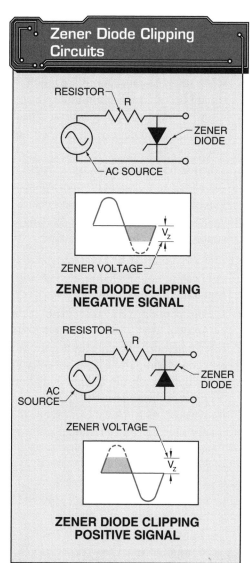

Figure 5-29. When supplied with alternating current and connected in parallel, a zener diode is capable of limiting either the positive or negative segments of an AC cycle.

Oscilloscope Calibration. A stable voltage source that is not affected by line voltage variations is typically necessary to calibrate an oscilloscope. A zener diode can be incorporated into existing oscilloscope circuitry to provide a calibration voltage. **See Figure 5-31.** A 10 V zener diode is used to provide a calibration of one volt per division. Resistor R_L establishes the basic zener test current. Variable resistor R_S is used for fine adjustment.

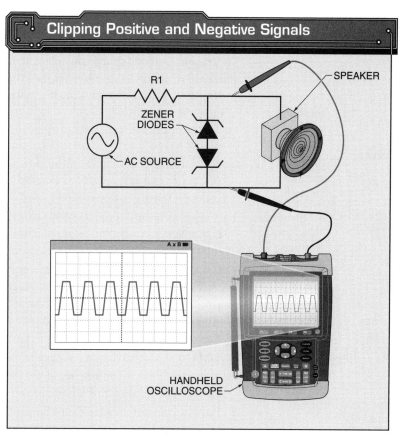

Figure 5-30. Two zener diodes must be wired back-to-back to clip the positive and negative portions of a signal.

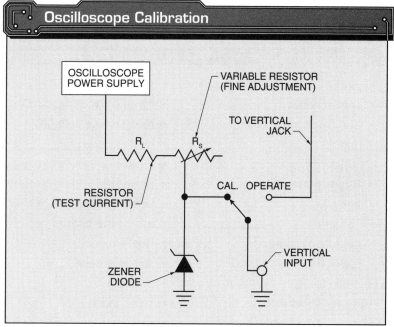

Figure 5-31. A zener diode can be incorporated into existing oscilloscope circuitry to provide a calibration voltage.

MAINTAINING SEMICONDUCTOR DEVICES

To prevent damage and avoid electrical shock, it is important to observe the following precautions when working with semiconductor devices:

- Test equipment and soldering irons should be checked to make certain there is no leakage current from the power source. If leakage current is detected, isolation transformers should be used.
- Always connect a ground between test equipment and a circuit before attempting to inject or monitor a signal.
- Ensure test voltages do not exceed maximum allowable voltage for circuit components and semiconductors. Never connect test equipment outputs directly to a semiconductor circuit.
- DMM ranges that require a current of more than 1 mA in the test circuit should not be used for testing semiconductors.
- Battery eliminators should not be used to furnish power for semiconductor equipment because they have poor voltage regulation and high-ripple voltage.
- The heat applied to a semiconductor when soldered connections are required should be kept to a minimum by using a low-wattage soldering iron and heat sinks, such as long-nose pliers, on the transistor leads.
- When it becomes necessary to replace semiconductors, do not pry transistors to loosen them from printed circuit boards.
- All circuits should be checked for defects before replacing a semiconductor.
- The power must be removed or disconnected from the equipment before replacing a semiconductor.
- Using conventional test probes on equipment with closely spaced parts often causes accidental shorts between adjacent terminals. These shorts can cause damage or ruin a semiconductor. To prevent these shorts, the probes can be covered with insulation, except for a very short length of the tips.

Refer to Chapter 5 Quick Quiz® on CD-ROM

KEY TERMS

- diode
- semiconductor
- free electrons (carriers)
- doping
- N-type material
- P-type material
- holes
- carriers
- depletion region
- anode
- cathode
- forward bias voltage
- forward current
- reverse current
- reverse operating current
- reverse bias voltage
- diode characteristic curve
- forward breakover voltage
- forward operating current
- peak inverse voltage (PIV)
- avalanche current
- heat sink
- silicon grease
- derating
- rectification
- rectifier
- half-wave rectifier
- voltage limiter
- clamping circuit
- avalanche diode
- biased clamping circuit
- zener diode
- voltage regulator
- temperature coefficient
- intermittent
- series regulator
- shunt regulator

REVIEW QUESTIONS

1. Why are diodes unique?
2. What is forward bias?
3. What is reverse bias?
4. How does current flow through a diode in reference to the anode and cathode?
5. What is the simplest form of a rectifier circuit?
6. What is a rectifier?
7. What is a semiconductor?
8. How is a semiconductor doped?
9. What is the difference between N-type material and P-type material?
10. What is a carrier?
11. Describe a diode characteristic curve and explain its use.
12. What is peak inverse voltage (PIV)?
13. What type of tests can be done on a diode using a DMM?
14. List the precautions that should be observed when using a DMM.
15. What special precautions should be observed when installing diodes?
16. Define the power capacity of a diode.
17. What is a zener diode?
18. How are zener diodes rated?
19. How are zener diodes tested?
20. What is the most common use for a zener diode?

DC POWER SUPPLY OPERATION AND TROUBLESHOOTING 6

Many devices used today contain electrical circuits that require DC power. Because AC power is easier to generate and transmit, and thus more common, a DC power supply is required to convert AC power to DC power. DC power supplies convert AC power into DC power using several electrical components.

OBJECTIVES

- Describe DC power supply operation.
- List and describe DC power supply components.
- Explain the operation of rectifiers.
- List and describe DC power supply filters.
- Explain the operation of voltage regulators.
- Explain the operation of voltage dividers.
- Explain the operation of voltage multipliers.
- Explain troubleshooting DC power supplies.
- List and describe power interruptions.

DC POWER SUPPLIES

Electrical circuits typically require DC power to operate properly. However, an AC power source is more practical because it is easier to generate and transmit than DC power. When DC power is needed, a DC power supply is used. A *DC power supply* is a device that converts alternating current (AC) to regulated direct current (DC) for use in electrical circuits.

A DC power supply consists of a rectifier, filter, voltage regulator, and voltage divider. **See Figure 6-1.** A DC power supply may also include a transformer and a voltage multiplier. A *transformer* is a device with two windings (primary and secondary) used to step up or step down AC voltage in an AC circuit to the proper operating voltage. Depending on the cost and application, a DC power supply may contain all or some of these components.

Rectifiers

A rectifier is an electrical circuit that changes AC into DC. Rectifiers typically consist of one or more diodes used to control the flow of current in a circuit. Three basic types of rectifiers used in single-phase DC power supplies are the half-wave, full-wave, and full-wave bridge rectifier. **See Figure 6-2.**

Half-Wave Rectifiers. A *half-wave rectifier* is a circuit containing an AC source, a load resistor (R_L), and a diode that permits only the positive half cycles of the AC sine wave to pass, creating pulsating DC. **See Figure 6-3.** Half-wave rectification is accomplished because current is allowed to flow only when the anode terminal of diode D1 is positive with respect to the cathode. Electrons are not allowed to flow through the rectifier when the cathode is positive with respect to the anode.

Direct Current Media Clip

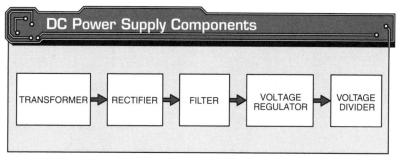
Figure 6-1. *A block diagram shows the basic components of a DC power supply.*

131

Rectifiers

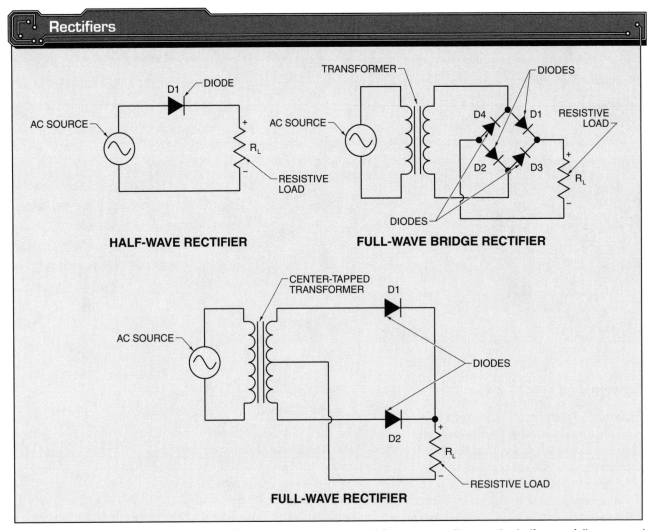

Figure 6-2. *The three basic types of rectifiers used in single-phase DC power supplies are the half-wave, full-wave, and full-wave bridge rectifiers.*

Half-Wave Rectifiers

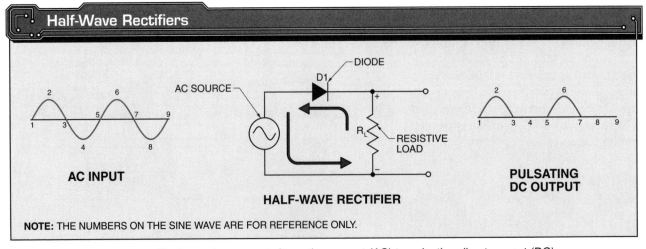

Figure 6-3. *A half-wave rectifier is used to convert alternating current (AC) to pulsating direct current (DC).*

The output voltage of the half-wave rectifier is considered pulsating DC with half of the AC sine wave cut off. *Pulsating DC* is direct current that varies in amplitude but does not change polarity. The rectifier can pass either the positive or negative half of the input AC cycle depending on the polarity of the diode in the circuit. Half-wave rectifiers are considered inefficient for many applications because one-half of the input cycle is not used.

One application that makes use of the reduced current of a half-wave rectifier is a light dimmer. **See Figure 6-4.** The circuit can be used for two-position light dimmers and high-intensity lamps that are switched from high (bright) to low (dim). When the dimmer switch is closed, the diode is shorted, which allows full current and full brightness. When the dimmer switch is open, the diode is placed in series with the lamp and blocks one-half of the current, which reduces the intensity of the lamp.

Full-Wave Rectifiers. A *full-wave rectifier* is an electrical circuit containing two diodes and a center-tapped transformer used to produce pulsating DC. The center-tapped transformer supplies out-of-phase voltages to the two diodes. When voltage is induced in the secondary from point A to B, point A is positive with respect to point N. Current flows from N to A, through the load R_L, and through diode D1. Diode D1 is forward-biased and allows electrons to flow. Diode D2 is reverse-biased and blocks current flow because point B is negative (−) and point A is positive (+). **See Figure 6-5.**

When the voltage across the secondary reverses during the negative half cycle of the AC sine wave, point B is positive with respect to point N. Current then flows from N to B, through the load, and through diode D2. Diode D2 is forward-biased and allows electrons to flow. Diode D1 is reverse-biased and blocks current because B is positive with respect to N. This is repeated every cycle of the AC sine wave, producing a full-wave DC output.

The output voltage of a full-wave rectifier has no off cycle. Electrons flow through the load during both half cycles. The constant electron flow results in a complete output signal. A full-wave rectifier is more efficient and has a smoother output than a half-wave rectifier.

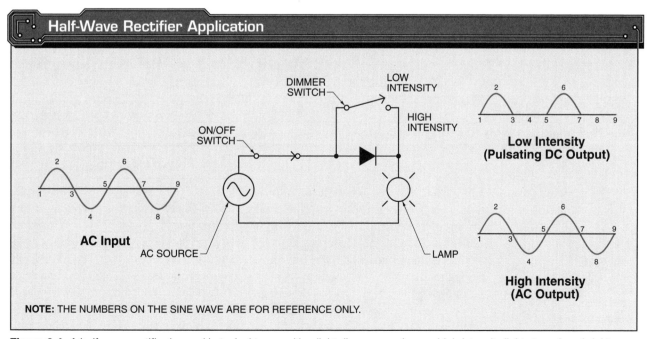

Figure 6-4. *A half-wave rectifier is used in typical two-position light dimmers and some high-intensity lights to reduce brightness.*

134 SOLID STATE DEVICES AND SYSTEMS

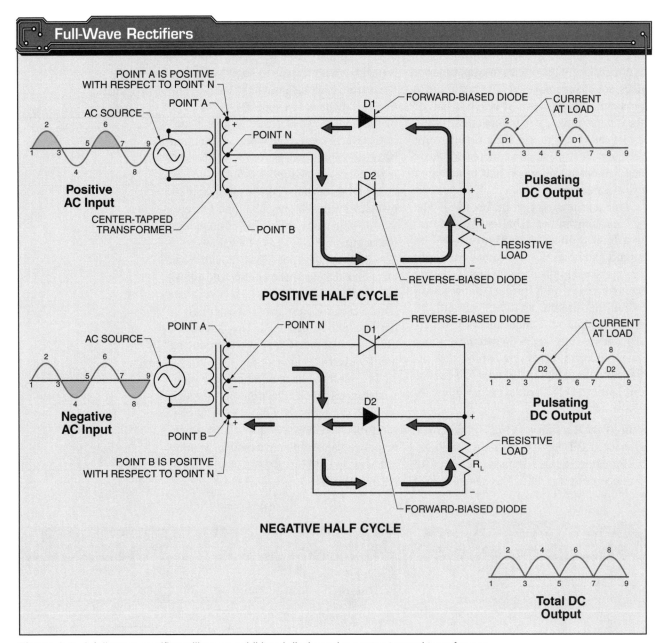

Figure 6-5. *A full-wave rectifier utilizes an additional diode and a center-tapped transformer.*

Full-Wave Bridge Rectifiers Media Clip

Full-Wave Bridge Rectifiers. A *full-wave bridge rectifier* is an electrical circuit that contains four diodes that allow both halves of a sine wave to be changed into pulsating DC. The full-wave bridge rectifier does not require a center-tapped transformer. It uses lower voltage diodes than the center-tapped circuit of a full-wave rectifier. The bridge diodes only need to block half as much reverse voltage as the center-tapped transformer diodes for the same output voltage. **See Figure 6-6.**

When the voltage is positive at point A and negative at point B, electrons flow from point B through diode D2, the load (R_L), and diode D1 to point A. When the AC supply voltage is negative at point A and positive at point B, electrons flow from point A through diode D4, the load (R_L), and diode D3 to point B.

Full-Wave Bridge Rectifiers

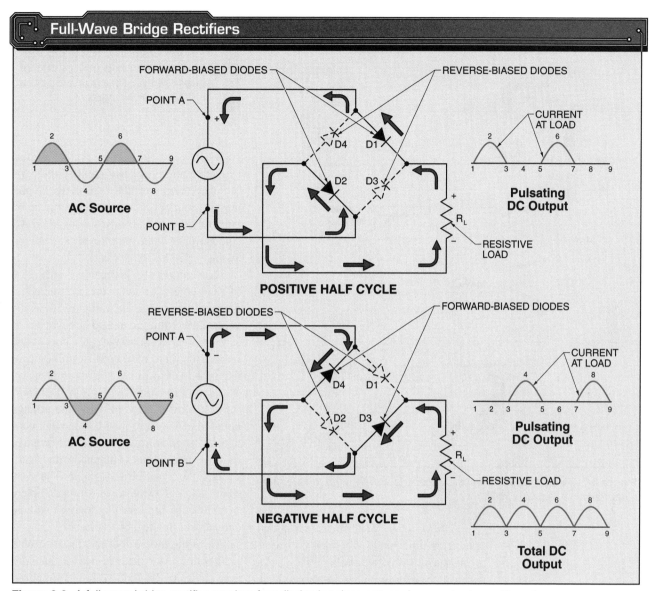

Figure 6-6. *A full-wave bridge rectifier requires four diodes but does not require a center-tapped transformer.*

One disadvantage of the full-wave bridge rectifier is that on each alternation, the direct current in the circuit must flow through two series-connected diodes. The forward DC voltage drop across the two rectifiers is, therefore, greater than the drop across a single rectifying diode. However, the voltage drop across silicon diodes is small (0.6 V), and the loss typically can be tolerated.

Rectifier Voltage Measurements. Measuring voltages in a rectifier requires knowledge of some of the terminology, math, and instruments necessary to take accurate measurements. A rectifier uses both AC and DC. Care must be exercised when measuring AC and DC and when converting or comparing any AC value to a DC value. For example, 100 VAC is not necessarily the same as 100 VDC. An AC voltage is constantly changing while a DC voltage remains the same. Comparing one value to another requires a mathematical conversion, depending on the test instrument used to take the measurement. Measurements in a rectifier are typically taken with either a DMM or an oscilloscope. **See Figure 6-7.**

Rectifier Measurements

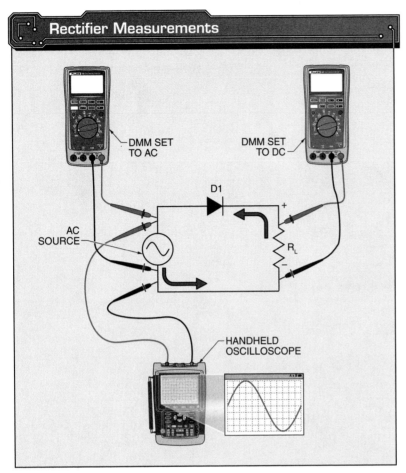

Figure 6-7. Either a DMM or an oscilloscope is usually used to take measurements in a rectifier circuit.

> A center tap is a connection in the middle of a winding. A center-tapped transformer has a secondary winding of two half cycles of equal voltage. One half is positive and the other is negative.

When using an oscilloscope, care must be exercised to properly identify the parts of a waveform so that the trace (scope pattern) is accurately converted for comparison purposes. Typical voltage values of an AC sine wave are peak, peak-to-peak, rms, and average. **See Figure 6-9.**

Peak voltage (V_p) is the maximum value of an alternating voltage that is reached on the positive or negative half cycle of a sine wave. Peak voltage can also be represented as V_{max}. *Peak-to-peak voltage* (V_{P-P}) is the maximum value measured from the negative peak to the positive peak of a sine wave. V_{P-P} is always twice the value of V_p. *Root-mean-square voltage* (V_{rms}) is the voltage equal to 0.707 multiplied by the peak voltage. *Average voltage* (V_{avg}) is the average of all instantaneous voltages during a half cycle, either positive or negative. The average voltage of a sine wave over a complete cycle is zero because the positive voltage cancels out the negative voltage.

A DMM set to VDC measures actual voltage. A DMM set to VAC measures root-mean-square (effective) voltage (V_{rms}). An oscilloscope measures peak-to-peak voltage (V_{P-P}). A table with conversion values may be used to convert peak, rms, and average voltage. **See Figure 6-8.**

All voltage conversions are done with peak values, not peak-to-peak values. To convert from V_{P-P} to V_{avg}, V_{P-P} is first converted to V_p and then converted to V_{avg}. To convert from V_{rms} to V_{P-P}, V_{rms} is first converted to V_p and then to V_{P-P}.

Voltage Conversions		
To Convert	**To**	**Multiply By**
rms	Average	0.9
rms	Peak	1.414
Average	rms	1.111
Average	Peak	1.567
Peak	rms	0.707
Peak	Average	0.637
Peak	Peak-to-peak	2.0

Figure 6-8. A table is often used to compare the relationship between peak, rms, and average voltage readings in a circuit.

Chapter 6—DC Power Supply Operation and Troubleshooting   **137**

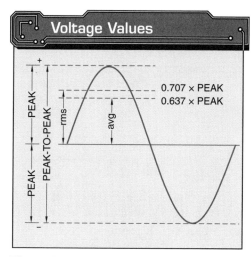

Figure 6-9. *The major values to recognize on an AC sine wave are peak, peak-to-peak, instantaneous, rms, and average.*

For example, to convert 311 V_{P-P} to V_{rms}, the value 311 V_{P-P} is first divided by 2 to obtain a value of 155.5 V_P (311 ÷ 2 = 155.5). Then, the value 155.5 V_P is multiplied by 0.707 to get 110 V_{rms} (155.5 × 0.707 = 110). **See Figure 6-10.**

Once the AC input voltage of a rectifier is determined, the output voltage can be found by taking a direct reading with a DMM set to measure VDC. An oscilloscope and several calculations may be used to find the output voltage. Connect an oscilloscope across the output. The V_P value of the AC output is one-half the V_{P-P} value of the input minus the 0.6 V drop across the diode (V_{D1}). To convert from V_P to V_{avg}, the voltage conversion values are used. In the circuit, peak voltage is equal to 154.9 V (155.5 – 0.6 = 154.9). Average voltage is equal to 98.67 V (154.9 × 0.637 = 98.67). Since every other pulse is missing, the average DC voltage is equal to 49.34 V (98.67 ÷ 2 = 49.34). **See Figure 6-11.**

When troubleshooting a rectifier, precise voltage measurements are typically not needed. Initially, it is only necessary to determine whether the device is operating. Approximated voltages provide enough information on the condition of a circuit for a technician to make intelligent decisions.

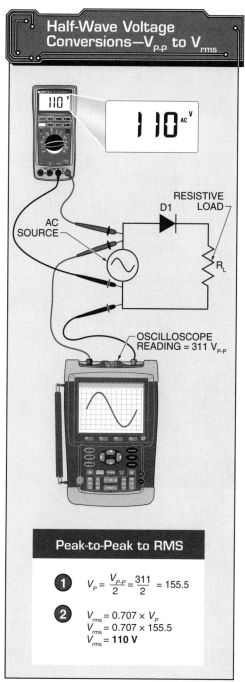

Figure 6-10. *Conversions from one value to another must always be made in relation to the peak voltage.*

A sine wave is a periodic waveform, the value of which varies over time according to the trigonometric sine function. Sine waves may shift in time or vary in amplitude while still retaining their sinusoidal form. They are characterized by smooth, repetitive oscillations between positive and negative peaks.

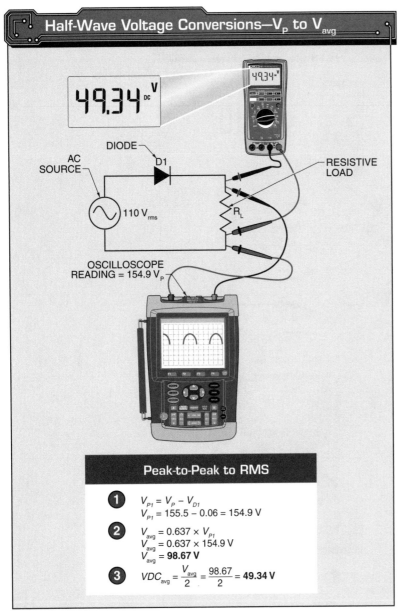

Figure 6-11. The output of a half-wave rectifier can be measured by taking a direct reading with a DC voltmeter. The output can also be found by using an oscilloscope and making calculations with conversion tables.

In a half-wave rectifier, one-half of the input sine wave is removed by the rectifier. Therefore, the VDC_{avg} is 45% of V_{rms} as follows:

$$VDC_{avg} = \frac{V_{rms} \times 0.9}{2}$$

$$VDC_{avg} = V_{rms} \times 0.45$$

$$VDC_{avg} = 110 \times 0.45$$

$$VDC_{avg} = \mathbf{49.5 \text{ V}}$$

Since one diode is in the circuit, a voltage drop of 0.6 V should be subtracted from the value, which results in 48.90 VDC_{avg} (49.5 − 0.6 = 48.90). This application is close to the calculated value of 49.34 V.

Since a full-wave rectifier uses both half cycles of an AC sine wave, it supplies twice the amount of DC voltage as a half-wave rectifier. Instead of using 45% of the rms value, the voltage output is 90% of the rms value. To determine the approximate output voltage, the line voltage (V_{rms}) from one side of the transformer to the center tap is measured at 18 VAC. The output voltage is found by multiplying the line voltage by 90% (18 × 0.9 = 16.2) and subtracting the voltage drop of one diode (0.6), which results in 15.6 VDC (16.2 − 0.6 = 15.6). The value of only one diode is subtracted because only one diode allows electron flow at a given time. **See Figure 6-12.**

Like the output from a full-wave rectifier, the output of the full-wave bridge rectifier is 90% of the rms value. Electrons in a full-wave bridge rectifier flow through two diodes at a given time. Therefore, when calculating the expected VDC output of a full-wave bridge rectifier, the voltage drop for two diodes must also be subtracted.

For example, to determine the approximate output voltage, the line voltage (V_{rms}) across a full-wave bridge rectifier is measured at 50 VAC. The output voltage is found by multiplying the line voltage by 90% (50 × 0.9 = 45) and subtracting the voltage drop of two diodes (2 × 0.6 = 1.2 V), which results in 43.8 V (45 − 1.2 = 43.8). **See Figure 6-13.** When replacing diodes, the peak inverse voltage (PIV) is critical.

To determine the expected average DC voltage (VDC_{avg}) of a full-wave rectifier, V_{rms} (measured at the rectifier input) is multiplied by 90%. The factor 90% can be found by dividing V_{avg} by V_{rms} as follows:

$$factor = \frac{V_{avg}}{V_{rms}}$$

$$factor = \frac{63.2}{70.7} = 0.9 \,(90\%)$$

For a full-wave bridge rectifier, the PIV is divided across the rectifiers. The PIV for each rectifier is the transformers peak voltage. In a conventional two-diode full-wave rectifier, the PIV is approximately twice the transformer voltage.

DC Power Supply Filters

A *filter* is a circuit in a power supply section that smoothes the pulsating DC to make it more consistent. A filter minimizes or removes ripple voltage from a rectified output by opposing changes in voltage and current. The filtering process is accomplished by connecting parallel capacitors and series resistors or inductors to the output of the rectifier. A capacitor smoothes voltage while an inductor smoothes current.

The capacitance of a capacitor determines its effectiveness in filtering voltages. With a larger capacitance, the capacitor filters more effectively. The effectiveness of an inductor (choke) in filtering voltages is also directly related to its inductance.

Capacitive Filters. A *capacitive filter* is a circuit consisting of a capacitor and resistor connected in parallel. The capacitive filter provides maximum voltage output to a load. Since a large capacitor is needed, an electrolytic capacitor is typically used.

As pulsating DC voltage from a rectifier is applied across capacitor C1, it charges to the peak voltage. **See Figure 6-14.** Between peaks, the capacitor discharges through the resistive load R_L, and the voltage gradually drops. *Ripple voltage* is the amount of varying voltage present in a DC power supply. In a capacitive filter, ripple voltage is the voltage drop before the capacitor begins to charge again. The amount of discharge between voltage peaks is controlled by the RC time constant of the capacitor and the load resistance. If the load resistance is large and the capacitance is large, the ripple voltage will be small, resulting in a smooth output. Ripple voltage increases when the load increases on the capacitive filter.

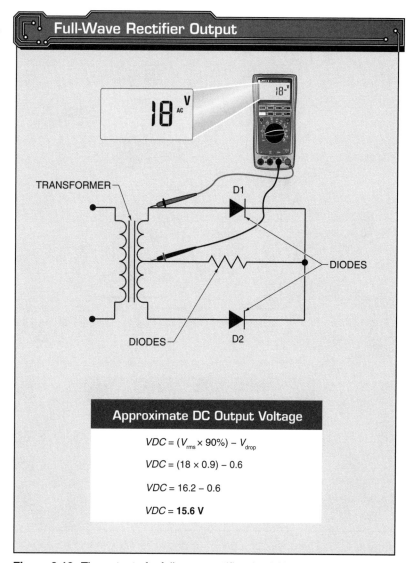

Figure 6-12. *The output of a full-wave rectifier should be approximately 90% of the effective AC input (V_{rms}) minus the 0.6 voltage drop across one diode.*

An integrated power module is a grouping of separate parts designed to deliver power to electronic devices.

140 SOLID STATE DEVICES AND SYSTEMS

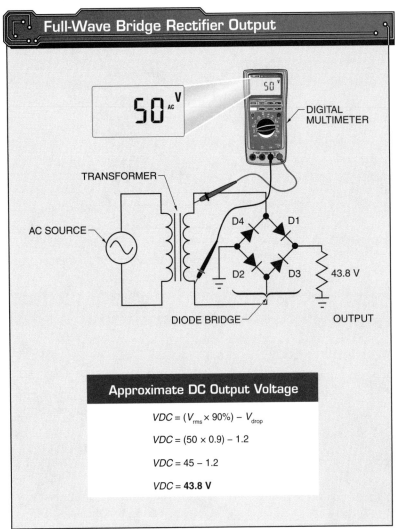

Figure 6-13. The output of a full-wave bridge rectifier should be approximately 90% of the effective AC input (V_{rms}) minus the voltage drop across two diodes.

An L-section resistive filter reduces surge currents by using a current-limiting resistor, R1. **See Figure 6-15.** R1 controls surge currents by limiting the current flow to slow the charging of the capacitor. R1 should always be used in series with the rectifier and the input capacitor of the filter system. This protects the rectifier from the high surge of charging current that flows through the rectifier from the input capacitor C1 when the circuit is first energized. A low-value resistor of about 50 Ω or less is typically used in the application. The filtering of the resistor is not as good as other filters, but it is less expensive.

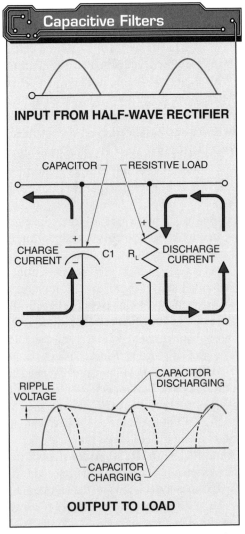

Figure 6-14. A capacitive filter is a capacitor connected in parallel with a load resistor.

Capacitive filters also tend to produce surge currents. *Surge current* is a larger than normal current that is created when a device is first turned ON. Over a period of time, surge currents may damage fuses and diodes in the circuit. Other power supply filters have components that reduce the effect of ripple-voltage variations and surge current.

L-Section Resistive Filters. An *L-section resistive filter* is a filter that reduces or eliminates the amount of DC ripple at the output of a circuit by using a resistor and capacitor as an RC time constant.

L-Section Inductive Filters. An *L-section inductive filter* is a filter that reduces surge currents by using a current-limiting inductor and a capacitor. **See Figure 6-16.** An inductor in series opposes a change in current by creating a counter electromotive force (CEMF) or counter voltage. Surge current is greatly reduced and the capacitor charges slowly. The inductor also aids the filtering effectiveness of the capacitor since the CEMF of the inductor tends to cancel out the effects of the ripple voltage.

The operation of the L-section inductive filter can also be seen through the effect that inductive reactance has on the circuit. When the pulsating DC voltage is applied to the inductor, the changing voltage produces a high inductive reactance. Therefore, the inductor tends to block the pulsating DC voltage. The DC portion of the signal is allowed to pass through the inductor. The pulses not blocked by the inductor are bypassed by the capacitor.

Pi-Section Filters. A *pi-section filter* is a filter made with two capacitors and an inductor or a resistor to smooth out the AC ripple in a rectified waveform. Pi-section filters get their name from the Greek letter pi (π) because the filter configuration resembles its symbol. The two types of pi-section filters are inductive and resistive.

A pi-section filter consists of three elements. In a pi-section inductive filter, there is a shunt input capacitor, C1; a series inductor (choke), L1; and a shunt output capacitor, C2. **See Figure 6-17.** As the input voltage reaches the first capacitor, C1, the capacitor shunts most of the AC ripple current to ground. This presents a much smoother current to L1. Since L1 presents a high inductive reactance to the remaining AC ripple, L1 tends to block the AC ripple. Finally, C2 shunts to ground any remaining AC ripple. The result is a relatively smooth DC voltage.

To save on cost, inductor L1 may be replaced by resistor R1. A pi-section filter that uses a resistor depends on the use of the RC time constant. Resistive filters are not as effective as inductors.

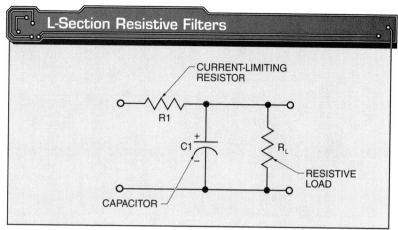

Figure 6-15. *An L-section resistive filter reduces surge currents by using a current-limiting resistor, R1.*

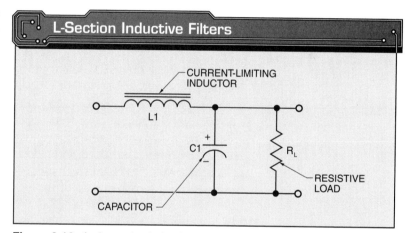

Figure 6-16. *An L-section inductive filter reduces surge currents by using a current-limiting inductor, L1.*

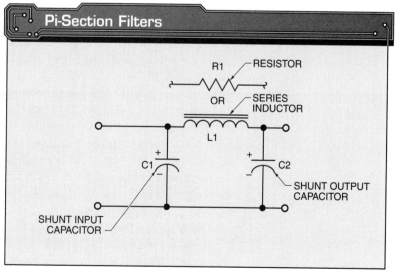

Figure 6-17. *A pi-section filter consists of two capacitors and an inductor, which may be substituted for a resistor.*

A building automation system controller power supply is normally 120 VAC to 24 VAC step-down transformer.

Filters and Peak Inverse Voltage (PIV). The peak inverse voltage (PIV) of a diode should not be exceeded. *Peak inverse voltage (PIV)* is the maximum reverse-bias voltage that a diode can withstand without breaking down and allowing electron flow. Peak inverse voltage is found by multiplying 1.414 by the rms voltage of the AC input. However, when a filter is added, the PIV becomes much higher. In a half-wave rectifier with a pi-section inductive filter added, when diode D1 conducts, capacitor C1 charges to almost the peak voltage of the AC input. C1 holds its charge for some time. Therefore, its voltage adds to the voltage already across the diode when the sine wave begins its negative half cycle. **See Figure 6-18.**

C1 voltage adds to the voltage across the transformer. C1 voltage increases the inverse voltage across the diode to almost twice the amount of peak voltage of the AC input. In a full-wave rectifier with a center-tapped transformer, the PIV is twice the peak voltage of the AC input on the secondary side of the transformer.

It may appear that the diode is rated much higher than necessary and that a lower-rated diode would appear to work. The circuit designer, however, has taken into account the additional PIV of the filter circuit. Therefore, a diode with a PIV rating equal to or higher than the one to be replaced must always be used.

Diodes are connected in series for higher PIV ratings and in parallel for higher current ratings. **See Figure 6-19.** In this case, if each diode were rated for 100 V, in series they would be 200 V. However, when silicon diodes are in series, each one consumes 0.6 V and must be added together as a voltage drop in the circuit. For example, two diodes in series have a combined voltage drop of 1.2 V. To increase current rating, the diodes are placed in parallel.

Voltage Regulators

A *voltage regulator* is an electrical circuit that is used to maintain a relatively constant value of output voltage over a wide range of operating situations. Power supplies are either regulated or unregulated. It depends on whether the final output voltage must be constant or if it can fluctuate over certain limits. A *regulated power supply* is a power supply that maintains a constant voltage across an output even when loads vary. An *unregulated power supply* is a power supply whose output varies, depending on changes of line voltage or load. Many electronic devices require regulated power supplies. These include motor controls, computers, and critical timing equipment. Voltage regulators include series and shunt regulators.

Series Regulators. A series regulator has a variable resistance that automatically changes as output voltage changes. The output voltage of a rectifier is held constant by a variable resistor, R1, in series with a load, R_L. **See Figure 6-20.** The load current flows through the variable resistor and causes a voltage drop across it. Therefore, the variable resistor and the load develop a DC voltage divider. The resistance can be adjusted to maintain a desired voltage across the load even when variations occur in the input voltage of the rectifier or when the load current changes.

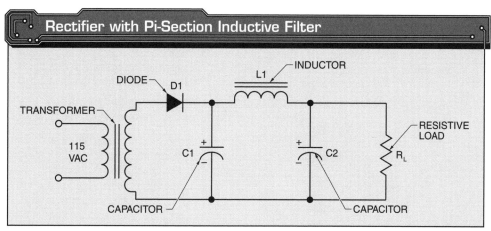

Figure 6-18. *When a pi-section inductive filter is added to a rectifier circuit, the peak inverse voltage (PIV) of the diode must be increased due to the voltage being held by capacitor C1 on the reverse alternation.*

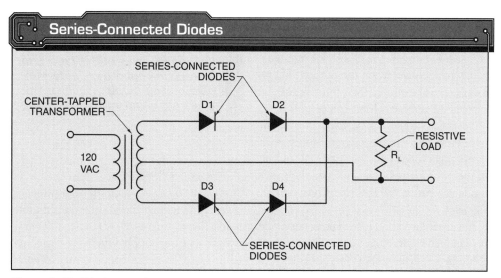

Figure 6-19. *Diodes can be connected in series to increase the peak inverse voltage (PIV) rating.*

As the load current increases, a larger voltage drop appears across the resistance. The voltage across R_L needs to be reduced to maintain voltage across R_L. The value of resistance R1 can be reduced. The reduction lowers the voltage across R1 until the voltage across R_L returns to its required value. *Note:* The circuit must be manually operated.

Shunt Regulators. A shunt regulator combined with the resistance of a power supply or with an additional resistor forms a voltage divider. **See Figure 6-21.** As the shunt resistance, R2, increases, more voltage appears across it as an output to the load. As R2 decreases, less voltage appears across it.

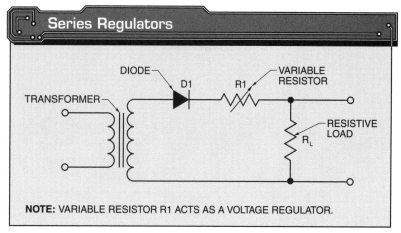

Figure 6-20. *The voltage regulator of a series-regulated power supply is a variable resistor in series. It can change as the output voltage changes.*

144 SOLID STATE DEVICES AND SYSTEMS

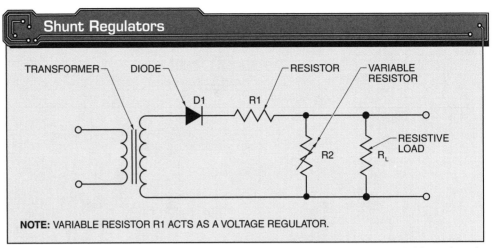

Figure 6-21. *The voltage regulator of a shunt-regulated power supply is a variable resistor in parallel. It automatically changes as the output voltage changes.*

The resistance of a shunt regulator needs to increase when the output voltage decreases. The resistance of a shunt regulator needs to decrease when the output voltage increases. Therefore, the shunt regulator returns the output voltage to normal.

A zener diode can be used in a shunt regulator. In a simple power supply with direct zener regulation, the "raw" DC output of the power supply must exceed the zener diode regulated voltage. **See Figure 6-22.** Resistor R1 supplies current electron flow to the zener diode and load. The zener diode is regulated by drawing more or less current through R1 as the voltage across it tries to increase or decrease. As a result, the voltage drop across R1 increases or decreases.

Potentiometers are variable resistors that function as voltage dividers. Voltage dividers of this type can function as control devices for variable analog outputs, such as volume controls on audio equipment.

Voltage Dividers

A *voltage divider* is a circuit that provides several different voltages for several different loads. **See Figure 6-23.** A voltage divider is used in a power supply when it is necessary to provide voltages for several different loads. It divides the available voltage into the voltages required by various loads.

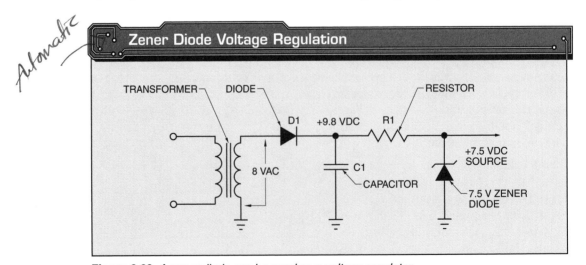

Figure 6-22. *A zener diode can be used as a voltage regulator.*

Voltage Dividers

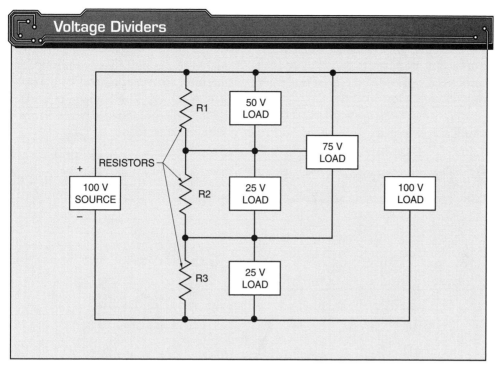

Figure 6-23. *A voltage divider distributes the available voltage to various loads.*

A voltage divider consists of a resistor, or series of resistors, connected across the output of a power supply. Sometimes it is tapped at different points to provide a selection of output voltages. The series of resistors are in parallel with the various loads. The total resistance of the voltage divider provides a network to decrease voltage to the desired value. The total resistance appears to the power supply as a fixed load and a discharge path for the capacitors when the power supply has been turned OFF. This fixed load provides a degree of voltage regulation.

Voltage Multipliers

A *voltage multiplier* is an electrical circuit designed to supply higher voltages than an AC source voltage without using a transformer. Voltage multipliers are designed to supply high voltages with a limited load current. Voltage multipliers are often referred to as transformerless power supplies. They are capable of delivering a DC voltage as high as 400 V from a 110 VAC source.

The basic operating principle behind voltage multipliers is that two capacitors are allowed to be charged individually from a 100 VAC source. **See Figure 6-24.** During the positive half cycle, diode D1 is forward-biased and allows capacitor C1 to be charged to 100 VDC. During the negative half cycle, diode D2 is forward-biased and allows capacitor C2 to be charged to 100 VDC.

Since capacitors C1 and C2 are in series, the voltage across the two capacitors is the sum of the two individual voltages. When a DMM is connected across the two capacitors, the reading is 200 VDC. Voltage multiplier circuits are series-connected capacitors. However, rectifiers and resistors are used to produce and control peak DC voltages.

Effect of RC Time on DC Output. The resistance in a transformerless half-wave rectifier can be varied to control the value of DC output. **See Figure 6-25.** During the positive half cycle of the AC source that diode D1 conducts, there is electron flow through variable resistor R1. With 110 VAC as the AC source, capacitor C1 charges to the peak value of 155.5 V (110 × 1.414 = 155.5).

During the negative half cycle, D1 is reverse-biased and does not conduct. During this time, C1 discharges through R1. This action repeats with each half cycle of input voltage. The result is a varying value of VDC across R1.

Every time D1 is forward-biased, C1 charges to the peak voltage. However, the extent to which the capacitor discharges during the negative half cycle depends on the RC time constant of the resistance-capacitance combination. Therefore, an average direct output voltage is obtained. This voltage is greater than the voltage obtained when using a simple resistive load.

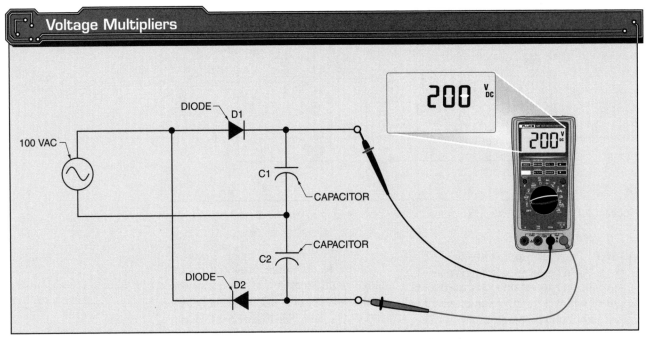

Figure 6-24. *A voltage multiplier uses capacitors to supply a voltage higher than the voltage source.*

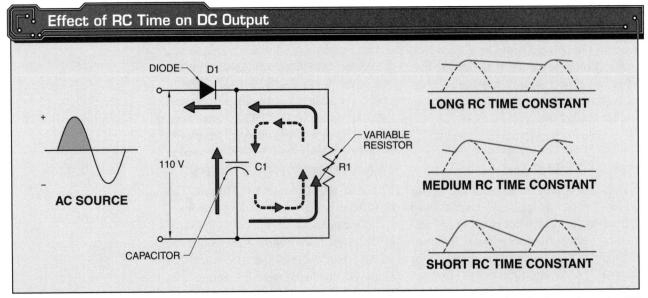

Figure 6-25. *In an RC circuit, the extent to which the capacitor discharges during a negative half cycle depends on the RC time constant of the resistance-capacitance combination.*

Voltage Doublers. A *voltage doubler* is a circuit designed to produce a DC output level that is approximately twice that of the peak AC value. The two most common types of voltage doublers are the half-wave doubler and the full-wave doubler.

A *half-wave doubler* is an electrical circuit containing two diodes, two capacitors, and a load resistor, where the output voltage is twice the peak input voltage. The circuit consists of two diodes, D1 and D2, two capacitors, C1 and C2, and a resistive load, R_L. The source is an AC voltage. During the negative half cycle of the AC source, diode D1 is forward-biased and allows capacitor C1 to be charged, while diode D2 is reverse-biased and does not allow current flow. **See Figure 6-26.**

During the positive half cycle of the AC source, diode D2 is forward-biased, and capacitor C2 is charged from the AC source and capacitor C1, since they are in series. Diode D1 is reverse-biased and does not allow electron flow. The voltages stored in capacitor C1 and the AC source are combined to charge capacitor C2 to twice the peak voltage. Capacitor C2 then discharges through the resistive load, R_L, at twice the input voltage.

Note: Capacitor C2 charges to the peak value of the AC source during the positive half cycle when diode D2 is forward-biased. During the negative half cycle when diode D1 is forward-biased, capacitor C1 charges to the peak value of the AC source and effectively doubles it for the duration of the time that diode D2 is forward-biased.

The disadvantage of this circuit is that the DC output contains a large 60 Hz ripple. In addition, capacitor C2 must be capable of operating at the full peak output voltage. This voltage is approximately three times the rms value of the AC source under no-load conditions. The higher voltage rating raises the cost of the capacitor.

A *full-wave doubler* is an electrical circuit that uses both halves of the AC source to charge two capacitors on alternate half cycles so that both capacitor voltages can be combined to double the output. **See Figure 6-27.** Both capacitors are charged on alternate cycles of the input so that the voltages can be added together in the output.

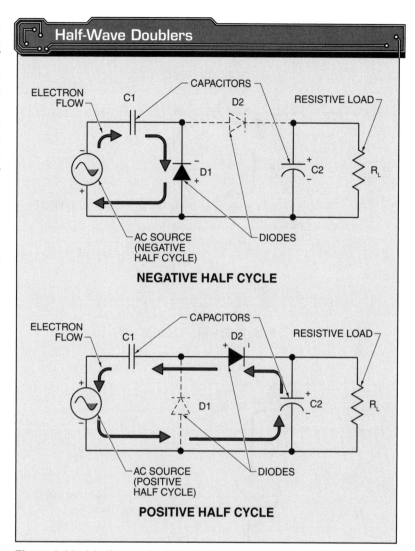

Figure 6-26. *A half-wave doubler circuit uses only one half of the AC source to double the input voltage.*

During the positive half cycle of the AC source, diode D1 is forward-biased, and capacitor C1 is charged to the peak value of the AC source, while diode D2 is reverse-biased and does not allow electron flow. During the negative half cycle of the AC source, diode D2 is forward-biased, and capacitor C2 is charged to the peak value of the AC source, while diode D1 is reverse-biased and does not allow electron flow. Because capacitors C1 and C2 are in series across the output, the voltages are combined. The regulation of the full-wave doubler is better than the half-wave doubler because one capacitor is always being charged during a given half cycle.

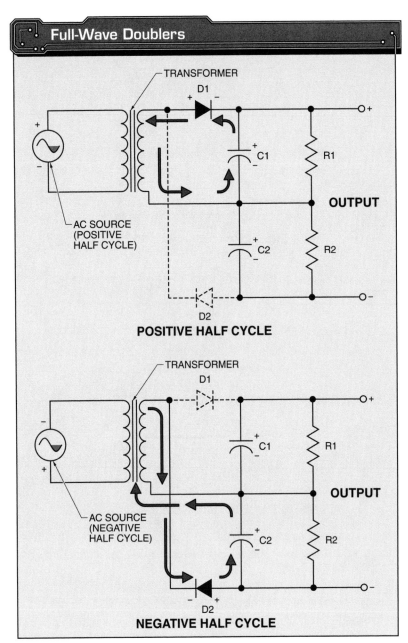

Figure 6-27. A full-wave doubler uses both halves of the AC source by charging two capacitors on alternate cycles to double the input voltage.

TROUBLESHOOTING DC POWER SUPPLIES

The operations of several types of basic power supplies including half-wave, full-wave, and full-wave bridge rectifiers all have certain traits in common. For example, all power supplies may exhibit similar problems and can be tested using a DMM or an oscilloscope. Remember that the tests used for problems in one power supply may not necessarily be the same tests used for another power supply.

Troubleshooting Half-Wave Rectifiers

In a simple half-wave power supply, AC power is provided by the power transformer. **See Figure 6-28.** Diode D1 allows current in one direction only during the positive half cycle of the AC source. The output of the diode is half of a sine wave, as shown on the oscilloscope screen. The output of the power supply is affected by capacitor C1. Capacitor C1 is charged from the output of diode D1 and then supplies power to the circuit when diode D1 is reverse-biased. This results in a smoothing process, as shown in the oscilloscope waveforms with a load and no load. *Note:* When capacitor C1 is open there is no smoothing of the waveform.

Some problems found in many types of power supplies can be explained by troubleshooting the half-wave rectifier. Common problems with half-wave rectifiers include blown fuses and tripped circuit breakers, no output voltage, and low power supply voltage.

Blown Fuses and Tripped Circuit Breakers. One of the most common problems in power supplies is a shorted diode. A shorted diode will cause a fuse to blow or a circuit breaker to trip. The reason is that with diode D1 shorted, capacitor C1 is effectively across the transformer secondary and has an excessive current. This excessive current is reflected into the primary of the transformer and the fuse blows or the circuit breaker trips. If the circuit is not overload protected, the transformer burns out. **See Figure 6-29.**

Another possible cause of a fuse blowing or circuit breaker tripping is a short in the load. For example, a shorted transistor or integrated circuit (IC) draws excessive current from the power supply. As with a diode, this problem can typically be isolated with a DMM.

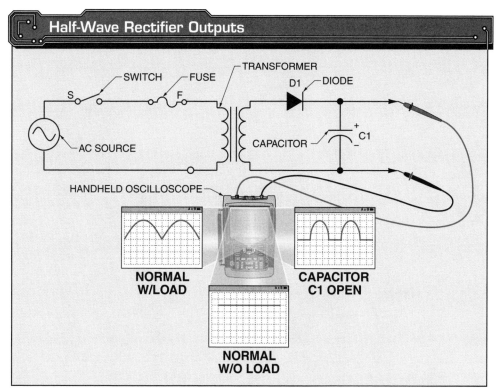

Figure 6-28. *The output of the half-wave rectifier circuit is affected by the capacitor.*

No Output Voltage. No output voltage, or a reading of 0 VDC, occurs when there is no AC input voltage. The lack of AC input voltage occurs when a fuse or circuit breaker is open, a switch does not close, the power transformer winding is open, or there is a loose connection such as a loose plug or fuse holder. All of the possible problems can be checked with a DMM. Maximum voltage is read across an open circuit, provided that it is the only open in the circuit. **See Figure 6-30.**

Low Power Supply Voltage. The most common cause of low power supply voltage is an open input-filter capacitor. A large increase in ripple can be observed on an oscilloscope when an input-filter capacitor is open.

A simple way to check a capacitor is to shunt another capacitor of the same value across it temporarily. It is important to make sure that the polarity of the capacitor is correct. If the voltage increases dramatically, the capacitor is bad. Any type of bypass jumping with a shunt should be used only as a temporary test.

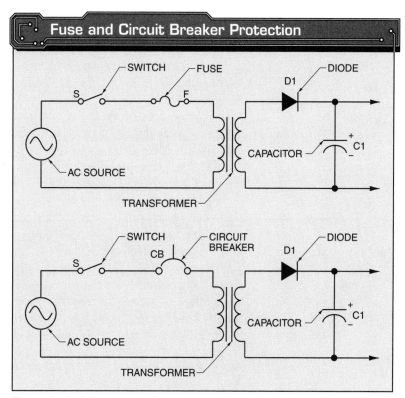

Figure 6-29. *If the primary of a transformer is not protected by a fuse or circuit breaker, the transformer will likely burn out when the diode shorts.*

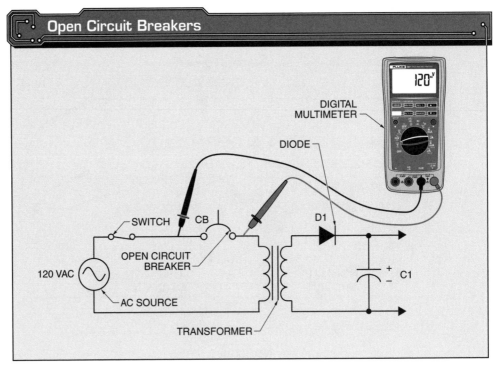

Figure 6-30. *Maximum voltage is read across an open circuit, provided that it is the only open in the circuit.*

Troubleshooting Full-Wave Rectifiers

When troubleshooting full-wave rectifiers, tracing the AC signal through the power supply is the most accurate method of troubleshooting problems that cannot be found by a visual check. The AC voltage is traced from the transformer where the AC sine wave changes to pulsating DC at the rectifier output. Then, the DC pulses are smoothed out by the filter. The point where the signal stops or becomes distorted is where the problem can be found. If there is no DC output voltage, there may be an open or short in the signal tracing process. If there is a low DC voltage, there may be a defective part.

An unregulated power supply converts and filters a power signal. **See Figure 6-31.** AC voltage is brought in from the power line by a line cord. AC voltage is connected to the primary of the transformer through switch S1. At the secondary winding of the transformer, points 1 and 2, the oscilloscope shows the stepped-up voltage developed across each half of the secondary winding as a complete sine wave. Each of the two stepped-up voltages is connected between a ground, point 3, and one of two diodes, points 4 and 5.

After passing through the diodes, point 6, the waveform is changed to pulsating DC. Pulsating DC is fed through inductor L1 and filter capacitor C1, which removes a large part of the ripple, point 7. Finally, the DC voltage is fed through inductor L2 and filter capacitor C2, which removes most of the remaining ripples, point 8.

Problems that may occur in full-wave rectifiers include the following:

- open fuse or circuit breaker in the primary or secondary, which causes a loss of voltage
- shorted or open transformer, which does not allow voltage across the circuit
- open or shorted diode D1 or D2, which causes the output voltage to drop by one-half
- both diodes open, which are indicated by 0 V at the load
- both diodes shorted, which short the entire transformer secondary

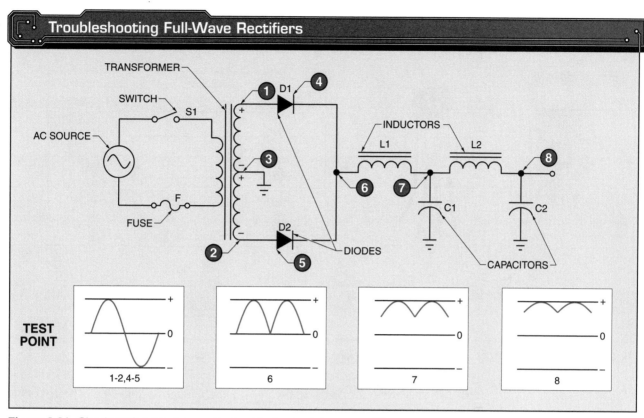

Figure 6-31. *Circuit tracing can be used to troubleshoot a full-wave rectifier.*

Troubleshooting Full-Wave Bridge Rectifiers

Problems in a full-wave bridge rectifier are the same as in half-wave and full-wave rectifiers. These problems can cause zero voltage, a fuse to blow, or a circuit breaker to trip. Low voltage may occur in a heavily loaded full-wave bridge rectifier if a diode opens. At the same time, ripple level increases, since a full-wave bridge rectifier with an open diode becomes a half-wave rectifier circuit.

It is difficult to check for an open diode with a DMM in a full-wave rectifier. This is because of a parallel path through the secondary winding and the complimentary diode. Disconnecting one end of either diode of a complementary pair breaks up the parallel path. Then, the pair is checked to see if one of the diodes is open. If the diodes are in one package, one transformer lead is disconnected and checked for a possible open diode from the AC input to the positive (+) and negative (−) terminals. **See Figure 6-32.** First, one lead of the power transformer is disconnected. Then, terminals 1 to 2, 1 to 3, 2 to 4, and 3 to 4 are measured. The DMM leads are reversed each time to check the diode. Any reading under 5 Ω, regardless of DMM lead direction, indicates a shorted diode.

Problems that may occur in a full-wave bridge rectifier include the following:

- open fuse or circuit breaker in the primary or secondary, which causes a loss of voltage
- shorted or open transformer, which allows no voltage to appear across the circuit
- open diode, which causes the DC output to equal a half-wave rectifier minus the voltage drop of the two diodes
- shorted diode, which causes the voltage output to be similar to a half-wave rectifier (the voltage should be 0 V)

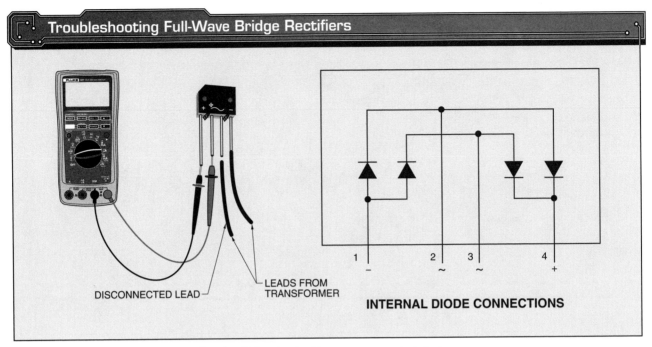

Figure 6-32. A full-wave bridge rectifier should be checked for open or shorted diodes.

Troubleshooting Zener Diode Voltage Regulators

Zener diodes that are not working properly may have too little or too much voltage across them. Too little voltage across the zener diode may be caused by a bad diode, low DC input voltage, increased value of a resistor, or a load that is too large. Too much voltage across the zener diode indicates an open.

A zener diode can be checked in its circuit using a DMM. However, a DMM only indicates whether a zener diode is open or shorted. The test does not tell whether the breakdown voltage of a zener diode is correct. For example, due to leakage, a 15 V zener diode may hold the voltage to 12 V. A DMM may not indicate any leakage because it only uses low test voltages. Like all diodes, the zener diode should have a low resistance in the forward-bias direction and a high resistance in the reverse-bias direction. **See Figure 6-33.**

A zener diode must be safely checked to confirm it is working properly. A DC source can be used to check a zener diode as long as it has a somewhat higher voltage than the zener diode to be checked. The limiting resistor value is not critical but should be large enough to limit the expected zener diode current to approximately 10 mA for zener diodes rated at 24 V or to 5 mA for higher voltage units. If a zener diode is rated at more than 1 W, higher test currents may be used. If an adjustable DC power supply is used, a 1000 Ω resistor is used. The supply voltage is increased until the voltage reading of the DMM across the zener diode does not increase. The DMM should indicate the zener diode rating.

POWER INTERRUPTIONS

All electrical equipment is rated for operation at a specific voltage. The rated voltage is accepted as a voltage range. The typical range is ±10%, but when components of an electrical system are derated to save energy and operating costs, the typical range is +5% to −10%. The +5% to −10% range is used because overvoltages are more damaging to loads than undervoltages. Equipment manufacturers, utility companies, and regulating agencies must routinely deal with changes in system voltage. Power interruptions are classified into standard industry categories. **See Figure 6-34.**

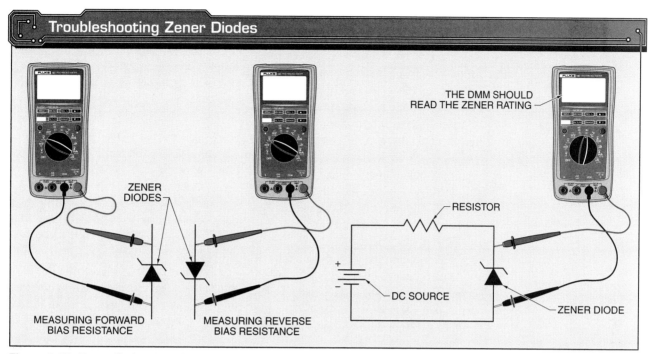

Figure 6-33. *Zener diodes should have a low resistance in the forward-bias direction and a high resistance in the reverse-bias direction.*

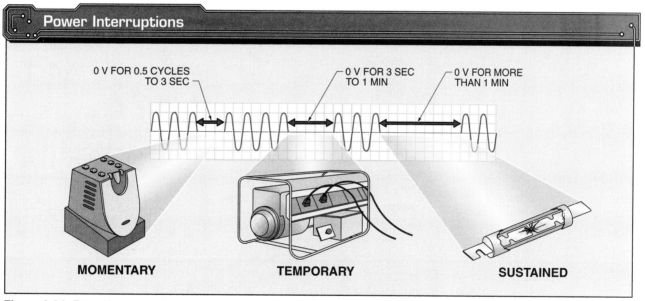

Figure 6-34. *Power interruptions to an electrical system can be momentary, temporary, or sustained.*

Momentary Power Interruptions

A *momentary power interruption* is a decrease to 0 V on one or more power lines lasting more than one half cycle to 3 sec. All power distribution systems encounter momentary power interruptions during normal operation. Momentary power interruptions are caused by lightning strikes, utility grid switching during a problem (short on a line), and open circuit transition switching. Open circuit transition switching is a process in which power is momentarily disconnected when switching a circuit from one voltage supply or level to another.

Temporary Power Interruptions

A *temporary power interruption* is a decrease to 0 V on one or more power lines lasting 3 sec to 1 min. Automatic circuit breakers and other circuit protection equipment protect power distribution systems by removing faults. An automatic circuit breaker typically requires 20 cycles to approximately 5 sec to restore power after a temporary power interruption.

A temporary power interruption can be caused by a gap in time between a power interruption and when a back-up power supply (generator) is started. Also, a temporary power interruption can occur if a circuit is accidentally opened by switching the wrong circuit breaker or gear switch. When power is not restored, a temporary power interruption becomes a sustained power interruption.

Sustained Power Interruptions

A *sustained power interruption (outage)* is a decrease to 0 V on all power lines lasting more than 1 min. Even the best power distribution systems have a complete loss of power at some time. Sustained power interruptions are commonly the result of storms or circuit breakers tripping due to damaged equipment.

Transient Voltages

A *transient voltage (voltage spike)* is temporary, unwanted voltage in an electrical circuit. Transient voltages range from a few volts to several thousand volts and last from a few microseconds to a few milliseconds. Transient voltages can occur at any point on an AC sine wave. Transient voltages are also referred to as voltage surges.

Research by the American Institute of Electrical Engineers (AIEE), service technicians, and equipment manufacturers has clearly shown that solid state circuits and devices do not tolerate transient voltages exceeding twice the normal operating voltage. Transient voltages are produced in all electrical systems from outside sources such as lightning. They are also produced within a system by turning off loads that include coils such as solenoids, magnetic motor starters, and motors.

Transient voltages are caused by the sudden release of stored energy due to lightning strikes, unfiltered electrical equipment, contact bounce, arcing, and generators being switched ON and OFF. Transient voltages are produced from stored energy contained within a system. The size and duration of transient voltages depend on the value of the inductance (L) and capacitance (C) of a system. Transient voltages on a 120 V power line reach several thousand volts or more. One high-voltage transient is all that is needed to damage circuits or electrical equipment.

Lightning is the most common source of transient voltages on utility power distribution systems. Lightning induces a transient voltage that travels in two directions from the point of contact. In most cases, the transient voltage is dissipated by the utility company grounding and protection systems after lightning has traveled through 10 to 20 utility poles (grounding points). Unprotected or improperly protected equipment is severely damaged when lightning strikes close to it. All transient voltages are attenuated, or dampened, as they travel through an electrical system.

High-level transient voltages caused by lightning strikes must be considered in all electrical applications because lightning may strike at any location. Low-level transient voltages are produced on the distribution system of a facility when loads are switched OFF. Transient voltages are produced continuously when using loads that include the following:
- solid state power supplies
- magnetic motor starters
- fluorescent lighting using electronic ballasts
- variable-frequency drives
- DC motor drives
- solenoids
- soft-start motor starters
- electronically controlled welding equipment
- motors that include brushes

Surge Suppressors

A *surge suppressor* is an electrical device that provides protection from high-level transients by limiting the level of voltage allowed downstream. Surge suppressors should be installed at service entrance panels, distribution panels, and individual loads. **See Figure 6-35.** Computers, electrical circuits, and specialized equipment require protection against transient voltages. Protection from transient voltages commonly includes proper wiring, grounding, shielding of power lines, and surge suppressors.

Transient voltages differ from voltage swells and voltage sags. Transient voltages are typically large in magnitude, have a short duration, have a steep (short) rise time, and are very erratic. As with voltage sags and voltage swells, a power quality meter should be used to monitor for transient voltages over a period of time. The size, duration, and time of transient voltages can be displayed at a later time. If transient voltages are identified as the problem, a surge suppressor should be used.

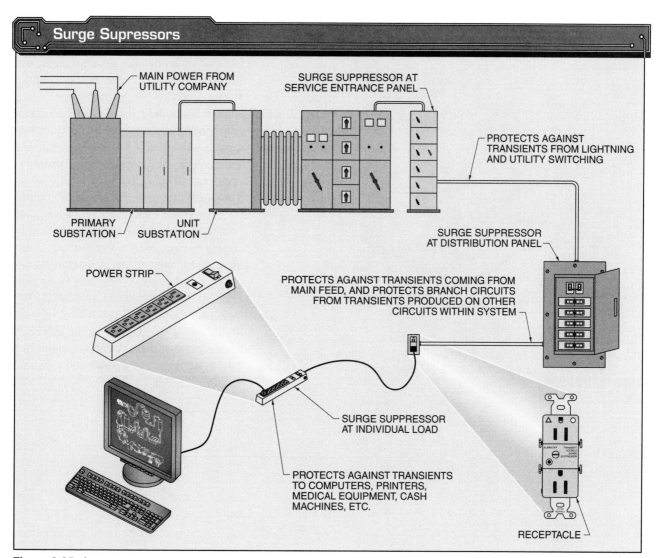

Figure 6-35. *A surge suppressor is an electrical device that provides protection from high-level transients by limiting the level of voltage allowed downstream from the surge suppressor.*

Uninterruptible Power Supplies (UPSs)

An *uninterruptible power supply (UPS)* is a battery-based system that includes all of the additional power conditioning equipment to make a complete self-contained power source. **See Figure 6-36.** UPS devices use batteries and inverters to keep electronic devices running for a period of time during a power interruption. UPS devices typically provide protection against power surges, brownouts, and line noise. A UPS consists of a rectifier, battery, and inverter.

Refer to Chapter 6 Quick Quiz® on CD-ROM

A rectifier converts AC power into DC power, which is used to charge a battery. A battery stores power for use during a power interruption and determines the length of time a UPS can support equipment. An inverter converts DC power into AC power and supplies continuous power to equipment.

UPS systems are classified into two categories, off-line and line interactive. An off-line UPS uses AC as the primary power source. If a drop in current or voltage is detected by a UPS, it automatically switches to a backup battery. When AC power is restored, a UPS switches back to the primary power source. The switchover time is typically no longer than a few milliseconds. If a power interruption is longer, the electronic equipment will turn off. Off-line UPS units are the least expensive variety of uninterruptible power supplies and are intended for residential use.

In a line interactive UPS, the battery charger, inverter, and source selection switch of the off-line UPS have been replaced with a combination inverter/converter. The inverter/converter charges the battery and converts DC to AC for use by the protected equipment. AC power is the primary power source and DC power (battery) is the secondary power source. When there is AC power, the inverter/converter charges the battery. When there is a power interruption, the transfer switch is automatically opened, and the device operates in reverse. These UPS systems generally cost more than off-line UPS systems and are used as backups for corporate and industrial electronic systems.

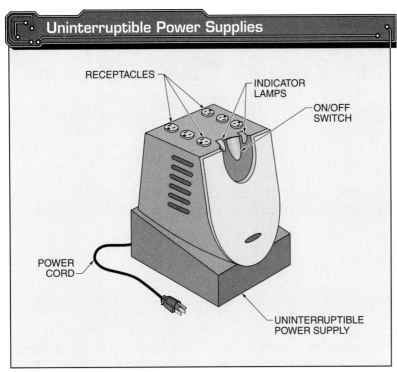

Figure 6-36. *An uninterruptible power supply (UPS) is a battery-based system that includes all of the additional power conditioning equipment to make a self-contained power source.*

KEY TERMS

- DC power supply
- DC voltage
- transformer
- half-wave rectifier
- pulsating DC
- full-wave rectifier
- full-wave bridge rectifier
- peak voltage
- peak-to-peak voltage
- root-mean-square voltage
- average voltage
- filter
- capacitive filter
- ripple voltage
- surge current
- L-section resistive filter
- L-section inductive filter
- pi-section filter
- peak inverse voltage
- voltage regulator
- regulated power supply
- unregulated power supply
- voltage divider
- voltage multiplier
- voltage doubler
- half-wave doubler
- full-wave doubler
- transient voltage
- surge suppressor
- uninterruptable power supply (UPS)

REVIEW QUESTIONS

1. What is the main function of a DC power supply?

2. What are the major parts of most DC power supplies?

3. Describe the three basic types of rectifier configurations used in single-phase DC power supplies.

4. How much voltage is usually dropped across a silicon diode?

5. What type of voltage will the following instruments read?
    - DC voltmeter
    - AC voltmeter
    - oscilloscope
    - multimeter

6. What is a rectifier?

7. What is ripple voltage?

8. What is pulsating DC?

9. What is peak voltage?

10. List four ways the output of *any* rectifier can be affected.

11. What is the function of the filter section in a DC power supply?

12. What are the four types of filters?

(continued)

REVIEW QUESTIONS (continued)

13. What are the disadvantages of capacitive filters?

14. What effect does a filter have on PIV?

15. What are the three sections (elements) of a pi-section filter?

16. How do the two types of regulators operate?

17. How can the PIV and current rating of a diode be increased?

18. What is a voltage divider and when is it used?

19. What are voltage multipliers and how do they work?

20. What effect does the RC time constant have on DC output?

POWER SOURCES AND RENEWABLE ENERGY

7

As the need for electrical energy grows, new sources of power will be necessary. Most electronic devices today are portable and are expected to work in remote locations. The power sources used to power electronic devices must be reliable and long-lasting. The main sources of electrical energy explored are based on the use of light, magnetism, and electrochemical reaction.

OBJECTIVES

- Explain how photovoltaic cells operate.
- Discuss photovoltaic cell output.
- Explain how photovoltaic systems operate.
- Discuss wind turbine systems.
- Explain how wind turbines operate.
- List and describe the various types of batteries.
- Describe fuel cells.

SOURCES OF ELECTRICAL ENERGY

Electrical energy is used to power electronic devices. Initially, electronic devices were required to be plugged into an outlet. Today, electronic devices are portable and can be taken almost anywhere. The common sources of electrical energy include sunlight, magnetism, batteries, and fuel cells. **See Figure 7-1.** Photovoltaic, wind turbine, battery, and fuel cell technology are commonly used to produce electrical energy. Technicians must know how to test, troubleshoot, and maintain these forms of energy producing technologies.

PHOTOVOLTAIC CELLS

A *photovoltaic cell* is a semiconductor device that directly converts solar energy into electrical energy. Photovoltaic cells are often called solar cells since solar energy is used to create electrical power. A photovoltaic cell is sensitive to light and produces a voltage without an external power source. **See Figure 7-2.**

Photovoltaic cells are ideal for portable applications and use in remote locations where it would be difficult to obtain electricity. The use of photovoltaic cells is increasing since they are becoming more efficient and economical to manufacture. Many manufacturers are designing photovoltaic cells for single- and multiple-cell applications. The schematic symbol for a photovoltaic cell indicates it is similar to a single-cell voltage source such as that found in a battery. A photovoltaic cell should have a long life span, provided it does not become damaged.

DOE/NREL, Robb Williamson

Photovoltaic arrays are used to convert solar energy into electrical energy and can be used in many applications.

160 SOLID STATE DEVICES AND SYSTEMS

Sources of Electrical Energy

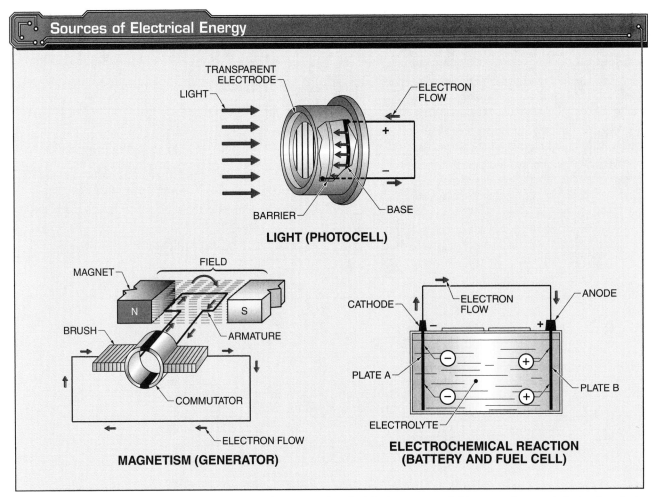

Figure 7-1. *Three sources of electrical energy used to power electrical devices are sunlight, magnetism, and electrochemicals.*

Photovoltaic Cells

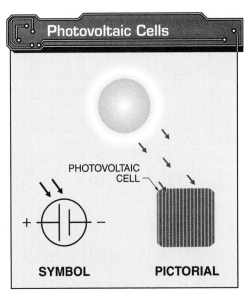

Figure 7-2. *A photovoltaic cell is sensitive to light and produces voltage without an external power source.*

Photovoltaic Effect Media Clip

Photovoltaic Cell Operation

Photovoltaic cells generate energy by using a PN junction to convert sunlight (solar energy) into electrical energy. **See Figure 7-3.** The photovoltaic cell produces a potential difference between a pair of terminals when exposed to sunlight. The voltage potential is caused by the absorption of photons across the PN junction of the semiconductor material. The *photovoltaic effect* is the production of electrical energy due to the absorption of light photons in a semiconductor material.

The photoelectric effect was first noted by Edmund Bequerel in 1839. He found that certain materials would produce small amounts of electric current when exposed to light.

Certain materials, such as cesium, selenium, and cadmium, emit electrons when exposed to light. Photons contain various amounts of energy depending on their wavelength. Higher energies are associated with shorter wavelengths (higher frequencies). The photons from sunlight are absorbed by the semiconductor material in photovoltaic cells, which cause electrons to flow through the semiconductor material and produce electricity.

Photovoltaic Cell Output

Photovoltaic cells are rated by the amount of energy they convert. Most manufacturers rate photovoltaic cell output in terms of voltage (V) and current (mA). Photovoltaic cells produce a limited amount of voltage and current. For example, each photovoltaic cell may produce up to 0.6 V. To increase the voltage output, cells are connected in series. **See Figure 7-4.**

In addition to the maximum voltage, each photovoltaic cell may produce up to 40 mA of current. To increase the current output, cells are connected in parallel. **See Figure 7-5.** To increase both voltage and current, the individual cells are connected both in series and parallel.

The photovoltaic effect is measured using a high-impedance voltage-measuring device such as a digital multimeter. In the dark, there is no open-circuit voltage present. When sunlight strikes the cell, the light is absorbed and, if the photon energy is large enough, it frees hole-electron pairs.

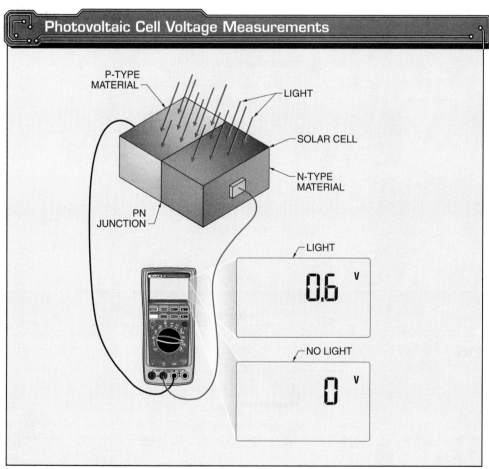

Figure 7-3. *A photovoltaic cell generates energy by using a PN junction to convert light energy into electrical energy.*

162 SOLID STATE DEVICES AND SYSTEMS

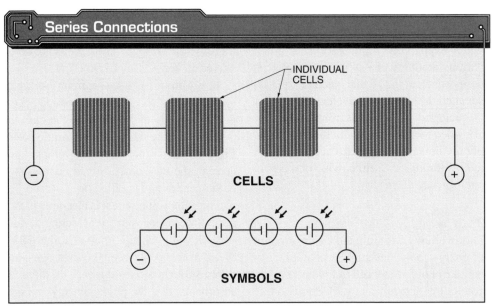

Figure 7-4. *Photovoltaic cells are placed in series to increase the voltage output from a set of photovoltaic cells.*

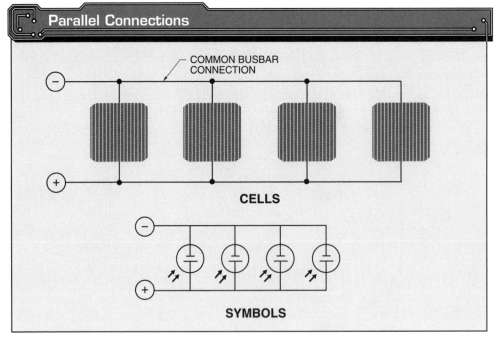

Figure 7-5. *Photovoltaic cells are placed in parallel to increase the current output from a set of photovoltaic cells.*

When sunlight passes through the transparent cover of a photovoltaic cell and strikes the barrier layer, electrons in the selenium atoms gain energy and move across the barrier layer to accumulate on the transparent cover. A voltage potential now exists between the collector post (–) in contact with the transparent cover and the base plate conductor (+) in contact with the selenium. The barrier layer allows electron flow in only one direction so that the electrons cannot return directly to the selenium compound. **See Figure 7-6.**

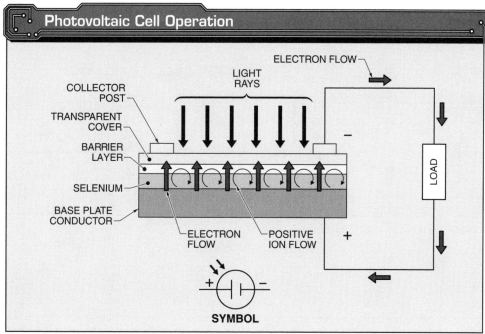

Figure 7-6. When light strikes the barrier layer in a photovoltaic cell, selenium electrons gain energy and move across the barrier layer to accumulate on the transparent cover.

At the PN junction, some recombination of the electrons and holes occurs, but the junction itself acts as a barrier between the two charges. The electrical field at the PN junction maintains the negative charges in the N-type material and the positive charges or holes in the P-type material.

When a load is connected across the PN junction, current flows. **See Figure 7-7.** When current flows through the load, electron-hole pairs formed by sunlight recombine and return to the normal condition prior to the application of sunlight. Consequently, there is no loss of electrons from or addition of electrons to the silicon during the process of converting absorbed sunlight into electrical energy.

Nanotechnology

Nanotechnology is technology that manipulates matter at the molecular level. Nanotechnology, such as a nanocoating, is used to increase the output of photovoltaic cells by increasing the ability of the cells to absorb sunlight.

A nanocoating treatment consists of several coatings made of silicon dioxide and titanium dioxide. Nanorods are vaporized and deposited on the photovoltaic material to form the coating. Conventional silicon photovoltaic modules, which are groups of individual cells, absorb approximately two-thirds of the sunlight that strikes them. Nanocoatings allow photovoltaic modules to absorb as much as 30% more sunlight. With the coating, solar energy can be absorbed from any angle. The nanocoating reduces the need for the modules to track the sun. Therefore, the costs associated with energy use are reduced.

Photovoltaic Systems

Photovoltaic cells produce a significant amount of electrical power when they are combined into large photovoltaic systems. A *photovoltaic system* is an electrical system consisting of photovoltaic cells, modules, and arrays used to convert light into electrical energy. A *module* is a photovoltaic device consisting of many electrically connected cells that form a panel. An *array* is a set of several connected modules. **See Figure 7-8.** A photovoltaic cell is the smallest component in a photovoltaic system.

164 SOLID STATE DEVICES AND SYSTEMS

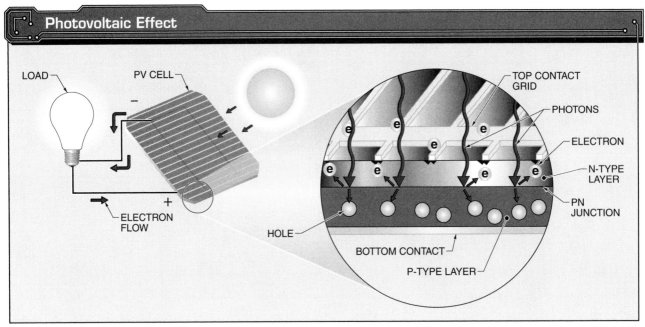

Figure 7-7. *The photovoltaic effect produces free electrons that must travel through conductors in order to recombine with electron holes.*

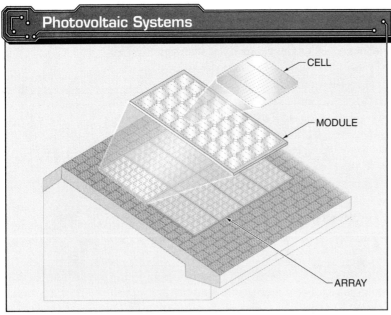

Figure 7-8. *The basic building blocks for photovoltaic systems include cells, modules, and arrays.*

One of the most common types of photovoltaic systems is the residential photovoltaic system. **See Figure 7-9.** A residential photovoltaic system is usually located on a rooftop. A residential photovoltaic system may include a battery storage feature.

Modules and arrays of these systems produce direct current (DC) power, which can be used to charge batteries or directly power DC loads. The DC power can also be converted to alternating current (AC) power to power AC loads. The AC power is created by using inverters.

Photovoltaic System Inverters. An *inverter* is an electronic device that changes direct current (DC) voltage to alternating current (AC) voltage. Inverters can be used to convert the DC output of a photovoltaic module, wind generator, or fuel cell to AC power. Inverters supply standard 120 V, 60 Hz AC power, similar to that supplied by a utility company. Inverters are broadly classified as either stand-alone or utility-interactive inverters. The only difference between a stand-alone inverter and a utility-interactive inverter is whether the inverter is connected to the energy source or directly to a battery bank.

A *stand-alone inverter* is an inverter connected to a battery bank for its source of power and that operates independently. **See Figure 7-10.** Stand-alone inverters can be used to convert DC from a battery to AC to run electronic equipment. In addition to obtaining power from an inverter, DC loads may be powered directly from the battery bank.

Typical Photovoltaic Systems

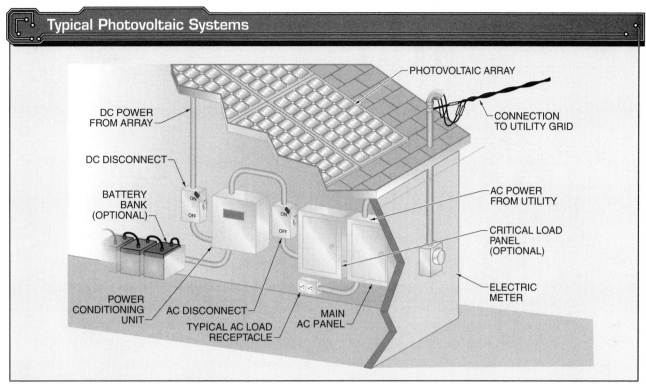

Figure 7-9. *A utility-connected, residential photovoltaic system is the most common system configuration. Various electrical components control, condition, and distribute the power to DC and/or AC loads.*

Stand-Alone Inverters

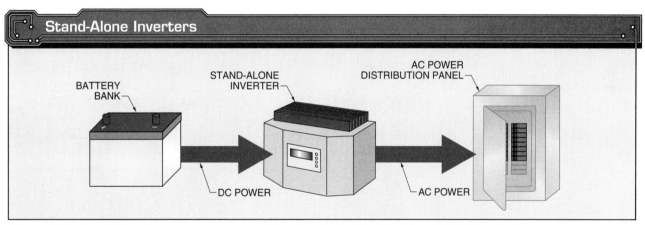

Figure 7-10. *Stand-alone inverters are connected directly to the battery bank in residential applications.*

A *utility-interactive inverter* is an inverter connected to, and operated in parallel with, an electric utility grid. These inverters interface the energy source with the utility grid. Inverters convert DC from the energy source to AC that is synchronized with the utility grid. Inverters also detect the loss of utility power and stop feeding the utility line to prevent electrocution of utility line workers. **See Figure 7-11.**

Photovoltaic Applications

Photovoltaic cells can be used as sensors in a wide variety of applications. They can also be used to power equipment that may be difficult to power with a conventional power source. For example, photovoltaic cells are used in light-intensity meters as sensors and satellite power sources.

Utility-Interactive Inverters

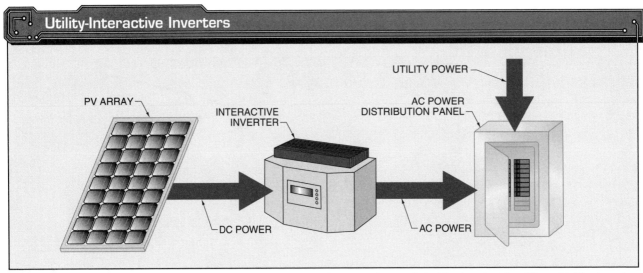

Figure 7-11. *Utility-interactive inverters are connected to, and operate in parallel with, the electric utility grid. The inverters convert DC output from a photovoltaic array to AC power that is synchronized with the utility grid.*

Light-Intensity Meters. A light-intensity meter, or exposure meter, uses photovoltaic cells to measure the amount of light necessary for good photographs. **See Figure 7-12.** The light reflected from the subject to be photographed passes through the window of the camera and strikes the photovoltaic cell. A sensitive meter is connected across the output of the photovoltaic cell, and current flows in proportion to the amount of light falling on it.

Satellite Power Sources. A weather satellite uses photovoltaic cells to power the transmitter so it can send signals to and from the surface of the earth. **See Figure 7-13.** The electrical systems of satellites are powered by solar energy. The photovoltaic cells convert light rays of the sun into electrical energy. Communication satellites have hundreds of photovoltaic cells that constantly convert solar energy into electrical energy.

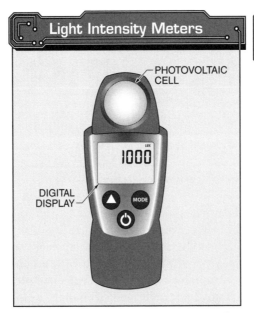

Figure 7-12. *A light-intensity meter uses a photovoltaic cell to measure the intensity of light.*

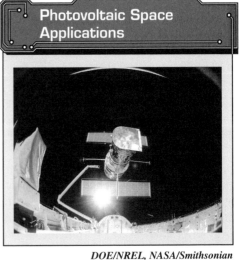

DOE/NREL, NASA/Smithsonian Institution/Lockheed Corp.

Figure 7-13. *Photovoltaic systems are used in space to recharge batteries in spacecraft and satellites.*

WIND TURBINES

A *wind turbine* is a power generation system that converts the kinetic energy of wind into mechanical energy, which is used to rotate a generator that produces electrical energy. Wind turbines can be used to produce power where no utility lines exist, or they can be connected to a utility grid to supplement grid power. Wind turbine outputs range from a few kilowatts to several megawatts. Wind turbines can be grouped in parallel to produce large amounts of power such as in wind farms.

Wind turbines are power plants that can be located almost anywhere a sustained wind system is feasible. In standard generating power plants (nuclear, coal, and hydro), up to 50% of the power generated is lost as line losses and heat before it reaches the loads. Adding wind turbines to a utility system increases system efficiency locally and reduces the need to transport and consume fuel from the utility generating station. With wind energy being generated locally, long distance distribution of power is eliminated along with its losses.

A commercial wind turbine primarily consists of a hub, rotor, nacelle, gearbox, generator, power converter, controller, tower, and a number of blades. **See Figure 7-14.** Wind turbines may be grid-connected or stand-alone systems.

Grid-Connected Wind Turbines

A *grid-connected wind turbine* is a wind energy system that is connected to an electric utility distribution system (grid). **See Figure 7-15.** A grid-connected wind turbine can reduce the consumption of utility-supplied electricity for lighting, appliances, and electric heating systems. When the turbine cannot deliver the amount of energy needed, the utility makes up the difference. When the wind system produces more electricity than required, the excess is sold to the utility.

A grid-connected wind turbine operates only when the utility grid is available. During power outages, the wind turbine is required to shut down due to safety concerns.

The Public Utility Regulatory Policies Act (PURPA) requires utilities to purchase power and allow connections from other energy producers such as those of wind energy. To address any power quality and safety concerns, the utility should be contacted before connecting to its distribution lines. Utilities typically provide a list of requirements for connecting a system to the grid.

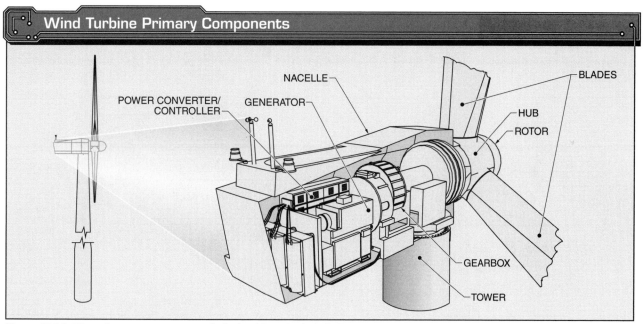

Figure 7-14. *The primary components of wind turbines include hubs, rotors, nacelles, gearboxes, generators, power converters, controllers, towers, and blades.*

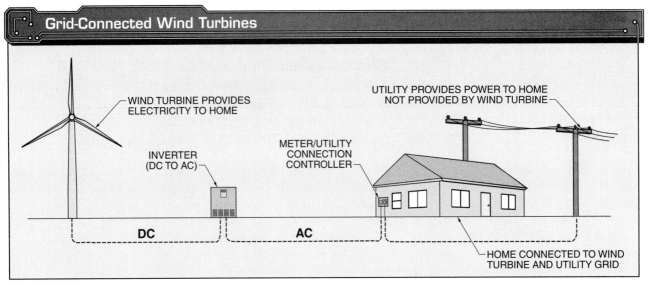

Figure 7-15. *A grid-connected wind turbine is connected to an electric utility distribution system (grid).*

Stand-Alone Wind Turbines

A *stand-alone (off-grid) wind turbine* is a wind energy system that is not connected to an electric utility distribution system (grid). **See Figure 7-16.** The storage capacity of stand-alone wind turbine systems must be large enough to supply electricity during noncharging periods. Stand-alone wind turbines usually provide power to a single residence.

The battery banks of these systems are typically sized to supply electric loads for one to three days. A wind turbine that produces 900 W to 1200 W of power may be sufficient for the operation of basic equipment. If more power is required, a turbine that produces 2 kW to 3 kW of power should be considered. All appliances and equipment that may be run at the same time determine the maximum load the system is required to operate.

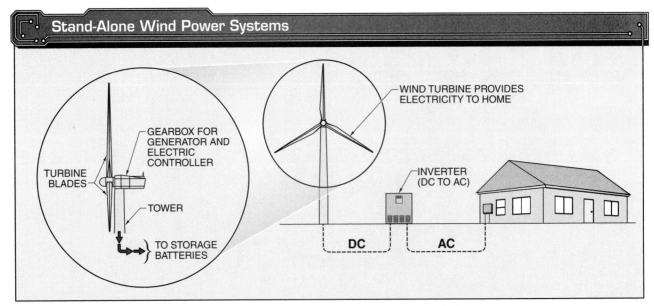

Figure 7-16. *A stand-alone wind turbine is not connected to an electric utility distribution system. The storage capacity of this system must be large enough to supply electrical needs during noncharging periods.*

Wind Turbine Converters and Inverters

Electrical loads and the utility power grid are designed to operate at a constant frequency of 60 Hz (50 Hz in some countries). The output of an AC variable-speed wind turbine generator is AC voltage at a varying AC frequency (Hz). In order to use the AC voltage on electrical loads designed to operate at a constant frequency or to add the voltage to the utility power grid, the voltage must be set to a constant frequency regardless of the wind or generator speed. For this reason, the output of an AC generator is converted from AC to DC, filtered, inverted to a constant-frequency AC voltage (60 Hz or 50 Hz), and filtered (conditioned). The voltage output is then sent to transformers to step up the low voltage to the higher voltage of the utility grid. **See Figure 7-17.**

A *converter* is an electronic device that changes alternating current (AC) voltage to direct current (DC) voltage. The power converter of a wind turbine uses diodes to rectify AC to DC.

An *inverter* is an electronic device that changes direct current (DC) voltage into alternating current (AC) voltage. Inverters used in wind turbines change DC voltage into AC voltage at a fixed frequency and voltage level. Inverters in wind turbines use fast-acting solid state switches that can control high current. The fast-acting solid state switches must also allow the lowest amount of power loss (voltage drop) possible to reduce the amount of heat produced and power loss. The solid state switches turn the DC voltage on and off to remake AC voltage. High-power transistors are used to switch the voltage on and off. The inverter changes the DC voltage into an AC voltage using pulse width modulation (PWM) technology.

> Wind is a form of solar energy. Winds are produced by the uneven heating of the atmosphere by the sun, the irregularities of the earth's surface, and the rotation of the earth.

Pulse width modulation (PWM) controls the amount of voltage output by converting the DC voltage into fixed values of individual DC pulses. **See Figure 7-18.** The fixed-value pulses are produced by the high-speed switching of transistors.

By varying the width of each pulse (time ON), the output voltage can be increased or decreased. The wider the individual pulses, the higher the DC output voltage. By varying the DC output levels, an equivalent AC voltage is produced. The inverter circuit reverses the DC voltage to produce the negative side of the equivalent AC output. Inductors and capacitors can be added as filters to help smooth the AC voltage.

U.S. Marine Corps

Wind turbines use wind energy to rotate generator rotors to convert kinetic energy into electricity.

Electronic Control Devices

Wind turbines are designed to generate electrical power to be used locally and/or to be added to a utility grid. In order to have a safe, economical, and efficient system, the electrical, mechanical, and environmental operating conditions must be measured and the data input into the control system.

170 SOLID STATE DEVICES AND SYSTEMS

Electrical Power Production and Conversion

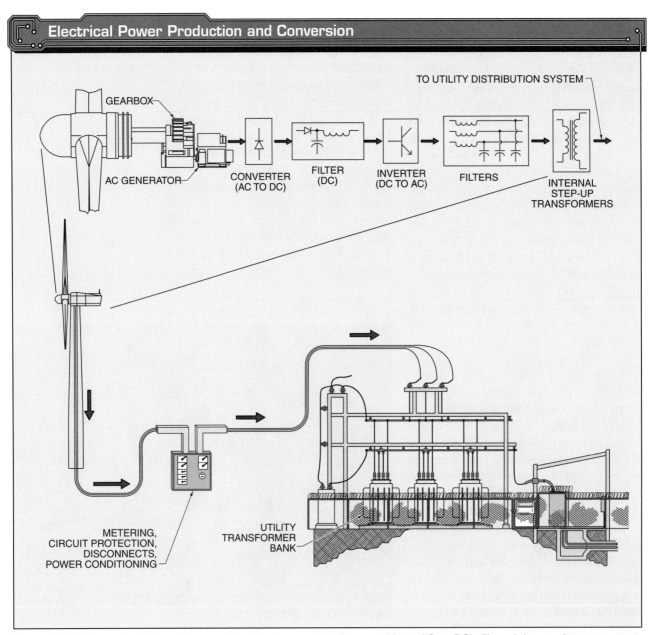

Figure 7-17. *The output of an AC generator is first converted (changed from AC to DC), filtered, inverted to a constant-frequency AC voltage (60 Hz or 50 Hz), filtered (conditioned), and sent to transformers to step up the low voltage to the high voltage of the utility grid.*

The wind turbine controller is the central electronic control device of a wind turbine system that receives signals based on operating conditions, which include voltage level, current level, power output, frequency, wind speed, wind direction, rotor speed, pitch position, vibration, operating temperature, and faults. This information is used to determine when the generated power can be connected or disconnected from the utility grid. Controllers can be standard microprocessors, PLCs, or manufacturer-built proprietary electronic control systems. The controller can be limited to the local wind turbine system only or connected to an external system that can monitor the wind turbine system from a remote location.

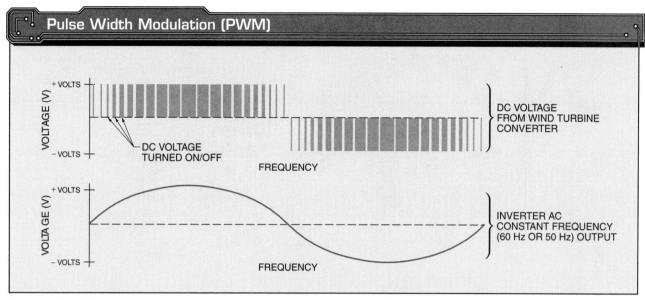

Figure 7-18. *PWM controls the amount of voltage output by converting the DC voltage into fixed values of individual DC pulses.*

Wind Turbine Instruments and Voltage Controllers

In addition to the primary components of a wind turbine, instruments and voltage controllers must be included as part of a wind turbine system. Wind turbines may include anemometers, wind vanes, and transformers. **See Figure 7-19.**

Anemometers. An *anemometer* is an instrument that measures wind speed. An anemometer is connected to the controller of a wind turbine and used to help determine when the wind turbine should rotate and when it should be stopped, such as during high winds and maintenance. The output signal from an anemometer may also be used to track wind speeds at all times and record, store, or send the wind speed information to a centralized location.

Wind Vanes. A *wind vane* is an instrument that measures wind direction. Like an anemometer, a wind vane is connected to the controller of a wind turbine and used to help determine when the wind turbine should rotate. The output signal of a wind vane may also be used to track wind direction and record, store, or send the wind direction information to a centralized location.

Transformers. A transformer is an electrical device that uses electromagnetism to change voltage from one level to another. Both step-up and step-down transformers are used in wind turbine systems.

Step-up transformers are used in wind turbine systems to increase the generator AC output voltage to the high voltage of the utility grid. For example, the output of a generator may be 690 VAC (common output voltage on 50 Hz turbines) or 600 VAC (common output voltage on 60 Hz turbines), but the utility grid voltage can be thousands or tens of thousands of volts. Transformers are the system interface that allows the generated power of wind turbines to be added to the utility grid.

Step-down transformers are used in wind turbine systems to reduce the generator AC output voltage to a low voltage level required by many system control and monitoring devices. Some of the low-voltage AC is also rectified to operate low-voltage DC loads and controls. The DC voltage can also be used to store power in, or recharge, DC batteries. The batteries in wind turbine systems are used to store power, sound an alarm, and/or send a system failure signal to operators when a power failure occurs.

172 SOLID STATE DEVICES AND SYSTEMS

Wind Turbine Supporting Components

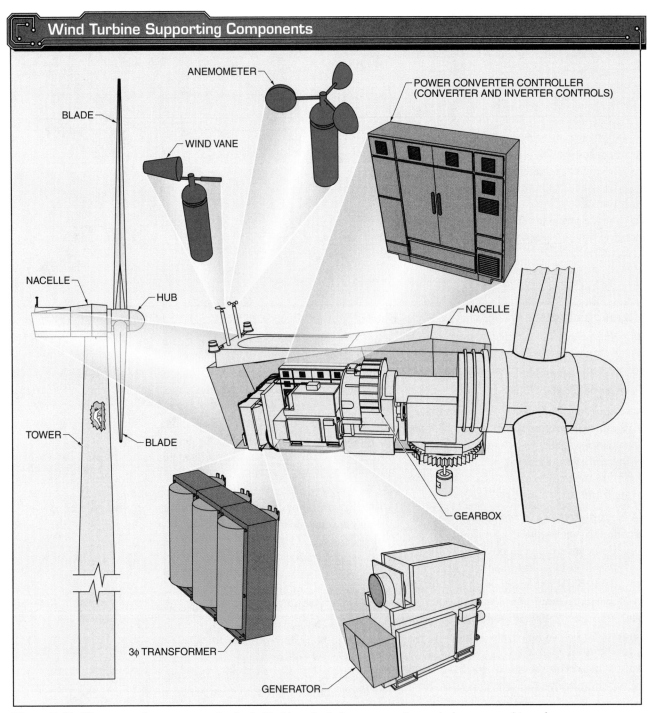

Figure 7-19. *The supporting components of wind turbines include anemometers, wind vanes, and transformers.*

Wind Turbine Power Output

All wind turbines are rated for the amount of power they can produce under normal operating conditions. The required power output depends on the intended use of a wind turbine. Wind turbines designed to feed all of their power to the utility grid are useful regardless of the amount of power they produce at any given time. Wind turbines designed as a primary power supply for individual homes, farms, and small businesses should be sized large enough to deliver enough power under normal operating conditions.

Between the power output location of a wind turbine and the utility grid, and at other points in the electrical system, there must be disconnect switches and overcurrent/short circuit protection devices (circuit breakers and/or fuses). Using disconnect switches is a safe method for turning the wind turbine off and on. The circuit breakers or fuses provide a point at which power is automatically removed when the system is overloaded or shorted and a known point for troubleshooting.

Electric power meters are connected to the wind turbine system to measure and record the amount of power output. Other meters, which include voltmeters, ammeters, frequency meters, and power factor meters, are connected to the system to monitor electrical quantities. Metering the system operating parameters helps to detect problems, provides data for developing system improvements, and provides documentation for billing power output and meeting regulatory requirements.

ELECTROCHEMICAL POWER SOURCES

An *electrochemical power source* is a device that uses an electrochemical reaction to produce electrical energy. An electrochemical power source serves as a back-up power source of DC voltage for devices such as electronic intercoms, paging systems, and emergency lighting systems. The most basic type of electrochemical power source is an electrochemical cell. An *electrochemical cell* is a single unit that uses an electrochemical reaction to produce electrical energy. A *battery* is a group of connected electrochemical cells that uses an electrochemical reaction to produce electrical energy.

Technicians should be aware of the different types of cells and batteries. It is also important for technicians to know how to maintain cells and batteries and to recognize problems when they occur and fix them.

Cells can be classified as either wet cells or dry cells. **See Figure 7-20.** A wet cell uses a liquid electrolyte. For example, the wet cells in automobile batteries use a liquid electrolyte of sulfuric acid. A dry cell uses an electrolyte in the form of paste. For example, the dry cells in flashlights use an electrolyte paste of ammonium chloride.

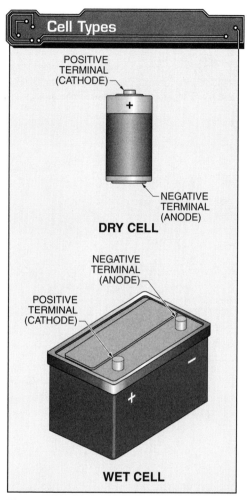

Figure 7-20. *Electrochemical power sources include wet cell and dry cell batteries.*

Wet Cells

A *wet cell* is an electrochemical power source consisting of two electrodes in a liquid electrolyte solution. **See Figure 7-21.** The electrolyte is usually a solution of acid in water. By electrochemical reaction, the electrolyte breaks down into particles called ions, which are groups of atoms. Certain ions have excess electrons and are negatively charged. Other ions have a shortage of electrons and are positively charged. The negative ions and the positive ions float in the electrolyte.

The positive ions are attracted to a positive electrode. The negative ions are attracted to a negative electrode. As one electrode takes on more electrons, it becomes negatively charged. As the other electrode loses electrons, it becomes positively charged.

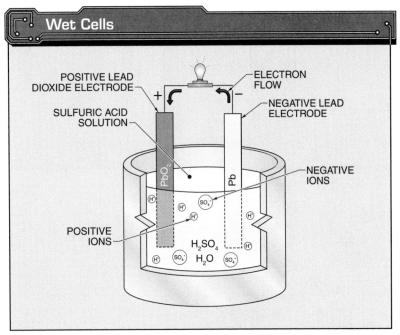

Figure 7-21. Wet cells consist of a copper electrode and zinc electrode immersed in a liquid electrolyte that allows electrons to flow from the zinc electrode to the copper electrode.

Each electrode in a practical cell connects to a terminal outside of the electrolyte. The cell has two terminals: positive (+) and negative (−). With one electrode negatively charged and the other electrode positively charged, a voltage, or difference of potential, exists across the electrodes.

Wet Cell Current Flow. When a conductor and a lamp are connected between the two electrodes, electrons flow between the electrodes and the lamp lights. The voltage generated across the electrodes depends on the materials used in the electrodes and the electrolyte. A cell of certain materials generates a fixed value of voltage for most of its useful life. The amount of current the cell can supply depends largely on the size of the cell and its parts. The electrodes of the cell and the electrolyte change chemically as the cell supplies current. After a period of time, the electrodes change to the point where the cell is no longer able to supply current. When this happens, some types of cells may be recharged and used again, while other cells must be replaced.

Wet cell batteries can explode when batteries are being jumped or charged due to sparks from the cable connections. Battery acid is corrosive and can burn the skin. Sulfuric acid is commonly used in lead-acid batteries. If battery acid comes in contact with the skin, the area should be immediately washed with plenty of soap and water, and anyone exposed should seek medical attention. If battery acid enters the eyes, the eyes should be immediately flushed with running water, and medical attention should be sought.

Sulfuric acid spills or splashes on clothing can be neutralized by immediately rinsing with a solution of baking soda or household ammonia and water. It is important to remove all clothing with acid on it so that the acid does not contact or remain on the skin. In nickel-cadmium batteries, the electrolyte is a potassium hydroxide solution that can be neutralized with a mixture of vinegar and water.

Dry Cells

A *dry cell* is an electrochemical power source consisting of a cylindrical container made of zinc, a graphite rod in the center, and a paste electrolyte that allows electrons to flow. The electrodes are zinc and graphite, with an acidic paste between them that serves as the electrolyte. The zinc contains the electrolyte and acts as the negative terminal (electrode) of the cell. The positive terminal of the cell is a graphite rod with a brass contact cap. The rod is placed in the center of the cell. **See Figure 7-22.** The electrolyte that is used in the dry cell is actually a damp paste of concentrated ammonium chloride and zinc chloride.

When the zinc comes in contact with the electrolyte, positive zinc ions enter the electrolyte paste. Positive hydrogen and ammonium ions are driven toward the graphite (positive) electrode. This electrochemical reaction produces the positive and negative charges on the electrodes that are necessary to create voltage. When an external circuit is connected, current flows from the zinc electrode, through the external circuit, and to the graphite electrode. Zinc ions also move from the zinc to the graphite electrode. One of the electrodes is always decomposed in this process. The electrode that is decomposed in the common dry cell is the one that forms the case. The corrosive action of these types of cells can be very detrimental to solid state equipment when left in place after the useful life of the cell has ended.

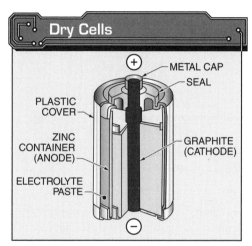

Figure 7-22. *Dry cells consist of a cylindrical container made of zinc, a graphite-center rod, and a paste electrolyte that allows electrons to flow.*

Dry cell batteries may explode when the polarity of the battery is reversed. Manufacturer's specifications and procedures for inserting batteries should always be followed. Dry cell batteries should be periodically checked for signs of leakage or corrosion. Leaking batteries should be replaced. If the equipment is not used for a long period of time, the batteries should be removed and stored separately.

Batteries can produce toxic and explosive mixtures of gases. Ventilation of the battery enclosure is sometimes required. For small battery systems charged at low rates, the accumulation of gases is minimal and natural ventilation may be adequate.

Cell and Battery Capacity

The capacity of a storage battery is stated in ampere-hours (Ah). An ampere-hour is the maximum current a battery can continuously deliver until it is completely discharged for a period of one hour. The capacity of a battery in ampere-hours is found simply by multiplying the number of amperes of current carried by the number of hours the current flows. A battery capable of discharging 2 A of current continuously for a period of 8 hr has a capacity of $2 \times 8 = 16$ Ah. Likewise, a battery that will deliver a current of 10 A for 12 hr has a capacity of 120 Ah. **See Figure 7-23.** Reserve capacity indicates how much time is available to keep equipment running when a charging system is not in place or has failed.

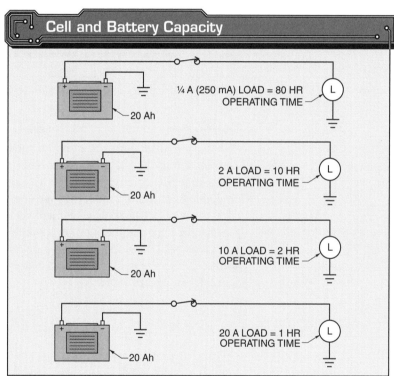

Figure 7-23. *Cell and battery capacity is listed in ampere-hours (Ah) and describes how long a battery can power a load.*

A battery delivers power for only a fixed amount of operating time. Once the battery power is used, the battery must be replaced or recharged. The operating time of a battery depends on the load. The more power the load requires, the less operating time. The less power the load requires, the greater the operating time. The amount of current the cell or battery can supply depends largely on the size of the cell or battery. **See Figure 7-24.**

Cell and Battery Current

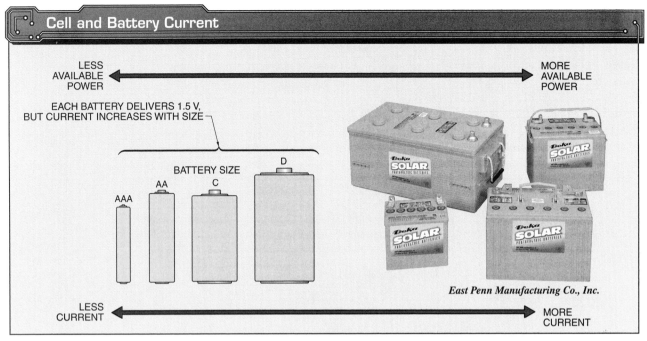

Figure 7-24. *The amount of current a cell or battery can supply depends on the cell or battery size.*

Connecting Cells and Batteries

A battery is made up of a group of cells connected together. Cells are connected in series to produce a higher voltage than is available from just one cell. They are connected in parallel to make available a greater current than can be provided by a single cell. They are connected in series-parallel to provide both higher voltage and higher current. **See Figure 7-25.**

Primary and Secondary Cells

A *primary cell,* also known as a single-use or nonrechargeable cell, is a cell that uses up its materials to the point where recharging is not practical. Most dry cells are primary cells. **See Figure 7-26.**

Americans purchase nearly three billion dry cell batteries every year to power radios, toys, cellular phones, watches, laptop computers, and portable power tools. Recycling batteries keeps heavy metals out of landfills. Recycling saves resources because the recovered plastics and metals can be used to make new batteries.

A *secondary (rechargeable) cell* is a cell that can be recharged. The electrodes and the electrolyte in secondary cells can be recharged or restored to their original, active states. The cells are recharged by passing current through the cells in the direction opposite to that of the load current. **See Figure 7-27.** Secondary cells are used in thin-film and lithium-ion batteries.

Surrette Battery Company
Batteries may be comprised of multiple electrochemical cells.

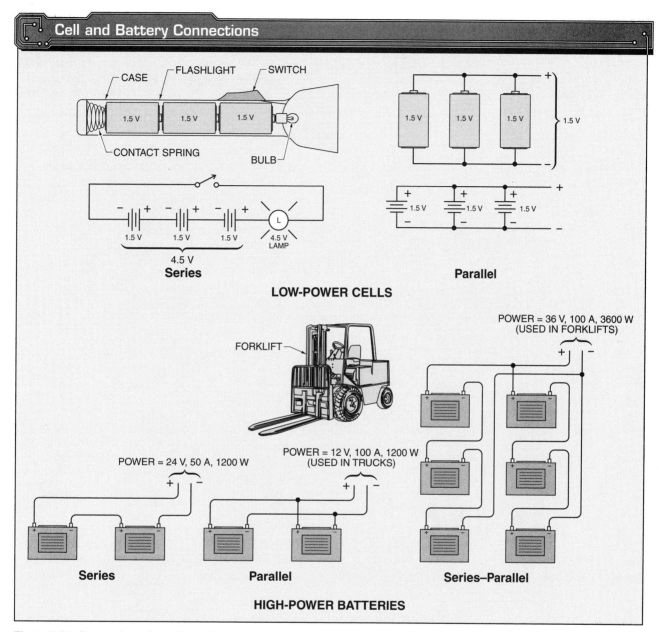

Figure 7-25. *Connecting cells and batteries in series increases voltage. Connecting cells and batteries in parallel increases current.*

Thin-Film Batteries. A *thin-film battery* is a rechargeable solid state battery that contains multiple secondary cells, is very thin, and may be flexible. Thin-film battery components are prepared by deposition techniques such as those used in the semiconductor industry. There is also a potential to use nanomaterials in thin-film batteries due to recent developments in nanotechnology. Once thin-film batteries can be manufactured at a low cost, the technology will most likely replace conventional batteries in many applications. **See Figure 7-28.**

Thin-film printing technology is now being used to apply solid state lithium polymer chemistry to create unique batteries for specialized applications. Thin-film batteries can be deposited directly onto printed circuit boards in any shape or size. Flexible thin-film batteries may also be made by printing onto plastics and thin metal foil.

Primary (Single-Use) Cells

Primary (Single-Use) Cells/Batteries	Case Type	Applications	Advantages	Disadvantages	Disposal
Carbon-zinc	• Cylindrical • Button	Clocks, calculators, flashlights, portable electronics, measuring instruments, electronics, remote control devices	• Least expensive • Readily available • Variety of sizes	• Cannot withstand heat or cold well • Being replaced by alkaline batteries	• Disposable
Alkaline-manganese-dioxide	• Cylindrical • Button	Flashlights, smoke detectors, clocks, calculators, cameras	• Environmentally safe • Maintain peak power output longer than carbon-zinc cells • Interchangeable with carbon-zinc cells • Available in sizes AAA, A, C, D, and 9 V	• About double the cost of carbon-zinc but more cost effective • Not rechargeable	• Disposable
Lithium-manganese dioxide	• Cylindrical • Button	Watches, medical equipment, electronic games, MP3 players, portable power tools, military applications	• Long shelf life • Small size • Lightweight	• Older batteries do not charge as well as new ones • Heat causes capacity loss	• Recyclable • Hazardous waste
Silver-zinc	• Button • Prismatic	Laptop computers, cellular phones, submarines, satellites, missiles	• Free from problems associated with thermal runaway, fire, and danger of explosion • Improved performance for notebooks and cellular phones	• Fluctuating cost of silver	• Recyclable • Hazardous waste
Zinc-air	• Button	Watches, hearing aids, emergency medical power, patient monitors	• Free from problems associated with thermal runaway, fire, and danger of explosion • Lightweight • Long shelf life	• Must be vented • Rechargeable only in certain forms • Sensitive to temperature and humidity	• Recyclable • Hazardous waste

Figure 7-26. *Primary cells are single-use cells because recharging them is not practical.*

These new batteries replace liquid or gel electrolyte with thin layers of solid glass-like or polymer materials, which are more stable. The batteries can survive a wide range of cold and hot temperatures. For example, they could be built into rubber tires to power air pressure sensors.

The different layers that make up a typical thin-film solid state battery are deposited using the sputtering or evaporation methods used to create semiconductor devices. Because thin-film batteries are manufactured using all solid state materials, these batteries are intrinsically safe. The risks, such as spilling, boiling, or gassing, associated with traditional batteries are not associated with thin-film solid state batteries.

Lithium-Ion Batteries. A *lithium-ion battery* is a rechargeable battery that consists of multiple secondary cells with a lithium compound as the electrolyte. Lithium-ion batteries produce approximately 4 V, which is almost three times the voltage of rechargeable nickel-cadmium batteries or nickel-metal hydride batteries. The larger voltage allows for smaller and lighter batteries and battery-powered equipment.

Secondary (Rechargeable) Cells

Secondary (Rechargeable) Cells/Batteries	Case Type	Applications	Advantages	Disadvantages	Disposal
Nickel-cadmium	• Cylindrical • Prismatic	Two-way radios, emergency medical equipment, power tools, hobby equipment, camcorders, toys, rechargeable flashlights, electric vehicles, standby power	• Long life • Low maintenance • Capacity retention • Rechargeable • Wide range of sizes	• Expensive • Limited availability • Contains toxic material • Needs recharging after storage	• Recyclable • Hazardous waste
Lead-acid	• Plastic	Communication equipment, security systems, UPS systems, car batteries, storage systems for alternate energy	• Withstand overcharging and overdischarging • Resists vibration • Inexpensive to manufacture	• Cannot be stored in discharged condition • Environmentally unsafe	• Recyclable • Hazardous waste
Sealed lead-acid (SLA)	• Plastic	Computer back-up power, telecommunication systems, emergency lighting, security alarms, cordless tools	• Can hold a charge for months • Cost-effective • Not subject to "memory" • Low maintenance	• Relatively low energy density • Thermal runaway with improper charging	• Recyclable • Hazardous waste
Nickel-metal-hydride	• Cylindrical • Prismatic	Cameras, cordless vacuum cleaners, power tools, laptop computers, electric plug-in vehicles	• Available in the same standard size of alkaline cells • Cost-effective • High capacity • Environmentally safe	• Limited useful life • High self-discharge • Requires full discharge three to four times per year to prevent crystal formation • High temperature reduces battery performance	• Recyclable • Disposable
Lithium-ion	• Prismatic • Cylindrical	High-end laptop computers, cellular phones, UPS systems, power tools, wheelchairs, electric vehicles	• Long shelf life • High capacity • Lightweight • Lasts two to three years with proper use	• More costly than alkaline	• Recyclable • Hazardous waste
Lithium-ion polymer	• Variety of custom shapes	Cellular phones, PDAs, mobile communicating devices	• Rechargeable in device	• Prone to burning if overcharged or corroded	• Recyclable • Hazardous waste
Thin-film	• Variety of custom shapes	RFID tags, smart cards, medical sensors	• No toxic materials used • No leakage • Rechargeable	• Low power output	• Recyclable

Figure 7-27. *Secondary cells are multiple-use cells because they are rechargeable.*

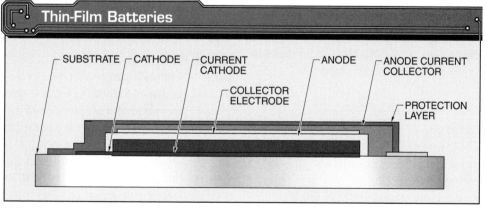

Figure 7-28. *Thin-film batteries are rechargeable solid state batteries that are very thin and may be flexible.*

Lithium-ion batteries should be charged per the manufacturer's specifications. The term "lithium" is used to describe the anode of lithium-ion batteries. However, batteries differ in cathode material, electrolyte, cell design, and other mechanical features. Each lithium system has its own characteristics.

Lithium-ion batteries may have a microprocessor or digital circuit that monitors the charge and discharge process. These safety circuits may also include reverse polarity protection so that the battery cannot be inserted backward.

Frequent full discharges of lithium batteries should be avoided. Several partial discharges with frequent recharges are better for lithium-ion batteries than one full discharge. Recharging a partially charged lithium-ion battery does not cause harm because the batteries do not suffer from memory effects. Lithium-ion batteries cannot be charged with a trickle charger. The lithium-ion chemistry cannot accept an overcharge without causing damage to the cell. Lithium-ion batteries should be kept cool. For prolonged storage, batteries should be kept at a 40% charge level. Lithium-ion batteries are typically designed to be recharged in the device they power rather than in an external charger. It is recommended to remove batteries from equipment when the equipment is running on fixed power or stored for long periods of time.

Lithium-ion-polymer batteries are the next stage in the development of batteries. With lithium-ion-polymer batteries, a polymer (plastic) electrolyte is used instead of a liquid electrolyte. The polymer electrolyte allows batteries to be made in a variety of shapes and sizes.

Cell Designs

Cells that are combined into batteries come in a variety of shapes and sizes to accommodate the flexible design criteria required for developing solid state equipment. Some equipment may require rectangular configurations and some require cylindrical designs. Camcorders may actually require the battery to be shaped around the electronics, requiring a pouch design. The most common battery cell designs include button, cylindrical, pouch, spiral-wound cylindrical, and prismatic. **See Figure 7-29.**

Button Cells. A *button cell* is a button-shaped electrochemical cell manufactured with a separator layer between circular electrodes. The assembly is placed into a nickel-plated container and electrolyte is added. Crimping a cap onto the container seals the cell.

Cylindrical Cells. A *cylindrical cell* is a cylindrically shaped electrochemical cell with a positive electrode, negative electrode, synthetic separator, and electrolyte. The cylindrical cell is the most common battery shape. It is easy to manufacture, offers high energy density, and provides good mechanical stability.

Pouch Cells. A *pouch cell* is an electrochemical cell enclosed in a flexible, heat-sealable foil with conductive foil tabs welded to the electrodes. The pouch cell uses a heat-sealable foil instead of a more expensive conventional metal casing. The electrical contacts consist of conductive foil tabs that are welded to the electrode and sealed to the pouch material. The use of foil makes it possible to use the available space for the battery more efficiently. Because of the absence of a metal case, the pouch pack is light. Therefore, pouch cells are vulnerable to external mechanical damage. Standardized pouch cells do not exist. Each manufacturer builds a pouch cell for a specific application.

Spiral-Wound Cylindrical Cells. A *spiral-wound cylindrical cell* is a cylindrically shaped electrochemical cell that contains an anode and a cathode wound together with a microporous separator interspaced between thin electrodes. Spiral-wound cylindrical cells are designed for high-current pulse capability as well as continuous high-rate operation. Therefore, high surface area is achieved and capacity is optimized.

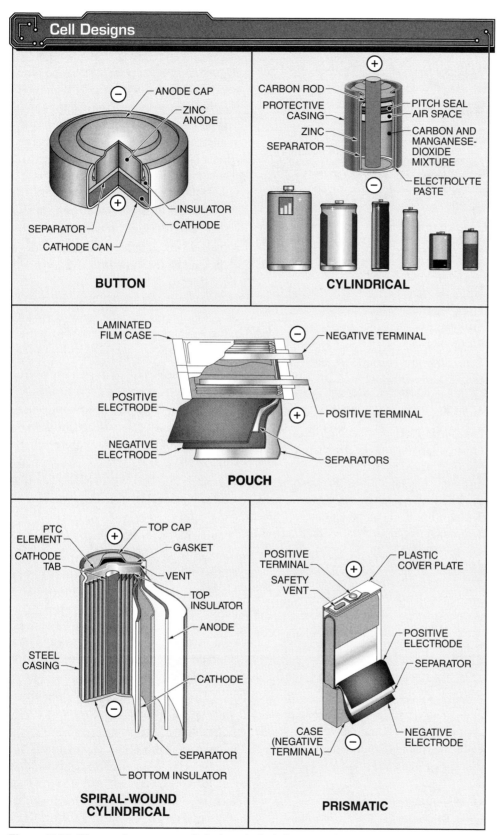

Figure 7-29. *The most common cell designs include button, cylindrical, pouch, spiral-wound cylindrical, and prismatic.*

182 SOLID STATE DEVICES AND SYSTEMS

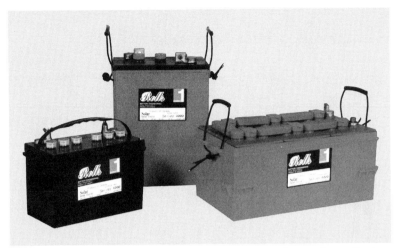

Surrette Battery Company
Deep-cycle batteries may be used for power storage in off-grid photovoltaic and wind applications.

Spiral-wound cells contain a safety vent mechanism to relieve internal pressure in the event of severe mechanical abuse. Spiral-wound cells also contain a resettable positive temperature coefficient (PTC) device, which limits current flow and prevents the cell from overheating if accidentally short-circuited.

Prismatic Cells. A *prismatic cell* is a rectangular-shaped electrochemical cell enclosed in a case that contains a flat, rectangular anode and a flat, rectangular cathode with a separator in between them. The electrodes are manufactured in a manner similar to those of the cylindrical cell, except that the finished electrodes are flat and rectangular in shape.

The basic differences between the prismatic cell and the cylindrical cell are the construction of the electrodes and the shapes of the containers. Prismatic cells are designed to meet the needs of compact equipment where space is limited. The rectangular shape of the prismatic cell permits more efficient battery assembly by eliminating the voids that appear in a battery constructed with cylindrical cells. Prismatic cells are typically found in mobile phones. There is no standard prismatic cell size.

Battery Maintenance and Storage

The dry cell is relatively easy to maintain as long as it is kept clean, dry, and recharged. General storage batteries, which are used to provide back-up power for large power systems, require much more attention. **See Figure 7-30.** There are many common issues associated with poorly maintained batteries. Most of these problems can be avoided with preventive maintenance. When storing batteries, the following rules should be applied:

- Do not recharge a battery unless it is specifically marked "rechargeable."
- Never allow a battery to remain discharged for any length of time.
- Avoid overcharging, which may result in excessive formation of gas or high electrolyte temperature.
- Never add extra acid to a cell.
- Replace only the amount of electrolyte spilled.
- Preserve battery life by switching off a device and removing the battery when it is not being used.
- Remove all batteries from equipment at the same time and replace them with new batteries of the same size and type.
- Remove and attach battery cables in the right order. The ground cable should be disconnected first and connected last.
- In cold weather, keep batteries as near full charge as possible. The freezing point of the electrolyte depends on its specific gravity.
- Wear eye protection when working near batteries.
- Service technicians should not wear watches, rings, or other jewelry that could potentially come in contact with battery terminals.
- All tools used near a battery should have insulated handles to prevent electrical contact.
- Inspect cables and connections for cracks. Replace the cables if needed. Clean the connectors with a wire brush.
- Do not apply significant force to connectors.
- Keep all external parts of the battery clean, dry, and tight.

Improperly Maintained Batteries

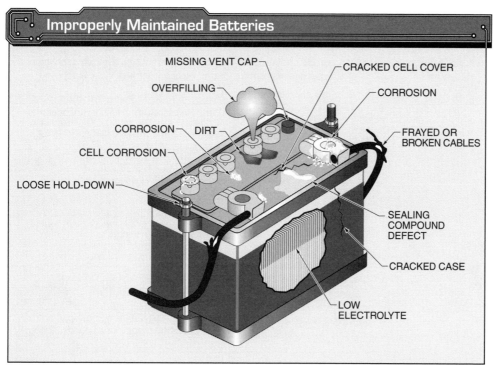

Figure 7-30. *Improperly maintained batteries will have problems powering equipment and are safety hazards.*

Battery Life

There are three factors to consider when determining overall battery life: shelf life, useful life, and recharge life. The shelf life of a battery is the length of time the battery can last just sitting before use. Shelf life varies depending on the battery type and temperature. Some battery manufacturers place a date code on the battery package. The useful life of a battery is the length of time the battery can last powering a particular device. The useful life of a battery depends on the current drain of the device and the capacity (mAh) rating of the battery. The recharge life of a battery is the number of times the battery can be discharged and recharged and remain effective. Over time, rechargeable batteries lose their capacity for recharging.

Battery Testing Methods

A battery is tested to determine whether it is good and stable, requires recharging, or should be replaced. The condition of a battery can be checked by performing various tests. Common battery tests include the specific-gravity test, light-load test, high-rate discharge test, and relative mode test. The high-rate discharge test is the only test that can be performed on a maintenance-free battery. The symptoms, causes, and remedies for many battery problems can be determined with proper troubleshooting. **See Figure 7-31.**

Specific Gravity Tests. With a specific gravity test, a sample of electrolyte is taken from each cell with an instrument called a hydrometer. The hydrometer is used to measure the specific gravity of the electrolyte in a cell. The hydrometer operates on a physics principle discovered by Archimedes. He found that a floating body sinks deeper into a light liquid (low specific gravity) than a heavy liquid (high specific gravity).

The specific gravity of a substance is a measure of how much a certain volume of that substance weighs in comparison to the same volume of water. For example, the specific gravity of water is unity, or 1.000. Concentrated sulfuric acid weighs 1.835 times as much as the same volume of water. Thus, the specific gravity of the acid is 1.835.

Troubleshooting Chart for Lead-Acid Batteries

Symptom	Cause	Remedy
Battery will not hold its charge	Battery worn out or deteriorated	Remove battery and test it for capacity and condition; replace if necessary
	Partial short circuit in the electrical system	Check electrical system for shorts or load; see that battery switch is turned OFF when not in use; remove and charge battery
	Battery switch left turned ON	
	Charging rate too low	Adjust voltage regulator for proper charging rate
	Electrical load too large for battery and generator	Install battery and generator with sufficient capacity for applied loads
	Battery left standing too long	Remove and recharge battery; when not in use for a long period of time, see that battery is recharged every 30 days
	Broken internal cell walls	Replace battery
	Plates "shorted" internally	Replace battery
Battery will not take charge	Battery is worn out	Replace battery
	Battery plates sulfated	Place battery on charger until specific gravity does not change for a period of 2 hr; then charge at 1 amp for 60 hr
Electrolyte runs out of vent cap	Electrolyte level too high	See that electrolyte is maintained at correct level; remove spilled electrolyte
	Charging rate too high, causing gassing and overcharge	Adjust voltage regulator for correct charging rate
Excessive corrosion in battery case and compartment	Overcharging	Adjust voltage regulator for correct charging rate
	Electrolyte level too high	Correct electrolyte level
	Vent lines clogged	Clear vent lines and see that they are properly connected

Figure 7-31. *Lead-acid batteries may require troubleshooting and testing to determine the sources of problems.*

Inside a hydrometer, there is a weighted bulb with a calibrated scale. The bulb floats free inside the hydrometer when a liquid is drawn up inside the cylinder. The specific gravity of the electrolyte can be read directly on the float scale. For an accurate reading, it is important to hold the hydrometer in a vertical position with the float bulb floating freely. The point on the scale provides the specific gravity reading. **See Figure 7-32.**

The specific gravity of an electrolyte varies with temperature. Changes in temperature affect the volume of any liquid. Fully charged specific gravity equals 1.27 to 1.38. Fifty percent (50%) discharged specific gravity equals 1.21 (needs recharging).

Hydrometer markings are calibrated to a specific temperature, typically 80°F. When specific gravity is measured at temperatures lower or higher than the reference temperature, a correction factor must be applied.

Light-Load Tests. A *light-load test* is a battery test that shows the ability of each cell to produce voltage under a light load. An instrument used for this test is the cell analyzer. The ability of a battery to maintain voltage while under load is a good indicator of battery condition. While a defective battery may accept a charge, it may not perform well under load. A specific-gravity test only reveals how much a battery is charged or discharged but does not indicate the condition and capacity of a battery.

Specific Gravity Testing

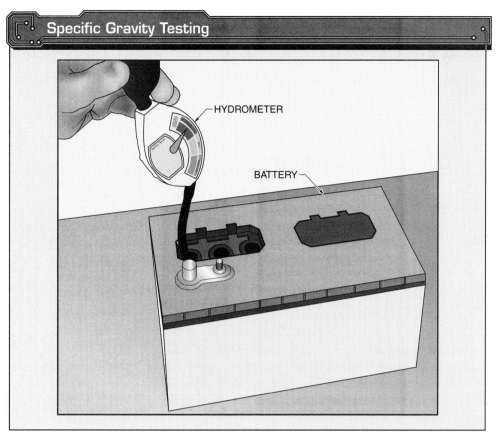

Figure 7-32. *Hydrometers can be used to measure the specific gravity of battery electrolytes.*

High-Rate Discharge Tests. A *high-rate discharge test* is a battery test that measures battery terminal voltage after 15 sec of discharge when a heavy load is placed on the battery. **See Figure 7-33.** The open-circuit voltage of a dry cell is misleading and should never be taken as an indication of the condition of the battery.

Relative Mode Tests. A *relative mode test* is a digital multimeter (DMM) test that records a voltage, current, or resistance measurement and displays the difference between the reading and subsequent measurements taken by a meter. For example, the relative mode is used when testing a forklift electrical system for the amount of current drawn by a load or circuit. A DMM set to measure DC using a DC clamp-on current probe accessory can measure total system current at the battery or any individual circuit current at the fuse box. **See Figure 7-34.**

Battery Load Testers

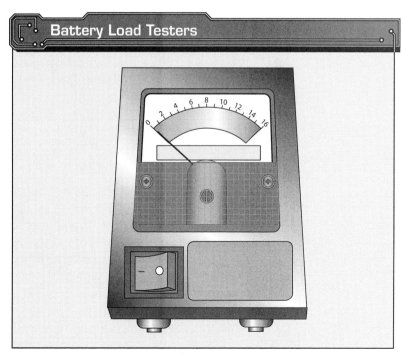

Figure 7-33. *A battery tester indicates the overall health of a battery by measuring the battery voltage under a high-current load.*

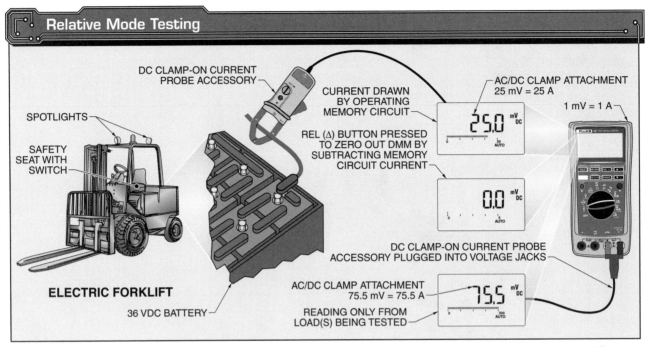

Figure 7-34. *During relative mode testing, voltage, current, or resistance measurements are recorded, and the difference between those readings and subsequent measurements taken by a meter are displayed.*

Relative mode can also be used to zero out any current that is constantly drawn by an electrical system during normal operation. For example, the memory circuits in computers, digital clocks, and cellular phones constantly draw a small amount of operating current. Pressing the REL button zeros out the DMM by subtracting the memory circuit current from the measurement being displayed. As an individual load is turned ON, current measurements are displayed for that load only. The relative mode allows for greater DMM accuracy when taking measurements.

Battery Charging Methods

Charging is the process of reversing the current of a cell or battery and converting the electrical energy back into chemical energy. In order for the battery to accept current, the voltage of the charging source must be higher than the battery voltage. For example, a 12 V lead-acid battery is charged with about 14.4 V and reads about 12.6 V when fully charged.

Charge rate is quantified the same way as discharge rate. For example, a charging rate of C/50 to a 100 Ah battery applies 2 A of current until the battery reaches a specified voltage. The maximum battery charging rate is specified by manufacturers as C where C refers to the ampere hour (Ah) rating of the battery. Using this terminology, a 10 Ah battery being charged at a C/5 rate corresponds to 2 A.

When charging a battery, the following guidelines should be followed:
- Check the charger nameplate information to determine whether the charger rating matches the battery.
- Turn OFF the charger prior to connecting or disconnecting cables from the battery.
- Set the charger to the correct voltage setting.
- Ensure the correct polarity of cables.
- Ensure proper ventilation when the battery is being charged.

WARNING: Overcharging any type of battery can damage the battery. During overcharging, excessive current causes the oxides of some batteries to loosen. Once removed, these oxides are no longer active in the battery, reducing the capacity of the battery. Overcharging also heats the battery and can remove water from the electrolyte.

The two basic methods used for charging batteries include constant-voltage and constant-current charging. Either method can be used to achieve a quick (high-rate and low-rate) charge or trickle charge. Regardless of the method used, the charging current must be DC current.

Constant-Voltage Charging. Constant-voltage chargers use direct-current sources that have constant voltage outputs, with ammeters to monitor the circuit currents. The polarity of the direct-current source is matched to the polarity of the battery terminals. When more than one battery is being charged, the additional batteries are connected in parallel.

When a discharged storage battery is placed in a constant-voltage charger circuit, the direct-current source furnishes a charging current of about 30 A to 50 A. With smaller batteries, the initial charging current is much less. The amount of current depends on whether the source is connected across one battery or divided among more than one battery.

As a battery begins to charge, terminal voltage increases and causes the current delivered to decrease. As the battery becomes more fully charged, less current is required to charge it further. Thus, the current in a constant-voltage charger tapers off during the charging period. This type of circuit is sometimes called a taper charger because of the gradual decrease in applied current.

The constant-voltage method is faster than the constant-current method. The constant-voltage method of battery charging requires approximately 6 hr to 8 hr to charge a battery fully.

Constant-Current Charging. With constant-current charging, the charger supplies a relatively uniform current, regardless of the battery state of charge or temperature. Constant-current charging helps to eliminate imbalances of cells and batteries connected in series. Constant-current chargers are most appropriate for cyclic operation where a battery is often required to obtain a full charge overnight. At these high rates of charge, there will be some venting of gases. Normally, the user of a cyclic application is instructed to remove the battery from a single-constant-current charger within a period of time that permits full charge yet prevents excessive grid oxidation.

Quick Charging. The quick charge method is a modification of the constant-voltage charging method. With quick charging, a large current passes through the battery for a short time. Then, a lower current completes the charge in approximately 1 hr to 2 hr.

Trickle Charging. A number of applications require the use of batteries that are maintained in a fully charged state. A constant fully charged battery state is accomplished by trickle charging the battery at a rate that will replace the capacity loss due to self-discharge. The preferred temperature range for trickle charging is between 50°F and 95°F.

Battery Chargers

The more features added to a battery charger, the more expensive the battery charger. Some manufacturers may supply a specific charger with the original equipment order or to the consumer at an additional cost. The rates at which batteries should be charged are determined by the manufacturer.

One feature that is very helpful when charging batteries is charger control. A more expensive battery charger will have an intelligent microprocessor that switches the charger OFF when the battery is fully charged. It can also recognize how much charge was originally in the battery and indicate, for liquid-based batteries, when adding water is necessary. Less expensive chargers charge batteries for a fixed length of time and may overcharge the battery, shortening its life. A simple light-emitting diode (LED) is typically used to indicate when the charge cycle is complete.

There are three basic types of chargers: continuous low-rate chargers, timer-controlled chargers, and smart chargers.

> By combining hydrogen fuel with oxygen, fuel cells can produce plenty of electric power while emitting only pure water as exhaust. Fuel cells are so clean that astronauts drink the water produced by fuel cells.

Continuous Low-Rate Chargers. A continuous low-rate charger relies on technician management of the charge time. The charger will charge the batteries as long as the charger is plugged in and the batteries are installed. Although slow charge is generally good for battery life, batteries can be overcharged if not properly managed by the user.

Timer-Controlled Chargers. A timer-controlled charger stops the main charge at a predetermined time. The timer reduces the need for user monitoring of charge time, and the low charge rates limit the effects of overcharging.

Smart Chargers. A smart charger uses a microprocessor to monitor the battery voltage characteristics to determine when the battery is fully charged. Typically, a simple program in the microprocessor monitors the change of voltage. The microprocessor may also have a program that monitors temperature. Sensors monitor battery temperature during battery charging. When the batteries reach the maximum temperature limit, the charge is terminated. This protection is important when fast charging methods are used. Once charged, a low-rate trickle charge is used to maintain a full charge on the battery. When compared to timer-controlled chargers, smart chargers can typically charge batteries faster without impacting performance.

Cell and Battery Disposal

Because of the large number of batteries being used with computers and other electronics, it is important to dispose of the batteries properly and safely to minimize any impact on the environment. In addition, there are over 500 recycling laws in the United States, and failure to dispose of batteries properly may result in fines.

The disposal of waste products in the United States is regulated by the Environmental Protection Agency (EPA). EPA Regulations are listed in the Code of Federal Regulations, or CFR 40. These regulations are entitled "Protection of Environment." Individual states and local communities also may establish regulations covering the disposal of products. These may actually be more stringent than the federal regulations and cover the disposal of household waste, which is not included in federal regulations. State and local agencies should be contacted regarding their battery disposal guidelines.

FUEL CELLS

A *fuel cell* is an energy source that transforms the chemical energy from fuel into electrical energy. All fuel cells operate quite similarly. Each fuel cell has one positive electrode called the cathode and one negative electrode called the anode. The chemical reaction that creates electricity takes place at the electrodes. All fuel cells also have an electrolyte that carries charged particles from one electrode to another. A catalyst, which increases the reaction of the electrodes, is also usually present. **See Figure 7-35.**

A fuel cell may only generate a small amount of DC electricity. To be useful, many fuel cells are assembled into stacks. For higher voltages, fuel cells are connected in series. For higher current, fuel cells are connected in parallel.

One of the main advantages of fuel cells is that they generate electricity with very little pollution. This is possible because most of the hydrogen and oxygen used as fuel recombine to form a by-product (waste) of pure water. In addition to hydrogen fuel cells, there are hydrocarbon fuel cells and chemical hybrids. The by-products of these types of cells are carbon dioxide and water.

Fuel Cells

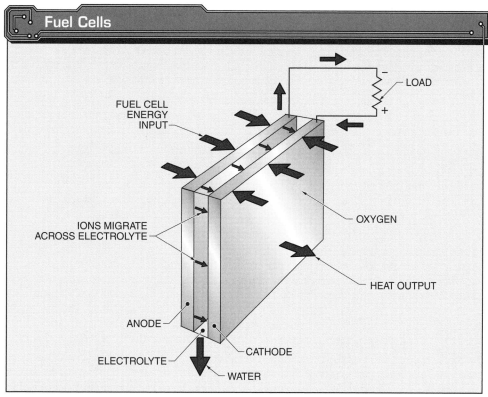

Figure 7-35. *A fuel cell is an energy source that transforms chemical energy from fuel into electrical energy.*

Fuel cells are environmentally friendly. Some parts of the United States have exempted power plants operating on fuel cells from some of the more stringent air permit requirements.

Fuel cells are also rapidly replacing batteries in applications in which a continuous source of electricity must be maintained. For example, the fuel cells in video equipment last longer than batteries and can be replaced or refurbished without having to shut OFF the equipment during a film shooting sequence.

The high reliability of fuel cells provides certain businesses a level of security during power outages. Organizations, such as hospitals, security companies, and 911 emergency centers, could use fuel cells during power outages. Through the use of fuel cells, a facility can continue to operate during a power outage because an independent system continues to power the facility.

A fuel cell consists of two electrodes separated by an electrolyte. Hydrogen fuel is fed into the anode of the fuel cell. Oxygen from the air enters the fuel cell at the cathode. Using a catalyst, the hydrogen splits into protons and electrons. The protons pass directly through the electrolyte. The electrons create a separate current which can be used before returning to the cathode. Once returned to the cathode, the electrons reunite with the hydrogen and oxygen to form a molecule of water.

Fuel cells can be made in a variety of sizes and with varying power outputs. Stationary fuel-cell applications provide power for buildings such as residential and commercial property. Transportation applications range from utility vehicles, golf carts, cars, boats, and buses. Portable applications include laptop computers, cellular phones, and digital cameras. The fuel cell technologies most commonly used include alkaline fuel cell (AFC), phosphoric acid fuel cell (PAFC), proton-exchange-membrane fuel cell (PEMFC), solid oxide fuel cell (SOFC), and zinc-air fuel cell (ZAFC) technology. **See Figure 7-36.**

Fuel Cell Technologies

Fuel Cell	Electrolyte	Operating Temperature (°F)	Efficiency	Applications
AFC	Aqueous potassium-hydroxide	• Low-temp cells 75°F to 170°F • High-temp cells 200°F to 480°F	60% to 70%	Space equipment
PAFC	Molten phosphoric-acid	300°F to 500°F	35% to 45%	Small power plants
PEMFC	Thin, solid polymer membrane	160°F to 180°F	45% to 60%	Auto industry
SOFC	Solid zirconium-oxide	1000°F to 1800°F	60% to 65%	Power plants
ZAFC	Zinc	175°F	35% to 45%	Hearing aids, cameras, auto industry

Figure 7-36. *The most common fuel cells include alkaline fuel cells (AFCs), phosphoric acid fuel cells (PAFCs), proton-exchange-membrane fuel cells (PEMFCs), solid oxide fuel cells (SOFCs), and zinc-air fuel cells (ZAFCs).*

Proton-Exchange-Membrane Fuel Cell Technology

A *proton-exchange-membrane fuel cell (PEMFC)* is a fuel cell that transforms chemical energy from a fuel into electrical energy and consists of a polymer electrolyte membrane that conducts protons (hydrogen ions) but not electrons. PEMFC technology works with a polymer electrolyte in the form of a thin, permeable sheet.

PEMFCs are believed to be the best type of fuel cell as the vehicular power source that will eventually replace gasoline and diesel internal combustion engines. Compared to other types of fuel cells, PEMFCs generate more power for a given volume or weight of fuel cell.

To speed the reaction of a PEMFC, a platinum catalyst is used on both sides of the membrane. Hydrogen atoms are stripped of their electrons, or "ionized," at the anode, and the positively charged protons diffuse through one side of the porous membrane and migrate toward the cathode. The electrons pass from the anode to the cathode through an exterior circuit and provide electric power along the way. At the cathode, the electrons, hydrogen protons, and oxygen from the air combine to form water.

The heart of the cell is the solid proton exchange membrane. The membrane is surrounded by a diffusion layer and a reaction layer. Under constant supply of hydrogen and oxygen, the hydrogen diffuses through the anode and the diffusion layer up to the platinum catalyst, the reaction layer.

One disadvantage of PEMFCs is that because the fuel used must be purified and a platinum catalyst is used on both sides of the cell membranes, the cost of PEMFCs is high. An advantage is that the solid, flexible electrolyte will not leak or crack. Also, PEMFCs operate at a low enough temperature to make them suitable for homes and cars.

Cogeneration

Cogeneration, also known as combined heat and power (CHP), is the use of a power source to generate electricity and useful heat at the same time. Compared to conventional generation, in which an electrical grid supplies the electricity, on-site cogeneration provided by fuel cells offers greater efficiency by producing both heat and electricity. Systems, such as residential fuel cell systems and cogeneration systems for office buildings and factories, are now in mass production. The fuel cell system generates constant electric power and produces hot air and water from the waste heat.

Refer to Chapter 7 Quick Quiz® on CD-ROM

KEY TERMS

- photovoltaic cell
- photovoltaic effect
- nanotechnology
- photovoltaic system
- inverter
- stand-alone inverter
- utility-interactive inverter
- wind turbine
- grid-connected wind turbine
- stand-alone wind turbine
- converter
- pulse width modulation
- anemometer
- wind vane
- transformer
- electrochemical power sources
- electrochemical cell
- battery
- fuel cell
- cogeneration

REVIEW QUESTIONS

1. What is a photovoltaic cell?
2. What is the photovoltaic effect?
3. Define nanotechnology.
4. What is a photovoltaic system?
5. What is an inverter?
6. What is a stand-alone inverter?
7. What is a utility-interactive inverter?
8. What is a wind turbine?
9. What is a grid-connected wind turbine?
10. What is a stand-alone (off-grid) wind turbine?
11. What is a converter?
12. How does an inverter use pulse width modulation?
13. What is an anemometer?
14. What is a wind vane?
15. What is a transformer?
16. Describe electrochemical power sources.

(continued)

REVIEW QUESTIONS (continued)

17. What is an electrochemical cell?

18. What is a battery?

19. What is a fuel cell?

20. Explain cogeneration.

TRANSDUCER APPLICATIONS AND TROUBLESHOOTING 8

Transducers convert mechanical, magnetic, thermal, optical, and chemical variations into the electrical voltages and currents that drive control systems. Transducers include sensing and detection devices, such as thermistors, piezoelectric sensors, ultrasonic sensors, radiation sensors, photoconductive cells, Hall effect sensors, and gas detectors.

OBJECTIVES
- List and describe the types of thermistors.
- Explain piezoelectricity.
- Explain Hall effect sensors.
- Describe ultrasonic sensors.
- List and describe the types of radars.
- Discuss gas detectors.
- Describe radiation detectors.

TRANSDUCERS

A *transducer* is a device that converts various forms of energy into electrical energy. Transducers provide a valuable link in industrial control systems. They convert mechanical, magnetic, thermal, electrical, light, and chemical variations into electrical voltages and currents. These voltages and currents are used directly or indirectly to drive other control systems. Types of transducers include thermistors, piezoelectric sensors, Hall effect sensors, ultrasonic sensors, radars, gas detectors, and radiation detectors.

THERMISTORS

A *thermistor* is a type of transducer that acts as a temperature-sensitive resistor. The resistance of a thermistor changes with a change of temperature. **See Figure 8-1.** When a match is placed under the thermistor, the resistance of the thermistor decreases and current flow increases. When the match is removed, the resistance increases to its original state (resistance value).

The operation of a thermistor is based on the electron-hole pair theory. As the temperature of the semiconductor increases, the generation of electron-hole pairs increases due to thermal agitation. Increased electron-hole pairs cause a drop in resistance.

Thermistor Types

Thermistors are popular because of their small size. They can be mounted in places that are inaccessible to other temperature-sensing devices. For example, a thermistor can be embedded in the coil of a motor to provide additional thermal protection. Thermistors can also be made into the shapes of beads, disks, washers, and rods. The sizes and shapes of thermistors vary, depending on their application. **See Figure 8-2.**

Transducers are used in electrical systems for residential, commercial, and industrial applications. Common applications include antennas that convert waves of electromagnetic energy into electrical energy and microphones that convert sound energy into signals of electrical energy.

Thermistors

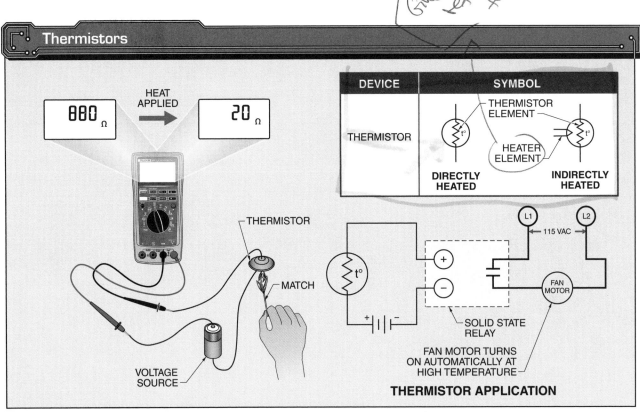

Figure 8-1. A thermistor is a transducer that acts as a temperature-sensitive resistor whose resistance changes with a change in temperature.

Thermistor Types

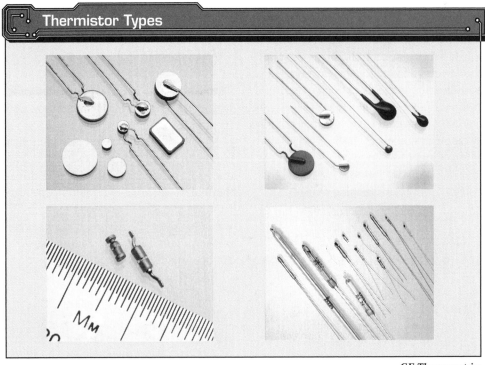

GE Thermometrics

Figure 8-2. Thermistors are tiny temperature-sensitive resistors that are available in many different shapes and sizes.

Thermistor Resistance Values

The resistance value of a thermistor varies from a few ohms (Ω) to the megohm (MΩ) range. Thermistors are linear or nonlinear. In a linear thermistor, the resistance changes the same amount for each degree of temperature change. In a nonlinear thermistor, the resistance varies dramatically for different temperature ranges.

Thermistors are sensitive and can provide fractional degree temperature control. Some thermistors can be accurate to a ±0.1°C change in temperature. Since thermistors are resistive devices, they can also operate on AC or DC.

Thermistor Schematic Symbols

Schematic symbols may be used for a directly heated and an indirectly (externally) heated thermistor. Directly heated thermistor symbols consist of a resistor labeled "t°" with a circle around the resistor and label. Indirectly heated thermistor symbols have an additional element that enters the circle.

Thermistor Classes

The two classes of thermistors are positive temperature coefficient (PTC) and negative temperature coefficient (NTC) thermistors. Although most thermistors operate on an NTC, some applications require operation on a PTC. With a PTC, an increase in temperature causes the resistance of the thermistor to increase. With the NTC, an increase in temperature causes the resistance of the thermistor to decrease.

Cold and Hot Resistance

Cold resistance and hot resistance refer to the operating resistance of a thermistor at temperature extremes. Cold resistance is measured at 25°C, or room temperature. However, some manufacturers specify lower temperatures. The specification sheet should always be checked.

Hot resistance is the resistance of a heated thermistor. In a directly heated thermistor, the heat is generated from the ambient temperature and the current passing through the device. In an indirectly heated thermistor, the heat is generated from the ambient temperature, current, and heating element of the thermistor.

Directly heated thermistors are used in voltage regulators, vacuum gauges, and electronic time-delay circuits. Indirectly heated thermistors are used for precision temperature measurement and temperature compensation. Each type of thermistor is represented by a separate schematic symbol.

NTC Thermistor Applications

The resistance temperature characteristic curve for an NTC shows that as temperature increases, resistance decreases. **See Figure 8-3.** The large swing of resistance in relation to temperature enables the thermistor to become a versatile solid state device. The resistance temperature plot for an NTC thermistor is approximately exponential (nonlinear). Formulas and tables are available from manufacturers that can be used to make specific calculations.

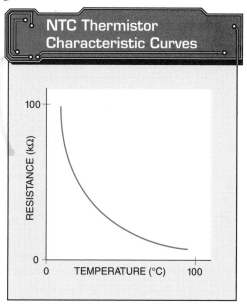

Figure 8-3. *The resistance temperature characteristic curve for an NTC thermistor is approximately exponential (nonlinear). It has a large enough swing of resistance to become a versatile solid state device.*

NTC Thermistors for Fire Protection. A fire alarm circuit is typical of a circuit that requires an NTC thermistor. **See Figure 8-4.** The purpose of this circuit is to detect a fire and sound an alarm. In normal operating environments, thermistor resistance is high because ambient temperatures are relatively low. The high resistance keeps the current to the control circuit low. The alarm remains OFF. However, the increased temperature of a fire lowers the resistance of the thermistor. The lowered resistance allows current to increase, and the alarm is activated.

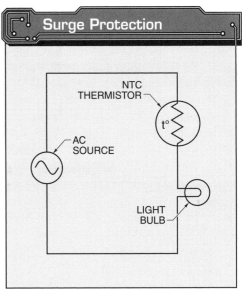

Figure 8-5. *An NTC thermistor acts as a shock absorber for a light bulb by dropping a high percentage of the available voltage across the thermistor. As the thermistor heats, its resistance drops, and more current is available to the light bulb.*

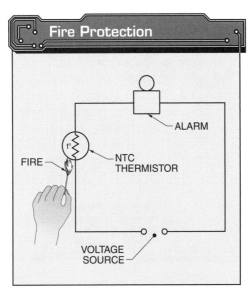

Figure 8-4. *In the presence of fire, the increase in temperature lowers the resistance of an NTC thermistor, which increases current and activates an alarm.*

NTC Thermistors for Surge Protection. An NTC thermistor may be used as a shock absorber for a light bulb (incandescent lamp). **See Figure 8-5.** When the circuit is energized, a high percentage of the available voltage is dropped across the thermistor. As the thermistor heats, its resistance drops, and more current is available to the light bulb. Using this circuitry, many hours of life are added to a light bulb.

NTC Thermistors for Flow Measurement. An NTC thermistor dissipates heat at a constant rate under different environmental conditions and can be used to measure flow. **See Figure 8-6.** Two thermistors may be directly heated in a circuit: one thermistor in the path of flow and the other shielded from flow.

If the flow is rapid, the thermistor in the path of the flow is much cooler than the other thermistor. Because the thermistor is in parallel with R2, the output signal will be large. If the flow is slow, heat is not carried away from the thermistor in the flow path as rapidly. Therefore, the output signal would be smaller. Flow rate sensors may be fixed systems, as in the case of an exhaust system, or portable systems that can be held in the hands of HVAC technicians. **See Figure 8-7.**

NTC Thermistors for Temperature Compensation. An NTC thermistor is commonly used for temperature compensation. When sensing outside ambient air temperatures, the thermistor is wired in series with resistor R1 located inside a thermostat. **See Figure 8-8.** With a decrease in outdoor temperature, the NTC thermistor increases its resistance, allowing less current to flow in the circuit. This reduces thermostat variations to improve the operating characteristics of a typical heat or heat pump system.

Flow Measurement

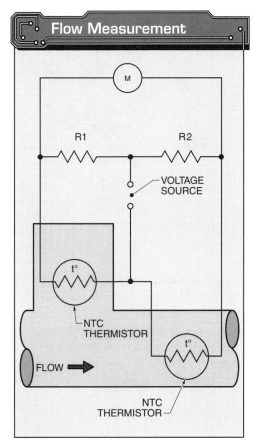

Figure 8-6. *The amount of flow in the circuit determines how rapidly heat is removed from the NTC thermistor. Rapid flow dissipates heat rapidly. Less flow dissipates heat slowly. With a larger flow, there is more resistance and the signal from R2 is larger.*

Most thermistors are made of oxidized metals, which include manganese, nickel, cobalt, and copper. Thermistors are manufactured by heating metal oxide powder in a process called sintering. The sintering process solidifies the powder without melting it.

Flow-Rate Sensors

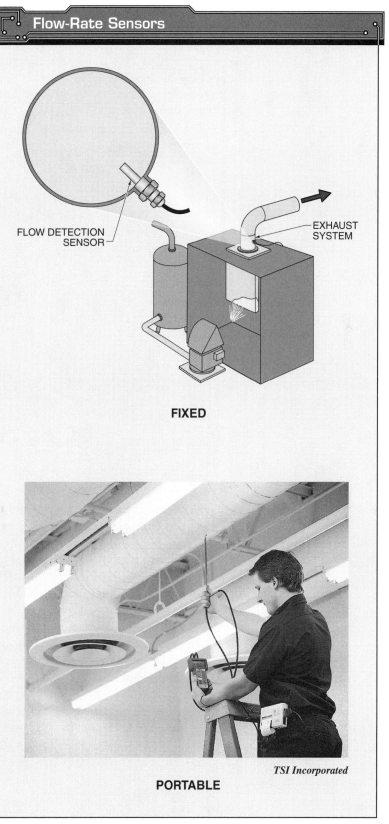

Figure 8-7. *Flow-rate sensors may be fixed, as in the case of an exhaust system, or portable.*

Figure 8-8. *An NTC thermistor used for temperature compensation in a thermostat can reduce the thermostat's variation in control temperatures to very low levels.*

PTC Thermistor Applications

A PTC thermistor is characterized by an extremely large resistance change for a small temperature span. **See Figure 8-9.** The temperature at which the resistance begins to increase rapidly is the switch temperature. This point can be changed from below zero to above 160°C.

The PTC thermistor is similar to a switch that is not perfect but practical for many applications. A thermistor is not a perfect switch because it has some resistance when closed and a large amount of resistance when open. An ideal switch has no resistance when closed and infinite resistance when open.

The PTC thermistor, depending upon the application, usually exhibits relatively low resistance levels during the ON state and relatively high resistance levels during the OFF state. Although current is always flowing in the circuit during the OFF state, the current is usually so low that it is negligible.

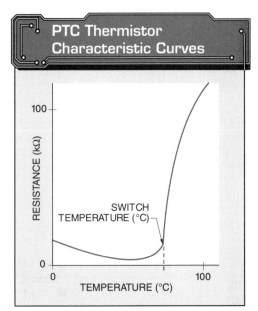

Figure 8-9. *A PTC thermistor has an extremely large resistance change for a small temperature span. The temperature at which the resistance begins to increase rapidly is the switch temperature.*

PTC Thermistors for Motor Starting. A PTC thermistor can replace the start switch in a single-phase (1ϕ) motor. **See Figure 8-10.** When the circuit is energized by switching the ON/OFF switch to the ON position, the PTC thermistor has a low resistance and permits most of the line voltage (AC source) to be applied to the starter winding. As the motor starts, the PTC thermistor heats up until the switch temperature is reached. The temperature to be reached is determined by the thermal inertia of the PTC thermistor and current flowing through the starter winding. Then, the thermistor rapidly changes from a low-resistance device to a high-resistance device.

PTC Thermistors as Current Limiters. The increase in resistance of a PTC thermistor at the switch temperature makes it suitable for current-limiting applications. **See Figure 8-11.** For currents lower than the limiting current, the power generated in the unit is insufficient to heat the PTC thermistor to its switch temperature. However, as the current increases to the critical level, the resistance of the PTC thermistor increases at a rapid rate so that any further increase in power dissipation results in a current reduction. The time required for the PTC thermistor to get into the current-limiting mode is controlled by the heat capacity of the PTC thermistor, its dissipation constant, and the ambient temperature.

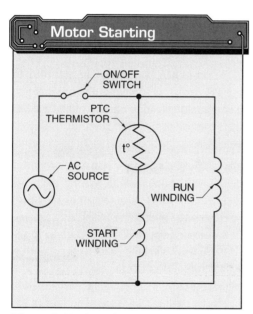

Figure 8-10. A PTC thermistor can be used to replace the troublesome start switch in a single-phase (1ϕ) motor. As the resistance of the PTC thermistor increases, it causes a low current in the start winding to remove the starter winding from the circuit.

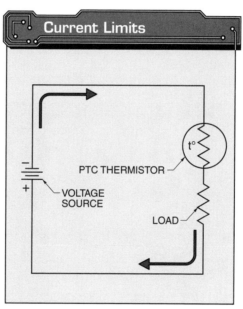

Figure 8-11. When a large current is presented to a load, the resistance of the PTC thermistor increases at a rapid rate so that a reduction in current occurs.

Once the thermistor reaches its high-resistance state, virtually no current flows through the thermistor or the starter winding. Since the current through the starter is negligible, the starter winding can be considered removed from the circuit. *Note:* The PTC thermistor becomes less effective on successive starts because its temperature is higher than the ambient temperature levels.

PTC Thermistors for Arc Suppression. A PTC thermistor can switch from a low resistance to a high resistance when the switch is open. **See Figure 8-12.** The low resistance of the PTC provides effective arc suppression. In addition, the PTC switching action transfers essentially all of the power supply voltage from the load to the PTC thermistor itself.

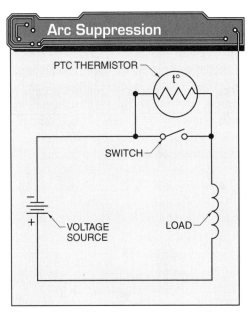

Figure 8-12. *A PTC thermistor provides arc suppression. It absorbs the voltage from an inductive load by rapidly switching from a low resistance to a high resistance once the switch is open.*

Boilers use transducers to measure the steam pressure and temperature, feed water flow and conductivity, boiler water level, fuel flow and pressure, and composition of the stack gas.

Troubleshooting Thermistors

A thermistor must be properly connected to an electronic circuit. A loose or corroded connection creates a high resistance in series with the thermistor resistance. The control circuit may sense the additional resistance and display a false temperature reading. The hot and cold resistance of a thermistor can be checked with a DMM. **See Figure 8-13.** To test the hot and cold resistance of a thermistor, apply the following procedure:

1. Remove the thermistor from the circuit.
2. Connect the DMM leads to the thermistor leads, and place the thermistor and a thermometer in ice water. Record the temperature and resistance readings.
3. Place the thermistor and thermometer in hot water (not boiling). Record the temperature and resistance readings.
4. Compare the hot and cold readings with the manufacturer specification sheet or with a similar thermistor that is known to be good.

Note: When water may be harmful to the PC board or surrounding components, a temperature-testing unit should be used as the standard temperature source.

Temperature Problems. Heat is produced in all electrical systems. Heat may be produced deliberately, such as with heating elements and heat lamps. Heat may also be the result of electricity flowing through electrical distribution equipment, such as conductors, transformers, and switching gear. Heat is produced when current flows through a resistance. The higher the current, the greater the temperature produced at the resistance. The higher the resistance of the conductors of switching gear, the greater the temperature produced. Technicians can use temperature measurements to identify problems or potential problems in electrical systems by measuring the heat produced by undersized conductors, switching gear, and transformers.

> *Thermistors can be encapsulated in glass beads for added protection. Glass-encapsulated thermistors are rugged, low-cost thermistors that operate in a wide range of temperatures and are sealed for protection against moisture.*

Testing Thermistors

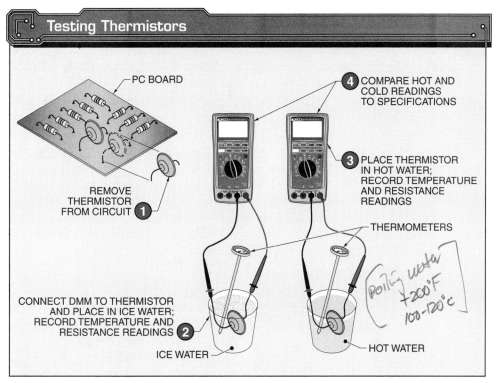

Figure 8-13. *The hot and cold resistance of a thermistor can be checked with a DMM.*

Surface Leakage Current. Surface leakage current is current that flows from areas on conductors where insulation is removed to allow electrical connections. Conductors are terminated to wire nuts, splices, spade lugs, terminal posts, and other fastening devices at different points along an electrical circuit. The point at which the insulation is removed provides a low resistance path for surface leakage current. Dirt and moisture allow additional surface leakage current to flow. Surface leakage current results in increased heat at a connection. Increased heat contributes to insulation deterioration and makes a conductor brittle. Surface leakage current is minimized by making good, tight, clean connections.

PIEZOELECTRIC SENSORS

Piezoelectricity is the voltage generated from the application of pressure. The *piezoelectric effect* is the electrical polarization of some materials when mechanically strained. A *piezoelectric sensor* is a transducer that operates based on the interaction between the deformation of certain materials and an electric charge. For example, an electric conversion transducer may change an electric signal from one frequency to another frequency. Specific frequencies can either be passed to a load or rejected.

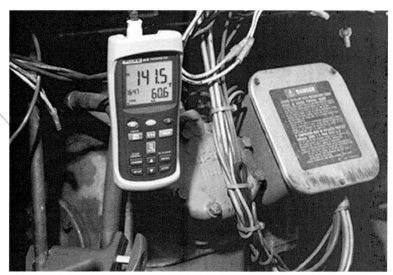

Fluke Corporation

Thermocouples are used in applications that require direct-contact temperature measurement. For example, a digital thermometer uses an external thermocouple as a temperature sensor.

The voltage potential produced by the pressure on a crystal is directly proportional to the amount of strain applied. If a crystal is stretched, the voltage potential across the crystal changes polarity. For example, the points on a crystal that are positive when pressure is applied become negative when the crystal is stretched. **See Figure 8-14.** Electrons flow in one direction in a circuit connected to a crystal when the crystal is decompressed. Also, the electron flow within the crystal in both cases is from the positive terminal to the negative terminal.

The positive terminal has a deficiency of electrons. The negative terminal has an excess of electrons. If either the compression or decompression force is held constant, electron flow continues until these two charges are equalized, and then it ceases.

Pressure can be applied to a crystal by compression, stretching, twisting, or bending. The best method to use for applying pressure depends on the crystal material. Some materials respond best to bending pressure, while others respond best to twisting pressure. The force of the pressure acts on the atoms of the crystal to force the electrons out of their orbits.

Piezoelectric Sensor Applications

Piezoelectric sensors are used in microphones, radio station transmitters and receivers, ultrasound equipment, strain and pressure gauges, and piezoelectric motors. For example, a piezoelectric (crystal) microphone converts the pressure of sound waves into electric signals. **See Figure 8-15.** The pressure from the sound waves causes the crystal material within the microphone to compress and decompress, producing the signals. If an audio electric signal is applied to the crystal, the microphone becomes an earphone. In the earphone application, the applied electric signal causes the crystal to compress and decompress, producing sound.

Piezoelectricity is also used in strain and pressure gauges. Strain and pressure gauges consist of very fine electrical wires attached to the crystal. When the crystal is compressed or stretched, the potential pressure is proportional to the force and can be measured with a voltmeter. Strain and pressure gauges are used to measure the vibration of machinery and structures.

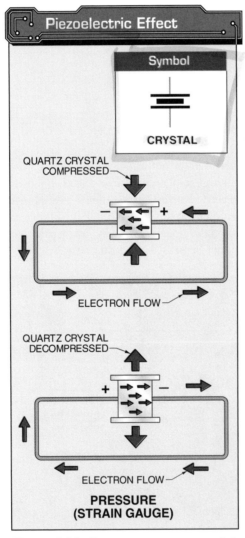

Figure 8-14. *Pressure on some crystals generates a surface charge on the crystal. The amount of charge is proportional to the amount of pressure.*

> *In accordance with the RoHS and WEEE directives, lead-free piezoelectric ceramics are being considered as alternatives to lead-based piezoelectric materials for high-frequency ultrasonic transducer applications.*

Chapter 8—Transducer Applications and Troubleshooting   **203**

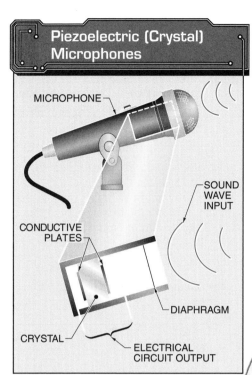

Figure 8-15. *A piezoelectric (crystal) microphone converts the pressure of sound waves into electric signals.*

HALL EFFECT SENSORS

A *Hall effect sensor* is a transducer that detects the proximity of a magnetic field. The output of a Hall generator depends on the presence of a magnetic field and the current flow in the Hall generator. Hall effect sensors are used in many products, such as computers, sewing machines, automobiles, aircraft, machine tools, and medical equipment.

Hall Effect Principle

A constant control current passes through a thin strip of semiconductor material (Hall generator). When a permanent magnet is brought near, a small voltage called Hall voltage appears at the contacts that are placed across the narrow dimension of the strip. As the magnet is removed, the Hall voltage is reduced to zero. Thus, the amount of Hall voltage is dependent on the presence of a magnetic field and on the current flowing through the Hall generator. If either the current or the magnetic field is removed, the output of the Hall generator is zero. In most Hall effect sensors, the control current is held constant, and the flux density is changed by the movement of a permanent magnet. **See Figure 8-16.**

Note: The Hall generator must be combined with a group of electronic circuits to form a Hall effect sensor. Since all of this circuitry is usually on an integrated circuit (IC), the Hall effect sensor can be considered a single device with a voltage output.

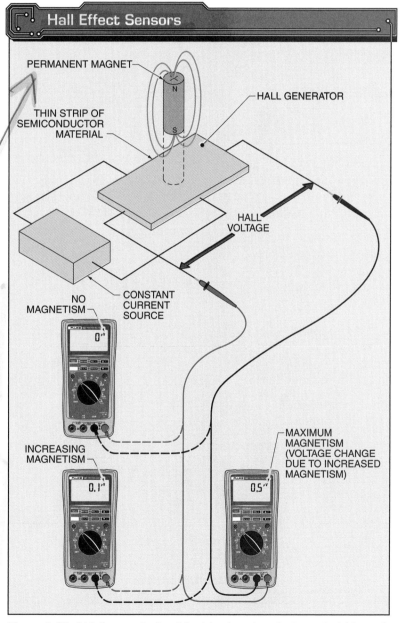

Figure 8-16. *A Hall generator is a thin strip of semiconductor material through which a constant control current is passed.*

Hall Effect Sensor Packaging

Hall effect sensors are packaged in different ways and chosen according to the type of job they are to perform. Five of the most common packaging arrangements include vane, plunger, current sensor, proximity, and cylindrical. **See Figure 8-17.**

Hall Effect Sensor Applications

Hall effect sensors are used in a variety of commercial and industrial applications. Their small size and ruggedness make them ideal for many sensing jobs. For example, Hall effect sensors can be embedded in the human heart as a timing element. Depending on the output requirements for each application, the following Hall effect sensor functions can be accomplished by using either a digital or linear output transducer:

- under- or overspeed detection
- disk speed detection
- automobile or tractor transmission control
- shaft rotation counting
- bottle counting
- camera shutter positioning
- rotary position sensing

RPM Sensors. An RPM sensor is one of the most common applications of a Hall effect sensor. The magnetic flux required to operate the sensor may be furnished by a ring magnet or by multiple magnets mounted on the shaft or hub. Each change in polarity results in an output signal. **See Figure 8-18.**

Liquid Level Sensors. There are two methods for using a Hall-effect sensor to measure the level of liquids in a tank. One method uses a notched tube with a cork floater inserted into a tank. **See Figure 8-19.** The magnet is mounted in the float assembly, which is forced to move in one plane (up and down). As the liquid level goes down, the magnet passes the digital output sensor. When the sensor is actuated, it could indicate a low level condition.

The second method uses a linear output sensor and a float in a tank made of aluminum. As the liquid level goes down, the magnet moves closer to the sensor, causing an increase in output voltage. This method allows the measuring of liquid levels without any electrical connections inside the tank.

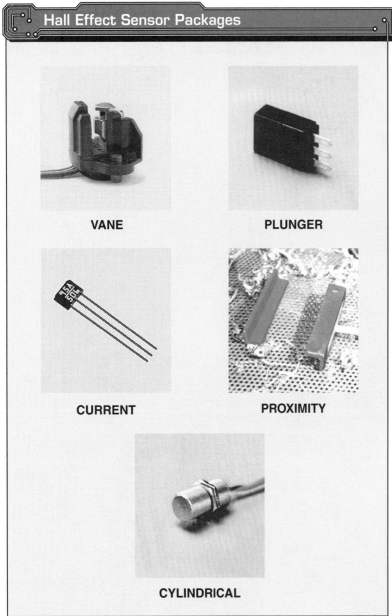

Honeywell

Figure 8-17. Hall effect sensors are available in a variety of packages for different applications.

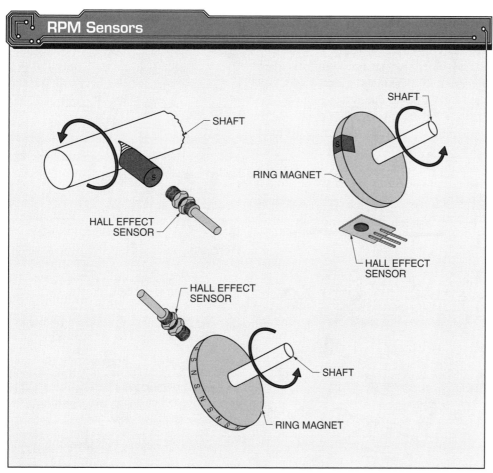

Figure 8-18. *Each change in polarity results in an output from the Hall effect sensor used in an RPM sensor application.*

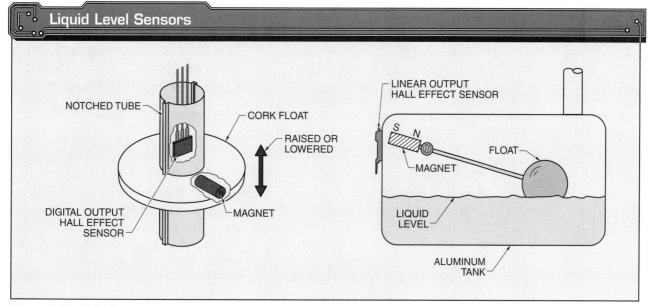

Figure 8-19. *Hall effect sensors can be used to measure liquids in a tank.*

Magnetic Card Readers. A door-interlock security system can be designed using a Hall effect sensor, a magnetic card, and the associated electronic circuitry. **See Figure 8-20.** The magnetic card slides by the sensor and produces an analog output signal. This analog signal is converted to a crisp digital signal to energize the door latch relay. When the solenoid of the relay pulls in, the door is opened. For systems that require additional security measures, a series of magnets may be molded into the card.

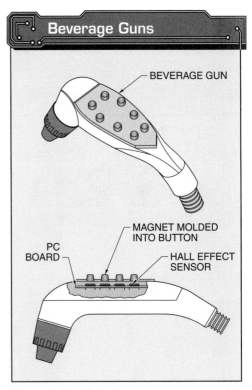

Figure 8-21. *Hall effect sensors are used in beverage gun applications because of their small size and reliability.*

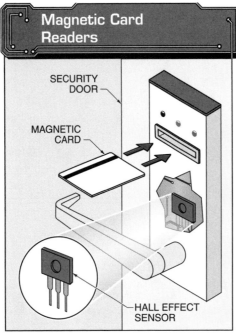

Figure 8-20. *A door-interlock system can be designed using a Hall effect sensor, a magnetic card, and the associated electronic circuitry.*

Beverage Guns. Hall effect sensors are used in beverage gun applications because of their small size and reliability. **See Figure 8-21.** Hall effect sensors cannot be contaminated by solid foods, syrups, or other liquids because they are completely enclosed in the beverage gun. The beverage gun is completely submersible in water for easy cleaning and requires little maintenance.

Tilt Sensors. Hall effect sensors may be installed in the base of a machine to indicate its level or degree of tilt. **See Figure 8-22.** Magnets are installed above the Hall effect sensors in a pendulum fashion. The machine is level as long as the magnets remain directly over the sensors. A change in the state of output (when a magnet swings away from a sensor) is indication that the machine is not level. The sensor/magnet combination may also be installed in such a manner as to indicate the degree of tilt.

Joysticks. Hall effect sensors may be used in a joystick application. **See Figure 8-23.** In this application, the Hall effect sensors inside the joystick housing are actuated by a magnet on the joystick. The proximity of the magnet to the sensor controls the activation of different outputs used to control cranes, operators, motor control circuits, wheelchairs, etc. The use of an analog device also achieves degree of movement measurements, such as speed.

Tilt Sensors

David White Instruments
Figure 8-22. Hall effect sensors may be installed in the base of a machine to indicate the level or degree of tilt.

Joysticks

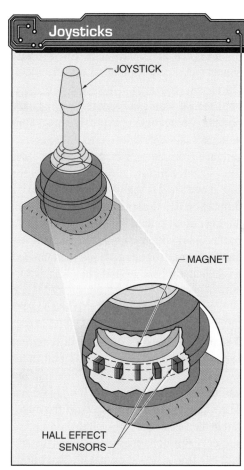

Figure 8-23. Hall effect sensors may be used in a joystick application.

Paper Sensors. A Hall effect plunger sensor can be used in printers to detect the level of paper in the metal tray as it drops. **See Figure 8-24.** Hall effect sensors have extremely long life and low maintenance costs, have no contacts to become gummy or corroded, and may be directly interfaced with logic circuitry.

Paper Sensors

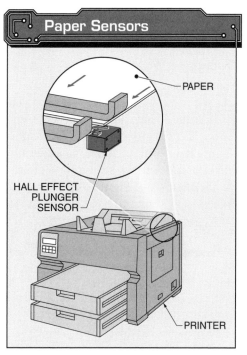

Figure 8-24. A Hall effect plunger sensor can be used in printers to detect paper flow.

ULTRASONIC SENSORS

An *ultrasonic sensor* is a transducer that can detect the presence of an object by emitting and receiving high-frequency sound waves. Ultrasonic sensors can provide an analog voltage output or a digital voltage output (switched output). The high-frequency sound waves are typically in the 200 kHz range. Ultrasonic sensors are used to detect solid and liquid objects at a distance of up to approximately 1 m (3.3′). Ultrasonics used in the security field for motion detection operate in the 40 kHz frequency band.

Ultrasonic sensors can be used to monitor the level in a tank, detect metallic and nonmetallic objects, and detect other objects that easily reflect sound waves. Soft materials, such as foam, fabric, and rubber, are difficult for ultrasonic sensors to detect. They are better detected by photoelectric or proximity sensors.

Ultrasonic sensors are used to detect clear objects (glass and plastic), which are difficult to detect with photoelectric sensors. For this reason, ultrasonic sensors are ideal for applications in the food and beverage industry or for any application that uses clear glass or plastic containers.

Operating Modes

The two basic operating modes of ultrasonic sensors are the direct mode and the diffused mode. In the direct mode, an ultrasonic sensor operates like a direct scan photoelectric sensor. In the diffused mode, an ultrasonic sensor operates like a scan diffuse photoelectric sensor. **See Figure 8-25.**

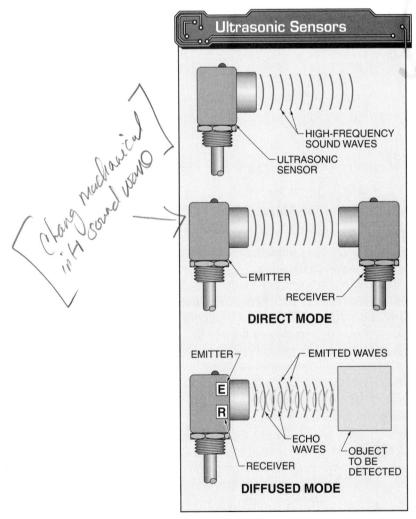

Figure 8-25. *Ultrasonic sensors detect objects by bouncing high-frequency sound waves off an object.*

Direct mode is a method of ultrasonic sensor operation in which the emitter and receiver are placed opposite each other so that the sound waves from the emitter are received directly by the receiver. Ultrasonic sensors used in the direct mode usually include an output that is activated when a target is detected. The output is normally a transistor that can be used to switch a DC circuit. Outputs are available in normally open and normally closed switching modes. Ultrasonic sensors include a means of adjustment for tuning the sensing distance of the sensor. Tuning the receiver to the emitter minimizes interference from ambient noise.

Diffused mode is a method of ultrasonic sensor operation in which the emitter and receiver are housed in the same enclosure. In the diffused mode, the emitter sends out a sound wave and the receiver listens for the sound wave echo bouncing off an object. Ultrasonic sensors used in the diffused mode may include a digital output or an analog output. The analog output provides an output voltage that varies linearly with the target's distance from the sensor. **See Figure 8-26.** The sensor typically includes a light emitting diode (LED) that glows with intensity proportional to the strength of the echo. The analog output sensor includes an adjustable background-suppression feature that allows the sensor to better detect only the intended target and not background objects.

RADAR SYSTEMS

Radar level sensors are transducers that use high-frequency (about 10 GHz) radio waves. These radio waves are aimed at the surface of the material in the storage vessel. The radio waves are reflected off the material in the vessel and returned to the emitting source. Common types of radar systems are pulsed, frequency modulated continuous wave, and guided wave radar systems.

Pulsed Radar

A *pulsed radar level sensor* is a level-measuring sensor consisting of a radar generator that directs an intermittent pulse with a constant frequency toward the surface of the material in a vessel. The time it takes for the pulse to travel to the surface of the material, reflect off the material, and return to the source is a function of the distance from the sensor to the material surface. **See Figure 8-27.** The two common types of antennae used to emit either pulses or continuous waves are in the form of a cone or a rod. A cone antenna is larger and sturdier than a rod and less subject to material buildup or condensation. Rod antennae are smaller and less expensive, making them more suitable for use in smaller vessels.

varies its frequency between a minimum and maximum value, a receiver that detects the signal, and electronics that measure the frequency difference between the signal and the echo. Bandwidth is the range of frequencies from the minimum to the maximum value in an FMCW radar level sensor. Sweeptime is the constant time for an FMCW emitter to vary the frequency from the lowest frequency to the highest.

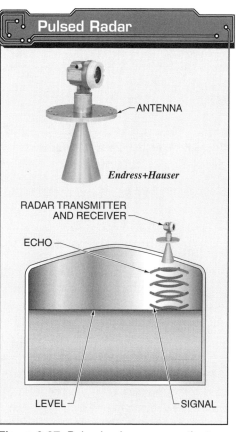

Figure 8-27. Pulsed radar measures the transit time from the transmitter to the surface of a material to determine level.

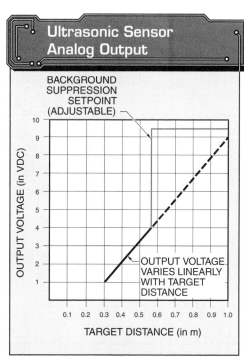

Figure 8-26. An ultrasonic sensor used in the diffused mode can provide an analog output that varies linearly with the target's distance from the sensor.

Frequency Modulated Continuous Wave Radar

A *frequency modulated continuous wave (FMCW) radar* is a level-measuring sensor consisting of an oscillator that emits a continuous microwave signal that repeatedly

The microwave signal travels to the surface of the material in the vessel and is reflected back to the emitter. The reflected echo signal has a different frequency than the emitted signal that is being generated at that instant. **See Figure 8-28.** These differences vary directly with the distance between the emitter and the surface of the material in the vessel. This permits the radar detector to be calibrated in units of level.

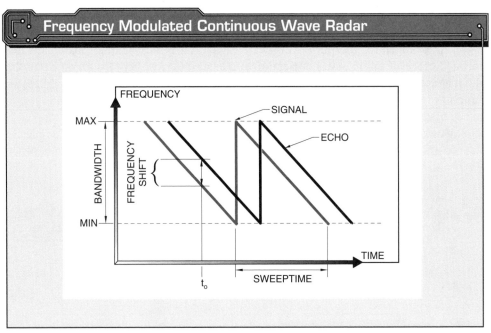

Figure 8-28. *Frequency modulated continuous wave (FMCW) radar measures the frequency difference between the radar signal and the echo to determine level.*

Guided Wave Radar

A *guided wave radar* is a level-measuring detector consisting of a cable or rod as a wave carrier extending from an emitter down to the bottom of a vessel and electronics to measure the transit time. A time domain reflectometer (TDR) is another name for guided wave radar. The material in the vessel reflects, or echoes, some of the microwave energy at the point where the carrier and material make contact. **See Figure 8-29.** The transit time is measured and used to calculate the level. This guiding reduces the effect of dust above granular solids as well as the turbulence in some liquids.

Radar Applications

Radar level-measuring instruments depend on the signals being reflected from the surface of materials back to the receivers for proper operation. It is difficult to use radars to measure the levels of materials that absorb, rather than reflect, a radar signal. Water and most other liquids are good reflectors. They do not absorb much microwave energy. Many other materials do absorb this energy and are harder to measure.

The dielectric constant of a material determines the degree of absorption and therefore, the strength of the reflected wave. For materials with a dielectric constant less than 1.5, it is difficult to measure level with a radar sensor. For materials with a dielectric constant between 1.5 and 2.0, it is usually possible to measure level with a radar sensor. Some materials with a low dielectric constant require a stilling well to help focus the beam.

Radar level sensors must be calibrated upon installation and periodically thereafter. Calibration is fairly simple since the sensors measure the time of flight or a frequency shift of the signal. The distance and the velocity of the radar signal determine the time of flight. Since radar is an electromagnetic pulse, the velocity of the radar signal is fixed at the speed of light. Any changes in the time of flight are related to the distance the signal has to travel.

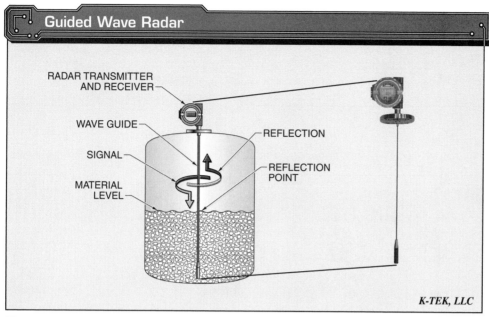

Figure 8-29. *Guided wave radar uses a wave guide to direct a radar signal down a wire.*

Since radar units are mounted at the top of a vessel or tank, there are few problems with tank integrity and product leakage, buildup on the sensors, or errors due to changes in the product. However, mixer blades and other objects in the vessel can cause false echoes, while dust in the air above a granular solids surface may absorb the radar signal. It is important to install and aim the radar continuous level sensor correctly to eliminate or minimize the possibility of error.

GAS DETECTORS

A *gas detector* is a transducer that detects gases or vapors. Technicians are sometimes required to work in confined spaces for certain tasks. A confined space is an area that has limited openings for entry and/or unfavorable natural ventilation that could cause a dangerous air contamination condition. The dangers in a confined space include excessively hot/cold temperatures and elevated noise, toxic gas, combustible gas, and dust levels. When a technician is required to work in a confined space (or any area of concern), test instruments that can measure the environmental conditions in a confined space must be used. **See Figure 8-30.**

Gas detectors are available that measure in parts per million (ppm) and display (with an LCD and alarms) the levels of gases. Gases that are typically measured include the following:

- carbon monoxide (CO)
- oxygen (O_2)
- hydrogen sulfide (H_2S)
- ammonia (NH_3)
- sulfur dioxide (SO_2)
- methane (CH_4)
- chlorine (Cl_2)
- nitrogen dioxide (NO_2)
- hydrogen cyanide (HCN)

Gas detectors are available as single or multiple gas detectors. Single gas detectors typically detect the more common gases that cause problems, such as carbon monoxide (CO). Today, gas detectors that measure three, four, or five gases are becoming more popular and can be used for uncommon gases.

Gas Detectors

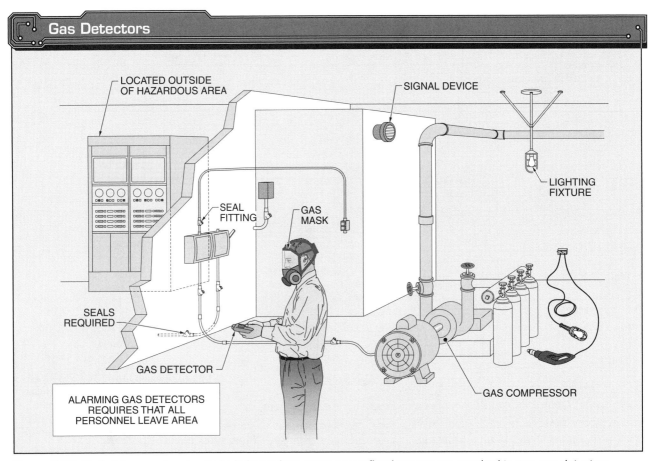

Figure 8-30. *Technicians performing tasks in hazardous areas or confined spaces are required to use gas detectors.*

A *toxic gas detector* is a hazardous atmosphere detector used to measure toxic gases or vapors that are not combustible but are harmful to people. **See Figure 8-31.** This category of harmful gases and vapors includes chlorine, phosgene, ammonia, and carbon monoxide. The sensor must be selected and installed to be responsive to a specific chemical. Toxic gas detectors are used in and around areas in a plant where there is a likelihood of an inadvertent release of a toxic gas or vapor to the atmosphere. Early detection is imperative in order to determine the size of the release and the direction of drift so that corrective actions can be taken.

A *permissible exposure limit (PEL)* is a regulatory limit on the amount of workplace exposure to a hazardous chemical. The Occupational Safety and Health Administration (OSHA) establishes a PEL based on an 8-hour, time-weighted average exposure. Each chemical is evaluated for toxicity and a PEL is established.

Toxic gas detectors are available in several designs. A simple and relatively inexpensive toxic gas analyzer uses a sensing cell with an inert dispersal barrier or flame arrestor with an internal chemical sensitive to the specific hazardous gas or vapor. The toxic gas or vapor can migrate through the barrier and activate the cell. The cell is set for a specific concentration range of the toxic material and the output is proportional to the concentration. This design can be permanently disabled by exposure to high concentrations of the toxic material and then the cell must be replaced.

Chapter 8—Transducer Applications and Troubleshooting 213

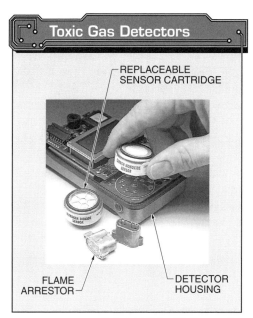

Industrial Scientific Corporation

Figure 8-31. *A toxic gas detector uses replaceable sensor cartridges to detect different gases.*

air conditioning, refrigeration, or process systems. **See Figure 8-32.** Leak detection is performed by using leak detection methods or one of many different types of leak detectors available, such as electronic, fluorescent, ultrasonic, fixed, or halide torch leak detectors.

Another design uses a chemical reagent that is placed in contact with a sample gas or vapor in a detector cell. A pump is used to carry an atmospheric sample into the analyzer. The reagent is specific for the selected hazardous gas or vapor and reacts with the gas if it is present. The cell generates an electrical signal dependent on the concentration of the toxic gas or vapor. The reagent needs to be replaced periodically to ensure proper operation of the analyzer.

Gas and Refrigerant Leak Detection

Gases are kept under pressure in storage tanks when not being used in a system. In any system under pressure, leaks can develop. Large leaks are quickly detected by smell or loss of process requirements (cooling). Small leaks can be much harder to detect.

Leak detection test instruments can be used to detect large and small leaks. A leak detector is a device that is used to detect refrigerant or other gas leaks in pressurized

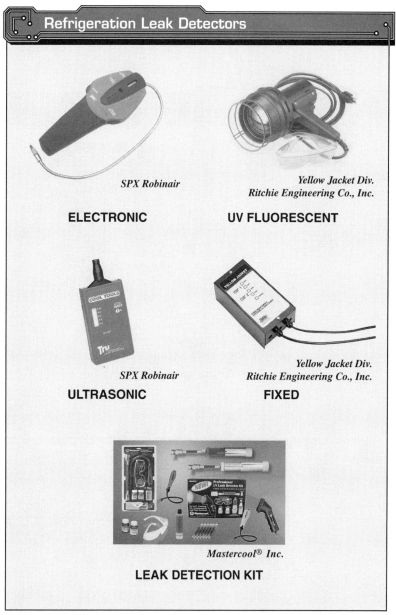

Figure 8-32. *A leak detection device detects refrigerant leaks in air conditioning, refrigeration, or process systems.*

Smoke and Gas Switches

Smoke and gas switches detect vapor. **See Figure 8-33.** *Vapor* is a gas that can be liquefied by compression without lowering the temperature. A *smoke switch (smoke detector)* is a switch that detects a set amount of smoke caused by smoldering or burning material and activates a set of electrical contacts. A smoke switch (smoke detector) is used as an early-warning device in fire protection systems. It is available as a self-contained unit, such as the common unit used in most houses, or as an industrial unit used as part of a large fire protection system.

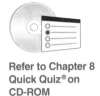

Refer to Chapter 8 Quick Quiz® on CD-ROM

A *gas switch* is a switch that detects a set amount of a specified gas and activates a set of electrical contacts. Many gas detectors have interchangeable sensor units that can detect different groups of gases. For example, one common gas sensor is designed to detect gases such as propane and butane but not carbon monoxide and smoke. This gas detector is used in applications such as detecting leaks in areas that have (or fill) propane tanks. Another sensor is designed to detect toxic gases such as carbon monoxide and ammonia. This gas detector is used in applications such as detecting high carbon monoxide levels in an area with operating combustion engines.

RADIATION DETECTORS

A *radiation detector* is a transducer that detects radioactive materials. Technicians are sometimes required to work in areas that contain radioactive materials. Radioactive materials are found in facilities such as nuclear power plants, hospitals, research labs, and nondestructive inspection areas. **See Figure 8-34.** All technicians in areas containing radioactive materials are required to monitor radiation levels at all times. Emergency response agencies must also have radiation detectors and personnel must have knowledge of their use.

The types of radiation that can be found in facilities are electromagnetic radiation, background radiation, and nuclear radiation. Nuclear radiation is radiation created by radioactive atoms. High levels of nuclear radiation are known to cause radiation sickness or death in humans. Radiation detectors are used to measure the levels of radiation, record radiation levels over time, and warn of danger when levels exceed a specific threshold.

> *Common smoke switches include photoelectric detectors and ionization detectors. An alarm is triggered when smoke disrupts a light source in a photoelectric detector or current between two electrodes in an ionization detector.*

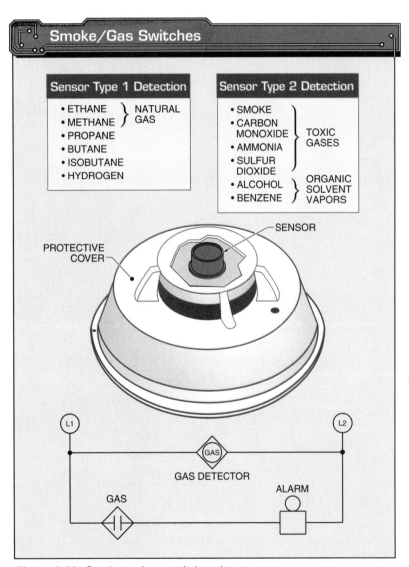

Figure 8-33. *Smoke and gas switches detect vapor.*

Chapter 8—Transducer Applications and Troubleshooting

Radiation Detectors and Meter Measurement

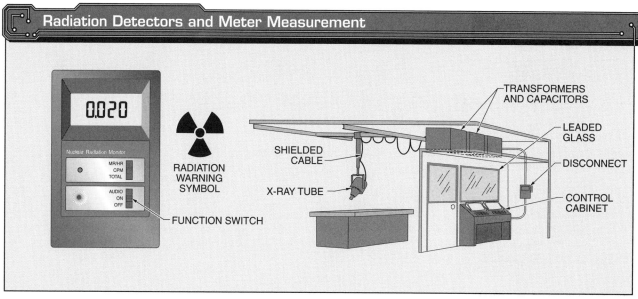

Figure 8-34. *Radiation detectors are required in facilities having radioactive materials or when performing tasks on instrumentation containing radioactive materials.*

KEY TERMS

- transducer
- thermistor
- piezoelectricity
- piezoelectric effect
- piezoelectric sensor
- Hall effect sensor
- ultrasonic sensor
- direct mode
- diffused mode
- pulsed radar level sensor
- frequency modulated continuous wave (FMCW) radar
- guided wave radar
- toxic gas detector
- permissible exposure limit (PEL)
- vapor
- smoke switch (smoke detector)
- gas switch (gas detector)
- radiation detector

REVIEW QUESTIONS

1. What is a transducer?
2. What is a thermistor?
3. Describe the types of thermistors.
4. Describe the thermistor classes.
5. What is cold and hot resistance?
6. Explain how thermistors are tested.
7. What is piezoelectricity?
8. What is the piezoelectric effect?
9. What is a piezoelectric sensor?
10. What is a Hall effect sensor?
11. What is an ultrasonic sensor?
12. Explain direct and diffused modes.
13. What is a pulsed radar level sensor?
14. What is a frequency modulated continuous wave (FMCW) radar?
15. What is a guided wave radar?
16. Describe toxic gas detectors.
17. What is a smoke switch (smoke detector)?
18. What is a gas switch (gas detector)?

BIPOLAR JUNCTION TRANSISTORS (BJTs) 9

Bipolar junction transistors are three-terminal electronic devices constructed of semiconductor material and may be used in switching or amplifying applications. The term "bipolar" is used because the operation of the transistor involves both electrons and holes. The bipolar junction transistor was invented at Bell Labs by William Shockley in 1948.

OBJECTIVES

- Explain the operation of bipolar junction transistors (BJTs).
- Describe biasing transistor junctions.
- Explain transistor operating characteristic curves.
- Explain transistors as DC switches.
- Discuss biasing transistors.
- Explain power dissipation.
- Describe procedures for testing transistors.
- List and describe transistor switching applications.

BIPOLAR JUNCTION TRANSISTORS

A *bipolar junction transistor (BJT)* is a three-terminal electronic device that switches or amplifies electrical signals. The two types of BJTs are NPN transistors and PNP transistors. **See Figure 9-1.** An NPN transistor is formed by sandwiching a thin layer of P-type material between two layers of N-type material. A PNP transistor is formed by sandwiching a thin layer of N-type material between two layers of P-type material.

The three terminals are the emitter (E), base (B), and collector (C). The terminals are in the same location for both types of BJTs. The only difference between the two symbols for the types of transistors is the direction in which the emitter arrow points. The emitter arrow points from the P-type material toward the N-type material in both cases. This means that the emitter arrow points toward the base in a PNP transistor and the emitter arrow points away from the base in an NPN transistor. BJTs are referred to as bipolar because both the holes and electrons are used as internal carriers for maintaining electron flow.

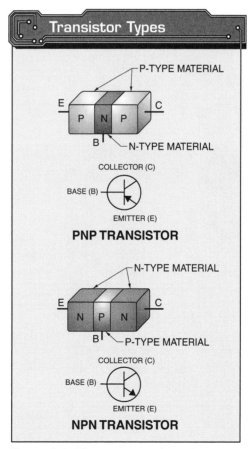

Figure 9-1. *The two types of transistors are PNP and NPN.*

Bipolar Junction Transistor Operation Media Clip

217

Transistors are manufactured with either two or three terminals extending from their case. **See Figure 9-2.** These packages are accepted industry-wide regardless of manufacturer. When a transistor with a specific shape must be used, a transistor outline (TO) number is used for reference. A *transistor outline (TO) number* is a number determined by the manufacturer that represents the shape and configuration of a transistor. **See Figure 9-3.** *Note:* The bottom view of transistor TO-3 shows only two leads (terminals). Typically, transistors use the metal case as the collector-pin lead.

Spacing can also be used to identify leads. Typically, the emitter and base leads are close together, and the collector lead is farther away. The base lead is usually in the middle. A transistor with an index pin must be viewed from the bottom. The leads are identified clockwise from the index pin. The emitter is closest to the index pin. **See Figure 9-4.** *Note:* For detailed information on transistor construction and identification, refer to a transistor manual or the manufacturer's specification sheets.

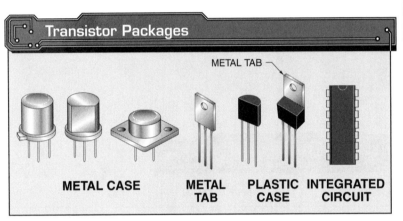

Figure 9-2. *Transistors have either two or three leads extending from their case.*

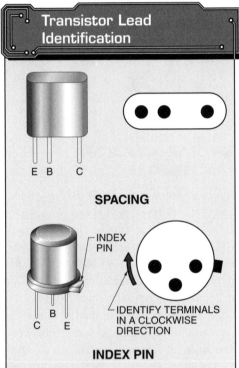

Figure 9-4. *Transistor leads can be identified by lead spacing or an index pin.*

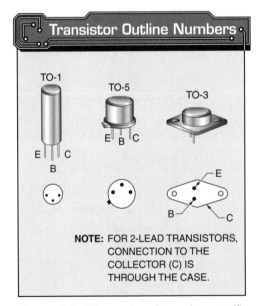

Figure 9-3. *When a transistor of a specific shape must be used, its TO number is used as a reference.*

The NPN transistor is by far the most popular type of transistor in use. One of the main reasons that the NPN transistor is popular is because a majority of power supplies have an output voltage which is positive with respect to ground. The schematic symbol and manufacturer specification sheets can help a technician identify NPN and PNP transistors.

BIASING TRANSISTOR JUNCTIONS

In any transistor circuit, the base-emitter junction must always be forward biased and the collector-base junction must always be reverse biased. The external voltage (bias voltage) is connected so that the positive terminal connects to the P-type material (the base), and the negative terminal connects to the N-type material (the emitter). **See Figure 9-5.** This configuration forward biases the base-emitter junction. Electrons flow in the external circuit as indicated by the arrows. The action that takes place is the same action that occurs for the forward-biased semiconductor diode.

The external voltage is connected so that the negative terminal connects to the P-type material (the base) and the positive terminal connects to the N-type material (the collector). **See Figure 9-6.** This configuration reverse biases the collector-base junction. Only a very small amount of current electrons (leakage current) flows in the external circuit as indicated by the dashed arrows. The action that takes place is the same action that occurs for a reverse-biased diode.

Modern building automation system controllers use different transistors for circuit operation.

Transistor Current

Individual PN junctions can be used in combination with two bias configurations. The base-emitter junction is forward biased while the collector-base junction is reverse biased. This circuit configuration results in an entirely different current path from the path that occurs with the individual circuits only. The main current path is directly through the transistor. It has a small amount of current flowing out of the base. **See Figure 9-7.**

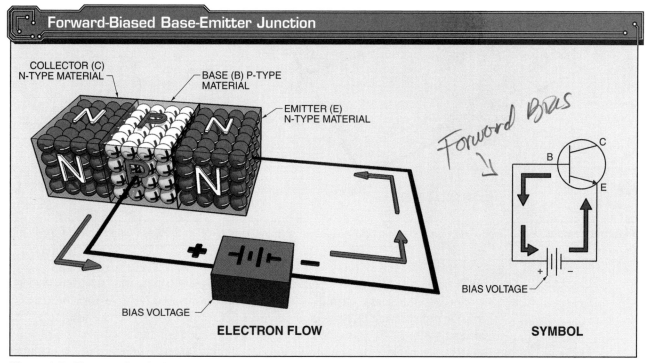

Figure 9-5. *Electrons flow in the external circuit when the base-emitter junction of an NPN transistor is forward biased.*

220 SOLID STATE DEVICES AND SYSTEMS

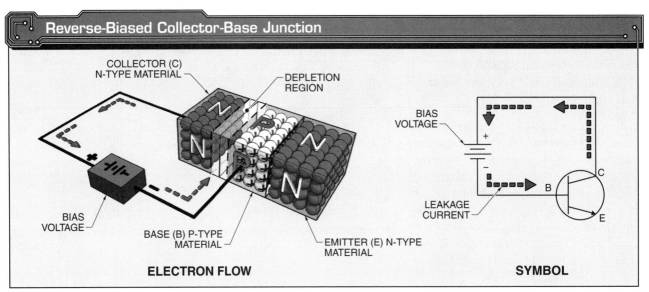

Figure 9-6. *Only a very small amount of electrons (leakage current) flows in an external circuit when the collector-base junction of an NPN transistor is reverse biased.*

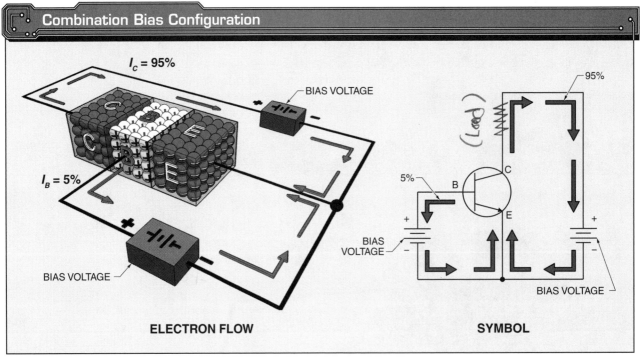

Figure 9-7. *Most electrons flow through the collector and a small amount flows through the base when both junctions are biased.*

The forward bias of the base-emitter circuit causes the emitter to inject electrons into the depletion region between the emitter and the base. Because the base (B) is thin (less than 0.001″ for most transistors), the much more positive potential of the collector (C) pulls the electrons through the thin base. As a result, the greater percentage (95%) of the available free electrons from the emitter passes directly through the base into the N-type material, which is the collector of the transistor.

Control of Base Current

The *base current* (I_B) is a critical factor in determining the amount of current flow in a transistor. It is critical because the forward-biased junction has a very low resistance and could be destroyed by large current flow. Therefore, the base current must be limited and controlled.

One method of limiting the base current in a transistor is to use a series-limiting resistor, R1. **See Figure 9-8.** Resistor R1 is extremely important because it keeps the base-emitter junction voltage and current from getting too high. The voltage applied to the base-emitter junction must be very small (0.6 V for silicon and 0.3 V for germanium). Resistor R1 forms a voltage divider with the resistance of the base-emitter junction to accomplish this voltage reduction. Since resistor R1 is considerably larger in resistance than the base-emitter junction, only a small fraction of the voltage applied will drop across the base-emitter junction. This type of voltage biasing is called fixed biasing, since resistor R1 does not change in value.

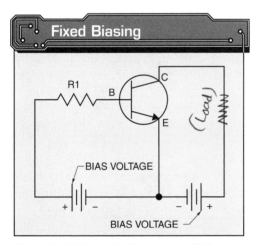

Figure 9-8. *Series-limiting resistor R1 forms a voltage divider with the resistance of the base-emitter junction, which is called fixed biasing.*

Voltage in a Transistor

In the base circuit, V_{BB} indicates bias voltage applied to the base-emitter circuit. V_{BE} indicates the voltage drop across the base-emitter junction. V_{R1} indicates the voltage drop across the series-limiting resistor, R1. Since the resistance of the base-emitter junction is less than resistor R1, most of the applied voltage (V_{BB}) will drop across R1. **See Figure 9-9.**

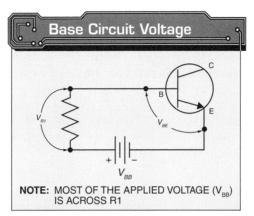

NOTE: MOST OF THE APPLIED VOLTAGE (V_{BB}) IS ACROSS R1

Figure 9-9. *Voltage in the base circuit is indicated by V_{BB}, V_{BE}, and V_{R1}.*

In the collector circuit, V_{CC} indicates the bias voltage applied across the collector-emitter circuit. Since the base-emitter and collector-base junctions are in series, the applied voltage (V_{CC}) will be divided between the junctions based on their individual resistances. Since the collector-base junction has a much greater resistance than the base-emitter junction, the collector-emitter voltage (V_{CE}) is larger than the collector-base voltage (V_{CB}). The collector-base voltage (V_{CB}), however, is greater than the base-emitter voltage (V_{BE}). **See Figure 9-10.**

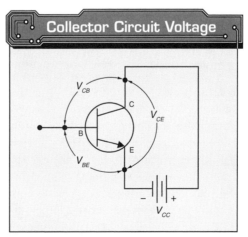

Figure 9-10. *Voltage in the collector circuit is indicated by V_{CB}, V_{CC}, and V_{CE}.*

TRANSISTOR OPERATING CHARACTERISTIC CURVES

Transistors are manufactured to perform different functions under a variety of operating conditions. The transistor can be made to operate with different combinations of voltages and currents to obtain a certain performance. To obtain as much information as possible about a transistor, a set of operating characteristic curves can be used. Operating characteristic curves are provided by the manufacturer. **See Figure 9-11.**

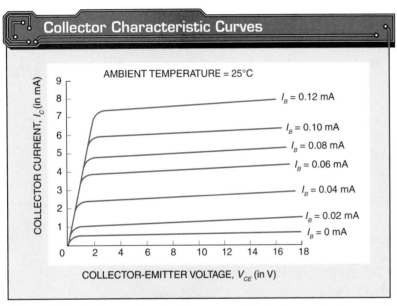

Figure 9-11. *Collector characteristic curves indicate how collector current changes when there is a change in base current or collector-emitter voltage.*

The curves on the graph are known as collector characteristic curves. These curves indicate how collector current (I_C) changes when there is a change in the base current (I_B) or collector-emitter voltage (V_{CE}). *Note:* The collector current (I_C) on the vertical axis is measured in milliamps (mA). The collector-emitter voltage (V_{CE}) on the horizontal axis is measured in volts (V).

Before using the entire set of curves, it should be understood how a curve is generated. A test circuit can be used to create the collector characteristic curve. The circuit is divided into two major sections: the base test circuit and the collector test circuit. **See Figure 9-12.**

In the base test circuit, the voltage (V_{BB}) provides the forward bias for the base-emitter junction. Potentiometer R1 is used to vary this bias to any specific value needed. Ammeter A1 is connected in series with base B, which is used to determine the value of I_B.

In the collector test circuit, the voltage (V_{CC}) provides the reverse bias for the collector-emitter junction. Potentiometer R2 is used to vary this bias to any specific value needed. Ammeter A2 is connected in series with collector C. It is used to determine the value of I_C. To determine V_{CE}, voltmeter V1 is placed across the collector-emitter junction.

A single collector characteristic curve for the transistor can be plotted. The vertical axis indicates I_C and is measured in milliamps (mA). The horizontal axis indicates V_{CE} and is measured in volts (V). **See Figure 9-13.**

The first step in plotting the curve is to adjust R2 so that no voltage exists between the collector and emitter. This is done by moving the potentiometer to its lowest value or grounded condition. Next, R1 is adjusted to produce a forward bias on the base-emitter circuit so that 0.1 mA (100 µA) flows in the base circuit. A1 is used to determine the value. At this point, with an I_B of 0.1 mA and no collector voltage, the I_C is zero. Point A represents a V_{CE} of zero and an I_C of zero with the base held constant at 0.1 mA.

If R2 were adjusted away from ground, a voltage would be applied between the collector and emitter. In this case, the potentiometer is adjusted to produce a V_{CE} of 1 V. Since the I_B must remain the same, R1 can be adjusted if necessary to maintain an I_B of 0.1 mA. Depending on the transistor being tested, a certain amount of I_C will begin to flow. In this example, A2 indicates 8 mA, which is shown on the graph. Point B represents a V_{CE} of 1 V and an I_C of 8 mA with the I_B held constant.

Test Circuits

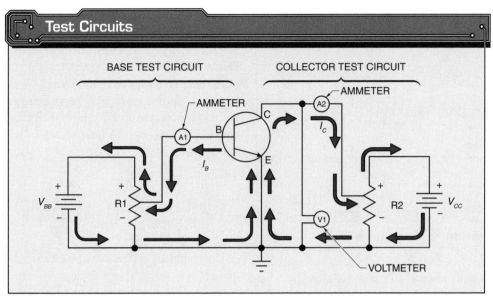

Figure 9-12. *A test circuit for plotting characteristic curves consists of two major sections: the base test circuit and the collector test circuit.*

Next, R2 is adjusted to produce a V_{CE} of 3 V while the base circuit remains at 0.1 mA. This increase in voltage causes the collector current to rise to 9 mA. Point C represents a V_{CE} of 3 V and an I_C of 9 mA with the I_B held constant. If the procedure is repeated, R2 is adjusted to a V_{CE} of 6 V at a base current of 0.1 mA. The I_C rises to 10 mA. Point D represents a V_{CE} of 6 V and an I_C of 10 mA with the I_B held constant.

If more information is needed on this transistor, additional curves can be plotted. When R1 is adjusted to produce more or less base current, it causes the curves to be above or below the original curve. The additional curves are denoted by the dotted lines. **See Figure 9-14.**

All manufacturers produce collector characteristic curves for their devices. To make their curves very accurate, many test points are established and the curves are accurately drawn. Collector characteristic curves are essential in predicting the operating condition of a transistor in a circuit.

Collector Characteristic Curves

With accurate curves, reasonable predictions can be made about the operation of a transistor. For example, to determine the amount of I_C that would flow if the V_{CE} is 4 V and the I_B is 0.4 mA, a vertical line is run from 4 V until it crosses the 0.4 mA base current line. This point is marked. Another line is drawn horizontally across to the collector current axis. At this point, the I_C is approximately 33 mA. Therefore, for a V_{CE} of 4 V and a constant I_B of 0.4 mA, a 33 mA current in the collector circuit should be expected. **See Figure 9-15.**

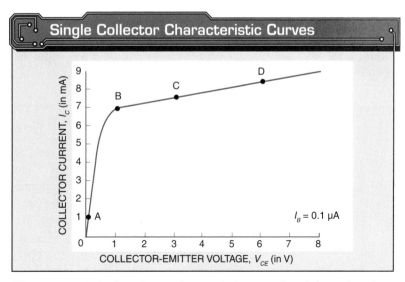

Figure 9-13. *A single collector characteristic curve gives information about the operation of a transistor.*

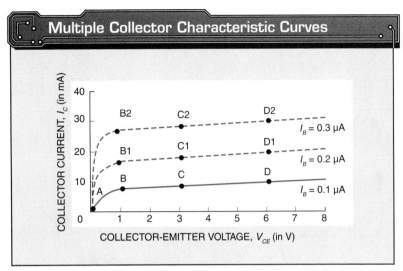

Figure 9-14. Multiple collector characteristic curves are plotted when additional information is needed on a transistor.

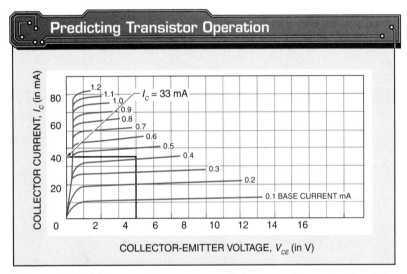

Figure 9-15. With accurate curves, predictions can be made about the operation of a transistor.

Current Gain (β) to Describe Transistor Operation

Transistor current gain (β) is the ratio of collector current (I_C) to base current (I_B) of a transistor with a constant collector-emitter voltage (V_{CE}). Current gain in a transistor is symbolized by the Greek letter beta (β).

Note: Designations beginning with "h" are also used to describe the operating characteristics of transistors. In the case of h_{fe}, it is used to designate current gain. Both β and h_{fe} are used interchangeably.

Current gain is used to describe the operation of a transistor because current changes are more dramatic in a transistor than voltage changes. In a formula, the Greek letter delta (Δ) means "a change in value." When the I_B is increased, the I_C increases, depending on the voltage applied to the collector-emitter circuit. The following formula is used to find the current gain of a transistor:

$$\beta = \frac{\Delta I_C}{\Delta I_B}$$

β = current gain
I_C = collector current
I_B = base current

The input of a transistor is a forward-biased PN junction. Therefore, input voltages across this junction are small and difficult to measure. With small voltage changes, it is difficult to compare transistor input voltage with transistor output voltage. The PN junction does, however, draw substantial current flow for a small signal input voltage, and the current can be easily measured. Current flow in the base controls I_C in the output. Therefore, most transistors should be compared on the basis of their ratio of current output to current input rather than their voltage. By using characteristic curves, it can be shown precisely how powerful the control of I_B is and how to predict its behavior.

> Multiple collector characteristic curves, commonly known as a family, are plotted on one graph to describe the operation of a working transistor.

Characteristic Curves to Find Current Gain

On a collector characteristic curve, a variety of base currents ranging from 0.1 mA to 1.2 mA are available. To determine what type of current change takes place in the collector circuit, it is necessary to find how much change takes place in the I_B. **See Figure 9-16.** For example, if the I_B changes from 0.1 mA to 0.7 mA, there is a change of 0.6 mA (ΔI_B = 0.6 mA). When the I_B is 0.1 mA, the I_C is 10 mA, and when the I_B is 0.7 mA, the I_C is 50 mA. The change in I_C is 40 mA (ΔI_C = 40 mA) while the V_{CE}

remains constant at 3 V. With the change in I_B known and the change in I_C known, current gain can be calculated as follows:

$$\beta = \frac{\Delta I_C}{\Delta I_B}, \text{ with a constant } V_{CE}$$

$$\beta = \frac{40}{0.6}$$

$$\beta = 66$$

A current gain of 66 means that any change in I_B results in a change 66 times greater in I_C. In terms of control, this device is extremely powerful. For each input, the output will be 66 times greater.

TRANSISTORS AS DC SWITCHES

The primary motivation behind the rapid development of the transistor was to replace mechanical switching. The transistor has no moving parts and can switch on and off quickly.

Mechanical switches have two conditions: open and closed or ON and OFF. The switch has a very high resistance when open and a very low resistance when closed. A transistor can be made to operate much like a switch. For example, it can be used to turn a pilot light PL1 ON or OFF. In this circuit, the resistance between the collector (C) and the emitter (E) is determined by the current flow between the base (B) and E. When no current flows between B and E, the collector-to-emitter resistance is high like that of an open switch. The pilot light does not glow because there is no current flow. **See Figure 9-17.**

If a small current flows between B and E, the collector-to-emitter resistance is reduced to a very low value like that of a closed switch. Therefore, the pilot light will be switched ON and begin to glow.

A transistor switched ON is usually operating in the saturation region of a characteristic curve. A *saturation region* is the region of a characteristic curve in which maximum current can flow in a transistor circuit. At saturation, the collector resistance is considered zero, and the current is limited only by the resistance of the load. Mathematically, saturation current is expressed as follows:

$$I_{SAT} = \frac{V_{CC}}{R_L}$$

where
I_{SAT} = saturation current
V_{CC} = supply voltage
R_L = resistive load

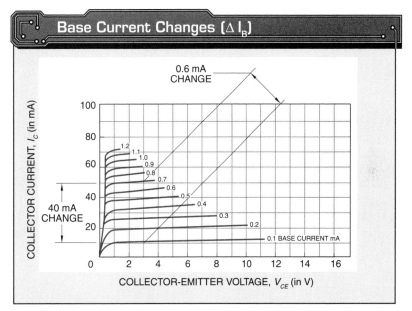

Figure 9-16. *On a collector characteristic curve, a 0.6 mA change in base current (I_B) results in a 40 mA change in collector current (I_C).*

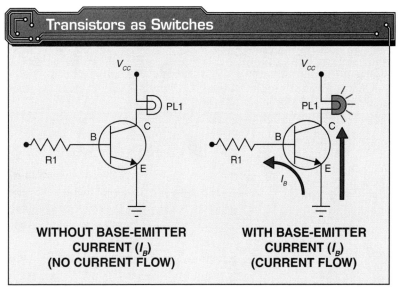

Figure 9-17. *A transistor can operate as a switch by manipulating its base-emitter current (I_B).*

When the circuit reaches saturation, the resistance of pilot light PL1 is the only current-limiting factor in the light circuit. When the transistor is switched OFF, it is operating in the cutoff region. The *cutoff region* is the region of a characteristic curve in which a transistor is turned OFF and no current flows through the collector. At cutoff, all the voltage is across the open switch (transistor) and the collector-to-emitter voltage is equal to the supply voltage. The characteristic curve shows the saturation point (A) and the cutoff point (B). **See Figure 9-18.** A *load line* is a line drawn on a collector characteristic curve that shows the relationship between the collector current (I_C) and collector-emitter voltage (V_{CE}) in a circuit.

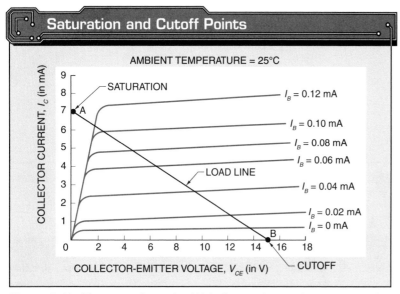

Figure 9-18. *The saturation point (A) indicates the maximum current that can flow in a collector at that voltage. The cutoff point (B) indicates the point at which there is no collector current.*

ESTABLISHING A LOAD LINE

A load line is drawn between two extremes of operation. It is assumed that a transistor is first saturated and then cut off. Saturation indicates the maximum current that can flow in the collector at a particular voltage. At saturation, a transistor is at the maximum output. At the cutoff, there is no I_C and the applied voltage source is dropped across the collector-emitter junction.

The output of the transistor is completely shut down. Therefore, the load line shows the maximum and minimum capabilities of a particular transistor. With this information, an operating point can be selected to give the outputs needed. **See Figure 9-19.**

The saturation for a load line is calculated by using the formula for saturation current. The voltage source (V_{CC}) is divided by the resistive load (R_L) to determine saturation (maximum) current. The cutoff voltage is equal to V_{CC}. For example, what is the saturation current of a transistor with an R_L of 2150 Ω and a V_{CC} of 15 V?

$$I_{SAT} = \frac{V_{CC}}{R_L}$$

$$I_{SAT} = \frac{15}{2150}$$

$$I_{SAT} = \mathbf{0.007\ A\ or\ 7\ mA}$$

Note: To find the cutoff, the desired voltage source (V_{CC}) is selected. In this case V_{CC} = 15 V.

BIASING TRANSISTORS

In order to achieve proper operation, a transistor must be properly biased. A transistor biased for saturation has the collector-base junction reverse biased. A transistor biased for cutoff has both junctions reverse biased.

If a transistor is biased for cutoff, it can be turned ON again only by applying a forward-bias voltage (signal) to the base-emitter. However, the signal must be strong enough to overcome the reverse bias and produce a sufficient current flow for the transistor to reach saturation. When the signal is removed from the transistor, it returns to reverse bias and cutoff. Using this technique, the transistor can be switched OFF and ON rapidly.

Base Bias

Increasing forward-bias voltage on a base causes the current through a transistor to increase. When the forward-bias voltage on the base decreases, current through the transistor decreases. A base-biasing configuration can use a photoconductive cell as part of the base circuit.

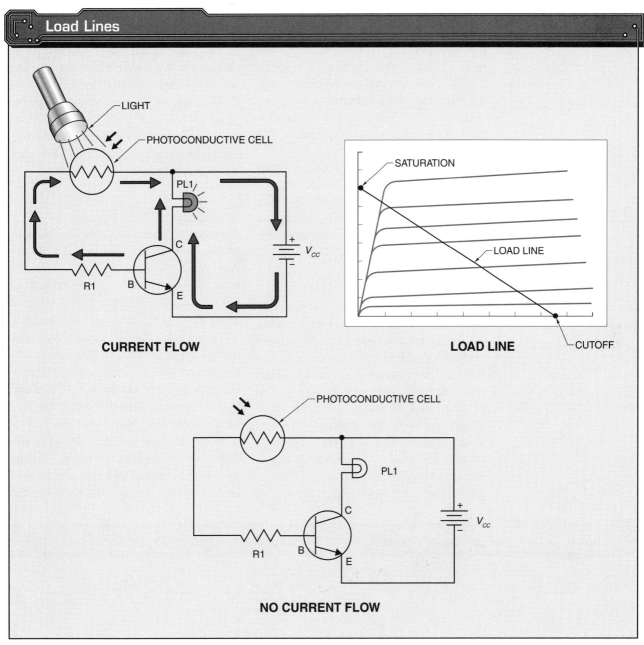

Figure 9-19. *A load line shows the relationship between the collector current and the collector-emitter voltage.*

Light shining on the photoconductive cell causes the resistance of the cell to decrease. With reduced resistance, the current into the base increases. This increase in current at the base-to-emitter junction causes the I_C to increase. The I_B is limited only by the bias-limiting resistor R1. If this bias value has been properly selected, the transistor will approach saturation and the pilot light turns on.

Removal of light from the photoconductive cell causes the resistance of the cell to increase. With the increased resistance, the current into the base decreases. The decrease in current at the base-to-emitter junction causes the I_C to decrease. If the bias value has been properly selected, the transistor will reach cutoff and the pilot light shuts off.

To make the transistor more flexible, an electromechanical relay can be added to the output circuit. With the addition of a relay with normally closed (NC) contacts, a photoconductive cell can be used to shut off a pilot light when light strikes the photoconductive cell. This type of circuit is often used in streetlight switching circuits. **See Figure 9-20.**

Note: When a transistor is used to switch a relay, a diode should be connected across the relay coil. This diode keeps the reverse-inductive surge of current of the relay coil from damaging the transistor when the relay is switched OFF.

Fixed Bias Configurations

A transistor can be biased from a single fixed-voltage source. The voltage source (V_{CC}) may be a definite block or section in the circuit. However, in most cases, the emitter (E) is connected to ground, which is the negative terminal of the source. The collector and R1 are connected to V_{CC}, which represents the positive terminal of the source. **See Figure 9-21.** This configuration is used because it is easier to represent a transistor circuit when several transistors are using power from a common source. Electrically, both circuits are the same.

The same circuit may be shown in yet another way for convenience of layout. The only difference is that the negative terminal of the voltage source is represented by the ground symbol, and the positive terminal of the source is represented by the positive sign and V_{CC} designation. Electrically, all circuits are exactly the same.

Clamping Diodes

A *clamping diode* is a diode that prevents voltage in one part of a circuit from exceeding the voltage in another part. A clamping diode protects a transistor from high-voltage surges. Clamping diodes are used when transistors control inductive loads.

Coils, such as those found in relays, have inductance. *Inductance* is the property of a circuit to oppose a change in current due to energy stored in a magnetic field. The collapsing magnetic field of a coil produces a high-voltage spike when switched OFF. A spike may damage or destroy a transistor. A diode prevents a spike from passing through the transistor because the diode conducts the excessive voltage away from the transistor.

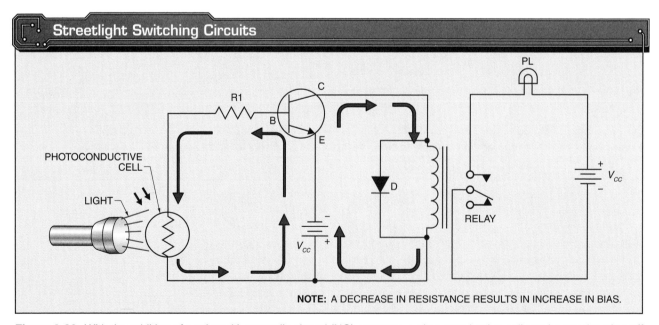

Figure 9-20. *With the addition of a relay with normally closed (NC) contacts, a photoconductive cell can be used to shut off a lamp when light strikes the cell.*

Fixed Bias Configurations

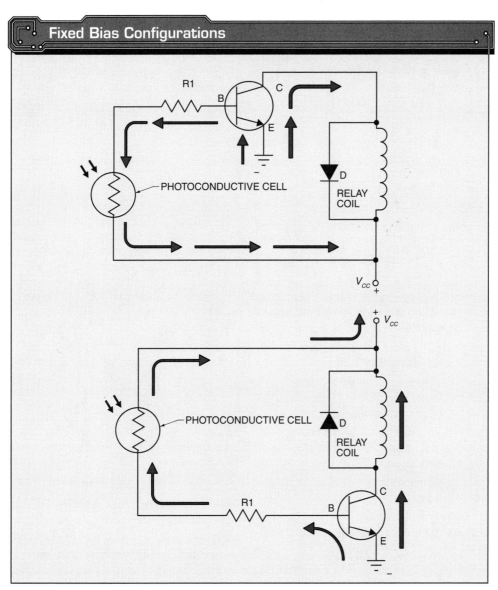

Figure 9-21. *Electrically, different fixed bias configurations are the same.*

Base Bias Instability

Base bias is one of the simplest methods of transistor biasing. With this type of bias, the amount of base current is dependent upon the value of the limiting resistor and the resistance of the base-emitter junction. As a signal reaches the limiting resistor, the signal will add to or subtract from the bias value to establish a variation in the output current of the device. Under normal stable temperatures, a base bias circuit will operate as it should. If the ambient temperature changes, the junction resistance also changes, and the bias current is affected.

Any change in bias shifts the operating point on the load line. **See Figure 9-22.** A shift of the operating bias due to heat results in moving the operating point closer to saturation or cutoff. *Thermal instability* is a change in bias due to heat. Base bias is easily affected by thermal instability. Bias stabilization circuits must be used to protect transistor circuits from thermal instability.

230 SOLID STATE DEVICES AND SYSTEMS

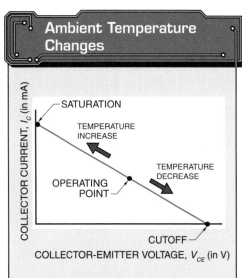

Figure 9-22. Any change in ambient temperature causes an operating point to shift up or down a load line.

Bias Stabilization Circuits

Bias stabilization circuits can be used to protect transistors in addition to heat sinking. A change in ambient temperature may cause thermal instability, which, if uncontrolled, results in thermal runaway. Thermal runaway can destroy a transistor. Bias stabilization circuits are designed to counteract the cumulative current increase due to a rise in temperature. The most common bias stabilization circuits are emitter-feedback, collector-feedback, and combination bias. Also, thermistors and diodes may be used as part of the circuit stabilization to provide thermal stabilization.

Emitter-Feedback Bias Stabilization Circuits. An emitter-feedback bias stabilization circuit is created by placing a resistor (R_E) in series with an emitter (E). A voltage develops at the emitter (E) and opposes the bias voltage at the base (B). The resistance values of R_B and R_E must be chosen so that the proper base-emitter bias current flows under ordinary operating conditions. Then, if the transistor temperature changes to increase bias currents, the voltage drop across R_E increases to increase the emitter voltage. The increased emitter voltage opposes the input bias voltage and reduces the base current. The collector current is then brought back to normal. **See Figure 9-23.**

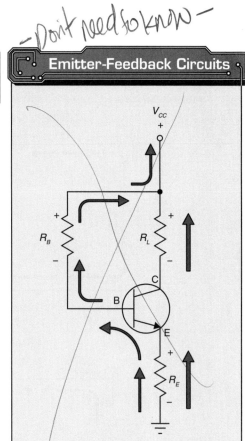

Figure 9-23. Emitter-feedback bias is accomplished by placing a resistor (R_E) in series with an emitter.

Collector-Feedback Bias Stabilization Circuits. A collector-feedback bias stabilization circuit is created by taking part of the output voltage and feeding it back into the input to return the circuit to normal operation. Collector feedback is also referred to as degenerative feedback or negative feedback.

In a collector-feedback stabilization circuit, the base resistor (R_B) limits input current and is connected directly to the collector (C) rather than directly to the voltage source (V_{CC}). The bias voltage now consists of a base voltage minus the drop across R_B. Therefore, if higher temperatures cause the transistor bias current to increase, the increased collector current causes a larger drop across the resistive load (R_L). The voltage at C decreases. This reduces the input bias current, causing the collector current to decrease and return to normal. The opposite happens if the transistor temperature decreases. **See Figure 9-24.**

Collector-Feedback Circuits

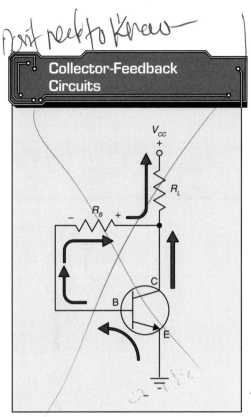

Figure 9-24. Collector feedback is accomplished by taking part of an output voltage and feeding it back into the input to return a circuit to normal operation.

Combination Circuits

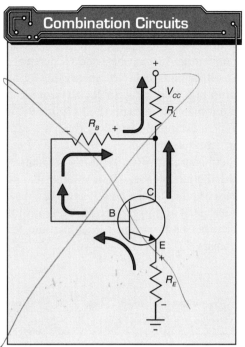

Figure 9-25. Combination bias requires the use of both emitter feedback and collector feedback.

Combination Bias Stabilization Circuits. A combination bias stabilization circuit is created by both emitter-feedback and collector-feedback circuits. The use of both bias voltages results in maintenance of good thermal stability. **See Figure 9-25.**

Voltage Divider with Emitter Bias. The bias for a transistor is often set by a voltage divider network. Resistors R1 and R2 form a voltage divider. They also set the fixed bias at the base terminal. Their resistance values are selected to provide the required bias current. Emitter resistor R_E is a temperature stabilizing resistor that determines the voltage on an emitter. The emitter current (I_E) is equal to the sum of the base current (I_B) and collector current (I_C). Current flow through R_E develops a voltage drop that opposes the base bias. Thus, any increase in the voltage drop across R_E due to temperature changes opposes the input bias and reduces the I_C. **See Figure 9-26.**

Voltage Divider— Emitter Bias

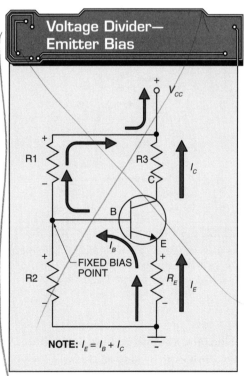

NOTE: $I_E = I_B + I_C$

Figure 9-26. By establishing a voltage divider network, the voltage drop across R_E (due to temperature changes) opposes the input bias and reduces collector current.

Thermistor Bias. Thermistors are also used for bias stabilization. Since a thermistor changes resistance with a change in temperature, it can be used as part of a stabilization circuit. The thermistor forms a voltage divider with R1 to provide a fixed bias for the base. **See Figure 9-27.** As temperature increases with an increase in base bias current, the resistance of the thermistor decreases. With less resistance, the thermistor drops less voltage for the base, and the base bias decreases. With a properly designed circuit, a thermistor changes base current to compensate for changes in temperature.

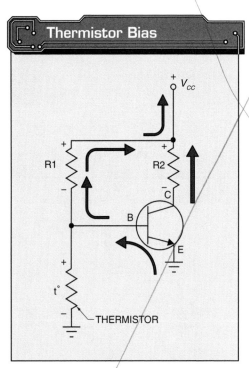

Figure 9-27. *A thermistor can be used to form a voltage divider circuit. As ambient temperature changes, the thermistor changes resistance, helping to stabilize the circuit.*

Thermistor/Emitter Feedback Bias. A thermistor can be used as part of a voltage divider with an emitter resistor (R_E) and a base resistor (R_B). The base is supplied with fixed bias through resistor R_B. R_E provides normal stabilization. By adding the thermistor, however, the current stabilization process becomes more effective. **See Figure 9-28.**

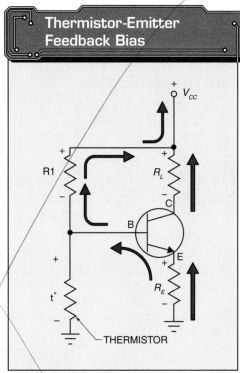

Figure 9-28. *A thermistor can be used with other types of stabilization circuits to provide more effective control.*

A change in temperature causes the resistance of the thermistor to change, producing a change in the bias. The change in the bias current offsets the change in the transistor, keeping the base current constant. The advantage of using a thermistor is that stabilization tends to occur before bias currents can change.

Diode Stabilization. A diode is often used for stabilization instead of a thermistor because its PN junction closely approximates that of a transistor. With thermistor stabilization, matching a thermistor to the operating characteristic of a transistor is often a problem. When a diode is used, it is forward biased. In stabilizing a circuit, a diode causes the same current changes as a thermistor. **See Figure 9-29.**

Diode Stabilization

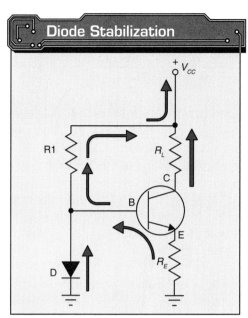

Figure 9-29. *By using a diode instead of a thermistor, the operating characteristics of a transistor can be more closely matched.*

POWER DISSIPATION

Because a transistor is extremely small, it can dissipate only a certain amount of heat (power dissipation factor) before it begins to change operating characteristics. If an overload continues for even a short period of time, the transistor junction may be destroyed by thermal runaway. Transistors are rated for wattage at 25°C. If the temperature drops below 25°C, more power can be handled. If the temperature exceeds 25°C, the transistor must be derated, or a heat sink must be added to dissipate the additional heat. Characteristic curves can be used to determine the power output at any given point. The power output of a transistor can also be determined by using the following formula:

$P = E \times I$

where

P = power (in W)
$E = V_{CE}$ = voltage (in V)
$I = I_C$ = current (in A)

For example, what is the power of a transistor with a V_{CE} of 10 V and an I_C of 10 mA?

$P = E \times I$
$P = 10 \times 10$
$P = \mathbf{100\ mW}$

Maximum Power Dissipation Curves

In order to protect the transistor, the maximum power dissipation curve must be considered. **See Figure 9-30.** A maximum power dissipation curve represents the maximum power that a transistor can handle under normal ambient temperatures. It shows the safe operating ranges of a transistor. The shaded areas indicate where the transistor may have problems. Under no circumstance should the load line move to the right of the maximum power dissipation curve. Circuit designers typically try to have operation well within the maximum power dissipation curve to help compensate for temperature variations. This allows for situations in which there is no control of temperatures such as in rooms without air conditioning during the summer months.

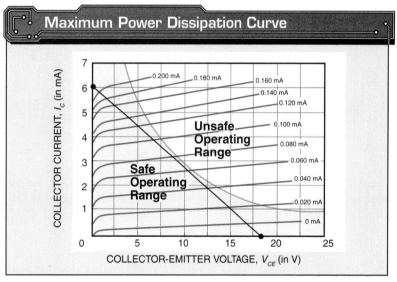

Figure 9-30. *A maximum power dissipation curve shows the safe operating range of a transistor.*

[Listed power rating]

Heat Sinks

Transistors, like diodes, use heat sinks for thermal protection. The cases of certain power transistors are designed specifically for ease in cooling. Some power transistors use radial fins to conduct heat away. Other power transistors are designed for use with heat sinks. **See Figure 9-31.** In some cases, the chassis of the equipment is used as a heat sink.

Heat Sinks

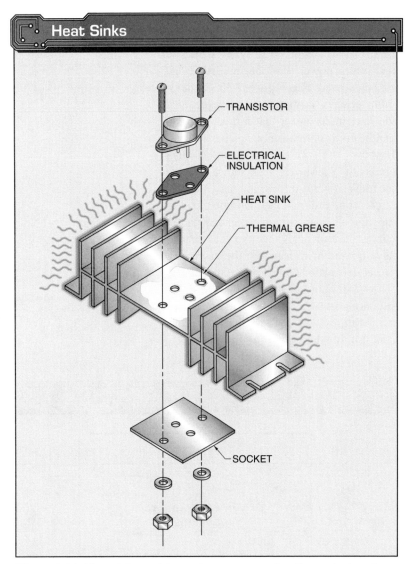

Figure 9-31. *Heat sinks and thermal grease spread on the contact surfaces of a transistor aid in the transfer of heat.*

When transistors are required to operate in high ambient temperatures, forced cooling by a fan or by air conditioning is used. In some precision solid state equipment, a unit shuts down if ambient temperatures exceed safe levels. Also, a unit is shut down if the forced cooling system is shut down.

Correct heat sink selection and mounting can be confirmed by taking an actual temperature measurement under worst-case conditions and comparing the reading to the allowable temperature chart that is provided by the manufacturer.

TESTING TRANSISTORS

Although there are a variety of transistor testers available, a DMM set to measure resistance (ohms) provides much of the information necessary to determine if a transistor is operating properly. Transistors are very rugged and are expected to be relatively trouble-free. Encapsulation and conformal coating techniques now in use ensure extremely long life expectancies.

In theory, a transistor should last indefinitely. However, if transistors are subjected to current overloads, the junctions will be damaged or destroyed. In addition, the application of excessively high operating voltages can damage or destroy the junctions through arc-overs or excessive reverse currents. One of the greatest dangers to a transistor is heat, which causes excessive current flow and damage.

To determine if a transistor is good or bad, a digital multimeter (DMM) or a transistor tester can be used. In many cases, a good transistor can be substituted for a transistor that is questionable to determine its condition. This method of testing is highly accurate and sometimes the quickest. However, it should only be used after it is certain that there are no circuit defects that might damage a replacement transistor. If more than one defective transistor is present in equipment, this testing method becomes cumbersome, as several transistors may have to be replaced before the problem is corrected.

To determine which transistors are not defective and which stages failed, all removed transistors must be tested. This test can be done by using a DMM, a transistor tester, or by observing whether the equipment operates correctly as each of the transistors is reinserted into the equipment.

Testing Transistors for Opens and Shorts

High voltages, improper connections, and overheating can damage a transistor. A technician is responsible for determining the condition of a transistor. Although many types of transistor testers are available, a technician may not have one when it is needed. Therefore, a DMM can be substituted for a transistor tester when finding opens and shorts. Transistors may be considered back-to-back diodes when conducting tests for opens and shorts. **See Figure 9-32.**

Note: A diode passes current when it is forward biased. It blocks current when it is reverse biased.

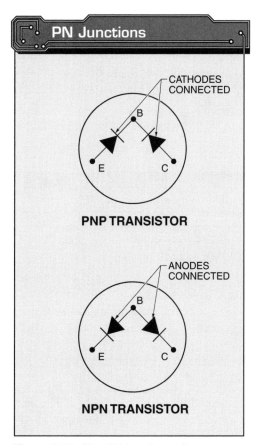

Figure 9-32. *The PN junctions of a transistor can be considered back-to-back diodes when testing for opens and shorts.*

In 1947, John Bardeen, Walter Brattain, and William Shockley invented the first point-contact transistor with a bipolar junction. They were awarded the 1956 Nobel Prize in Physics for their achievement.

A transistor may become defective from excessive current or temperature. A transistor typically fails due to an open or shorted junction. The two junctions of a transistor may be tested with a DMM set to diode test mode on the multimeter for accurate readings. **See Figure 9-33.** To test an NPN transistor for an open or shorted junction, apply the following procedure:

1. Connect a DMM to the emitter and base of the transistor. Measure the resistance.
2. Reverse the DMM leads and measure the resistance. The base-emitter junction is good when the resistance is high in one direction and low in the opposite direction.

Note: The ratio of high to low resistance should be greater than 100:1. Typical resistance values are 1 kΩ with the positive lead of the DMM on the base, and 100 kΩ with the positive lead of the DMM on the emitter. The junction is shorted when both readings are low. The junction is open when both readings are high.

3. Connect the DMM to the collector and base of the transistor. Measure the resistance.
4. Reverse the DMM leads and measure the resistance. The collector-base junction is good when the resistance is high in one direction and low in the opposite direction.

Note: The ratio of high to low resistance should be greater than 100:1. Typical resistance values are 1 kΩ with the positive lead of the DMM on the base, and 100 kΩ with the positive lead of the DMM on the collector.

5. Connect the DMM to the collector and emitter of the transistor. Measure the resistance.
6. Reverse the DMM leads and measure the resistance. The collector-emitter junction is good when the resistance reading is high in both directions.

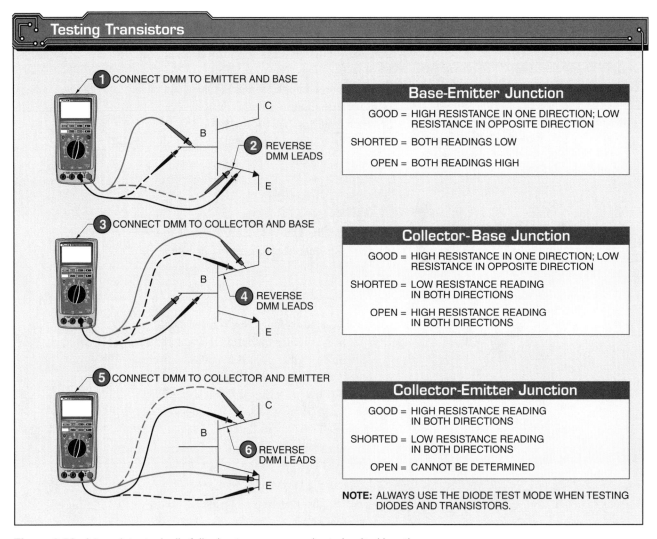

Figure 9-33. *A transistor typically fails due to an open or short-circuited junction.*

Testing Transistor Switches

It is impossible to determine whether a transistor is switched ON or OFF just by visual inspection. However, if an indicator light is in the collector circuit, the ON/OFF status of a transistor can be determined. Typically, a DMM set to measure voltage is used to determine the status of a transistor. **See Figure 9-34.** To check a transistor switch, apply the following procedure:

1. When the transistor switch is closed (ON), V_{CE} reads approximately 0 V. A transistor in this condition has virtually no resistance just as an electromechanical switch would have no resistance. Therefore, the transistor has no voltage drop.

2. When the switch is open (OFF), V_{CE} equals approximately V_{CC}. The particular V_{CC} reading in the open condition occurs because R_L has no current flowing through it. Approximately all the voltage is dropped across the collector-emitter.

3. If the load fails to energize, check the voltage input to see that it is high enough to turn the transistor ON. If the voltage input is correct, a silicon transistor should have a V_{BE} reading of approximately 0.6 V when turned ON. The remaining input voltage is dropped across the current-limiting resistor.

4. If the voltage input is correct but voltage V_{BE} is considerably higher or lower than 0.6 V, the base-emitter junction is bad and the transistor must be replaced.
5. If V_{BE} is correct and V_{CE} is approximately equal to V_{CC}, the transistor has an open collector-base junction and the transistor must be replaced.
6. If the transistor fails to shut off, check for a short circuit from the collector to base or from the collector to emitter. In either case, the transistor must be replaced if there is a short.
7. If V_{CE} is near 0 V but the load does not energize, the load probably has an open circuit and must be replaced.

Fluke Corporation
A multimeter can be used to take test measurements of solid state components.

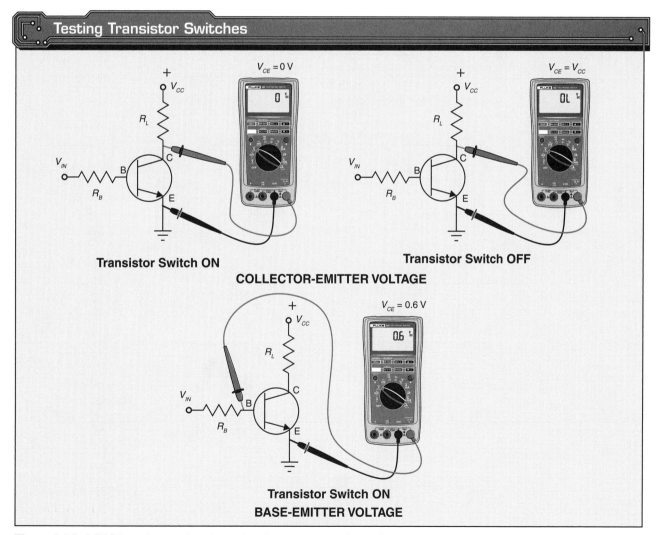

Figure 9-34. *A DMM can be used to determine if a transistor is ON or OFF.*

TRANSISTOR SWITCHING APPLICATIONS

Transistors are used for switching because of their reliability and speed. In certain situations, transistors are also integrated with other solid state components to form more complex devices. In each application, however, the fundamental operating principle of the transistor remains the same.

Seven-Segment Displays

By switching various combinations of transistors ON or OFF, different numbers can be created on a seven-segment display. **See Figure 9-35.** For example, if all transistors (A through G) are switched ON, an "8" should appear on the display. If all transistors are switched ON except E and D, a "9" should appear. There is typically circuitry in addition to the seven-segment transistor devices to help in decoding the proper signals for the display. When all circuitry is present, it is called a seven-segment decoder/driver display, or readout device.

Electronic Timers

An electronic timer can be used for the time control of a machine or process. **See Figure 9-36.** An electronic timer circuit uses an RC time constant to provide timing. This is developed by resistors R1 and R2 in addition to capacitor C1. The circuit is triggered by momentarily closing N.O. switch S1 shorting out capacitor C1 allowing it to discharge. With capacitor C1 discharged, transistor Q1 no longer conducts and the relay drops out causing the LED to turn OFF. Once switch S1 is released to its N.O. condition, capacitor C1 will begin to charge to the point where transistor Q1 is forward biased and begins to conduct. At this point, the relay is activated and the LED turns ON.

To change the time interval, the resistance of potentiometer R2 is changed. If the resistance is increased, the time interval is increased. If the resistance is decreased, the time interval is decreased. The light emitting diode (LED) indicates when the load is ON.

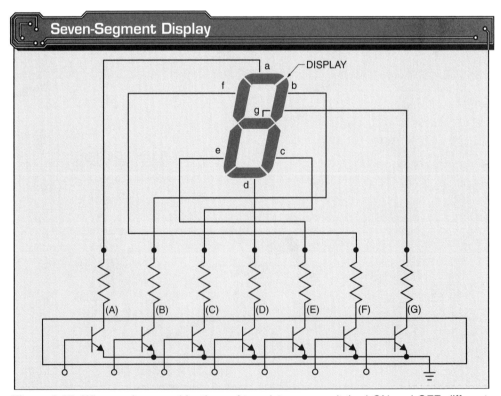

Figure 9-35. When various combinations of transistors are switched ON and OFF, different numbers appear on a seven-segment display.

Water Level Detectors

A water level detector can be used to detect the level of water in a tank. The main components of a water level detector circuit consist of a simple, low-cost transistor and a set of probes. **See Figure 9-37.** The advantage of the circuit is its ability to use a very small amount of current (in mA) to control a much larger output such as a sump pump motor.

In the circuit, the amount of collector current determines when the relay will trip. The base is held negative with respect to the emitter by the voltage divider. The voltage divider consists of resistors R1 and R2. Current stabilization is maintained by resistor R3.

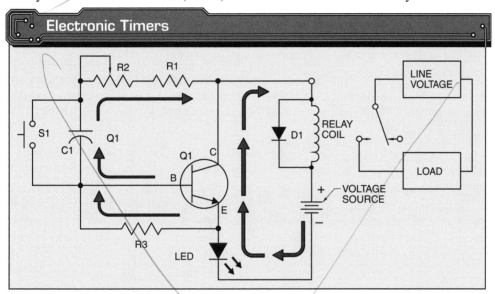

Figure 9-36. *An electronic timer consists of a transistor with an RC time constant to control a relay.*

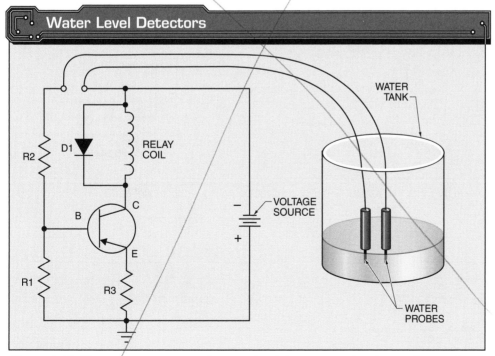

Figure 9-37. *The main components of a water level detector consist of a transistor and a set of probes.*

240 SOLID STATE DEVICES AND SYSTEMS

Refer to Chapter 9 Quick Quiz® on CD-ROM

As the water reaches a predetermined level set by the mounting height of the water probes, the bias circuit completes. With the bias circuit complete, voltage is applied to the circuit, causing base current to flow. With a base current established, the collector current will flow through the relay. If bias current has been properly established, the relay should trip and activate the circuit. Once the water level is below the probes, the relay is de-energized.

Beverage Dispensing Guns

A beverage dispensing gun uses transistors to switch a high load voltage and current to dispense beverages. Current flows through both the base-emitter and the collector-emitter when the transistor is switched ON. The base-emitter current is very low (approximately 5%) compared to the collector-emitter current. A transistor allows a small control current to control a high load current.

When a transistor is used as a switch, it acts as a relay. A relay allows a small amount of coil voltage and current to control a high load voltage and current. For example, a pushbutton on a beverage gun may be used to apply a low current to the base of a transistor. **See Figure 9-38.**

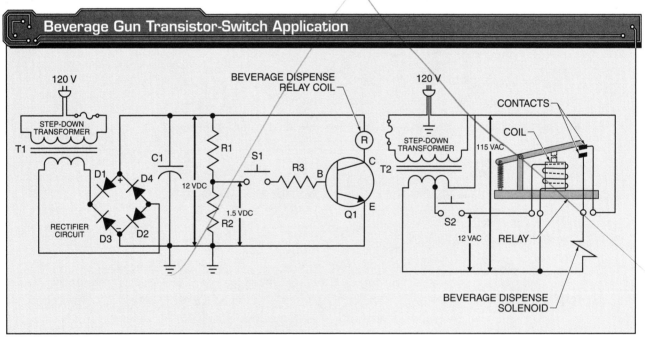

Figure 9-38. *A beverage dispensing gun uses transistors to switch a high load voltage and current to dispense beverages.*

KEY TERMS

- bipolar junction transistor (BJT)
- bipolar
- base
- collector
- emitter
- transistor outline (TO) number
- base current (I_B)
- base-emitter junction
- base-emitter voltage (V_{BE})
- collector current (I_C)
- collector-base junction
- collector-base voltage (V_{CB})
- collector-emitter junction
- collector-emitter voltage (V_{CE})
- collector feedback
- emitter current (I_E)
- emitter feedback
- current gain (β)
- Beta (β or h_{fe})
- saturation region
- saturation current (I_{SAT})
- applied voltage (V_{BB} or V_{CC})
- cutoff region
- load line
- clamping diode
- inductance
- thermal instability
- seven-segment display
- conductivity probe
- sink circuit

REVIEW QUESTIONS

1. What are the two types of transistors?

2. What are the three main parts of a transistor?

3. How are transistors referenced?

4. What is the bias configuration for the normal operation of any transistor?

5. What is the critical factor in determining current flow through a transistor?

6. Define the following abbreviations: V_{BB}, V_{BE}, V_{CE}, and V_{CB}.

7. What is used to determine the different voltage and possible current combinations in a transistor?

8. Explain why current gain is used to describe the operation of a transistor.

9. How is beta determined mathematically?

10. In what region is a transistor operating when it is ON and when it is OFF?

11. What line is drawn between the two extremes of operation on a characteristic curve?

12. What should be placed across the relay coil of a relay that is used with a transistor?

13. Explain thermal instability.

14. What are the three most common types of base bias stabilization circuits?

15. What is a maximum power dissipation curve?

16. At what temperature is the power of a transistor rated?

17. In transistor operation, what is the function of a heat sink?

18. How is a transistor tested to determine if it is an NPN or a PNP?

19. How could it be determined that a transistor junction is open?

20. What is the V_{CE} of a transistor that is switched OFF?

TRANSISTORS AS AMPLIFIERS 10

Transistors can be used as AC amplification devices. Amplification is the process of taking a small signal and making it larger. It is accomplished by using a small input signal to control the energy output from a larger source, such as a power supply. In control systems, transistor AC amplifiers are used to increase small signal currents and voltages so that they can do useful work.

OBJECTIVES
- Explain amplifier gain.
- Describe bandwidth.
- Explain how a decibel is used to describe changes in sound.
- List and describe the types of transistor amplifiers.
- Explain how operating points are set on load lines.
- Describe the classes of amplifier operation.
- List the types of transistor testers.
- List and describe the types of multistage amplifiers.

AMPLIFIER GAIN

The primary purpose of an amplifier is to produce gain. *Gain* is the ratio of the amplitude of an output signal to the amplitude of an input signal. In determining gain, an amplifier can be thought of as a "black box." **See Figure 10-1.** A signal applied to the input of the black box gives the output of the box. Mathematically, gain can be found by dividing output by input as follows:

$$gain = \frac{output}{input}$$

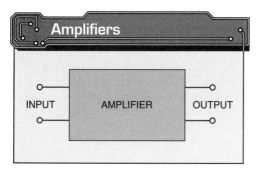

Figure 10-1. An amplifier can be thought of as a "black box" with an input and an output.

Sometimes a single amplifier does not provide enough gain to increase the amplitude of the output signal as needed. In such a case, a cascade amplifier can be used to obtain the required gain. A *cascade amplifier* is a series of connected amplifiers in which the output of one amplifier is connected to the input of another amplifier. For example, if amplifier A has a gain of 10, and amplifier B has a gain of 10, the total gain of the two amplifiers is 100 ($10 \times 10 = 100$). **See Figure 10-2.** If the gain of one amplifier were 8 and the other were 9, the total gain would be 72 ($8 \times 9 = 72$). For many amplifiers, gain is in the hundreds and even thousands.

Note: Gain is a ratio of output to input and has no unit of measure, such as voltage or amperes, attached to it. Therefore, the term "gain" can be used to describe amplifier gain, current gain, voltage gain, and power gain. In each case, the output is merely being compared to the input.

> An amplifier is commonly used in an audio system where it functions as an electric device that turns medium-level audio signals into strong audio signals that are then sent to speakers.

243

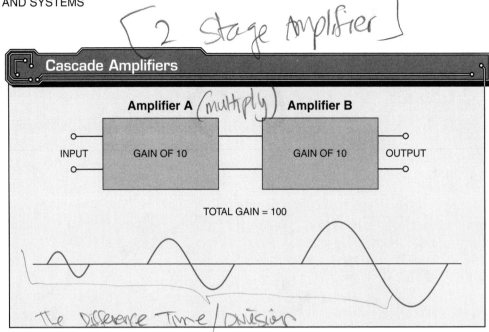

Figure 10-2. A cascade amplifier is a series of amplifiers connected so that the output of each amplifier is connected to the input of the next amplifier.

Some audio systems have cascade amplifiers that include a series of amplifiers and preamplifiers. Preamplifiers accept low-level audio signals, such as those from microphones and turntables, and amplify them into medium-level audio signals that are then sent to amplifiers.

Computing Current Gain

Current gain is the ratio of output current to input current. Mathematically, current gain is expressed as follows:

$$\text{current gain} = \frac{I_{out}}{I_{in}}$$

where
I_{out} = output current
I_{in} = input current

Note: The current gain for an amplifier is not the same as the current gain for an individual transistor, which is shown as beta (β). Beta refers to the transistor as a single device. It does not refer to the entire amplifier circuit containing the transistor.

The transistor current gain (β) is the maximum gain possible for a single-stage (one-transistor) amplifier circuit. The actual circuit gain depends on the particular values chosen for a load and the type of bias. The amplifier gain for a particular transistor circuit should always be less than the β of the single transistor in the circuit.

Computing Voltage Gain

Voltage gain is the ratio of output voltage to input voltage. To calculate the voltage gain of an amplifier, the RMS output voltage and input voltage of the amplifier are first measured. Then, the output is divided by the input. Mathematically, voltage gain is expressed as follows:

$$\text{voltage gain} = \frac{V_{rms\text{-}out}}{V_{rms\text{-}in}}$$

where
$V_{rms\text{-}out}$ = output voltage
$V_{rms\text{-}in}$ = input voltage

Computing Power Gain

Power gain is the ratio of output power to input power. Mathematically, power gain is expressed as follows:

$$\text{power gain} = \frac{P_{out}}{P_{in}}$$

where
P_{out} = output power
P_{in} = input power

BANDWIDTH (range of frequency)

The gain of an amplifier is not the same at all frequencies. Amplifiers are designed to operate within a given frequency range. If an amplifier is operated outside of this frequency range, the gain may decrease. *Bandwidth* is the range of frequencies over which the gain of an amplifier is maximum and relatively constant. The bandwidth for a given amplifier is often shown by a graph called a frequency-response curve. For example, a frequency-response curve for an amplifier that has a working frequency range starting at 2000 Hz and stopping at 20,000 Hz has a bandwidth. **See Figure 10-3.** Mathematically, bandwidth is expressed as the difference between the upper half-power frequency and the lower half-power frequency.

The upper and lower frequency limits are expressed as those points on the frequency-response curve where the gain drops to 0.707 times its maximum gain. These points are called half-power points. A *half-power point* is the point on a frequency-response curve where the power output is one-half the maximum value. Mathematically, a half-power point is expressed as follows:

$$\text{power output} = (0.707 \times V_{max}) \times (0.707 \times I_{max})$$
$$\text{power output} = 0.707 \times 0.707 \times V_{max} \times I_{max}$$
$$\text{power output} = 0.5 \times V_{max} \times I_{max}$$
$$\text{power output} = 0.50\ P_{max}$$

where
V_{max} = maximum voltage
I_{max} = maximum current
P_{max} = maximum power

Bandwidths can be wide or narrow. A *wide bandwidth* is a large range of frequencies over which an amplifier operates. A *narrow bandwidth* is a small range of frequencies over which an amplifier operates.

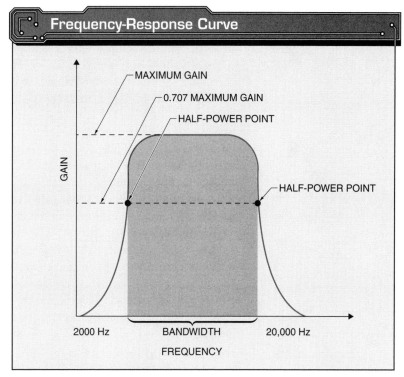

Figure 10-3. *The working frequency of an amplifier can be determined by using a frequency-response curve.*

DECIBELS

A *decibel (dB)* is a unit of measure used to express the relative intensity of sound. The human ear responds to the intensity, or loudness, of sound.

The human ear is not sensitive to the amount of change in sound but rather to the ratio of change that occurs. As the power changes from 1 W to 1.26 W (an increase of 0.26 W), a change in volume is evident. However, the power has to change from 79.4 W to 100 W (an increase of 20.6 W) before the same change is noticed. The ratio between the two powers is equal and can be calculated as follows:

$$\frac{1.26}{1} = 1.26 \quad \text{and} \quad \frac{100}{79.4} = 1.26$$

Since the human ear responds to the ratio of change rather than the amount of change, it is helpful to measure the output of audio amplifiers with that in mind. Each time the power of an amplifier increases, or the loudness of the sound increases, by a ratio of 1.26, it has increased by 1 dB.

The decibel is a unit of measure used to describe a change between two measurements. In electronics, this change is usually related to voltage or power. The decibel is a subunit of the bel. The bel is a larger unit than the decibel and was originally used to show the loss of signal in two miles of telephone wire. Use of the decibel is more common than use of the bel since most changes in small amplifiers are small.

Since decibels denote change, they are often used in describing amplifier gain and line signal losses. For example, an amplifier might have a +30 dB change. The plus sign in front of the number indicates there is an increase between input and output. If a minus sign were in front of a number, as in –10 dB, then there would be a signal reduction between input and output. **See Figure 10-4.**

For example, the microphone and cord of a tape-recording system provide a loss of –30 dB into the amplifier. The amplifier provides a gain of 30 dB to compensate for the loss. However, the cable and connectors introduce a loss of –5 dB prior to going into the tape recorder. If the recorder does not have amplification, it introduces another loss of –5 dB. The net result is a signal loss of –10 dB, which may have to be compensated for by an additional amplifier.

The decibel is widely used in the audio industry. Meters are marked, frequency response is graphed, and microphone outputs are rated in decibels. Amplifiers are rated in decibel gain. Noise levels are measured by the number of decibels they are below the level of the desired sound.

TYPES OF TRANSISTOR AMPLIFIERS

The three basic types of transistor amplifiers are the common-emitter, common-base, and common-collector transistor amplifier. **See Figure 10-5.** The amplifier is named after the transistor connection that is common to both the input and the load. For example, the input of a common-emitter circuit is across the base and emitter while the load is across the collector and emitter. Thus, the emitter is common to both the input and load.

Common-Emitter Amplifiers

A *common-emitter amplifier* is a bipolar junction transistor with both the input and output signals connected to the emitter. The transistor amplifies AC through the use of its input circuit and output, or load, circuit. The signal enters the amplifier through the input circuit and exits through the output circuit. **See Figure 10-6.**

Note: The emitter is connected to ground (reference or common reference point), and both input and output are common to the emitter. Because of this, the common-emitter circuit is sometimes

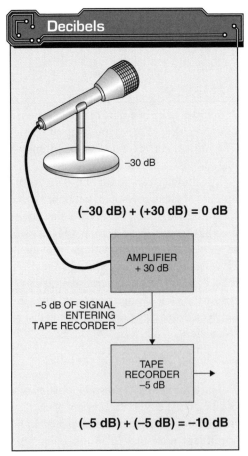

Figure 10-4. *A decibel is a unit of measure used to describe a change between two measurements.*

called a grounded-emitter circuit. Bias voltage for this circuit is provided by base resistor R_B.

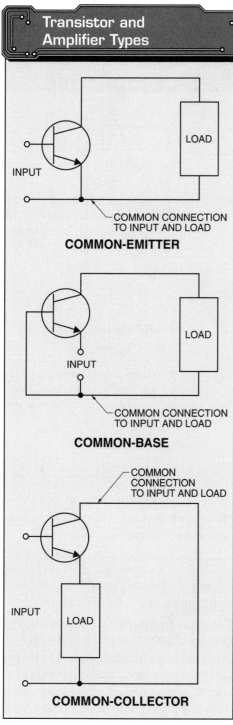

Figure 10-5. *The three basic types of transistor amplifiers are the common-emitter, common-base, and common-collector.*

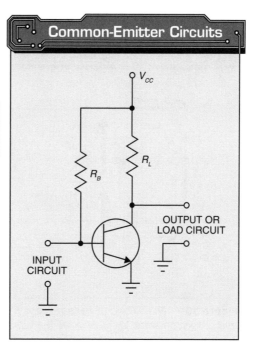

Figure 10-6. *In a common-emitter circuit, the emitter is connected to ground, and the input and output circuits are common to this ground.*

The values of the base resistor (R_B) and the collector voltage (V_{CC}) determine the base current (I_B). **See Figure 10-7.** The amount of base current can be found by using the following formula:

$$I_B = \frac{V_{CC}}{R_B}$$

where
I_B = base current
V_{CC} = collector voltage
R_B = base resistor voltage

Common-Emitter Amplifier Signals. An input signal must be applied to make an amplifier change from a static condition (at rest) to a dynamic condition (normal operation). **See Figure 10-8.** In this circuit, a square wave is applied to the amplifier. The signal rises to a given value and remains at that value for a period of time. It then reverses polarity and rises to a given value until it is time for the cycle to repeat. Because of the alternating nature of the signal, it will add to or subtract from the emitter-base bias voltage.

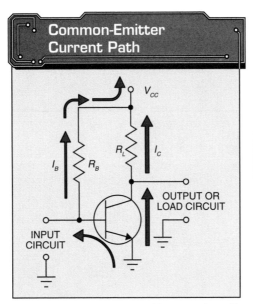

Figure 10-7. The values of base resistor R_B and collector voltage V_{CC} determine the base current.

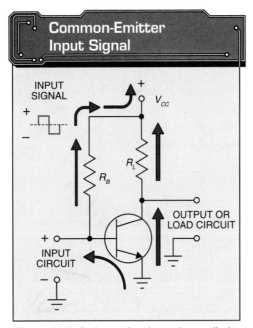

Figure 10-8. An input signal must be applied to make an amplifier change from a static condition (at rest) to a dynamic condition (normal operation).

> Sound energy is a sinusoidal waveform, the intensity of which can be measured in decibels to express a peak value, or amplitude. Amplitude is the distance a sine wave travels from its x-axis to its peak. The greater the amplitude, the louder the sound.

The input signal increasing in the positive direction adds to the emitter-base forward bias. **See Figure 10-9.** When the forward bias increases, the base current and collector current increase.

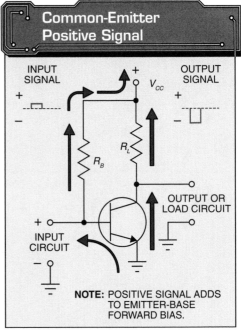

Figure 10-9. When the input signal increases in the positive direction, it adds to the emitter-base forward bias, which causes the base and collector currents to increase.

The input signal increasing in the negative direction subtracts from, or reduces, the emitter-base forward bias. **See Figure 10-10.** When the forward bias decreases, the base current and collector current decrease.

Note: As the input increases, the output signal decreases. As the input decreases, the output signal increases. This is called a phase inversion or 180° phase shift.

Common-Emitter Phase Inversion. The voltage divider concept can be used to easily explain phase inversion. The output circuit of a transistor can be considered as a series voltage-divider network composed of one fixed resistor (R_L) and one variable resistor (R_{CE}). It shows the effect of an input signal on an output signal. **See Figure 10-11.**

Chapter 10—Transistors as Amplifiers   249

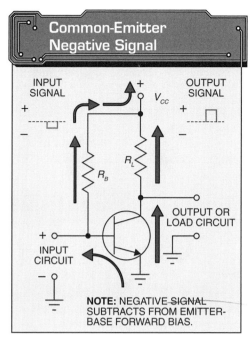

Figure 10-10. *When the input signal increases in the negative direction, it subtracts from the emitter-base forward bias, which causes the base and collector current to decrease.*

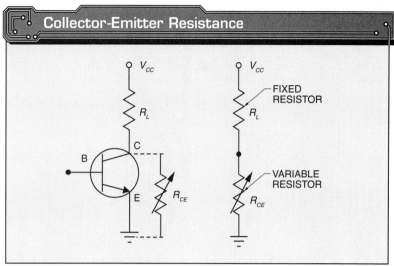

Figure 10-11. *As the base current changes, the collector-emitter junction acts similar to variable resistor R_{CE}.*

The total supply voltage (V_{CC}) of a common-emitter amplifier equals 10 V. In this circuit, no signal is applied, and V_{CC} is divided equally across R_{CE} and R_L. **See Figure 10-12.** *Note:* These voltages are used for illustrating a point and are relatively large for most transistors.

When a negative-going signal is coupled to the base bias, the forward bias is reduced. This causes the collector-emitter section of the transistor to increase in resistance. Since the resistance of R_{CE} has increased in a greater proportion relative to resistive load R_L, more voltage will be dropped across the output circuit resistance. The output signal (V_{CE}) will then increase. In this case, it will increase to 8 V.

When the signal is positive, the voltage aids the forward bias. This causes the collector-emitter section of the transistor to decrease in resistance. Since the resistance of R_{CE} now decreases in relation to R_L, less voltage will be dropped across the output circuit resistance. The output signal will then decrease. In this case, it will decrease to 2 V. The output signal of a common-emitter amplifier will be opposite, or an inversion, of the input signal.

Common-Base Amplifiers

A *common-base amplifier* is a bipolar junction transistor with both the input and output signals connected to the base. In this circuit, the input signal is applied to the emitter-base junction and the output signal is taken from the collector-base junction. The base is common to both the input and the output. It is the reference point in the circuit and is referred to as "ground". **See Figure 10-13.**

Milwaukee Electric Tool Corporation
DC electronic amplifier circuits are used in apllications such as portable radio/CD players.

Common-Emitter Phase Inversion

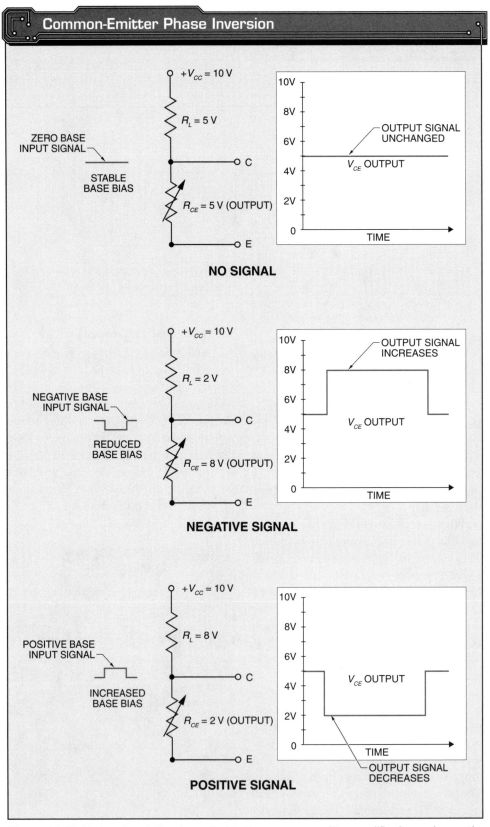

Figure 10-12. *Various input signals applied to the common-emitter amplifier base change the output signal (V_{CE}).*

Common-Base Circuits

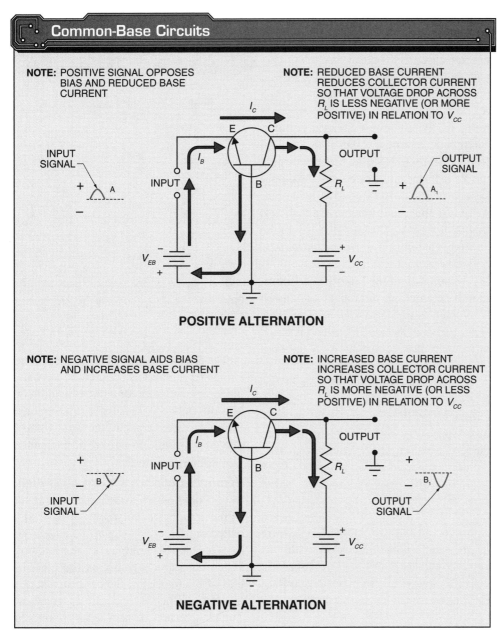

Figure 10-13. *In a common-base circuit, the input signal is applied to the emitter-base junction and the output signal is taken from the collector-base junction.*

Common-Base Circuit Operation. The input signal opposes the emitter-base voltage on the first alternation of the input signal when it is positive. It also reduces the emitter current accordingly. This reduces the collector current and thus, the voltage drop across output load resistor R_L decreases. The output voltage from collector to ground is equal to the algebraic sum of the voltage drop across R_L and the voltage of the collector source. Therefore, the reduction in voltage across R_L produces an increase in voltage from collector to ground. Thus, the output signal moves in a positive direction as the input signal moves in a positive direction.

Similarly, when the input signal swings in a negative direction, it adds to the emitter-base bias. Adding to the emitter-base bias increases collector current and produces an increased voltage drop across R_L. This increased voltage drop opposes more of the collector source voltage. Therefore, it reduces the output signal from collector to base. As the input signal moves in the negative direction, the output signal moves in the negative direction. *Note:* In the common-base amplifier, the input signal is applied directly to the emitter. The emitter current controls the collector current.

The amplification factor of all common-base amplifier circuits is designated by alpha (α), the current gain. *Note:* This is not the same as beta (β), which is based on a change in base current. Alpha is based on a change in emitter current.

Alpha is the ratio of the change in collector current (I_C) to the change in emitter current (I_E) with a constant collector voltage. Alpha is also often called the forward-current transfer ratio. Mathematically, alpha is expressed as follows:

$$\alpha = \frac{\Delta I_C}{\Delta I_E}$$

Since the emitter current is always larger than the collector current in this circuit, the current gain for a common-base circuit is always less than one. Therefore, the common-base circuit is rarely used.

Common-Collector Amplifiers

A *common-collector amplifier* is a bipolar junction transistor (BJT) with both the input and output signals connected to the collector. Because the output is taken from the emitter, this circuit is also called an emitter-follower circuit. It looks similar to the common-emitter circuit. However, the common-collector circuit has the input signal applied between the base and collector while the output is taken from the collector-emitter circuit. **See Figure 10-14.**

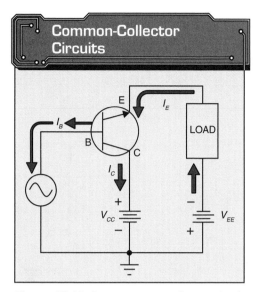

Figure 10-14. *In a common-collector circuit, the output is taken from the emitter.*

The biasing circuits of common-collector amplifiers are almost identical to the biasing circuits of common-emitter amplifiers. The difference is that the output voltage for the circuit is equal to the input voltage minus the voltage drop from the base of the transistor to the emitter.

Common-Collector Circuit Operation. In the common-collector circuit, the input signal is applied between the base and collector. The input signal either aids or opposes the forward bias of the transistor. When the input signal aids the forward bias of the transistor, the base current (I_B) increases. When it opposes, I_B decreases. The change in I_B causes a corresponding change in emitter current I_E and collector current I_C.

The output voltage of the circuit is developed across the load. The load is connected between the emitter and collector regions of the transistor. The emitter current (I_E) flowing through the load is much greater than the base current (I_B). Therefore, the circuit provides an increase in current between the input and output terminals.

The voltage developed across the load is always slightly lower than the voltage applied to the circuit. The slightly lower voltage appears at the emitter of the transistor.

This is because the transistor tends to maintain a relatively constant voltage drop across its emitter-base junction. The forward-bias voltage drop may be equal to approximately 0.2 V if the transistor is made of germanium or 0.6 V if the transistor is made of silicon. Therefore, the output voltage appearing at the emitter tends to track, or follow, the input voltage applied to the base of the transistor.

The common-collector circuit functions as a current amplifier, but it does not produce an increase in voltage. The increase in output current results in a moderate increase in power. The input resistance of any transistor connected in a common-collector arrangement is extremely high. This is because the input resistance is the same resistance that appears across the reverse-biased collector-base junction. The input resistance can be as high as several hundred thousand ohms in a typical low-power transistor. The output resistance appearing between the emitter and collector regions of the transistor will be much lower because of the relatively high emitter current (I_E) that flows through the output lead. It is often as low as several hundred ohms.

The common-collector circuit is used widely in applications where its high input resistance and low output resistance can perform useful functions. The circuit is often used to couple high-impedance sources to low-impedance loads. Therefore, it can perform the same basic function as an impedance-matching transformer. *Note:* Since the output signal corresponds to the input signal, there is no phase shift from input to output.

OPERATING POINTS

A common-emitter amplifier schematic and its corresponding characteristic curves are used to set the operating point. The operating point, or quiescent (Q) point, is established on a load line. The load line is where the transistor is biased when no signal is applied to the input. **See Figure 10-15.**

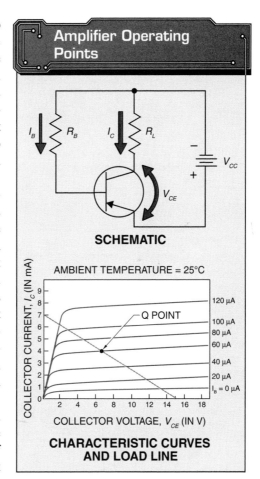

Figure 10-15. *The operating point, or quiescent (Q) point, of a typical common-emitter amplifier is established on a load line.*

Operating points should be chosen to accomplish the basic operation of the circuit. A signal swing should be set so that the signal never reaches the actual extremes of saturation or cutoff. Distortion of the signal would result if the signal reaches saturation or cutoff. **Figure 10-16.**

Linear Amplification

A *linear amplifier* is an amplifier that increases and maintains the exact duplicate of an input signal. In other words, the output signal is the same as the input signal, only it is larger. **See Figure 10-17.** Stereo amplifiers are linear so that music can be duplicated with the least amount of distortion. *Distortion* is any undesirable change in a signal.

254 SOLID STATE DEVICES AND SYSTEMS

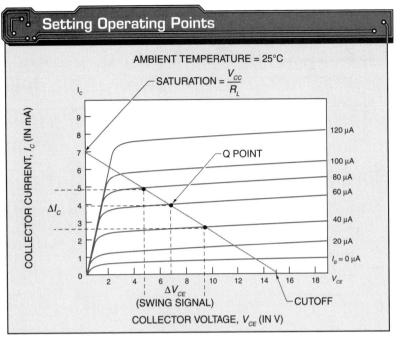

Figure 10-16. When selecting an operating point, the signal swing should never reach the extremes of saturation or cutoff.

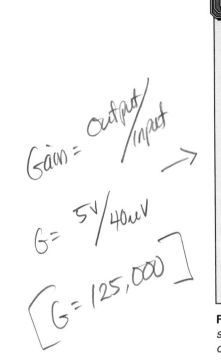

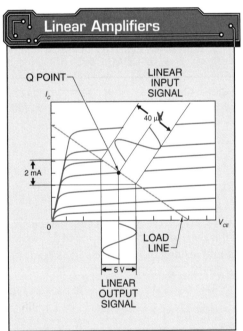

Figure 10-17. Linear amplifiers have an output signal that is the same as the input signal, only larger.

Amplifier distortion is often referred to as total harmonic distortion (THD) and is expressed as a percentage, such as 0.05% THD. The lower the percentage value, the higher the quality of the amplifier.

Signal Change Due to Saturation

A signal may be clipped or distorted by an operating point located too close to saturation. When the input signal swings positive, the transistor reaches its maximum output even though the signal continues to increase. There is no more output even with further signal increases. Any increase of signal is lost. In certain cases, clipping may be desirable, but in audio circuits, it is considered a form of distortion. **See Figure 10-18.**

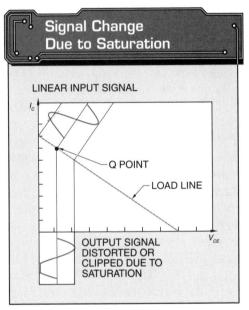

Figure 10-18. A signal may be clipped or distorted by operating at a point too close to saturation.

Signal Change Due to Cutoff

A signal may be clipped or distorted by an operating point located too close to cutoff. When the signal swings negative, the signal continues to be reproduced until the transistor reaches cutoff. Through cutoff, that part of the signal is also lost. Again, clipping of this nature is unacceptable for linear operation. However, it may be used in certain types of circuits, such as in a radio transmitter circuit. **See Figure 10-19.**

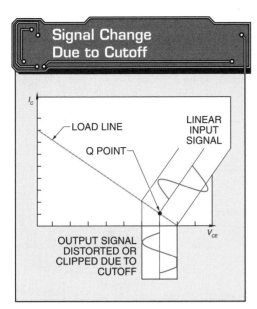

Figure 10-19. *A signal may be clipped or distorted by operating at a point too close to cutoff.*

Signal Change Due to Saturation and Cutoff

When an input signal for a normal operating amplifier is too large, clipping or distortion may occur on both ends (alternations) of the signal. **See Figure 10-20.** A good example of this type of problem is when a volume control is turned up too high. An advantage of clipping or distortion on both ends of a signal is that either can be used to develop switching pulses to trigger a circuit.

CLASSES OF OPERATION

The classes of amplifier operation are designated by letters that reference the level of amplifier operation in relation to the cutoff condition. The cutoff condition is the point at which all collector current is stopped due to the absence of base current. The four main classes of operation for an amplifier are designated as A, B, AB, and C.

Class A Operation

Class A amplifier operation is well above the cutoff condition for an amplifier. It is near the center of the straight portion of the characteristic curve on the load line. With Class A operation, collector current flows during the entire input signal. The output should duplicate the input and increase it in amplitude. Class A amplifiers are also called linear amplifiers. Class A amplifiers are used as audio amplifiers where true reproduction of the waveform is demanded. **See Figure 10-21.**

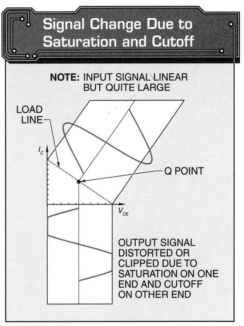

Figure 10-20. *When an input signal is too large, clipping and distortion may occur at both ends (alternations) of the signal.*

Class B Operation

Class B operation is located near the lower end of the load line closer to cutoff. A Class B amplifier reproduces only 180° of the input signal sine wave. Collector current flows only during one-half of the input signal. Operation during the negative half cycle does not produce base current or collector current. Class B amplifiers are used mostly with a pair of transistors in the power output of audio amplifiers.

Class AB Operation

Class AB operation is somewhere between Class A and Class B amplifiers on the load line. With Class AB operation, collector current flows for more than half the input signal but less than the full input signal.

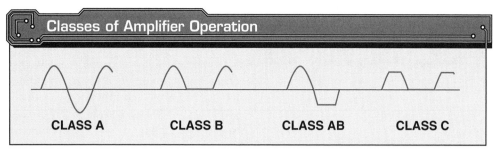

Figure 10-21. *The four main classes of operation for an amplifier are designated as A, B, AB, and C.*

Class C Operation

Class C operation is located closest to cutoff. With Class C operation, collector current flows for less than one-half the input signal. Class C operation is usually found in radio-frequency transmitters.

INPUT AND OUTPUT IMPEDANCES

Input impedance is the loading effect an amplifier presents to an incoming signal. **See Figure 10-22.** In other words, the signal sees the amplifier as a load. The amount of load or resistance is the input impedance.

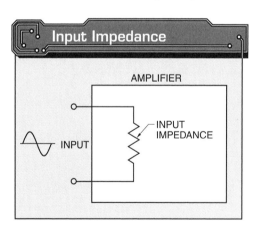

Figure 10-22. *Input impedance is the loading effect an amplifier presents to an incoming signal.*

Output impedance is the loading effect an amplifier presents to another device. **See Figure 10-23.** The device could be another stage of amplification or an output device, such as a speaker. The amount of input impedance and output impedance varies with different types of amplifier configurations.

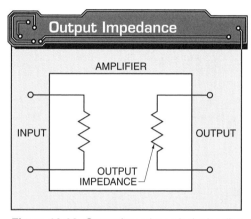

Figure 10-23. *Output impedance is the loading effect an amplifier presents to another device.*

Impedance matching is the process of setting the input impedance of a load equal to the output impedance of a signal source. When connecting microphones, antennas, speakers, or other types of devices to an amplifier, each must be impedance-matched for proper operation. **See Figure 10-24.**

Common-Emitter Amplifier Impedances

A typical PN junction has some resistance because of the presence of impurities and minority carriers. This resistance is relatively low (usually a few hundred ohms). It is called the base-emitter impedance, or the input impedance. *Note:* Base-emitter impedance is a low value and is the input impedance of the common-emitter amplifier.

The reverse-biased collector-base circuit with a relatively large voltage applied to the circuit has very little current flow because of the reverse-bias connection. In other words, it appears as a large value of resistance. A typical resistance value for a transistor is 100,000 Ω. The collector-emitter circuit represents the output for a common-emitter amplifier and has high impedance.

Common-Base Amplifier Impedances

The amount of change in emitter-base impedance needed to cause an appreciable change in emitter current is very small (usually 100 Ω or less.) Hence, the input circuit presents low impedance to the source. However, the output impedance of the transistor is very high (usually exceeding 100,000 Ω). It is high because the collector current is independent of the collector-base voltage. This large ratio of output impedance to input impedance makes it possible for the amplifier to produce a large voltage gain. A large voltage gain is produced despite the absence of current gain. Power gain for this amplifier is moderate.

Common-Collector Amplifier Impedances

The input impedance of a common-collector circuit is high (approximately 100,000 Ω). The output impedance is low (1000 Ω or less). The common-collector transistor circuit is used primarily for impedance matching or for isolation of coupling transistors. The common-collector circuit also has the ability to pass signals in either direction (bilateral operation).

The common-collector circuit is often used as an isolation amplifier because of its high input impedance. The high impedance loads the input signal and allows very little current flow. This in turn causes the signal to be isolated from the other stages. Generally, when a common-collector circuit is used in this way, a common-emitter amplifier follows to provide good gain. **See Figure 10-25.**

> *For signal transfer to occur with minimal loss, an amplifier should have extremely high input impedance (preferably infinite) and extremely low output impedance (preferably zero).*

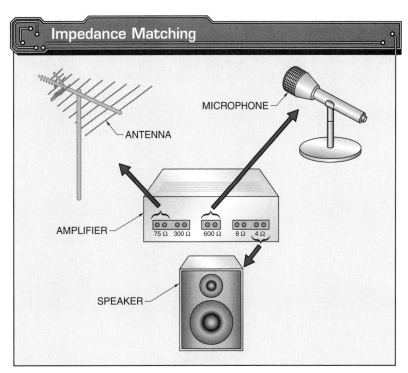

Figure 10-24. *When connecting microphones, antennas, speakers, or other types of devices to an amplifier, each must be impedance-matched for proper operation.*

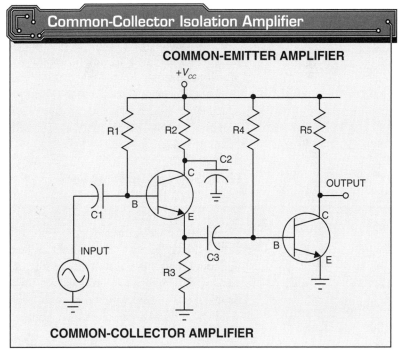

Figure 10-25. *The common-collector transistor circuit can be used for impedance matching or for the isolation of coupling stages. When a common collector is used in this manner, a common-emitter amplifier follows to provide good gain.*

TRANSISTOR SPECIFICATION SHEETS

Semiconductors have unique characteristics. The characteristics of these devices are presented on specification sheets. A typical specification sheet for a transistor includes maximum ratings, mechanical data, heat sinking information, and characteristic curves.

Maximum Ratings

The maximum ratings for a transistor must not be exceeded under any operating condition. Permanent damage to the device would result if the maximum ratings were exceeded. Typically, transistors are operated well below these ratings to ensure long life and proper performance.

The maximum ratings, as specified in data sheets, have been defined by the Joint Electronic Device Engineering Councils (JEDEC) and standardized by the Electronics Industries Association (EIA). The maximum ratings include voltage, current, power dissipation, and temperature.

Mechanical Data

The mechanical data for a transistor usually includes items such as weight, mounting position, dimensions, lead length, and type of case. Mechanical data can be used to specify the dimensions and pin layout of transistors when designing a printed circuit board using design software.

Heat Sinking Information

Transistor collector heat dissipation is dependent on ambient temperatures. As the temperature increases, the permissible wattage dissipation decreases. When replacing transistors, the wattage rating at various temperatures is critical and must be taken into consideration for proper heat sinking. Heat sinking information is usually provided on a chart or graph.

Characteristic Curves

Most manufacturers provide a set of characteristic curves on their specification sheets. These characteristic curves can be used in designing and developing new circuits. Generally, an electrician or technician has no need to use these curves.

TRANSISTOR TESTERS

A *transistor tester* is an electric device that tests the electrical characteristics of transistors and diodes. Transistor testers are typically used to test transistors for proper operation, current gain, and leakage current. They are also used to identify transistors as NPN or PNP. The three types of transistor testers are the in-circuit, field-service, and laboratory-standard transistor testers.

In-Circuit Transistor Testers

An in-circuit transistor tester is used to quickly determine whether a transistor is still operating. The primary advantage of this type is that the transistor does not have to be removed from the circuit. It should be noted that the information gained from this instrument is limited. The same information could be obtained just as readily by using a high-impedance voltmeter.

Field-Service Transistor Testers

A field-service transistor tester gives more information about the operating condition of a transistor than an in-circuit transistor tester. The field-service transistor tester detects current gain, leakage current, and any shorts or opens in a PN junction. Again, it should be noted that with the proper voltage measurements and a few simple calculations, all this information is available to the technician.

Laboratory-Standard Transistor Testers

A laboratory-standard transistor tester is found in laboratories and quality assurance stations. Laboratory-standard transistor testers give information that voltage checks cannot give. They measure transistor characteristics under actual operating conditions. This type of transistor tester is expensive and generally not used for service work.

TRANSISTOR SERVICE

When servicing a transistor circuit, the following points should be observed to prevent damage to the transistor:

- The voltage should be OFF before removing or installing a transistor. This prevents surge currents from damaging the transistor.
- Before the voltage is reconnected to a circuit, the transistor must be firmly and correctly inserted into its socket.
- Because transistors are very sensitive to improper bias voltages, testing transistors by short-circuiting various points to ground should be avoided. A short circuit from base to collector would immediately destroy the transistor.
- When making voltage measurements, meter probes must not be allowed to short-circuit terminals that are close together.
- A voltmeter should have a sensitivity rating of 20,000 Ω/V or more on all ranges when used with transistor circuits. A high-impedance voltmeter, such as an FET meter, is preferred.
- When making resistance measurements on a transistor, the transistor must be removed from its socket or unsoldered for testing.
- When bench testing transistor circuits, it is important to use well-regulated power supplies. Semiconductors respond to very minute changes in current. Poorly regulated power supplies may introduce other problems in addition to the one that is being corrected.

MULTISTAGE AMPLIFIERS

A single stage of amplification is often not sufficient to drive a load to the required amount of output from any given input source. Therefore, various stages of amplification must be coupled together, or connected, to build up the output to the required level. Coupling transfers a signal from the output of one stage to the input of the next without distortion or loss.

Amplifier Coupling

Amplifier coupling is the joining of two or more circuits so that power can be transferred from one amplifier stage to another. The three basic techniques for coupling amplifier stages together and to loads are capacitive coupling, transformer coupling, and direct coupling. When choosing a coupling technique for different applications, frequency, impedance matching, cost, size, and weight must be considered. Therefore, no one coupling technique is always best.

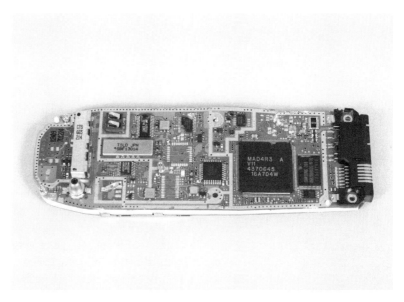

Amplifiers are sometimes used with small electronic devices such as cellular telephones.

Capacitive Coupling. Capacitive coupling has the disadvantage of poor impedance matching. It also limits the lower frequency response of the amplifier. Despite these drawbacks, capacitive coupling is often used because it is inexpensive and few components are required. If signal losses are encountered, they are generally offset by adding another transistor stage. Another reason why capacitive coupling is used is that it can block or isolate the bias circuits of each stage. The series coupling capacitor presents an open circuit to the flow of DC.

When a 5 V source is attached to a series coupling capacitor circuit, the capacitor (C1) is charged to 5 V. Current flows through resistor R1 only while capacitor C1 is charging. A voltage appears momentarily across resistor R1. Once capacitor C1 is charged, the voltage across it should be 5 V, and the voltage across resistor R1 should be 0 V. If the source voltage were increased to 10 V, capacitor C1 would be charged to 10 V. The charging current should produce another momentary 5 V pulse of voltage across resistor R1. If the voltage source were reduced to 3 V, capacitor C1 would be discharged to 3 V. The discharge current would produce a momentary pulse of 7 V across resistor R1 because it would be discharging to a lower voltage. **See Figure 10-26.**

Both AC and DC voltage can be present in the circuit. The input signal of 2 VAC varies around a DC level of 5 V. The AC signal causes the input to vary between 7 V and 3 V. **See Figure 10-27.** The capacitor charges and discharges according to the varying AC signal, assuming that the capacitance value has been selected to pass the lowest frequency. It should be noted that the output varies at around 0 V. The DC is effectively removed because the capacitor blocks the DC component.

Because a capacitor can block a DC component, the capacitor is called a blocking capacitor, or coupling capacitor. Blocking capacitors are used to connect one section of a multistage amplifier to another. **See Figure 10-28.**

Transformer Coupling. For maximum power transfer between amplifier stages, or between a stage and the load, correct impedance matching is required. Special interstage transformers are used to match input and output impedances between stages. Transformers are also used to match load impedance to amplifier output impedance. In the transformer coupling circuit, capacitor C2 is required to prevent the fixed bias of transistor Q1 from passing through transformer secondary T1. **See Figure 10-29.**

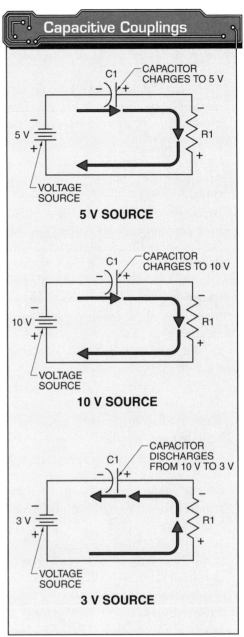

Figure 10-26. *The voltage source attached to the series coupling capacitor circuit charges or discharges capacitor C1 to the same voltage as the source.*

Although transformer coupling can provide a high level of circuit efficiency, the weight, size, and cost of transformers rule out this technique for many applications. Also, frequency response is better with capacitive coupling circuits, especially at higher frequencies.

Blocking DC Sources

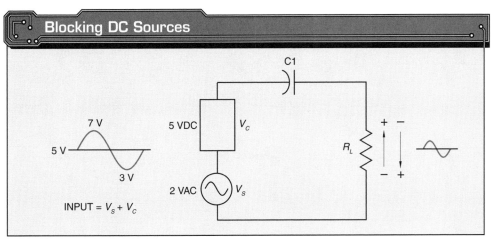

Figure 10-27. *When the 5 VDC source is effectively blocked, the 2 VAC source is allowed to pass on to load resistor R_L.*

Blocking Capacitors

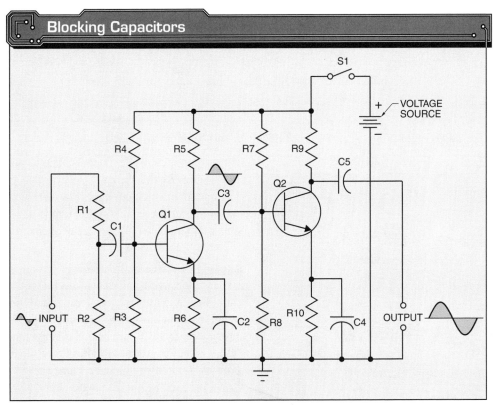

Figure 10-28. *Blocking capacitors, or coupling capacitors, are used to keep DC voltage from passing from one stage to another but allow AC signals to pass easily.*

Direct Coupling. For many industrial circuits, it is necessary to amplify very low frequency signals. Capacitive and transformer coupled amplifiers have poor frequency response at low frequencies because both of them block low frequency signals. Therefore, direct coupling must be used when low frequencies are involved. Direct coupling is also used where the DC value and the AC value of a signal must be retained. The directly coupled amplifier provides a frequency response that ranges from zero hertz (DC) to several thousand hertz.

262 SOLID STATE DEVICES AND SYSTEMS

Transformer Coupling

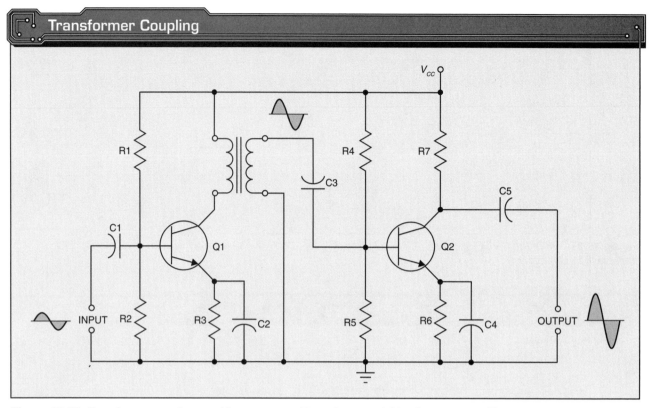

Figure 10-29. *Transformer coupling provides a means of impedance matching between amplifier stages.*

IBM

Transistors may be used in a wide array of electronic equipment, including computers and TVs.

In a direct-coupled amplifier circuit, the collector of transistor Q1 is connected directly to the base of transistor Q2. **See Figure 10-30.** The collector load resistor (R2) also acts as a bias resistor for transistor Q2. Any change in bias current is amplified by the direct-coupled circuit, which is very sensitive to temperature changes. This disadvantage can be overcome with bias stabilization.

Darlington Circuits. It is often desirable to have a wide frequency range in a transistor amplifier. A wide frequency range can be achieved by using a Darlington circuit. A Darlington circuit uses two bipolar transistors and a direct connection without any other type of coupling circuit. **See Figure 10-31.** The advantages of the Darlington circuit are extensive. For example, fewer components are required and there is no loss incurred in the coupling. The Darlington circuit can be used as an output stage of a transistor amplifier that is driven by a stage of low or intermediate power. In a typical audio amplifier, coupling of this type of circuit requires very large capacitors. Therefore, it is inefficient. By using the Darlington circuit, smaller and lighter bipolar transistors may be used and less power is dissipated.

Direct Coupling

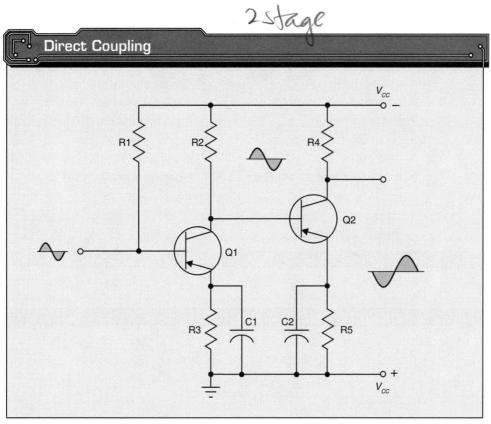

Figure 10-30. *Direct coupling should be used when low frequencies are involved. It should also be used when the DC value and the AC value of a signal must be retained.*

Darlington Circuits

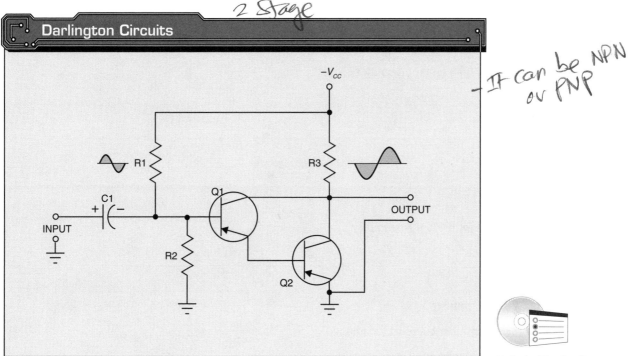

Figure 10-31. *An advantage of the Darlington circuit is that it uses fewer components. Also, there is no loss incurred in the coupling.*

Refer to Chapter 10 Quick Quiz® on CD-ROM

KEY TERMS

- gain
- cascade amplifier
- current gain
- voltage gain
- power gain
- single-stage amplifier
- bandwidth
- frequency-response curve
- half-power point
- wide bandwidth
- narrow bandwidth
- decibel (dB)
- common-emitter amplifier
- phase inversion
- common-base amplifier
- alpha (α)
- common-collector amplifier
- linear amplifier
- distortion
- class A operation
- class AB operation
- class B operation
- class C operation
- input impedance
- output impedance
- impedance matching
- isolation amplifier
- transistor tester

REVIEW QUESTIONS

1. What is amplification?
2. What is gain?
3. Explain the difference between amplifier current gain and beta (β).
4. Define bandwidth.
5. What is a decibel?
6. Describe the three basic types of transistor amplifiers.
7. What is phase inversion?
8. What are the forward voltage drops usually associated with germanium and silicon?
9. What is another name for a common-emitter circuit?
10. Where are common-collector circuits used?
11. What is a linear amplifier?
12. Define distortion.
13. What causes clipping?
14. What are the four main classes of amplifier operation?
15. What is another name for a Class A amplifier?
16. Define input impedance.
17. Define output impedance.
18. What is a multistage amplifier?
19. What are the major amplifier coupling techniques?
20. What are the advantages of using Darlington circuits?

JFETs, MOSFETs, AND IGBTs

Field-effect transistors (FETs) are the cornerstone of new developments in semiconductor technology. Common FETs include the junction field-effect transistor (JFET), metal-oxide semiconductor field-effect transistor (MOSFET), and insulated gate bipolar transistor (IGBT). The IGBT combines the advantages of a power MOSFET and a bipolar junction transistor. The characteristics of FETs make them ideal for amplification. Many amplifiers have low input impedance. By using an FET with a multistage amplifier, high impedance can be obtained.

OBJECTIVES

- Describe the types of field-effect transistors (FETs).
- Explain the operation of junction field-effect transistors (JFETs).
- Describe JFET circuit configurations.
- List JFET Applications.
- Explain the operation of metal-oxide semiconductor field-effect transistors (MOSFETs).
- Explain power MOSFETs.
- Describe insulated gate bipolar transistors (IGBTs).
- Explain troubleshooting IGBTs.
- Describe multistage amplifiers.

FIELD-EFFECT TRANSISTORS (FETs)

Field-effect transistors have been the cornerstone of new developments in semiconductor technology. A *field-effect transistor (FET)* is a three- or four-terminal device in which output current is controlled by an input voltage. FET terminals include a gate, drain, and source.

Two of the most common types of FETs include the junction field-effect transistor (JFET) and the metal-oxide semiconductor field-effect transistor (MOSFET). The newest type of FET is the insulated gate bipolar transistor (IGBT). The IGBT combines the advantages of a power MOSFET and a bipolar junction transistor.

JUNCTION FIELD-EFFECT TRANSISTORS (JFETs)

A *junction field-effect transistor (JFET)* is a simple field-effect transistor (FET) with a PN junction in which output current is controlled by an input voltage. The two types of JFETs include the N-channel and P-channel. A JFET, like all FETs, contains a gate (G), drain (D), and source (S). The gate is a control element, while the drain and source provide the same function as the emitter and collector on a bipolar junction transistor. **See Figure 11-1.**

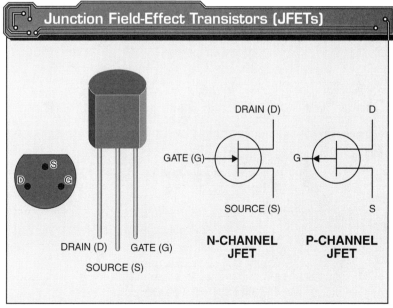

Figure 11-1. *The arrow of a schematic symbol for an N-channel JFET points inward. The arrow points outward for a P-channel JFET.*

265

JFETs may be used as direct current (DC), audio frequency (AF), and radio frequency (RF) amplifiers. They can also be used as switches and gates. JFETs are even found in voltage regulators and current limiters. JFETs are fully compatible with other semiconductor devices, such as standard bipolar transistors, silicon-controlled rectifiers, SCRs, triacs, and ICs.

JFET Operation

A JFET is a unipolar device, which differs in operation from a bipolar junction transistor. The output current of a JFET is controlled by the voltage on the input. The input voltage creates an electrical field, or depletion region, within the device. **See Figure 11-2.** The JFET is considered a voltage-driven device rather than a current-driven device like the bipolar junction transistor.

JFET Operation Media Clip

The source and drain of the JFET are connected to a common N-type material. This common material constitutes the channel of the JFET. If a DC potential is connected between the source and the drain, current should flow in the external circuit and through the channel. At zero gate voltage, channel height is maximum and channel resistance is minimum, resulting in current flow. With a slight positive voltage, the channel height opens further, allowing maximum current flow.

In 1926, the Polish-American physicist Julius E. Lilienfeld submitted a patent, "Method and Apparatus for Controlling Electric Current," which described a three-terminal device and semiconductor material, or the first FET.

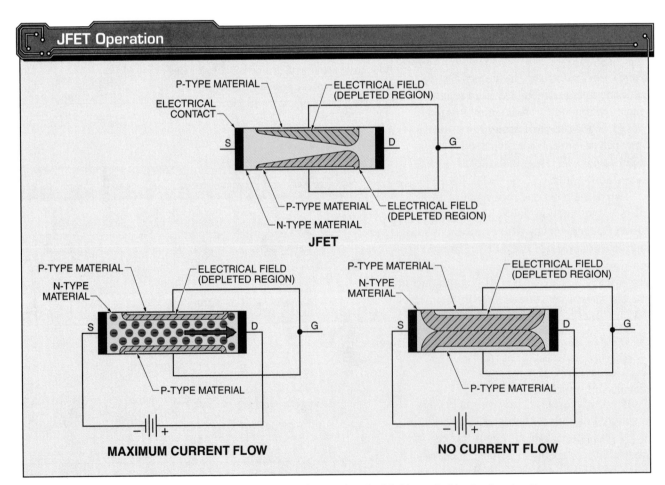

Figure 11-2. *The output current of a JFET is controlled by an electrical field created by the input voltage.*

The two sections of P-type material constitute the gate. Each section has its associated electrical field, or depletion region. If the gate is made negative with respect to the channel, the diodes (formed by the P-type and N-type materials) become reverse-biased and the depletion regions increase. At a large enough gate-to-source voltage (V_{GS}), the channel is effectively "pinched off" because the depletion regions touch. The depletion regions merge at a particular gate-to-source voltage of somewhere between 1 V and 8 V.

The size of the depletion region is controlled by the gate-to-source voltage, or V_{GS}. When V_{GS} increases, the depletion region increases. When V_{GS} decreases, the depletion region decreases.

This gate-to-source voltage controls the drain current and must always provide a reverse-bias voltage. This is quite different from a bipolar transistor. In a bipolar transistor, the junction must be forward-biased so that the junction impedance is extremely low, and there is current flow in the junction. In a JFET, the junction must be reverse-biased so that the junction impedance is high, and there is little current flow in the junction. The major advantage of this is good control using voltage rather than current. The power consumption of the JFET (in standby operation) is thousands of times less than that of a bipolar transistor controlling the same function.

JFET Output Characteristic Curves

A JFET characteristic curve shows the operating characteristics of a JFET. **See Figure 11-3.** The drain source voltage (V_{DS}) is plotted along the horizontal axis, and the drain current (I_D) is plotted along the vertical axis on the characteristic curve for a typical N-channel JFET. By using one curve, the detailed information produced may be more readily obtained.

The ohmic region is the first portion of the characteristic curve of a JFET where the drain current rises rapidly. In the ohmic region, the drain current is controlled by the drain-source voltage and the resistance of the channel. The knee of the curve is the pinch-off voltage. On the flattened portion of the graph, the JFET is at the pinch-off region. In this region, the drain current is controlled primarily by the width of the channel. When the drain current begins another sharp increase, avalanche (breakdown) begins. At this point, a large amount of current begins to pass through the channel. If this current is not limited by an external resistance or load, the JFET may be damaged.

Note: A JFET should never be forward-biased. If the voltage applied to the gate source were forward-biased, the drain current would increase to very high levels. Also, forward biasing the gate source causes the input impedance to drop to a low value.

JFET Power Dissipation

The JFET, like most semiconductor devices, can dissipate only a certain amount of power. The safe dissipation limits for a JFET are listed on the manufacturer specification sheet, which usually has a power dissipation curve.

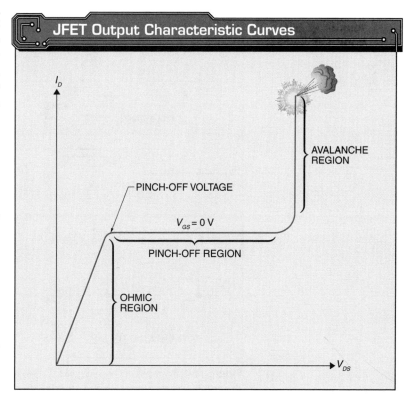

Figure 11-3. *A JFET characteristic curve shows the operating characteristics of a JFET.*

The normal dissipation rating applies to ambient temperatures of 25°C or less. If a JFET must operate at a high temperature, it must be derated according to manufacturer specifications.

JFET Circuit Configurations

JFETs can be connected into three basic circuit configurations: common source, common gate, and common drain. Common-gate and common-drain circuits are constructed in a similar common element arrangement. Furthermore, the only difference between an N-channel and P-channel circuit configuration is the direction of electron flow in the external circuit and the polarity of the bias voltages. **See Figure 11-4.**

Common-Source Amplifier Configuration. The common-source JFET can be biased from the voltage drop across a source resistor (R_s) in series with the source terminal. **See Figure 11-5.** In each circuit configuration (N-channel and P-channel JFET), the gate bias voltages have opposite polarities and opposite directions of current flow. For the N-channel JFET, the current flow is from ground through R_s. This makes the source side of R_s positive with respect to ground.

> JFETs do not have actual anodes and cathodes because the channel is only either N or P, which determines electron flow and bias voltage polarity.

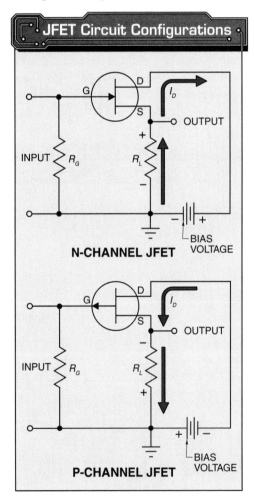

Figure 11-4. *In a common-source JFET circuit, the input is across the gate and source terminals, and the output is across the drain and source terminals.*

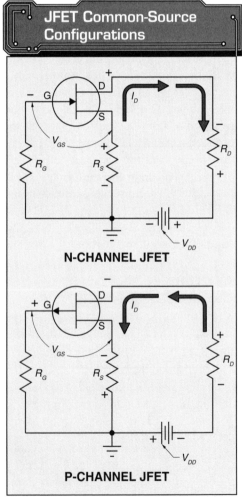

Figure 11-5. *A common-source JFET can be biased from the voltage drop across resistor R_s in series with the source terminal.*

In the N-channel JFET circuit, the voltage across the gate resistor (R_G) has a polarity that correctly biases the gate negative with respect to the source since R_G is also returned to ground. The opposite relationship exists in the P-channel JFET. In the P-channel JFET circuit, the current direction is such that the source side of R_S is negative with respect to ground. Hence, the gate is made positive with respect to the source. This is the correct polarity for reverse biasing the junction when the channel is made of P-type semiconductor material.

Common-Source Amplifier Operation. In a common-source amplifier, an AC signal is applied across gate resistor R_G. The DC bias is set by the DC voltage drop across source resistor R_S. To reduce the effect of AC on the DC bias, capacitor C is used as a bypass to shunt any AC signals around resistor R_S. Thus, a constant DC gate-to-source voltage level is maintained. With the input signal applied across resistor R_G, the signal varies the gate voltage around the DC operating bias, causing a variation in I_D.

In a P-channel JFET circuit, the positive swing of the input signal increases the gate bias. The drain current decreases and the drain voltage becomes more negative. This produces a negative swing in the output signal. A negative swing of the input signal decreases the gate bias. The drain current increases, and the drain voltage becomes more positive, producing a positive swing in the output signal. **See Figure 11-6.**

In N-channel JFET circuit, a positive swing of the input signal decreases the gate bias. This causes an increase in the drain current and a less positive (more negative) swing of the drain voltage and output signal. The result is a drop in the positive drain voltage and a negative swing of the output signal. The output signal at the drain then is 180° out of phase with the input signal. The common-source circuit is the only JFET configuration that produces an inverted output signal. **See Figure 11-7.**

The input resistance of a JFET can be greater than 100 MΩ. The gate resistor greatly influences the input resistance. In most cases, the value of the gate resistor determines the input impedance of the stage. Therefore, the gate resistor usually has a very high resistance (1 MΩ or greater).

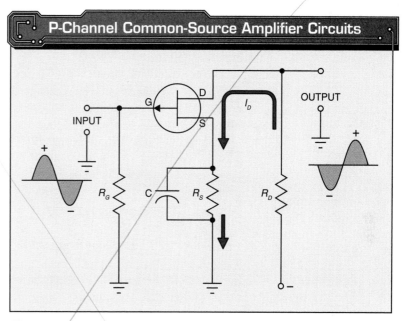

Figure 11-6. In a P-channel JFET circuit, a positive swing of the input signal results in an increase in the drain current and the drain voltage becomes more positive, producing a positive swing in the output signal.

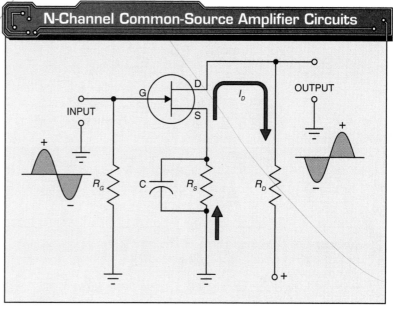

Figure 11-7. In an N-channel common-source amplifier, a positive swing of the input signal results in a drop of the positive drain voltage and a negative swing of the output signal.

Common-Gate Amplifier Configuration.
In common-gate amplifier circuits using N-channel and P-channel JFETs, the AC input signal is applied to the source and the gate, with the gate at ground potential. The input impedance of the common-gate configuration is the lowest of the three basic configurations. The output signal is present between the drain and the gate and has a high output impedance.

The common-gate circuit is used in applications where the source of the signal requires a low-to-high impedance match. The common-gate is often used in communication circuits. This is because it is capable of good voltage gain and reasonable power gain over a wide range of frequencies (particularly those in the higher frequency ranges). Thus, the common-gate is often found in UHF and VHF amplifiers.

Common-Gate Amplifier Operation.
The input signal and output signal are in phase for the common-gate amplifier. For the N-channel, a negative swing of the input signal decreases the gate source bias, increases the drain current, and decreases the drain voltage. Conversely, a rise in the input signal voltage increases the gate bias, decreases the drain current, and increases the drain and output voltages. The output voltage is amplified and is in phase with the input voltage. **See Figure 11-8.**

In the P-channel, the positive swing of the input signal decreases the gate bias. Therefore, the drain current increases, and the drain voltage becomes less negative (positive swing of the output signal). Conversely, with the negative swing of the input signal, there is a higher gate-to-source bias. Also, the drain current drops, and the drain voltage becomes more negative.

Junction field-effect transistors (JFETs) are also known as junction unipolar gate field-effect transistors (JUGFETs). When drawing JFET symbols, the symbol should include a circle around the component to indicate that it is a discrete device; however, in schematic diagrams, the JFET symbol is typically drawn without the circle. Recently, the JFET symbol has been drawn without a circle for discrete devices.

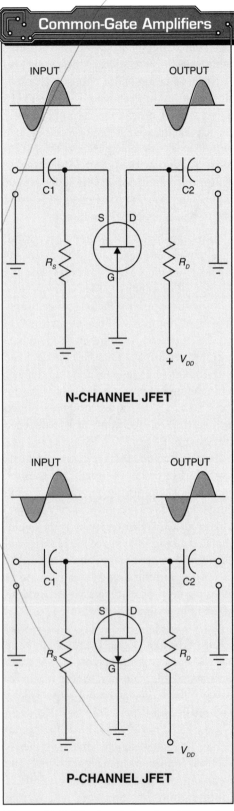

Figure 11-8. *The input signal and the output signal are in phase for the common-gate amplifier.*

Common-Drain Amplifier Configuration. In common-drain circuits for N-channel and P-channel FETs, the input resistance is high. The input resistance is high because of the degenerative effect of the AC output voltage developed across source resistor R_S. Like the emitter-follower stage, the output voltage from the common-drain is always less than the input voltage. Also, the circuit is not capable of voltage gain. However, the common-drain does produce good current gain and power gain. **See Figure 11-9.**

The output impedance is low for the common-drain circuit because the signal is taken from the source terminal. Since the output impedance is low, the common-drain stage is often used when a high-impedance signal source must be matched to a low-impedance load.

Common-Drain Amplifier Operation. In the N-channel circuit, a positive swing of the input signal decreases the gate bias. As a result, the drain-to-source current rises and the source and output voltages swing more positive. Conversely, a negative swing of the input signal increases the gate bias and decreases the drain-to-source current. The source voltage becomes less positive because the current decreases in the source load resistance (negative swing of the output signal).

For the P-channel circuit, the positive swing of the input signal increases the gate bias. Therefore, the drain-to-source current decreases. Also, the voltage drop across the source resistor is less negative (positive swing of the output signal). During the negative swing of the input signal, the gate bias decreases. Therefore, the drain-to-source current rises in the source resistor. Because of the direction of the current, this rise makes the source more negative, and there is a negative swing of the output signal voltage.

The voltage change at the output has the same direction as the input change, so the signals are in phase. Both common-gate and common-drain circuits have in-phase signal outputs. The common-source amplifier is the only JFET configuration that inverts the phase of the signal.

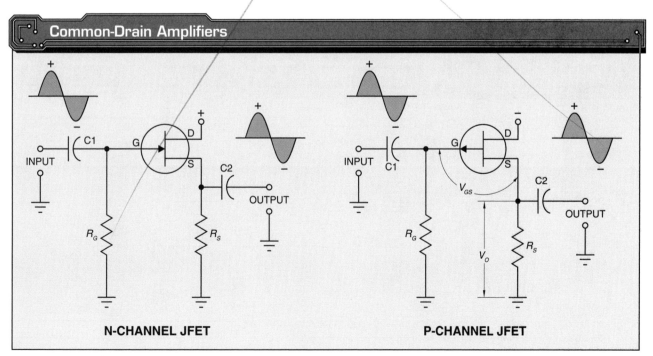

Figure 11-9. *The common-drain amplifier produces good current gain and power gain.*

JFET Applications

The JFET is used extensively in circuits where low power and high impedance are factors. JFETs need very little power to produce a large output at the load. JFETs are also easily matched to other semiconductor devices such as SCRs, triacs, and ICs.

JFETs are used in many amplifier configurations because they can amplify a fairly wide range of frequencies and have high input impedance. High input impedance is a definite advantage in the transfer of power. Most bipolar transistors have low input impedance and require a relatively powerful input signal to produce a large output at the load.

Preamplifiers. The JFET common-source amplifier can be used as a low-level preamplifier. A *preamplifier* is a circuit that provides gain for a weak signal before the signal goes through the normal stages of amplification. **See Figure 11-10.** This circuit permits direct input from a high-impedance, small-signal device, such as a crystal microphone. It then matches the microphone to a low-impedance power amplifier. Source resistor R_S provides gate bias. To obtain a higher AC gain from this circuit, a bypass capacitor (C1) connected across source resistor R_S maintains the bias on the gate at the desired operating point.

Impedance Matching. An impedance-matching circuit can step down impedance levels and preserve the bandwidth and linearity of the signals being amplified. In this case, a high-impedance microphone is coupled to a low-impedance coaxial cable to avoid losing frequency response. A common-drain JFET circuit (Q1) can also step up the input impedance of a bipolar transistor (Q2). In effect, the common drain serves as an impedance transformer with power gain. **See Figure 11-11.**

Multimeters. FET multimeters use JFETs to provide high input impedances with good sensitivity. The JFET is capable of producing an input of 11 MΩ or more with sensitivities exceeding those of a vacuum tube voltmeter (VTVM). *Note:* The "J" in "JFET" is usually dropped when referring to an FET multimeter.

In a simple FET multimeter circuit, the JFET is one "leg" in the DC bridge. Source resistor R_S provides negative feedback for high linearity in the response. The circuit requires good regulation, which is provided by the zener diode. **See Figure 11-12.**

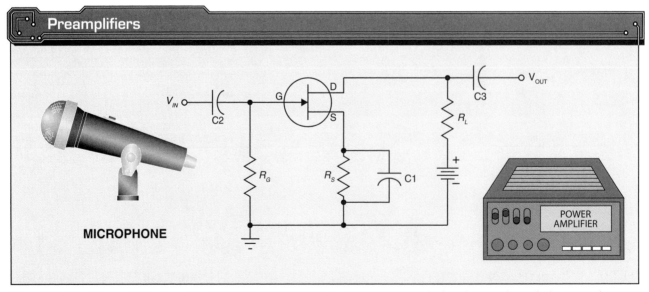

Figure 11-10. *A preamplifier is a circuit that provides gain for a very weak signal before it goes through the normal stages of amplification.*

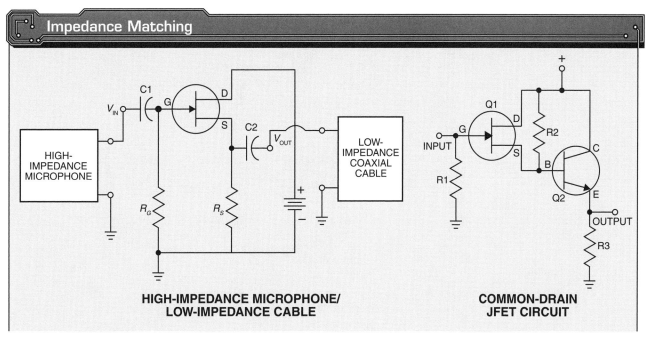

Figure 11-11. *An impedance-matching circuit can step down impedance levels and preserve the bandwidth linearity of the signals being amplified.*

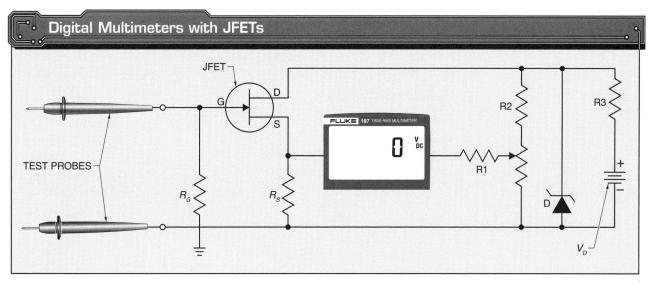

Figure 11-12. *A multimeter can provide high input impedance with good sensitivity by using a JFET in its front-end circuit.*

Voice-Operated Relays. In the circuit, a JFET provides a high impedance match of the microphone to the low impedance of the bipolar transistor Q2. When the microphone receives a signal, it is amplified through the JFET and the bipolar transistor. It is then applied to the JK flip-flop. The JK flip-flop has the unique feature of changing its state when a voltage, or high state, is applied to its J and K inputs. Thus, as each amplified pulse hits the JK flip-flop, the JK flip-flop should be either high or low. If the JK flip-flop is high, the relay should close because transistor Q3 is turned ON. If the JK flip-flop is low, the relay should drop out because Q3 is turned OFF. **See Figure 11-13.**

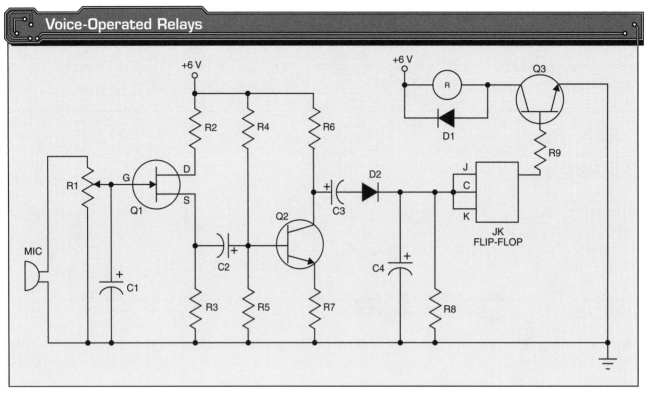

Figure 11-13. *A voice-operated relay uses a JFET as a high-impedance match between the microphone and the bipolar transistor.*

METAL-OXIDE SEMICONDUCTOR FIELD-EFFECT TRANSISTORS (MOSFETs)

A *metal-oxide semiconductor field-effect transistor (MOSFET)* is a three-terminal or four-terminal electronic switching device with metal-oxide or polysilicon insulating material that can be used for amplification. MOSFETs have the same terminal designations (gate, drain, and source) as JFETs. In addition, the four-terminal MOSFET has a designation for the material (substrate). Using the word "metal" in the description of this device can sometimes be misleading since the metal gate material is now being replaced by polysilicon. Polysilicon gates are made from polycrystalline silicon.

Polycrystalline silicon is a pure substance produced from raw quartzite (silica sand). Before silicon gate technology, the gate control of the MOSFET was made from aluminum. Polysilicon gates are three to four times faster than aluminum gates. They also consume half as much power as aluminum gates, have lower leakage current, and are more reliable.

The gate voltage of the MOSFET that controls the drain current is different from that of a JFET. The main difference in operation between a JFET and MOSFET is that voltage is applied between the gate and the P and N regions of the MOSFET structure. An electric field is generated, penetrates through the oxide layer, and creates an inversion layer, or channel, at the semiconductor-insulator layers. Varying the voltage between the gate and the P and N layers controls the conductivity of this layer, allowing current to flow between the drain and source. MOSFETs can operate in either an enhancement mode or a depletion mode.

MOSFET Operation Media Clip

Due to the vulnerability of a MOSFET to electrostatic discharge, electricians and other personnel should wear wrist grounding straps or other types of grounding devices to protect the MOSFET when handling or soldering it.

MOSFET Schematic Symbols and Lead Identification

MOSFETs are available as N-channel and P-channel devices. With an N-channel MOSFET, the arrow on the substrate points toward the channel. With a P-channel MOSFET, the arrow on the substrate points away from the channel. **See Figure 11-14.** *Note:* On schematic symbols for four-terminal MOSFETS, the substrate is not connected to the source.

Enhancement MOSFETs

The main body of the enhancement MOSFET is composed of a highly resistive P-type material. Two low-resistance N-type regions are diffused in the P-type region, forming the source and drain. When the source and drain are completely diffused, the surface of the MOSFET is covered with a layer of insulating material. The insulative layer of the MOSFET is silicon dioxide (SiO_2).

Holes are cut into the silicon dioxide insulating material, allowing contact with the N-type regions, thereby connecting the source and drain leads. The MOSFET construction is complete when a polysilicon contact (gate) is placed over the insulating material in a position to cover the channel from source to drain. **See Figure 11-15.**

Note: The metal contact is placed on top of the insulating material. There is no physical contact between the gate and the P-type material. A MOSFET is sometimes called an insulated (isolated) gate FET, or IGFET, because of this insulation.

Enhancement MOSFET Operation. A review of capacitor operation helps explain the enhancement operation of a MOSFET. The metal contact of the gate, the insulating material, and the P-type material are essentially a capacitor. The metal gate and the P-type material can be considered the plates of a capacitor. The oxide insulator is its dielectric. A voltage applied to the plates of a capacitor distorts the electrons that are in orbit in the dielectric. In this case, a positive voltage is applied to the upper plate, and the electrons move in the direction of the positive plate. **See Figure 11-16.**

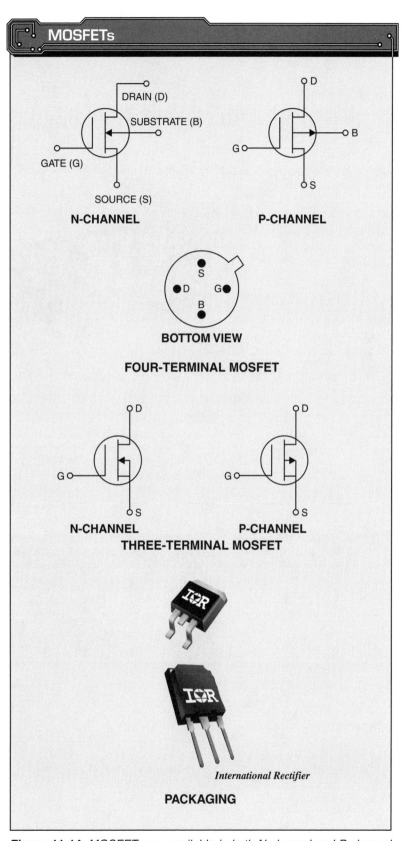

Figure 11-14. *MOSFETs are available in both N-channel and P-channel constructions.*

276 SOLID STATE DEVICES AND SYSTEMS

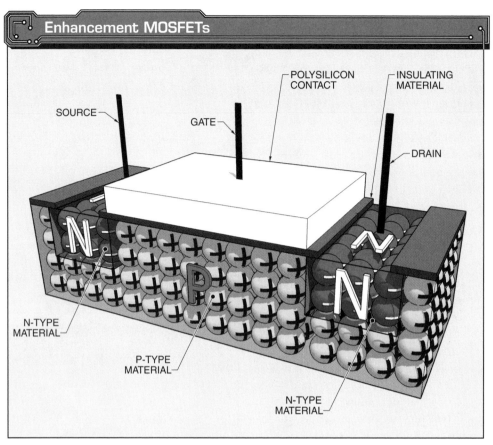

Figure 11-15. *An enhancement MOSFET contains a polysilicon contact placed over insulating material to cover a channel from source to drain.*

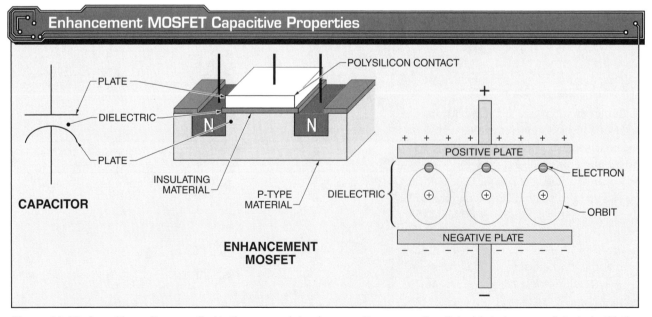

Figure 11-16. *A positive voltage applied to the upper plate of a capacitor causes the dielectric to become distorted, with the electrons moving in the direction of the positive plate.*

Applying this principle to the enhancement MOSFET, the gate can be used to produce a conductive channel from the source to the drain. The positive charge of a positive voltage placed on the gate induces a negative charge on the P-type material.

With increasing positive voltage, the holes in the P-type material are repelled until the region between the source and drain becomes an N-channel. Once the N-channel is forward-biased between the source and gate, current begins to flow. **See Figure 11-17.**

Since electrons have been added to form the N-channel, this MOSFET has enhanced current flow resulting from the application of a positive gate voltage. With a more positive gate voltage, the channel becomes wider. Therefore, current flows from source to drain due to the decreased channel resistance.

When the gate has zero voltage or a negative voltage, no enhancement effect is possible, and the MOSFET does not conduct. Since the enhancement MOSFET does not conduct at zero gate voltage, it is often called a normally off MOSFET. In the schematic, a broken line between terminals indicates the channel. The broken line signifies that the MOSFET is normally off. **See Figure 11-18.**

Note: A P-channel enhancement MOSFET is constructed similarly to an N-channel device, however, all P and N material are reversed. For the P-channel device, a negative gate voltage induces a P-channel and enhances current flow through the use of holes in the channel.

Depletion-Enhancement MOSFETs

The depletion-enhancement MOSFET is constructed similarly to the enhancement MOSFET. The main difference is the addition of a physical conducting channel between the source and drain. **See Figure 11-19.** The presence of the channel allows current to flow from source to drain, even without a gate voltage.

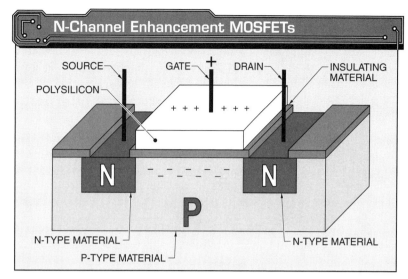

Figure 11-17. *The gate of an enhancement MOSFET can be used to produce an N-channel from source to drain.*

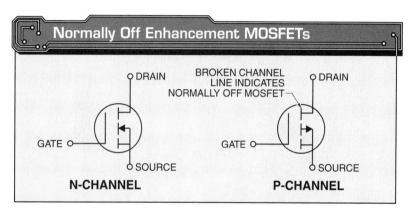

Figure 11-18. *The broken line in the schematic indicates that the MOSFET is normally off.*

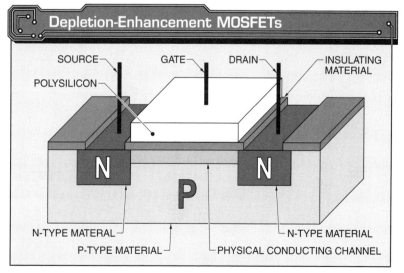

Figure 11-19. *A depletion-enhancement MOSFET has an added physical conducting channel between the source and the drain.*

The depletion-enhancement MOSFET has the same capacitive effect as the enhancement MOSFET. However, when a negative voltage is applied to the gate, holes from the P-type material are attracted into the N-channel. The holes neutralize the free electrons. **See Figure 11-20.** The result is that the N-channel is depleted, or reduced, in the number of carriers. The depletion of carriers increases channel resistance and reduces current. *Depletion mode* is the operation of a MOSFET with a negative gate voltage.

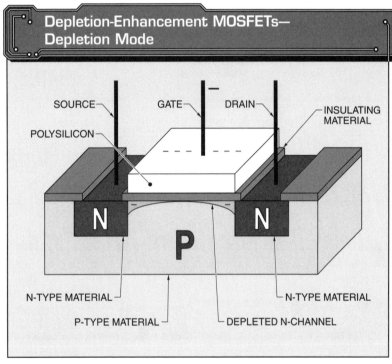

Figure 11-20. *A depletion-enhancement MOSFET operates in the depletion mode when a negative voltage is applied to the gate.*

With a positive gate voltage, the N-channel MOSFET can operate in the enhancement mode. *Enhancement mode* is the operation of a MOSFET with a positive gate voltage. **See Figure 11-21.** The positive gate voltage widens the N-channel, causing an increase in channel current. Since the depletion-enhancement MOSFET conducts a significant current even when V_{GS} is zero, it is often called a normally on MOSFET.

Dual-Gate MOSFETs

A MOSFET can be constructed with two gates such as in a dual-gate MOSFET arrangement. Current through the MOSFET can be cut off by either gate. A MOSFET also operates on the capacitive effect. **See Figure 11-22.**

The dual-gate arrangement allows the MOSFET to be used in a variety of circuits. For example, in a gain control circuit, the audio signal is applied to Gate 1 and the gain control voltage is applied to Gate 2. The gain control voltage can then be used to control the output from Gate 2.

Installation and Removal of MOSFETS

Care must be exercised when handling a MOSFET since the gate insulation is very thin. Any static charge introduced at the gate can perforate the insulation and destroy the device. Manufacturers of some MOSFETs wrap them in metal foil or short their leads with a metal eyelet or spring for protection. The shorting eyelet should be removed only after the device is installed in its circuit.

> *In 1905 Albert Einstein described the nature of light and the photoelectric effect on which photovoltaic technology is based, for which he later won a Nobel prize in physics.*

Power MOSFETs

Power MOSFETs exhibit the properties of small-signal MOSFETs but are designed to handle higher currents. Power MOSFETs were designed primarily for switching applications. Power MOSFETs can switch faster than bipolar transistors.

Unlike the small-signal MOSFET, the power MOSFET is fabricated with a vertical rather than lateral structure. MOSFETs are made using the double diffused metal-oxide substrate (DMOS) process and use polysilicon gates. The gate of this device is isolated from the source by a layer of insulating silicon oxide.

When voltage is applied between the gate and source terminals, an electric field is set up within the MOSFET. This field alters the resistance between the drain and source terminals.

The DMOS power MOSFET contains an inherent PN-junction diode. Its equivalent circuit can be considered as a diode in parallel with the source-to-drain channel, as shown in the schematic symbol. **See Figure 11-23.**

Power MOSFETs as Switches

There are a variety of solid state switch technologies available to perform switching functions. Each switch technology, however, has strong and weak points. The ideal switch would have zero resistance in the ON state, infinite resistance in the OFF state, switch instantaneously, and would require minimum input power to make it switch. The primary characteristics that are most desirable in a solid state switch are fast switching speed, simple drive requirements, and low conduction loss. For low voltage applications, power MOSFETs offer extremely low ON resistance and approach characteristics of the desired ideal switch.

Power MOSFETs have a wide range of specifications for high-frequency switching power supplies at frequencies above 100 kHz. The high power and high gain of power MOSFETs make them usable as power amplifiers in solid state transmitters for FM radio and TV broadcasting. The main disadvantage of power MOSFETs is the higher static drain-to-source ON-state resistance, which can cause unacceptable power losses in certain switching applications. The advantages of using power MOSFETs over power bipolar transistors include the following:

- faster switching
- lower switching losses
- wider safe operating area (SOA)
- simple drive circuitry
- ability to be paralleled easily (the forward voltage drop increases with increasing temperature, ensuring an even distribution of current among all components)

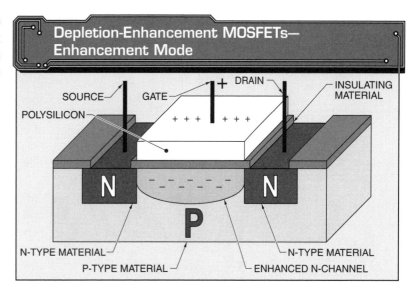

Figure 11-21. *A depletion-enhancement MOSFET operates in the enhancement mode when a positive voltage is applied to the gate.*

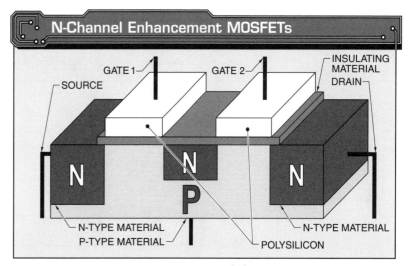

Figure 11-22. *Current through a dual-gate MOSFET can be cut off by either gate of a dual-gate MOSFET.*

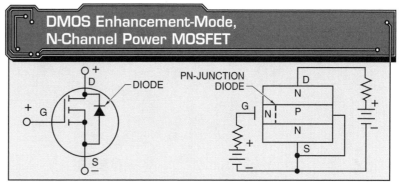

Figure 11-23. *The DMOS power MOSFET contains an inherent PN-junction diode, and its equivalent circuit can be considered as a diode in parallel with the source-to-drain channel.*

INSULATED GATE BIPOLAR TRANSISTORS (IGBTs)

An *insulated gate bipolar transistor (IGBT)* is a three-terminal switching device that combines an FET for control with a bipolar transistor for switching. **See Figure 11-24.** The IGBT combines the advantages of bipolar transistor and MOSFET technology.

The main difference in construction between the power MOSFET and IGBT is the addition of an injection layer in the IGBT. Due to the presence of the injection layer, holes are injected into the highly resistive N-layer and a carrier overflow is created. This increase in conductivity of the N-layer allows the reduction of the ON-state voltage of the IGBT. **See Figure 11-25.**

The silicon IGBT has become known as the power switch of high-voltage (greater than 500 V) and high-power (greater than 500 W) applications. The IGBT is a combination of the bipolar transistor and the MOSFET. It has the output switching and conduction characteristics of a bipolar transistor but is voltage-controlled like a MOSFET. This means it has the advantage of the high-current handling capability of a bipolar transistor with the ease of control of a MOSFET.

The IGBT is a power semiconductor device, noted for high efficiency and fast switching. The decision of whether to use an IGBT or MOSFET depends on the application. Cost, size, speed, and environmental requirements should all be considered when selecting an IGBT.

IGBT Operation

IGBTs are fast switching devices. IGBT operation consists of blocking, ON/OFF state, and latch-up operations. The safe operating area (SOA) of an IGBT protects against inductive shutoff. The two IGBT configurations include the punch-through (PT) and non-punch-through (NPT) configurations.

Blocking Operation. The ON/OFF state of an IGBT is determined by the gate voltage. If the voltage applied to the gate contact, with respect to the emitter, is less than the threshold voltage, then no MOSFET inversion layer is created, and the device is turned OFF. When this is the case, any applied forward voltage should fall across the reverse-biased junction. The only current flow should be a small leakage current.

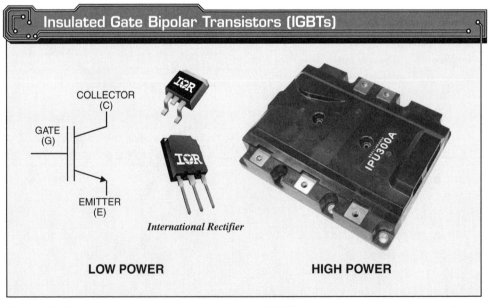

Figure 11-24. *An insulated gate bipolar transistor (IGBT) is a three-terminal switching device that combines an FET with a bipolar transistor.*

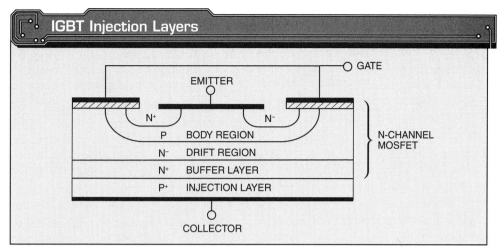

Figure 11-25. *The main difference between a power MOSFET and an IGBT is the addition of an injection layer in the IGBT.*

The forward breakdown voltage is, therefore, determined by the breakdown voltage of the junction. This is important for power devices with large voltages and currents. The breakdown voltage of the junction is dependent on the doping. A lower doping ratio results in a wider depletion region and a lower maximum electric field in the depletion region, which is why the drift region (N⁻) is doped much lighter than the body region (P⁻).

The buffer layer (N⁺) is present to prevent the depletion region from extending into the bipolar collector. The benefit of this buffer layer is that it allows the thickness of the drift region to be reduced, thus reducing ON-state losses.

ON-State Operation. The ON-state of the IGBT is achieved by increasing the gate voltage so that it is greater than the threshold voltage. This increase in voltage results in an inversion layer forming under the gate, which provides a channel that links the source to the drift region of the IGBT. Electrons are then injected from the source into the drift region. At the same time, holes are injected into the drift region.

This injection causes conductivity of the drift region, where both the electron and hole densities are higher than the original N⁻ doping. This conductivity gives the IGBT its low ON-state voltage. This is possible because of the reduced resistance of the drift region. Some of the injected holes should recombine in the drift region. Others may cross the region by drift and diffusion and reach the junction of the body region where they are collected.

OFF-State Operation. Either the gate must be shorted to the emitter or a negative bias must be applied to the gate. When the gate voltage falls below the threshold voltage, the inversion layer cannot be maintained, and the supply of electrons into the drift region is blocked. At this point, the shutoff process begins. The shutoff process cannot be completed as quickly as desired due to the high concentration of minority carriers injected into the drift region during forward conduction. The collector current rapidly decreases due to the termination of the electron current through the channel. Then, the collector current is reduced as the minority carriers recombine.

Latch-Up Operation. During ON-state operation, paths for current to flow in an IGBT allow holes to be injected into the drift region from the collector (P⁺). Parts of the holes disappear by recombination with electrons from the MOSFET channel. Other parts of the holes are attracted to the vicinity of the injection layer by the negative charge of electrons. These holes cross the body region and develop a voltage drop in the resistance of the body.

In addition to parasitic diodes, parasitic transistors are formed in an IGBT. These transistors can form the equivalent of an SCR within the parasitic molecular structure of the IGBT. Voltage can then forward bias the parasitic transistor, turning it ON. When this occurs, both NPN and PNP parasitic transistors turn on. The thyristor composed of these two transistors creates a latch-up condition for the IGBT. Once in a latch-up condition, the MOSFET gate has no control over the collector current. The only way to shut off the IGBT is to shut off the current, just as for a conventional SCR. *Note:* If latch-up is not terminated quickly, the IGBT may be destroyed by the excessive power dissipation.

IGBT Safe Operating Area. The safe operating area (SOA) is the current-voltage limit in which a power switching device like an IGBT can be operated without being destroyed. The area is defined by the maximum collector-emitter voltage (V_{CE}) and collector current (I_C) the IGBT operation must control to protect the IGBT from damage. The types of safe operating areas for IGBTs are the forward-biased safe operating area (FBSOA), reverse-biased safe operating area (RBSOA), and short-circuit safe operating area (SCSOA). The two primary conditions that could affect the SOA of an IGBT are operation during a short circuit and inductive shutoff.

Protection must be in place when switching inductive loads. This can be done by the use of regular diodes, zener diodes, or resistors. The method used depends on the application. IGBTs often need this type of protection from inductive loads to prevent inductive shutoff.

PT and NPT Configurations. The two types of structures used for IGBT construction are the punch-through (PT) structure and the non-punch-through (NPT) structure. An IGBT is called a PT IGBT when there is a buffer layer (N^+) between the injection layer (P^+) and the drift region (N^-). Otherwise, it is called an NPT IGBT. The buffer layer improves shutoff speed by reducing the minority-carrier injection quantity and by raising the recombination rate during the switching transition. The PT IGBT has similar characteristics as the NPT IGBT for switching speed and forward voltage drop. **See Figure 11-26.** Currently, most commercialized IGBTs are PT IGBTs.

Troubleshooting IGBTs

To troubleshoot an IGBT, the gate-to-emitter resistance and the collector-to-emitter resistance should be measured. The leakage resistance measurements are used to determine if the resistance is too low and the device is defective.

> *IGBTs are beginning to replace MOSFETs in high-voltage applications where conduction losses must be kept low. IGBTs have high input impedance and fast turn-on speeds similar to MOSFETs, but have better efficiency and lower costs.*

IGBT Configuration Comparison

	NPT	PT
Switching loss	Medium Moderate increase with temperature	Low Short tail current; significant increase with temperature
Conduction loss	Medium Moderate increase with temperature	Low No change to slight decrease with temperature
Paralleling	Easy	Difficult
Short-circuit rated	Yes	Limited

Figure 11-26. *Two different configurations, NPT and PT, are used for IGBT construction.*

Gate-to-Emitter Resistance. The leakage resistance is measured between the gate (G) and emitter (E), with a jumper between the collector (C) and emitter (E). **See Figure 11-27.** When a volt-ohm multimeter is used, it should be verified that the internal battery voltage is not higher than 20 V. A high voltage can damage the IGBT. The resistance reading for an IGBT in good condition ranges from several megohms (MΩ) to infinity. The device is considered defective if the resistance between the gate and emitter is very low.

Collector-to-Emitter Resistance. The leakage resistance between the collector (C) and emitter (E) is measured with a jumper between the gate (G) and emitter (E). **See Figure 11-28.** The collector should be connected to the positive (red) test lead and the emitter to the negative (black) test lead. The volt-ohm multimeter resistance reading would range from several megohms to infinity. The device is considered defective if the resistance between the collector and emitter is very low.

There are many advantages of IGBTs. The main advantages of an IGBT are the following:
- low ON-state voltage drop
- possibility of small chip size
- low driving power
- simple drive circuit
- wide SOA

IGBTs also include disadvantages. One disadvantage of an IGBT is that the switching speed is inferior to that of a power MOSFET. The tailing of the collector current due to the minority carrier causes the shutoff speed to be slow. However, the switching speed of an IGBT is superior to that of a BJT. Another disadvantage of the IGBT is the possibility of latch-up due to the internal PNPN thyristor structure.

IGBT Applications

IGBT applications include adjustable frequency drives and pulse width modulation. IGBTs are used for these applications because of their fast switching capabilities. They can also handle a fair amount of power.

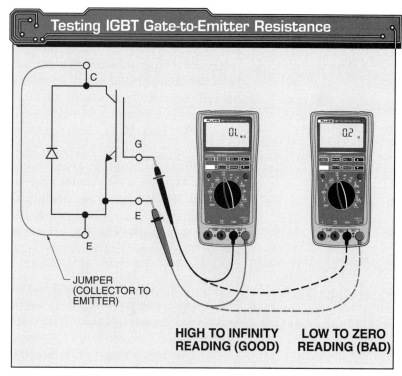

Figure 11-27. *The leakage resistance between the gate (G) and emitter (E) is measured with the collector (C) and emitter (E) shorted to each other.*

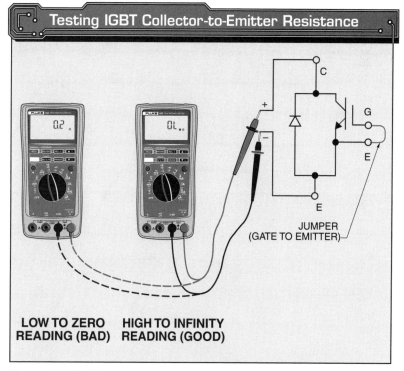

Figure 11-28. *The resistance between the collector (C) and emitter (E) is measured with a jumper between the gate (G) and emitter (E).*

Adjustable Frequency Drives. Adjustable frequency drives (AFDs) are a significant part of green technology. Energy savings, less wear on mechanical components, and better control are the primary factors that have advanced AFD technology.

AFD technology with pulse width modulation (PWM) uses IGBTs to generate the variable voltage and frequency required to control the speed of an AC motor. Since IGBTs can switch at high carrier frequencies (up to 15 kHz), there are several advantages of their use. The advantages of AFD technology with PWM include low-speed torque, quiet motor operation, and low-speed stability.

Pulse Width Modulation Motor Control. Pulse width modulation (PWM) is a type of adjustable frequency drive that achieves frequency and voltage control. By using PWM technology, a constant DC bus voltage is chopped into voltage pulses of fixed amplitude and variable width to approximate a sine wave output to an AC motor. **See Figure 11-29.**

Switching devices, like IGBTs, have the capacity to turn on and off much faster than devices using older technology. This ability results in increased performance and efficiency.

MULTISTAGE AMPLIFIERS

A *multistage amplifier* is an amplifier circuit that allows several stages of amplification. A multistage amplifier is needed to increase a signal if one amplifier does not provide enough gain. In multistage amplifiers, the signal is coupled, or connected, to several stages of amplification. Each stage is a complete amplifier and produces a specific amount of gain. The advantage of this system is the multiplication factor involved. If the first stage has a gain of 40 and the second has a gain of 50, the result is a gain of 2000 (40 × 50 = 2000).

> *Newer models of reliable IGBTs are widely used to control AC motors in electric and hybrid vehicles.*

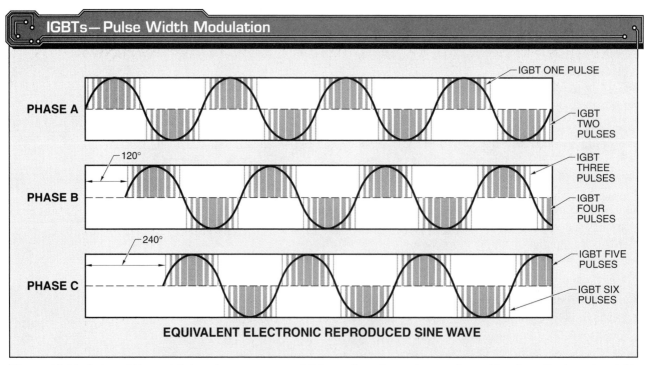

Figure 11-29. *Pulse width modulation chops a constant DC bus voltage into voltage pulses of fixed amplitude and variable width to approximate a sine wave output to an AC motor.*

In addition to gain, a multistage amplifier can provide protection by combining different types of transistors in sequence to obtain certain results. For example, many amplifiers have low input impedance. By using an FET on the front end of the multistage amplifier, high impedance can be obtained. The characteristics of FETs make them ideal for preventing excessive loading on the circuit.

Amplifier Coupling

Amplifier coupling is the joining of two or more circuits so that power can be transferred from one amplifier stage to another. A single stage of amplification is often not sufficient to drive a load to the required amount from any given input source. Therefore, various stages of amplification must be connected, or coupled together, to build up the output to the required level. Coupling transfers the signal from the output of one stage to the input of the next without distortion or loss.

The three basic techniques for coupling amplifier stages together and to loads are capacitive coupling, transformer coupling, and direct coupling. The frequency, impedance matching, cost, size, and weight must be considered when choosing a coupling technique for different applications. Therefore, there is not a coupling technique considered to be best for all applications.

Capacitive Coupling. *Capacitive coupling* is a method of amplifier coupling that uses an amplifier circuit consisting of two bipolar transistors connected by a capacitor. Capacitive coupling has the disadvantage of poor impedance matching. It also limits the lower frequency response of the amplifier. Despite these drawbacks, capacitive coupling is often used because it is inexpensive and few components are required. If signal losses are encountered, they are generally offset by adding another transistor stage. Another reason capacitive coupling is used is that it can block or isolate the bias circuits of each stage. The series coupling capacitor presents an open circuit to the flow of DC.

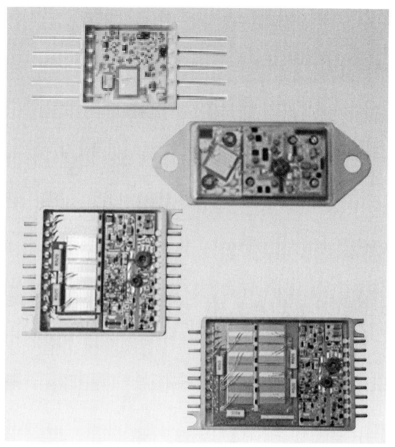

Micropac Industries, Inc.
Solid state power controllers can be used for motor switches, industrial automation, and test equipment.

A series coupling capacitor affects a circuit when transferring a signal from a source to a load. When a 5 V source is attached to the circuit, capacitor C charges to 5 V. Current flows through resistor R only during the charging of capacitor C, causing a voltage to appear momentarily across resistor R. Once capacitor C is charged, the voltage across it should be 5 V, and the voltage across resistor R should be 0 V. If the source voltage is increased to 10 V, capacitor C should increase its charge to 10 V. The charging current should momentarily produce 5 V across resistor R. If the source voltage is reduced to 3 V, capacitor C should discharge to 3 V. The discharge current should produce a momentary pulse of 7 V across resistor R because it is discharging to a lower voltage. **See Figure 11-30.**

286 SOLID STATE DEVICES AND SYSTEMS

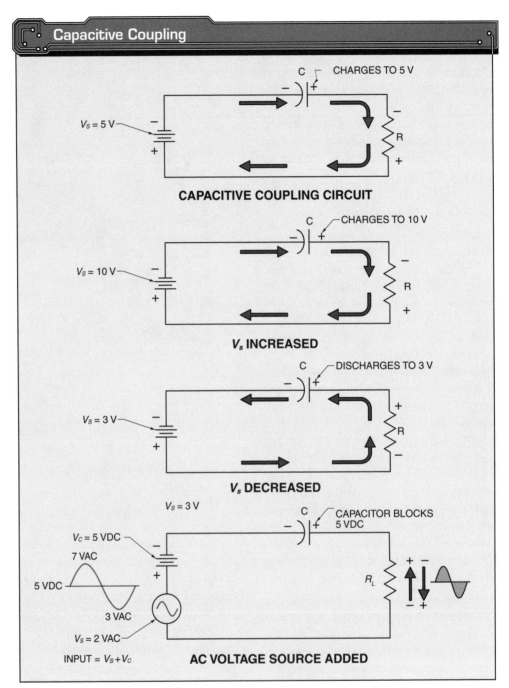

Figure 11-30. *Capacitive coupling requires a capacitor to amplify a signal.*

Both AC and DC voltage can be present in a capacitive coupling circuit. An input signal of 2 VAC varies around a level of 5 VDC. The AC signal causes the input to vary between 3 V and 7 V. The capacitor charges and discharges according to the varying AC signal, assuming the capacitance value has been selected to pass the lowest frequency. The output varies around zero level, and the DC is effectively removed because the capacitor blocks the DC component. Because of this effect, the capacitor is called a blocking capacitor, or coupling capacitor. Blocking capacitors are used to connect one section of a multistage amplifier to another. **See Figure 11-31.**

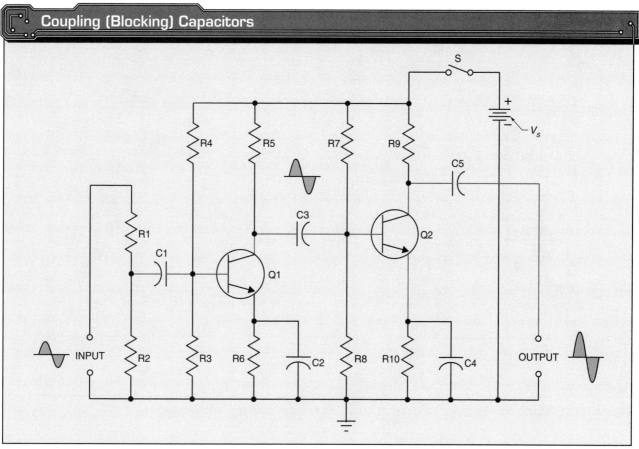

Figure 11-31. *Capacitive coupling circuits use blocking capacitors to connect one section of a multistage amplifier to another.*

Transformer Coupling. *Transformer coupling* is a method of amplifier coupling that uses an amplifier circuit consisting of two bipolar transistors connected by a transformer. For maximum power transfer between amplifier stages, or between a stage and a load, correct impedance matching is required. Special interstage transformers are used to match input and output impedances between stages. Transformers are also used to match load impedance to amplifier output impedance.

Transformer coupling can provide a high level of circuit efficiency. However, the weight, size, and cost of the transformers rule them out for many applications. Also, the frequency response is not as good as with capacitive coupling circuits, especially at higher frequencies. In the circuit, capacitor C2 is a bypass capacitor in the emitter circuit of transistor Q1. The purpose is to bypass AC signals present in the emitter to ground as the base will affect the emitter current. Bypassing these AC signals to ground will keep the emitter current at a constant level. **See Figure 11-32.**

Direct Coupling. *Direct coupling* is a method of amplifier coupling that uses an amplifier circuit consisting of two bipolar transistors connected so that the collector of the first transistor is directly connected to the base of the second transistor. In many industrial circuits, it is necessary to amplify very low frequency signals. Capacitive- and transformer-coupled amplifiers have poor frequency response at low frequencies because they both block low frequency signals. Therefore, direct coupling must be used when low frequencies are involved. Direct coupling is also used where the DC value and the AC value of a signal must be retained. The directly coupled amplifier provides a frequency response that ranges from zero hertz (DC) to several thousand hertz.

Transformer Coupling

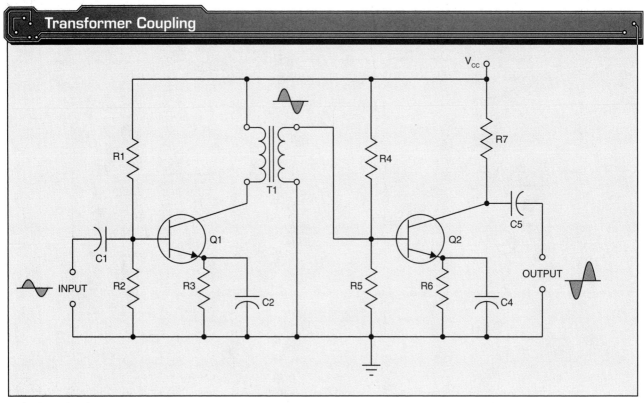

Figure 11-32. *Transformer coupling provides a means of impedance matching between stages.*

In a directly coupled amplifier, the collector of transistor Q1 is connected directly to the base of transistor Q2. In addition, the collector load resistor R2 acts as a bias resistor for transistor Q2. Any difference in bias current is amplified by the directly coupled circuit, which is very sensitive to temperature changes. This disadvantage can be overcome with bias stabilization circuits. Another disadvantage is that each stage may require different bias voltages for proper operation. **See Figure 11-33.**

Darlington Circuits

A *Darlington circuit* is an amplifier circuit that consists of two bipolar transistors connected so that the emitter of the first transistor is connected to the base of the second transistor. Current amplified by the first transistor is amplified by the second transistor. It is often desirable to have a broad frequency range in a transistor amplifier. The Darlington circuit uses two bipolar transistors and direct connection without using any other type of coupling circuit. **See Figure 11-34.**

The advantages of the Darlington circuit are extensive. A Darlington circuit uses fewer components than other amplifier circuits and there is no loss incurred in the coupling. The Darlington circuit can be used as an output stage of a transistor amplifier that is driven by a stage of low or intermediate power. In a typical audio amplifier, coupling of this type of circuit requires very large capacitors. Therefore, it is inefficient. By using the Darlington circuit, smaller and lighter bipolar transistors may be used, and less power is dissipated.

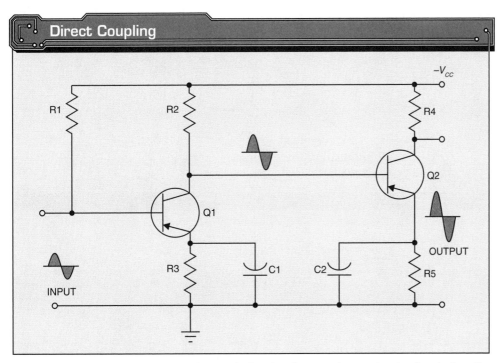

Figure 11-33. *Direct coupling must be used when low frequencies are involved and when the DC value and the AC value of a signal must be retained.*

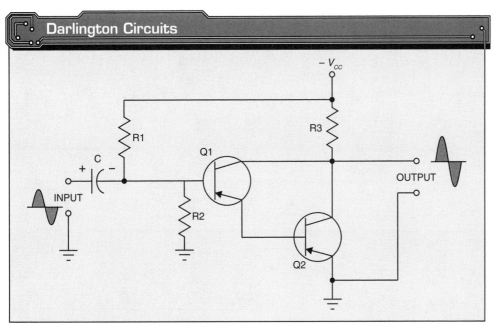

Figure 11-34. *A Darlington circuit uses fewer components than other amplifier circuits, and there is no loss in the coupling.*

Refer to Chapter 11 Quick Quiz® on CD-ROM

KEY TERMS

- field-effect transistor (FET)
- junction field-effect transistor (JFET)
- gate
- drain
- source
- JK flip-flop
- metal-oxide semiconductor field-effect transistor (MOSFET)
- depletion-enhancement MOSFET
- depletion mode
- enhancement mode
- insulated gate bipolar transistor (IGBT)
- multistage amplifier
- amplifier coupling
- capacitive coupling
- transformer coupling
- direct coupling
- Darlington circuit

REVIEW QUESTIONS

1. What are the two categories of field-effect transistors (FETs)?
2. What do the letters G, D, and S stand for in reference to FETs?
3. What is the procedure for troubleshooting an IGBT?
4. Define pinch-off voltage.
5. What is the control element of a junction field-effect transistor (JFET)?
6. What is the ohmic region of a JFET?
7. List the three basic configurations for JFETs.
8. What is the phase relationship between the input and the output of a common-source amplifier?
9. Where are common-gate amplifiers typically used?
10. Name three applications for JFETs.
11. What is a metal-oxide semiconductor field-effect transistor (MOSFET)?
12. What are the two MOSFET operating modes?
13. Explain the operation of an enhancement MOSFET.
14. Explain the operation of a depletion-enhancement MOSFET.
15. What precaution must be observed when installing or removing MOSFETs?
16. What is an insulated gate bipolar transistor (IGBT)?
17. List and describe common IGBT applications.
18. What is a junction field-effect transistor (JFET)?

SILICON-CONTROLLED RECTIFIERS (SCRs) 12

OBJECTIVES

- Define silicon-controlled rectifier (SCR).
- Explain SCR construction.
- Describe methods for troubleshooting an SCR.
- List and describe applications for SCRs.

Silicon-controlled rectifiers are used in high-power switching applications. For example, they may be used to control battery chargers, motor speed, and welding equipment. A silicon-controlled rectifier consists of four layers of semiconductor material and has an anode, cathode, and gate.

SILICON-CONTROLLED RECTIFIER PROPERTIES

A *silicon-controlled rectifier (SCR)* is a four-layer (PNPN) semiconductor device that uses three electrodes for normal operation. **See Figure 12-1.** The three electrodes are the anode, cathode, and gate. The anode and cathode of the SCR are similar to the anode and cathode of an ordinary diode. The gate serves as the control point for the SCR. An SCR is also known as a thyristor.

An SCR differs from an ordinary diode in that it does not pass significant current, even when forward-biased, unless the anode voltage equals or exceeds the forward breakover voltage. However, when forward breakover voltage is reached, the SCR switches on and becomes highly conductive. The gate current is used to reduce the level of breakover voltage necessary for the SCR to conduct or fire.

Low-current SCRs can operate with an anode current of less than 1 A. High-current SCRs can handle load currents in the hundreds of amperes. The size of an SCR increases with an increase in current rating. **See Figure 12-2.**

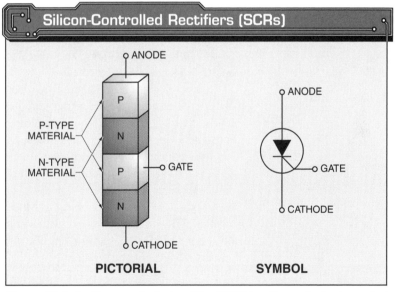

Figure 12-1. *A silicon-controlled rectifier (SCR) has three electrodes called the anode, cathode, and gate.*

A silicon-controlled rectifier (SCR) should not be confused with a silicon-controlled switch (SCS), which is similar to an SCR but is manufactured with two gates, rather than one. Both gates can be used for turning on an SCS.

SCR Operation
Media Clip

291

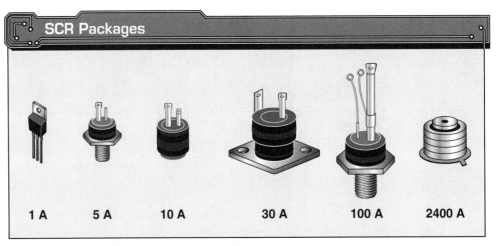

Figure 12-2. *The SCR comes in a variety of packages.*

Silicon-controlled rectifiers have voltage ratings of up to 2500 V and current ratings of up to 3000 A. Silicon-controlled rectifiers are used in power switching, phase control, battery charger, and inverter circuits. In industrial applications, they are applied to produce variable DC voltages for motors (from a few to several thousand horsepower) from AC line voltage. They can also be used in some electric vehicles.

Another common application involves phase control circuits used with inductive loads. SCRs are also found in welding equipment where they are used to maintain a constant output current or voltage. Large SCR assemblies with many individual devices connected in series are used in high-voltage DC converter stations. Much smaller SCRs are also used as electrostatic discharge (ESD) protection circuits on modern CMOS integrated circuits.

Characteristic Curves

A voltage-current characteristic curve shows the operating characteristics of an SCR when its gate is not connected. **See Figure 12-3.** When an SCR is reverse-biased, it operates similar to a regular semiconductor diode. While reverse-biased, there is a small amount of current until avalanche is reached. After avalanche is reached, the current increases dramatically. This current can cause damage if thermal runaway begins.

When an SCR is forward-biased, there is a small amount of forward leakage current called forward blocking current. This current stays relatively constant until the forward breakover voltage is reached. At that point, the current increases rapidly. The region is often called the forward avalanche region. In the forward avalanche region, the resistance of the SCR is very small. The SCR operates similar to a closed switch, and the current is limited only by the external load resistance. A short in the load circuit of an SCR can destroy the SCR if overload protection is not adequate.

Operating States. The operation of an SCR is similar to that of a mechanical switch. An SCR is either ON or OFF. When an applied voltage is above the forward breakover voltage (V_{BRF}), the SCR fires or turns on. The SCR remains on as long as the current stays above the holding current. When voltage across the SCR drops to a value too low to maintain the holding current, it returns to its OFF state.

Gate Control of Forward Breakover Voltage. When the gate is forward-biased and current begins to flow in the gate-cathode junction, the value of forward breakover voltage can be reduced. Increasing values of forward bias can be used to reduce the amount of forward breakover voltage necessary to get the SCR to conduct.

Voltage-Current Characteristic Curves

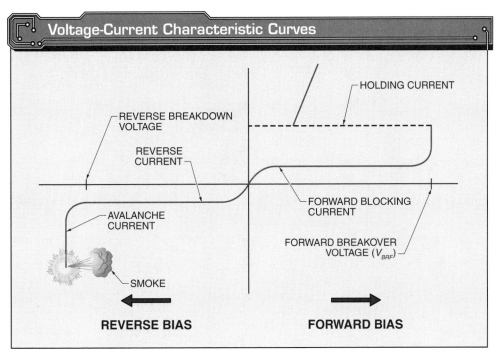

Figure 12-3. *The voltage-current characteristic curve of an SCR shows how an SCR operates.*

Once the SCR has been turned on by the gate current, the gate current loses control of the SCR forward current. Even if the gate current is completely removed, the SCR remains on until the anode voltage has been removed. The SCR also remains on until the anode voltage has been significantly reduced to a level where the current is not large enough to maintain the proper level of holding current.

Triggering Methods

An SCR normally can be triggered into conduction by applying a pulse of control current to the gate. Once turned on, an SCR remains on as long as there is a minimum level of holding current flowing through the load circuit. Once the current drops below the holding current value, the SCR turns off. **See Figure 12-4.**

The correct method of turning on an SCR is to apply a proper signal to the gate of the SCR (gate turn-on). Damaging methods of turning on an SCR are voltage breakover turn-on, static turn-on, and thermal turn-on.

When solid state switching devices are used to control current, the devices require heat sink mounting because of the large amounts of heat generated.

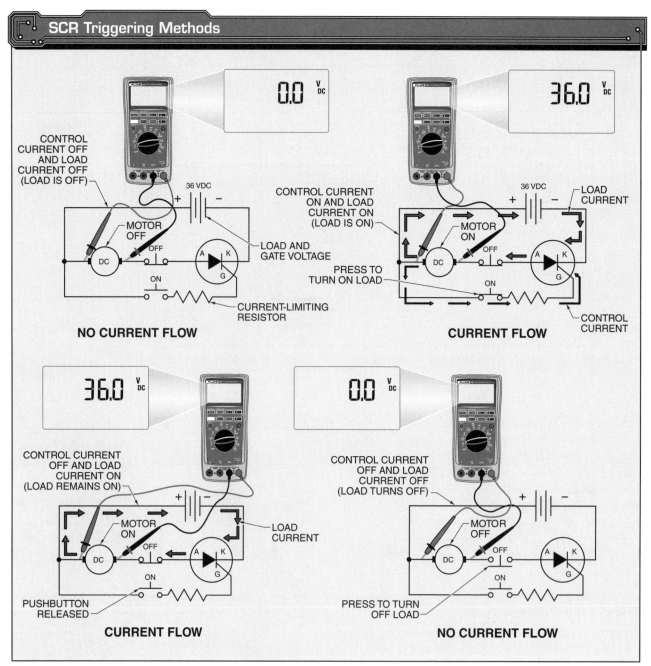

Figure 12-4. *An SCR is turned on by applying a pulse of control current to the gate. Once turned on, an SCR remains on as long as there is a minimum level of holding current flowing through the load circuit.*

Gate Turn-on. Gate turn-on is a method of turning on an SCR when the proper signal is applied to the gate at the correct time. Gate turn-on is the only correct way to turn ON an SCR. The gate signal must be positive with respect to the cathode polarity for the SCR to turn ON. **See Figure 12-5.**

Voltage Breakover Turn-on. Voltage breakover turn-on is a method of turning on an SCR when the voltage across the SCR terminals exceeds the maximum voltage rating of the device. Excessive voltage causes localized heating in an SCR and damages the SCR.

Turning on SCR Gates

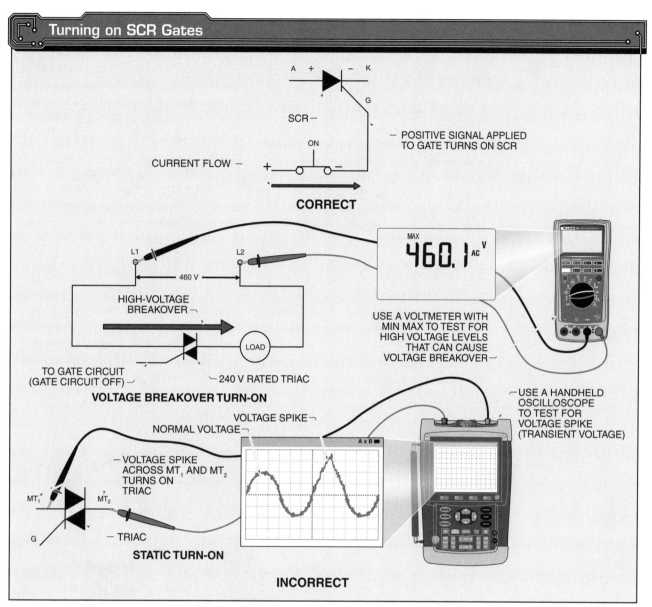

Figure 12-5. SCRs can be turned on by damaging methods, such as voltage breakover turn-on, static turn-on, and thermal turn-on.

Static Turn-on. Static turn-on is a method of turning on an SCR when a fast-rising voltage is applied across the terminals of a triac. Manufacturers refer to the point of turn-on as the dv/dt rating. The dv/dt rating defines the level of voltage over a given time period that causes the device to turn ON. For example, a typical rating for an SCR is 250 V/sec.

Static turn-on does not damage the SCR, provided the surge current is limited. A snubber circuit is typically added across the SCR terminals to protect the SCR when static turn-on is a problem.

Thermal Turn-on. All solid state components are heat sensitive. Thermal turn-on is a method of turning on an SCR when heat levels exceed the limit of the SCR, typically 230°F (110°C). When a solid state SCR is turned on by heat, the SCR is typically destroyed. Using the correct heat sinks for a circuit eliminates the thermal turn-on of SCRs.

SCR CONSTRUCTION

The internal structure of an SCR consists of two PN junctions joined together that form a PNPN layer. When reverse-bias voltage is applied, there is no current flow. Reverse bias exists when the positive lead of the voltage source is attached to the cathode and the negative lead is attached to the anode. The positive terminal of the voltage source at the cathode attracts electrons away from the PN junction nearest to the cathode. This creates a wide depletion region at that junction. The negative terminal of the voltage source at the anode attracts holes away from the PN junction nearest to the anode. Another wide depletion region is created at this junction. In the reverse-bias condition, there is not an appreciable amount of current flow. **See Figure 12-6.**

When forward-bias voltage is applied to the junctions, there is also no appreciable amount of current flow. In this case, the negative terminal (cathode) repels electrons through the N-type material toward the PN junction closest to the cathode. This attracts holes to the PN junction closest to the cathode away from the center PN junction. The positive terminal repels holes through the P-type material toward the PN junction nearest to the anode and away from the PN junction in the center. The net result is the creation of a larger depletion region at the center PN junction. Therefore, there is very little current flow. **See Figure 12-7.**

The SCR conducts in forward bias if the forward breakover voltage is reached or if a small forward bias voltage is applied to the gate. When the gate is pulsed or triggered and the proper forward bias is applied to the SCR, the depletion regions close and current begins to flow through the SCR. **See Figure 12-8.**

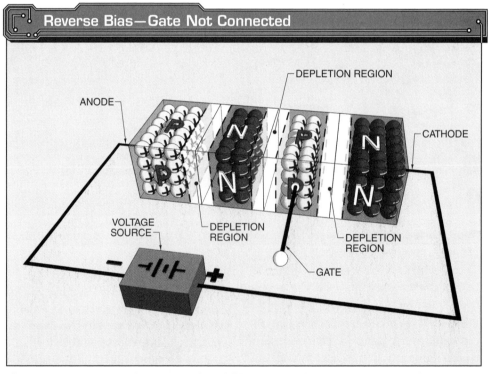

Figure 12-6. *The internal structure of an SCR consists of two PN junctions joined together that form a PNPN layer. With reverse-bias voltage applied to this layered structure, there is no noticeable current flow due to the formation of depletion regions.*

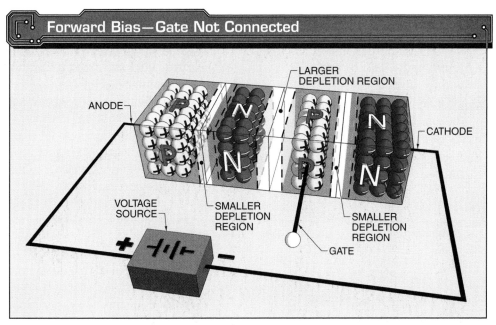

Figure 12-7. With forward-bias voltage applied to the layered structure, there is no noticeable current flow due to the formation of the larger depletion region in the center.

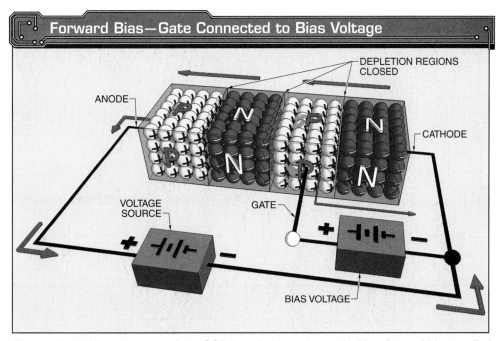

Figure 12-8. When the gate of the SCR is pulsed or triggered with a forward bias applied between the cathode and anode, the SCR conducts. This current continues to flow as long as current is supplied.

The positive terminal (anode) repels the holes in the P-type material toward the center junction closest to the anode. This attracts an equal number of electrons from the N-type material on the other side of the junction, resulting in the depletion region closing at the center junction. All three junctions now have closed depletion regions, and the SCR conducts current from the cathode to anode. This current continues to flow as long as current is being supplied.

The SCR turns off whenever the current is turned off or when the voltage is reduced or removed. Another gate pulse is required to restore conduction.

SCR Equivalent Circuits

One way to understand a complex device is to compare it to a simple device. This method is called the equivalent circuit analogy or comparison technique. The equivalent circuit of an SCR is a circuit consisting of interconnected PNP and NPN transistors. **See Figure 12-9.**

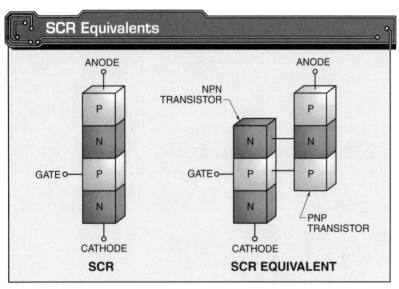

Figure 12-9. *An SCR equivalent can be constructed by interconnecting a PNP transistor and an NPN transistor.*

The emitters (E) of the PNP and NPN transistors represent the anode and cathode respectively. The base (B) and collector (C) of each transistor represent the gate connection. **See Figure 12-10.** When biasing the equivalent circuit (just as an SCR is biased), the anode is made positive and the cathode is made negative with no gate voltage applied. In this state, the NPN transistor does not conduct because its emitter-base junction is not forward-biased. Since the NPN transistor has no base current, the emitter-base junction of the PNP transistor cannot conduct. Both transistors are turned off and therefore, no current flows. Without a gate potential, the SCR equivalent circuit, like the SCR, cannot conduct in either direction, regardless of the polarity of the applied voltage.

To have the SCR turned on, the gate must be connected, and a voltage must be applied. This can be accomplished by applying a positive potential with respect to the cathode across the gate lead. With only a momentary pulse (as little as 500 µs) to the gate, the circuit starts operating. The momentary pulse causes the emitter-base junction of the NPN transistor to be forward-biased and to conduct. The current from the NPN transistor causes base current to flow through the PNP transistor.

The collector current through the PNP transistor supports the base current through the NPN transistor. The transistors hold each other in a state of constant conduction. The circuit was placed into a state of conduction by only a momentary pulse to the gate. Once in conduction, the circuit is self-sustaining and the gate signal can be removed. In order to switch the SCR circuit off, the anode-to-cathode voltage must be reduced to nearly zero, or the circuit must be momentarily opened.

DC Switches

The most basic application of an SCR is a DC switch. An SCR used as a DC switch has many advantages over mechanical DC switching. The SCR provides arcless switching, low forward voltage drop, and rapid switching time. It also lacks moving parts.

In the circuit, a single SCR is used to turn power to the load on and off. **See Figure 12-11.** The circuit is turned on by momentarily closing the gate control switch (S1). Closing S1 causes a very small gate current to flow through the current-limiting resistor (R1), which in turn causes the SCR to conduct. Once the SCR is ON, the gate signal can be removed and the SCR remains in conduction.

In order to turn off the SCR, switch S2 must be momentarily closed and switch S1 must be open. This shorts out the SCR for an instant and reduces the anode-to-cathode voltage to 0 V. With zero voltage, the forward current through the SCR drops below the holding value and the SCR turns off.

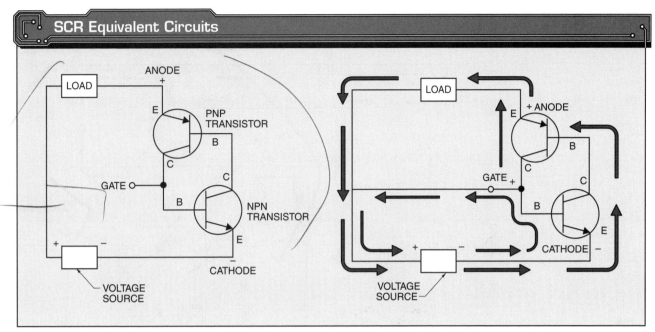

Figure 12-10. *In an SCR equivalent circuit, the emitters (E) of the PNP and NPN transistors represent the anode and cathode. The base (B) and collector (C) of each transistor are interconnected to represent the gate connection.*

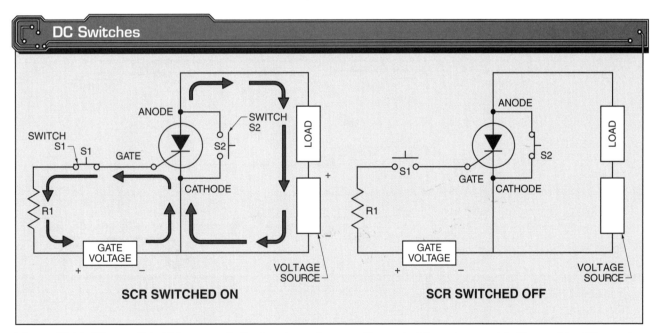

Figure 12-11. *When switch S1 is momentarily closed, a gate current is created through current-limiting resistor R1, causing the SCR to conduct. Once the SCR is ON, the gate signal can be removed and the SCR will remain in conduction.*

The drawback of using a circuit is that switch S2 must be closed and reopened across the high DC load current. Using a mechanical switch across an electronic switch defeats the purpose of electronic switching. This disadvantage can be overcome by using two SCRs and some additional circuitry. In the new circuit, SCR1 controls the DC power applied to the load. SCR2, along with capacitor C and resistor R1, turn SCR2 and the load off. **See Figure 12-12.**

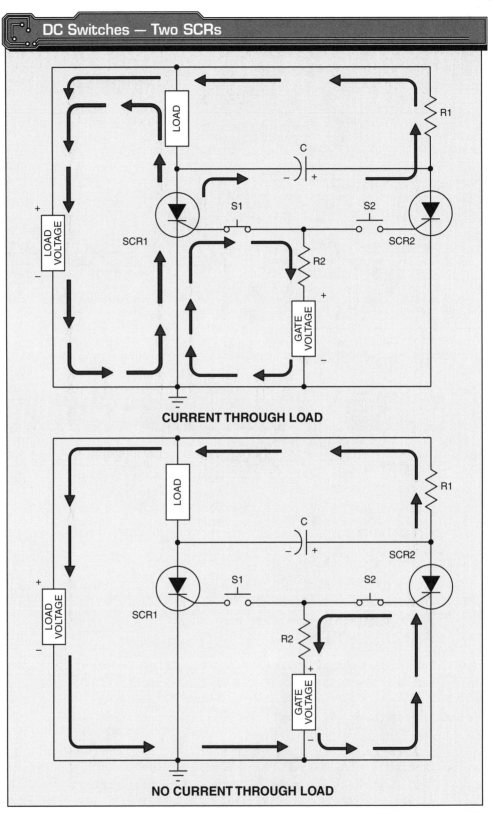

Figure 12-12. When switch S1 is closed, the gate current flows through SCR1 and resistor R2, turning on SCR1. This in turn allows current to flow to the load. Turning on SCR1 also grounds the left side of capacitor C so that it may charge through R1.

In the circuit, when switch S1 is momentarily closed, gate current flows through SCR1 and resistor R2, turning on SCR1. This in turn allows current to flow to the load. Turning on SCR1 also grounds the left side of capacitor C to a negative potential through SCR1. With one side of capacitor C grounded, the capacitor can charge through resistor R1.

To turn off SCR1, switch S2 must be momentarily closed with S1 open, energizing the gate of SCR2. Gate current now passes through SCR2 and R2, causing SCR2 to turn ON.

With SCR2 conducting, the right side of capacitor C is now brought to ground potential, placing the capacitor across SCR1. The polarity of the voltage across the capacitor then reverse biases SCR1, dropping its forward current below the holding level (current). **See Figure 12-13.** *Note:* The capacitor is actually across SCR1 with a completely opposite polarity.

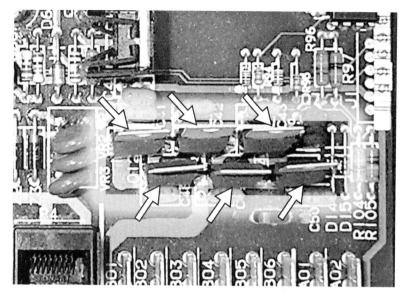

Building automation system controllers use thyristors as switches to turn equipment ON or OFF.

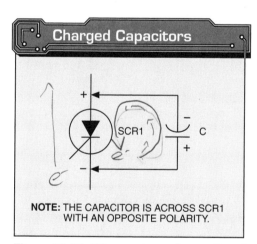

Figure 12-13. *When a charged capacitor of opposite polarity is applied across an SCR, the SCR will stop conducting. The polarity of the voltage across the capacitor reverse biases the SCR so that the forward current of the SCR drops below the holding level and stops conducting.*

Even if this reverse polarity takes place for only a moment, SCR1 stops conducting. To restart the circuit, S1 would again have to be momentarily closed.

AC Switches

Two SCRs are required for AC switching. **See Figure 12-14.** With switch S1 closed, the gate is connected for SCR1 through diode D1 and resistor R1. During the positive half of each AC cycle, SCR1 is forward-biased. With this forward bias, the gate current flows through D1 and R1, causing SCR1 to conduct. When the positive half cycle is complete, the voltage goes to zero and SCR1 shuts off. To switch the negative half cycle to the load, switch S2 is closed. With switch S2 closed, the gate is connected for SCR2 through diode D2 and resistor R2.

As the negative half cycle begins, SCR2 becomes forward-biased and gate current flows through D2 and R2, causing SCR2 to conduct. Once again, when the voltage crosses the zero voltage point, SCR2 shuts off.

Note: The entire cycle continues to produce full wave rectification until S1 and S2 are opened. Opening S1 and S2 breaks the conducting paths to both gate circuits.

Phase Control

Regular diodes cannot be used for AC and DC switching, while SCRs can be used. A regular diode is either completely ON or completely OFF. An SCR, however, can be turned on at different points in the conducting cycle.

AC Switches

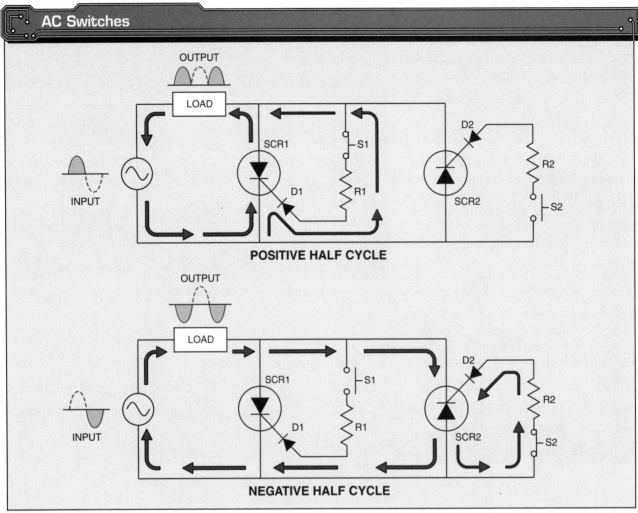

Figure 12-14. *During the positive half of each cycle, SCR1 is forward-biased. During the negative half of each cycle, SCR2 is forward-biased.*

The ability of an SCR to turn ON at different points in the conducting cycle can be useful for varying the amount of power delivered to a load. This type of variable control is called phase control. *Phase control* is the control of the time relationship between two events when dealing with voltage and current. In this case, it is the time relationship between the trigger pulse and the point in the conducting cycle when the pulse occurs. With phase control, the speed of a motor, the brightness of a lamp, and the output of an electric-resistance heating unit can be controlled.

The most basic control circuit using this principle is a half-wave phase control circuit. A *half-wave phase control circuit* is an SCR that has the ability to turn ON at different points of the conducting cycle of a half-wave rectifier. **See Figure 12-15.** In the circuit, the input voltage is a standard 60 Hz line voltage.

When a negative half cycle is applied to the circuit, the SCR is reverse-biased and should not conduct. In addition, diode D1 is reverse-biased and no gate current flows. Diode D2, however, is forward-biased and allows capacitor C to charge to the polarity shown. No current flows through R1 since D2 is in parallel with R1. The forward-biased resistance of D2 is so much lower than R1 that almost all current passes through D2 instead of through R1. Since no current is flowing through the load, the oscilloscope should not indicate a voltage drop.

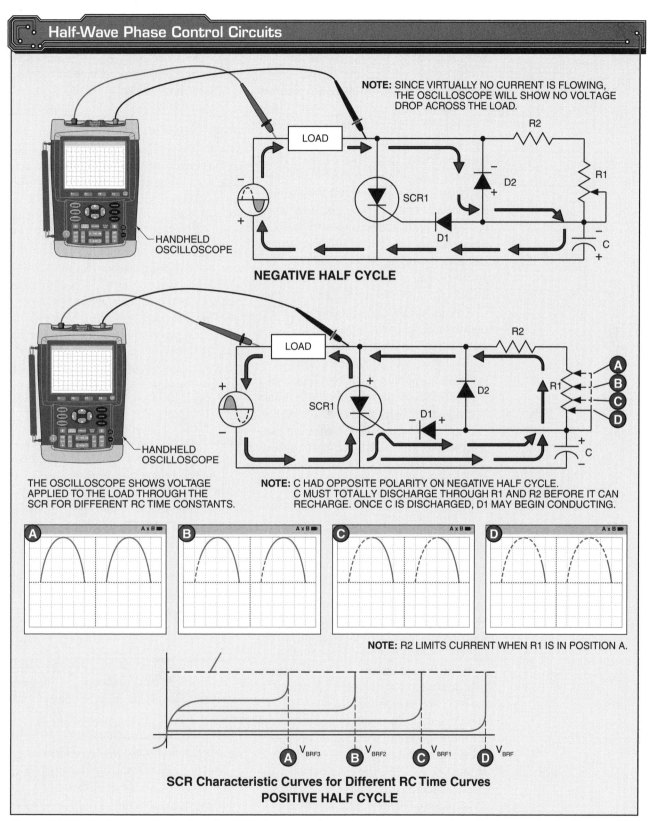

Figure 12-15. On the negative half cycle, SCR1 is reverse-biased and does not conduct. On the positive half cycle of the AC input, the SCR is forward-biased such that gate current will flow through diode D1 and variable resistor R1, causing the SCR to conduct.

Note: The oscilloscope on the negative half cycle should show only a slight movement of the trace. This is because the SCR is reverse-biased, and the current drawn through diode D2, when charging capacitor C, is small.

When the positive half cycle of the voltage source is applied, the circuit operation changes. When the positive half cycle is applied to the circuit, the SCR becomes forward-biased in such a way that it conducts forward current if a gate current of sufficient strength is present. The important factor is the amount of gate current. If the gate current is too small, the SCR does not fire. The amount of gate current in this circuit is controlled by variable resistor R1 and capacitor C. Resistor R1 and capacitor C form an RC network that determines the time of charge and discharge of the capacitor.

For capacitor C to charge on the positive half cycle, it must first discharge the polarities it received on the negative half cycle. Once discharged, it may recharge with the opposite polarity. The charging and discharging rates are determined by resistor R1. The discharge and charge time of the RC network determine the time sufficient current is allowed to flow through the gate circuit to fire the SCR.

If variable resistor R1 is in position A, or has no resistance, the capacitor will discharge its reverse polarity almost immediately. Current would then be allowed to pass through D1, which in turn fires the SCR. The result is that the entire positive half cycle is allowed to pass through the SCR to the load as shown by the oscilloscope display pattern correlated to position A.

If variable resistor R1 is moved to position B, where more resistance is placed in the RC network, the time of discharge for C is increased. The result is a delay in time for the gate current to reach its necessary value. Also, part of the positive half cycle is blocked as shown by the oscilloscope display pattern correlated to position B.

As more resistance is added in positions C and D, the time delay is increased. The result is that the positive half cycle is increasingly blocked while a decreasing amount of current is delivered to the load. Thus, the SCR then can deliver a varied output to the load on the positive half cycle. *Note:* To control the output in both the positive and negative half cycles, two SCRs must be used.

Mounting and Cooling

The proper operation of an SCR depends on the correct mounting and cooling of the device. The SCR may fail because of thermal runaway if the temperature of the SCR is allowed to rise too high. Circuits may also change characteristics because of insufficient cooling, which reduces the forward breakover voltage. For these reasons, most SCRs are designed with some type of heat transfer mechanism to dissipate internal heat loss.

A mounting surface, like a heat sink, is an integral part of the heat transfer path of an SCR. Proper mounting is always essential for successful SCR cooling. Incorrect mounting and cooling of the SCR result in the same problem and therefore, must be treated together.

Lead-Mounted SCRs. For small lead-mounted SCRs, cooling is maintained by radiation and convection from the surface of the SCR case. Cooling is also maintained by thermal conduction through the SCR leads. **See Figure 12-16.** The following practices for minimizing the SCR temperature should be observed:

- Minimum lead length to the terminal board, socket, or PC board should be used because it permits the mounting points to assist in the cooling of the SCR.
- Other heat-dissipating elements, such as power resistors, should not be connected directly to the SCR leads where avoidable.
- High-temperature devices, such as lamps, power transformers, and resistors, should be shielded from radiating their heat directly on the SCR case.
- In the final mounting, insulators should be in position and silicon paste should be applied to both sides for maximum heat transfer.

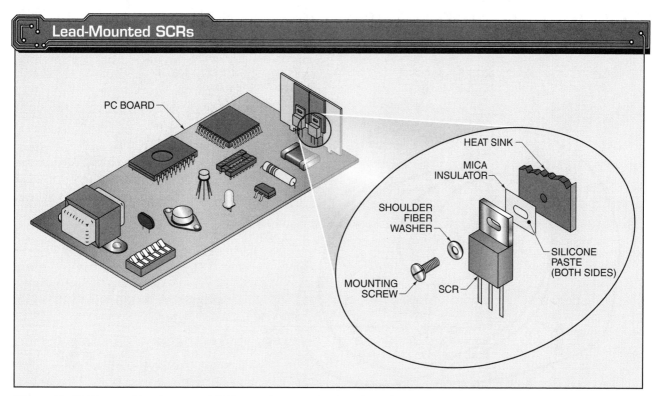

Figure 12-16. *For small lead-mounted SCRs, cooling is maintained by radiation and convection from the surface of the SCR case and by thermal conduction through the SCR leads.*

Press-Fit SCRs. Many medium-current SCRs use the press-fit package. **See Figure 12-17.** This package is designed primarily for forced insertion into a slightly undersized hole in the heat exchanger (heat sink). When properly mounted, the press-fit SCR has a lower thermal drop to the heat exchanger than the stud-mounted SCR. Also, in high-volume applications, a press-fit SCR generally costs less than a stud-mounted SCR.

To mount a press-fit SCR, a hole may be punched and reamed in a flat plate, extruded and sized in sheet metal, or drilled and reamed. A slight chamfer on the hole should be used to guide the housing. To ensure maximum heat transfer, the entire knurl should be in contact with the heat exchanger. The SCR must not be inserted into the heat exchanger past its knurl. This is to prevent the heat sink from taking pressure off the knurl in a deep hole. The insertion force must be limited as specified by the manufacturer. Following the manufacturer's specification helps prevent misalignment with the hole. Excessive SCR-to-hole interference is also avoided. Pressure must be uniformly applied to the face of the header when inserting the SCR.

Note: Heat exchanger materials may be (in order of preference) copper, aluminum, or steel. The heat exchanger thickness should be a minimum of 1/8″ the width of the knurl on the housing.

Stud-Mounted SCRs. Stud-mounted SCRs are installed using the same procedure used for mounting stud-mounted diodes. Manufacturer specifications should be followed for SCR installation. Typically, stud-mounted SCRs are fastened to heat sinks to ensure good heat flow. **See Figure 12-18.**

> *Stud mounting is one of the most common methods of mounting silicon-controlled rectifiers (SCRs) because it can ensure secure and reliable operation. The specifications available from the manufacturer specify how tight the threaded stud should be after installation.*

306 SOLID STATE DEVICES AND SYSTEMS

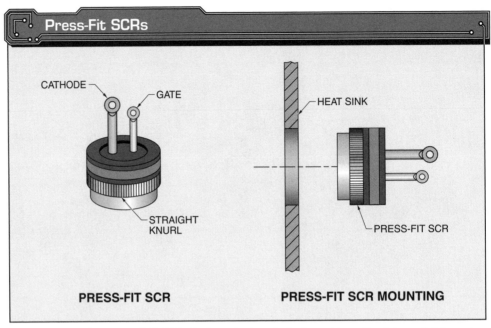

Figure 12-17. *The press-fit SCR is designed primarily for forced insertion into a slightly undersized hole in the heat exchanger (heat sink).*

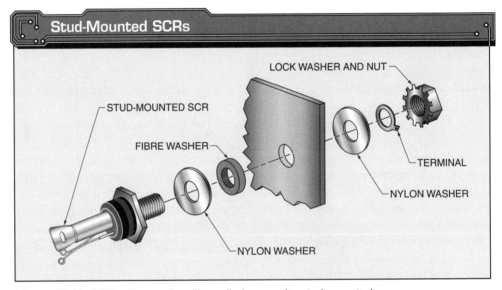

Figure 12-18. *SCRs, like regular silicon diodes, can be stud-mounted.*

Flat-Pack and Press-Pack SCRs. The advantage of flat-pack and press-pack SCRs is the increased surface area available for thermal conduction. **See Figure 12-19.** At a given temperature, their large, smooth, and flat surface allows greater power dissipation than the conventional stud-mounted SCRs. The major disadvantage is the cost of the more elaborate mounting procedure.

> *Mica and Teflon® washers are used as insulation between the PC board and SCR. Mica is a mineral that has high dielectric strength and can be sliced into thin layers to be used for washers. Teflon, also known as polytetrafluoroethylene (PTFE), is a synthetic plastic material that has low thermal conductivity and high dielectric strength.*

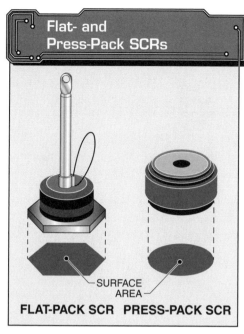

Figure 12-19. *The advantage of flat-pack and press-pack SCRs is the increased surface area available for thermal conduction.*

A press-pack SCR must be properly installed to provide maximum heat transfer. Uneven force distribution results in poor thermal and electrical contact, which seriously impairs SCR performance.

SCR TROUBLESHOOTING

In industry, a troubleshooter has many different kinds of electronic circuits with which to work. If a troubleshooter is well trained and knows the circuit, circuit disturbance tests can be used to easily collect the information needed to localize many of the problems that occur.

A circuit with an SCR could be used to spot an oversize box on a conveyer belt or determine whether someone has passed through a certain passageway in the time since the circuit was reset. The circuit could also be used to detect a power failure or when a single interruption of a light beam needs to be noted and recorded.

The circuit operates when a light source (L1) shines on the photoresistive cell (PR1). **See Figure 12-20.** The resistance of the cell lowers when the light shines on it, allowing more current to flow from the 12 V source to the base of transistor Q1. Transistor Q1 saturates and its collector-emitter voltage drops. This reduces the base voltage of Q2 to below cutoff, and no Q2 collector current flows. Therefore, the emitter voltage from Q2 drops to 0 V, and 0 V is applied to the gate of the SCR.

When the light source is interrupted, the resistance of photoresistive cell PR1 increases, thus dropping the base voltage of Q1 so that it no longer conducts. The collector voltage at Q1 rises and supplies "turn-on" bias for Q2. The emitter voltage of Q2 rises and supplies sufficient gate voltage to trigger the SCR into conduction, thus turning on the alarm. Once triggered, the SCR continues to conduct until the voltage from anode to cathode is either reduced to near zero or reversed. Once the source light is interrupted, the alarm lamp stays on until the reset button is pushed.

If an AC voltage is issued as a source for the warning lamp, the lamp turns off when the light beam is restored. This occurs because the AC voltage is reversing 60 times a second. Reversal from anode to cathode of the SCR causes the SCR to "unlatch" as soon as the gate signal is removed.

Troubleshooting Light-Sensing Circuits

A good technician can troubleshoot a light-sensing circuit by using signal substitution or circuit disturbances. The light could be turned off. The warning lamp should light immediately if the circuit is working properly. The technician could also block the light with an opaque object, such as a hand or piece of cardboard. Again, the warning lamp should light. The reset button may also be pushed to see if the warning lamp goes out.

If the warning lamp does not turn ON when the light source is lost, the warning lamp may be checked by pushing the reset button. If the lamp lights, the SCR is checked by temporarily shorting the anode to the gate (in higher voltage or current circuits, a 470 Ω resistor is used for anode-to-gate triggering). This triggers a good SCR, and the warning lamp should light and stay lit.

Warning Lamp Circuits

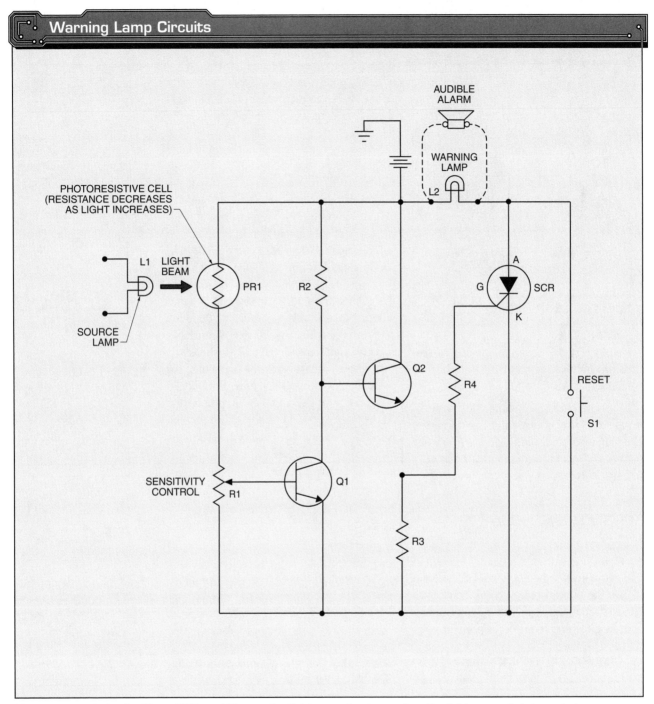

Figure 12-20. *A circuit with an SCR could be used to light a warning lamp when the light source is interrupted.*

If an SCR is working properly, the circuit should be studied to see whether a safe and simple method of checking transistor operation is possible. Since transistor Q2 must conduct to provide gate voltage for the SCR, the base is short-circuited to the emitter of transistor Q1. This zero-biases transistor Q1 and (if Q1 is working) its collector voltage rises, turns on transistor Q2 (if Q2 is working), and triggers the SCR. Transistor Q1 can be zero-biased by turning the sensitivity control to minimum.

If the warning lamp does not remain off after the reset button is pushed, the sensitivity control is turned toward maximum. If this does not help, photoresistive cell PR1 may be defective, transistor Q2 may have excessive collector-to-emitter leakage, or transistor Q1 may be open.

Photoresistive cell PR1 is checked with an ohmmeter. Its resistance should change from high to low when the light strikes it. Transistor Q2 is checked for excessive leakage by checking its base voltage. If it is below cutoff (approximately 0.55 V), transistor Q2 should be checked to ensure that the emitter voltage is 0 V. If it is not, transistor Q2 is leaky and must be replaced. If transistor Q1 is open, the base of transistor Q2, which is connected to the collector of transistor Q1, is high regardless of the base-bias voltage of transistor Q1.

CAUTION: Information provided by manufacturers on the installation of SCRs must be followed closely. The case of the SCR is often used as the anode or cathode connection.

Troubleshooting Procedures

An SCR must be tested under operating conditions using an oscilloscope. An oscilloscope shows exactly how the SCR is operating. Most high-power SCRs must be tested using a test circuit and an oscilloscope. Low-power and some high-power SCRs may be tested using an analog ohmmeter. To ensure that the analog ohmmeter delivers enough output voltage to fire the SCR being tested, a known working SCR should be tested first. **See Figure 12-21.**

Before taking any resistance measurements using an ohmmeter, it must be ensured that the meter is designed to take measurements on the circuit being tested. The operations manual of the test instrument should be consulted for all measuring precautions, limitations, and procedures. The required personal protective equipment should be worn and all safety rules should be followed when taking measurements. To troubleshoot a low-power SCR using an ohmmeter, apply the following procedure:

1. Set the selector switch of the ohmmeter to +DC and the range switch to the proper resistance range (R × 100 is typical).
2. Connect the test leads of the ohmmeter to the meter jacks.
3. Connect the negative test lead of the ohmmeter to the cathode of the SCR.
4. Connect the positive test lead of the ohmmeter to the anode of the SCR. The ohmmeter should read infinity.
5. Short-circuit the gate to the anode using a jumper wire. The ohmmeter should read almost 0 Ω. Remove the jumper wire. The low-resistance reading should remain.
6. Reverse the ohmmeter test leads so that the positive lead is on the cathode and the negative lead is on the anode. The ohmmeter should read almost infinity.
7. Short-circuit the gate to the anode with a jumper wire. The resistance displayed on the analog ohmmeter should remain high.
8. Remove the ohmmeter from the SCR.

SCR APPLICATIONS

In addition to being used in light dimmers and speed controls, SCRs can be used in industry to control the power in battery chargers, power supplies, and machine tools. Also, SCRs can be used as power control elements in welding equipment, power regulators, and temperature control systems.

Precision Heat Control

An SCR can be used in circuits to provide heat control. For example, an SCR control circuit can bring a chemical mixture stored in a vat up to a specific temperature and maintain that temperature. **See Figure 12-22.** With the proper circuitry, the temperature of the mixture can be precisely controlled. Using a bridge circuit, the temperature can be maintained within 1°F over a temperature range of 20°F to 150°F.

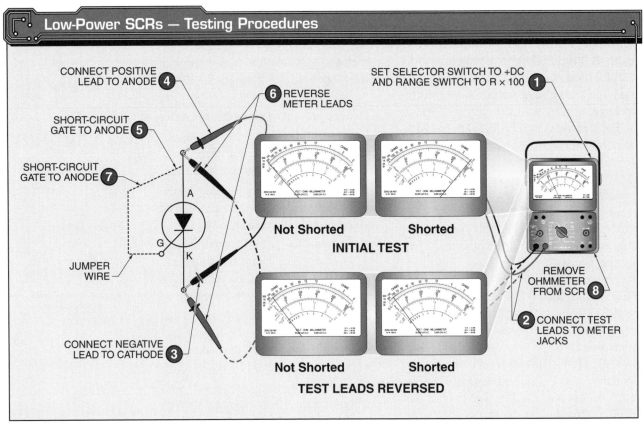

Figure 12-21. Low-power and some high-power SCRs may be tested using an analog ohmmeter. A good SCR will have readings of 0 Ω and infinity.

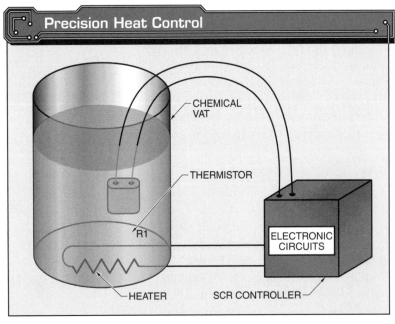

Figure 12-22. An SCR can be used in circuits to provide precision heat control.

In the circuit, transformer T has two secondary windings, W1 and W2. Winding W1 supplies voltage through the SCR to relay coil K. Winding W2 supplies AC voltage to the gate circuit of the SCR. Primary control over this circuit is accomplished through the use of the bridge circuit. The bridge circuit is formed by thermistor R1, fixed resistors R2 and R3, and variable resistor R4. Resistor R5 is a current-limiting resistor used to protect the bridge circuit. The fuse protects the primary of the transformer. When the resistance of R1 equals the resistance setting on R4, the bridge is balanced. The AC voltage introduced into the bridge by winding W2 is not applied to the gate of the SCR. Hence, the relay coil (K) remains de-energized and its normally closed contacts apply power to the heating elements. **See Figure 12-23.**

Precision Heat Control Circuits

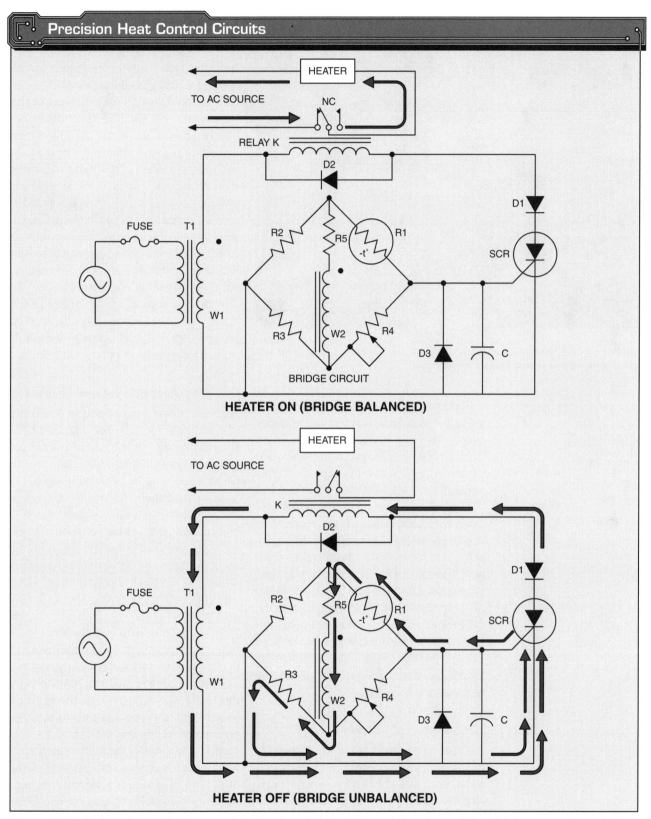

Figure 12-23. *Voltage is provided to a precision heat control circuit through transformer T1, which has secondary windings W1 and W2. A bridge is formed by thermistor R1 and fixed resistors R2, R3, and variable resistor R4. When the bridge is balanced, there is no gate current and relay K1 remains de-energized, allowing the heater to stay on.*

Saftronics, Inc.
Silicon controlled rectifiers are tested by connecting the PC board to a test circuit and using ohmmeters and oscilloscopes for diagnostics.

If the temperature increases above a preset level, the resistance of thermistor R1 decreases. The bridge becomes unbalanced such that current flows to the gate of the SCR while the anode of the SCR remains positive. This turns on the SCR and energizes the relay coil (K), thereby switching power from the load through the relay contact. If the temperature falls below the preset temperature setting, R1 unbalances the bridge in the opposite direction. Therefore, a negative signal is applied to the gate of the SCR when the anode of the SCR is positive. The negative signal stops the SCR from conducting and allows current to continue to flow to the heating elements. *Note:* Locating the thermistor in the vat where the temperature must be controlled provides the necessary feedback information.

Time-Delayed Relays

An SCR can be used in circuits to provide time delay for motor control, photo enlargers, and other timed or delayed functions. In many situations, an SCR can be combined with a relay to form a hybrid circuit. A *hybrid circuit* is a circuit that mixes mechanical devices, such as a relay, and electronic devices, such as an SCR, in the same circuit. The hybrid circuit turns a load on and off after a preset time delay. A hybrid circuit has an electronic component (SCR) and an electromechanical component (relay) in the same circuit.

The operation of a hybrid circuit begins with switch S in the RESET position. **See Figure 12-24.** Capacitor C quickly charges to the peak negative value of the supply voltage (approximately 170 V) and effectively shuts off any gate current to the SCR. In this position, the load is OFF. Switch S turned to the TIME position turns on the load. Capacitor C starts to discharge toward zero at a rate determined by the setting of variable resistor R2. Since a delay is built into this circuit, the RC time constant is long. The delay may be shortened by reducing the resistance of R2. When the voltage across C finally drops to zero, reverses, and reaches about 1 V, the SCR triggers relay coil K to turn off the load.

Security Alarm System Circuits

An SCR can be used as part of a security alarm system for cars. **See Figure 12-25.** By using diodes, switches, and an SCR firing circuit, visual and audible alarms can be activated to serve as warning signals in an attempted theft. By placing diodes in series with the ignition coil, radio, or CD player, each car accessory serves as a means for triggering the SCR when it is powered up. If normally open switches are installed as shown, the trunk, hood, or a pressure-sensitive switch under the floor mat may also be used.

In the circuit, when both the arm switch (S1) and the reset switch (S2) are closed and the ignition switch is turned on (by either a key or a jumper wire), current flows through diode D1 (associated with the ignition switch) and resistor R1. This puts a charge on capacitor C. When the charge on C is sufficient to turn ON the transistor, the current through the transistor also turns on the SCR. The SCR supplies current to a horn or siren. The SCR remains on regardless of the condition of the diodes or transistor. Only opening the reset switch (S2), which is concealed, turns off the SCR.

Time Delay Circuits

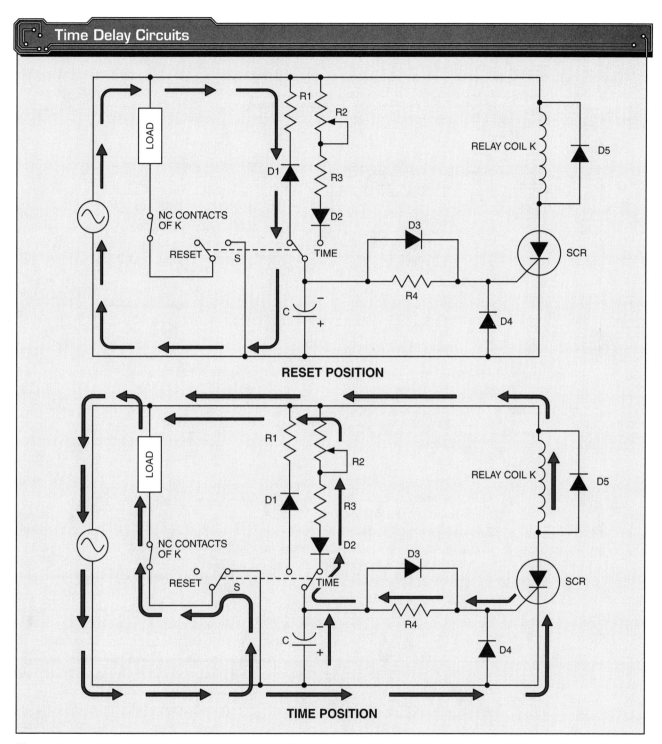

Figure 12-24. When the time delay circuit is in the RESET position, capacitor C quickly charges to the peak negative value of the supply voltage, effectively shutting off any gate current to the SCR.

The values of R1 and C in the timing network are selected to provide sufficient delay for the vehicle owner to enter the vehicle and open the reset switch to disarm the alarm. The arm switch is placed in the NORMAL position when the vehicle is in use. It is closed or armed when the vehicle is unattended.

Car Alarm Circuits

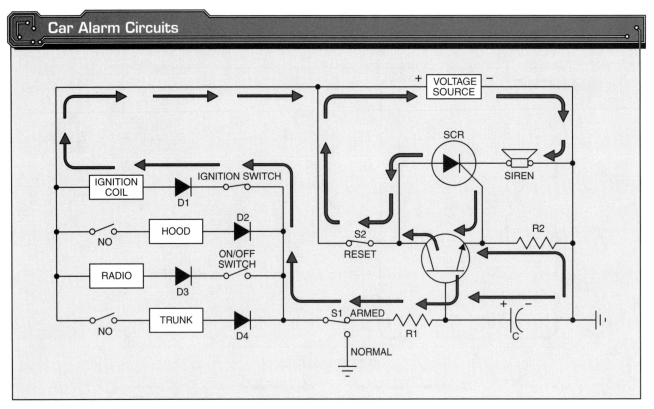

Figure 12-25. *An SCR can be used as part of a security alarm system for cars.*

This type of car alarm circuit has been improved greatly. It is no longer necessary to enter the vehicle to disarm the alarm. A car alarm may now be armed/disarmed with a remote control. However, a similar circuit is still used in some home alarm systems.

Adjustable Temperature Control

A variable heat source is required in many heating applications. With a variable heat source, the amount of heat output is adjustable. **See Figure 12-26.** Different temperatures are required for the proper heating of different materials and different sizes of materials. Too much or not enough heat prevents proper sealing.

To control the amount of heat output, the voltage to the heating element must be controlled. SCRs are used to control the amount of voltage output. The SCR is triggered into conduction in only one direction by a quick pulse of control current to its gate. An electronic triggering circuit is used to control the pulses to the gate. The triggering circuit may be set to allow for any voltage setting between completely on and completely off. The higher the applied voltage, the greater the heat output.

Solid State Starting

Solid state starting is a form of reduced-voltage starting for standard motors. The heart of the solid state starter is the SCR, which controls motor voltage, current, and torque during acceleration.

A reduced-voltage, solid state starter ramps up motor voltage as the motor accelerates instead of applying full voltage instantaneously. Unlike across-the-line starters, solid state starters reduce inrush current. Solid state starters also provide smooth acceleration and minimize starting torque. High starting torque can damage some loads connected to the motor. **See Figure 12-27.**

Chapter 12—Silicon-Controlled Rectifiers (SCRs) 315

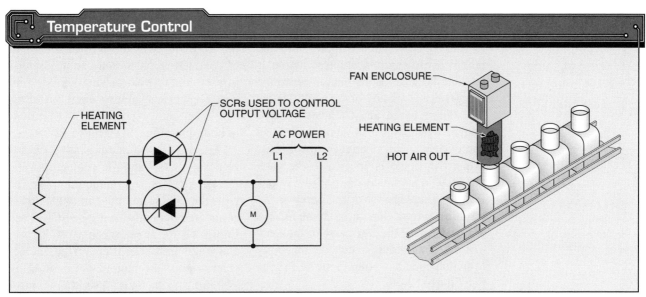

Figure 12-26. *SCRs can be used to control the amount of heat output by controlling the applied voltage.*

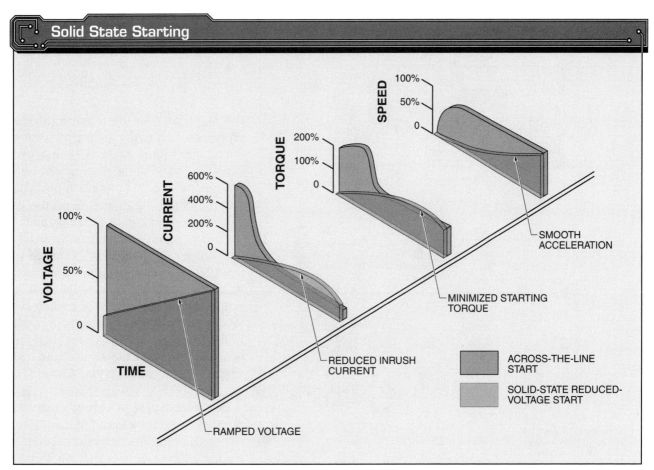

Figure 12-27. *A solid state starter ramps up voltage, reduces inrush current, smoothes acceleration, and minimizes starting torque.*

Solid state starting provides smooth, stepless acceleration in applications such as starting conveyors, compressors, and pumps. This smooth, stepless action is possible because of the unique switching (triggering) capability of the SCR. Unlike an ordinary diode, an SCR does not pass current from cathode to anode unless an appropriate signal is applied to the gate. When the signal is applied to the gate, the SCR is switched on, and the anode resistance decreases sharply. Therefore, the resulting current flow through the SCR is only limited by the resistance of the load. The advantage of this device is its ability to switch on at any point in the half cycle. **See Figure 12-28.**

The average amount of voltage and current can be reduced or increased by switching the SCRs, which controls the amount of conduction. SCRs may be used alone in a circuit to provide one-way current control. They may also be wired in reverse-parallel circuits to control AC line current in both directions. **See Figure 12-29.**

A typical solid state starting circuit consists of start and run contactors. The start contactor is connected in series with the SCRs, and the run contactor is connected in parallel with the SCRs. **See Figure 12-30.** The start contacts C1 close, and when the starter is energized, the acceleration of the motor is controlled by switching on the SCRs. The SCRs control the motor until it approaches full speed, at which time the run contacts C2 close, connecting the motor directly across the power line. At this point, the SCRs are turned off, and the motor runs with full power applied to the motor terminals.

Power Supply Protection Circuits

An SCR can be used for overvoltage protection as a power supply protection circuit. **See Figure 12-31.** A basic circuit consists of an SCR placed in parallel with the output of a DC power supply. The purpose of placing an SCR in parallel is to create a short circuit on the output of the power supply so that excessive voltage will not reach the load. Damage to the SCR and power supply is prevented by placing a fuse in front of the SCR to limit short-circuit current. A sensing device is connected to the gate of the SCR to detect excessive voltage between the gate and cathode. Excessive voltage will trigger enough current in the SCR to blow the fuse.

A power supply protection circuit is referred to as a "crowbar" circuit. This is because the effect of its operation is similar to the effect of placing a steel crowbar across the output terminals of a power supply.

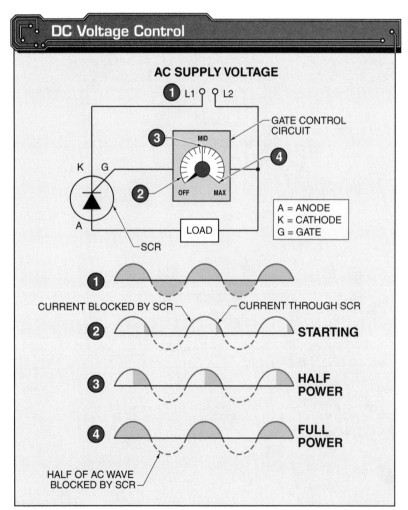

Figure 12-28. *When the signal is applied to the gate, the SCR is triggered on and the anode resistance decreases sharply.*

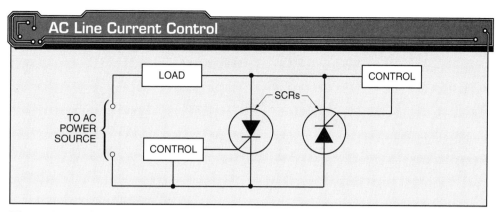

Figure 12-29. SCRs may be used alone in a circuit to provide one-way current control. They may also be wired in reverse-parallel circuits to control AC line current in both directions.

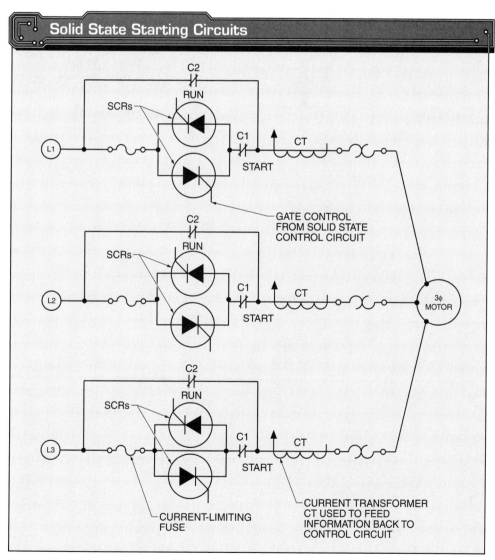

Figure 12-30. An SCR circuit with reverse-parallel wiring of SCRs provides maximum control of an AC load.

Refer to Chapter 12 Quick Quiz® on CD-ROM

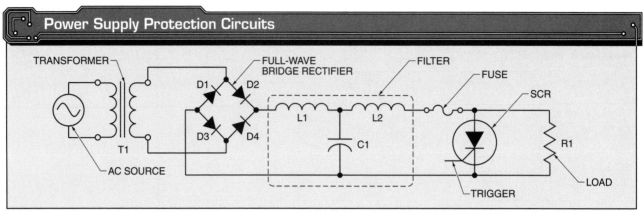

Figure 12-31. *Power supply protection (crowbar) circuits are placed in DC circuits to provide overvoltage protection.*

KEY TERMS

- silicon-controlled rectifier
- gate
- forward breakover voltage
- holding current
- phase control
- half-wave phase control circuit
- hybrid circuit
- gate turn-on
- voltage breaker turn-on

REVIEW QUESTIONS

1. How is the physical size of an SCR affected by an increase in current rating?
2. Which electrode on an SCR is the control point for the SCR?
3. What effect does gate current have on the level of breakover voltage?
4. Name the three electrodes of an SCR.
5. What is forward blocking current?
6. How many layers of P-type and N-type material does an SCR have?
7. What are the two operating states of an SCR?
8. What two events can cause an SCR to conduct in the forward direction?
9. What devices could be connected together to form an SCR equivalent circuit?
10. How is an SCR most often used?
11. What is a hybrid circuit?
12. How is an SCR turned off?
13. Define phase control.
14. What type of circuit can be used to provide a time delay in an SCR circuit?

TRIACS, DIACS, AND UNIJUNCTION TRANSISTORS 13

Triacs, diacs, and unijunction transistors (UJTs), along with silicon-controlled rectifiers (SCRs), are often found in the same circuitry. Triacs and SCRs are control devices. Diacs and UJTs form the triggering circuits for triacs and SCRs. Triacs, diacs, UJTs, and SCRs operate only as switches and may be used in a variety of switching applications.

OBJECTIVES

- Describe the operation of a triac.
- List and describe triac applications.
- Describe the operation of a diac.
- List and describe diac applications.
- Describe the operation of a unijunction transistor (UJT).
- List and describe UJT applications.

TRIACS

A *triac* is a three-electrode, bidirectional AC switch that allows electrons to flow in either direction. The word "triac" is derived from the phrase "triode for alternating current." Triacs are the equivalent of two silicon-controlled rectifiers (SCRs) connected in a reverse parallel arrangement with gates also connected to each other.

A triac is triggered into conduction in both directions by a gate signal in a manner similar to that of an SCR. The triac was developed to provide a means for producing improved controls for AC power. Triacs are available in a variety of packaging arrangements. **See Figure 13-1.** They can handle a wide range of current and voltage. Triacs generally have relatively low-current capabilities compared to SCRs. Triacs are usually limited to less than 50 A and cannot replace SCRs in high-current applications.

Triac Construction

Although triacs and SCRs look similar, their schematic symbols are different. The terminals of a triac are the gate, terminal 1 (T1), and terminal 2 (T2). **See Figure 13-2.**

There is no designation of anode and cathode. Current may flow in either direction through the main switch terminals, T1 and T2. Terminal 1 is the reference terminal for all voltages. Terminal 2 is the case or metal-mounting tab to which a heat sink can be attached.

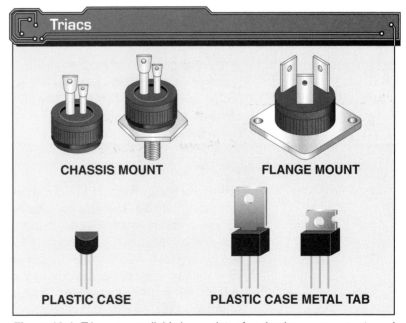

Figure 13-1. *Triacs are available in a variety of packaging arrangements and can handle a wide range of current and voltage.*

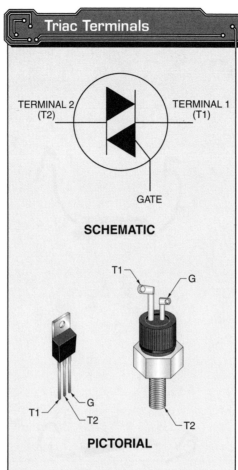

Figure 13-2. Triac terminals include a gate, terminal 1 (T1), and terminal 2 (T2).

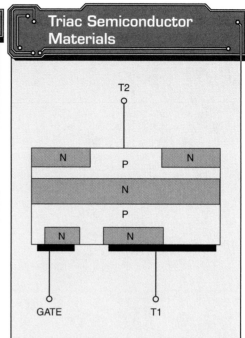

Figure 13-3. A triac can be considered two NPN switches sandwiched together on a single N-type material wafer.

The structure of the triac is complex. However, for all practical purposes, a triac can be considered as two NPN switches sandwiched together on a single N-type material wafer. **See Figure 13-3.**

Triac Operation

Triacs block current in either direction between T1 and T2. The triac operates in a manner similar to a pair of SCRs connected in a reverse parallel arrangement. **See Figure 13-4.** A triac can be triggered into conduction in either direction by a momentary positive or negative pulse supplied to the gate. If the appropriate signal is applied to the triac gate, it conducts electricity.

The triac remains off until the gate is triggered at point A. **See Figure 13-5.** At point A, the trigger circuit pulses the gate and the triac is turned ON, allowing current to flow. At point B, the forward current is reduced to zero and the triac is turned OFF. The trigger circuit can be designed to produce a pulse that varies at any point in the positive or negative half cycle. Therefore, the average current supplied to the load can vary.

One advantage of the triac is that virtually no power is wasted by being converted to heat. Heat is generated when current is impeded, not when current is switched OFF. The triac is either fully ON or fully OFF. It never partially limits current. Another important feature of the triac is the absence of a reverse breakdown condition of high voltages and high current, such as those found in diodes and SCRs. If the voltage across the triac goes too high, the triac is turned ON. When turned ON, the triac can conduct a reasonably high current.

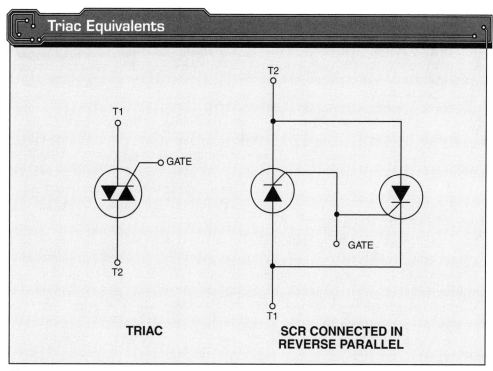

Figure 13-4. *A triac operates similar to a pair of SCRs connected in a reverse parallel arrangement.*

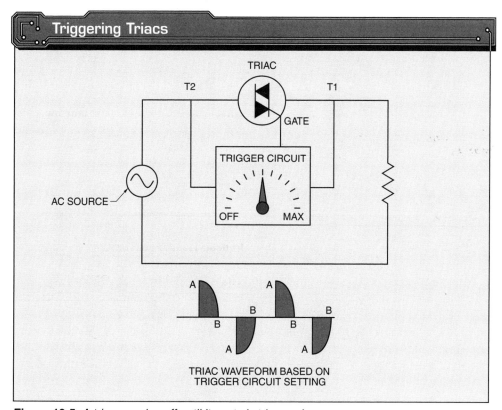

Figure 13-5. *A triac remains off until its gate is triggered.*

Triacs are considered versatile in part because of their ability to operate with positive or negative voltages across their terminals. Since SCRs have a disadvantage of conducting current in only one direction, controlling power in an AC circuit is better served with the use of a triac.

Triac Characteristic Curves

The characteristics of a triac are based on T1 as the voltage reference point. The polarities shown for voltage and current are the polarities of T2 with respect to T1. The polarities shown for the gate are also with respect to T1. **See Figure 13-6.** Again, the triac may be triggered into conduction in either direction by a gate current (I_G) of either polarity.

Triac Applications

Triacs are often used instead of mechanical switches because of their versatility. Also, where amperage is low, triacs are more economical than back-to-back SCRs.

Single-Phase Motor Starters. Often, a capacitor-start or split-phase motor must operate where arcing of a mechanical cut-out start switch is undesirable or even dangerous. In such cases, the mechanical cut-out start switch can be replaced by a triac. **See Figure 13-7.** A triac is able to operate in such dangerous environments because it does not create an arc. The gate and cut-out signal is given to the triac through a current transformer. As the motor speeds up, the current is reduced in the current transformer, and the transformer no longer triggers the triac. With the triac OFF, the start windings are removed from the circuit.

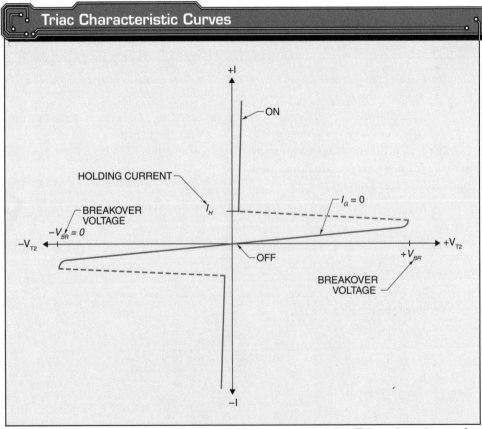

Figure 13-6. *The characteristics of a triac are based on terminal 1 (T1) as the voltage reference point.*

Solid state motor starting is associated with a host of benefits in applications such as starting conveyors, compressors, and pumps. Aside from an ability to minimize starting torque, solid state starting also creates smooth motor acceleration for an increase in overall efficiency.

Another method of sending signals to the triac is to use a tachometer generator. A tachometer generator produces a specific voltage for a certain amount of revolutions per minute (rpm). As the rpm increases, the voltage increases. As the rpm decreases, the voltage decreases. The signal would be sent to a trigger circuit, which would activate the triac at a preset rpm. With this arrangement, the dropout or cutout could be precisely set.

Replacement of Reversing Contactors. Triacs are also used to replace reversing contactors. When used with a reversing permanent split capacitor motor, a pair of triacs can provide a rapidly responding reversing motor control circuit. **See Figure 13-8.** Switches 1 and 2 (S1 and S2) may be reed switches. In this particular circuit, it is important to ensure that the triacs have a sufficient voltage rating. If one triac is turned ON, the other triac must block the voltage. The voltage is created by line voltage due to the inductance and capacitance between the motor winding and the capacitor. *Note:* The triac voltage rating should be 1.5 times the voltage rating of the capacitor.

Heater Control. A triac is commonly used to control the amount of current that reaches a load, such as an electric heater. The heating element becomes hotter when more of the sine wave reaches the heating element. **See Figure 13-9.**

A triac or two SCRs can transmit all or part of an AC sine wave. They can be triggered at any time during an AC cycle. A triac operates as if it contains a switch that turns on and off so quickly that it passes only a portion of the AC cycle. The result is that more or less of the sine wave is applied to the load.

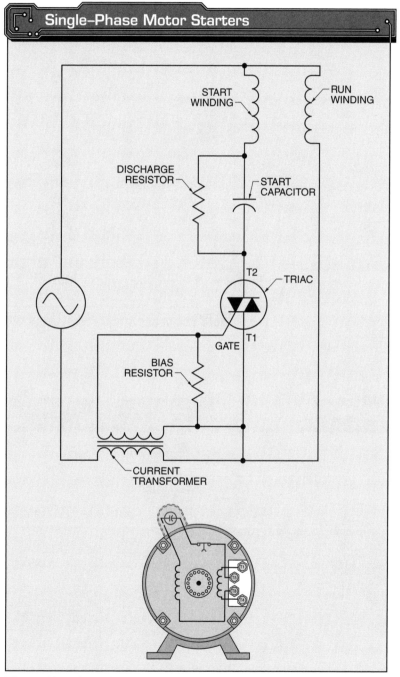

Figure 13-7. *A mechanical cut-out start switch may be replaced by a triac. As a motor speeds up, the current is reduced in the current transformer, and the transformer no longer triggers the triac. With the triac OFF, the start windings are removed from the circuit.*

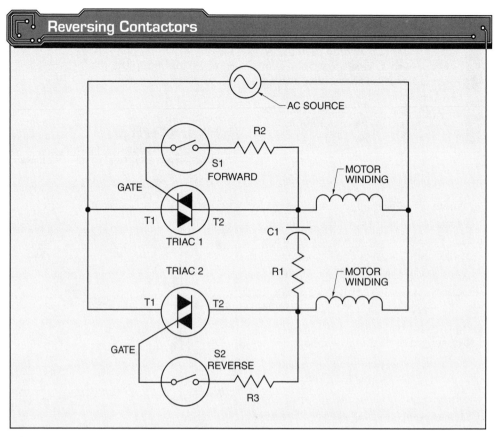

Figure 13-8. *A pair of triacs can replace reversing contactors on a permanent split capacitor motor.*

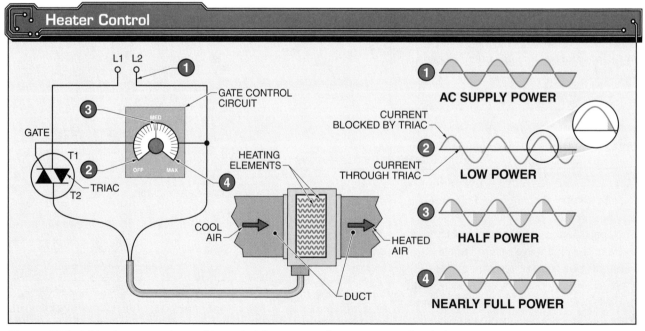

Figure 13-9. *A triac is commonly used to control the amount of current to an electric heater.*

Automatic Security Lights. A triac can be used in a residential or commercial security system circuit to turn lights ON at night and OFF in the morning. **See Figure 13-10.** A voltage-divider circuit divides the DC voltage between resistor R2 and the parallel combination of the photocell, with resistor R3 in series with inductor L1. When the photocell is exposed to light, its resistance drops to a few ohms. This deprives inductor L1 of enough current to close the reed switch S1. With switch S1 open, the triac does not conduct and the light will remain off. If no light falls on the photocell, its resistance increases to several megohms. The current from resistor R2 then passes through resistor R3 and inductor L1. This current causes the reed switch (S1) to close and energizes the triac by passing current to the gate. Once the triac activates, the light turns on.

Saftronics, Inc.
Sometimes electric motor drive components are removed from service so shop or benchtop troubleshooting can be performed.

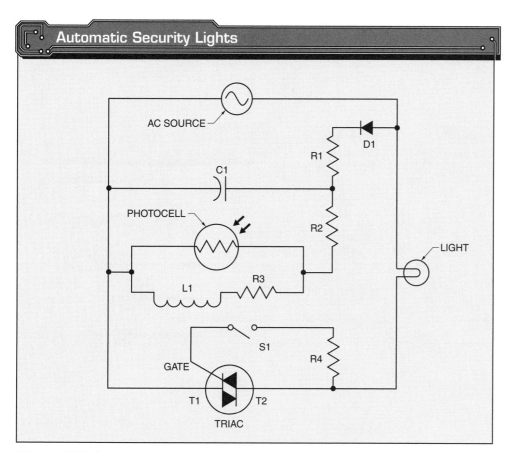

Figure 13-10. *An automatic security light is triggered by a photocell and switch.*

Triac Testing

Triacs should be tested while in operation using an oscilloscope. An oscilloscope shows exactly how a triac is operating. However, when a triac is out of a circuit, an ohmmeter can be used to perform a rough test. **See Figure 13-11.**

Note: If using a digital multimeter, it must first be ensured that the meter is designed to take resistance measurements on the circuit being tested. The operations manual of the test instrument must be referred to for all measuring precautions, limitations, and procedures. The proper PPE must always be worn and all safety precautions must be followed when taking measurements.

To test a triac using an ohmmeter, apply the following procedure:

1. Set the selector switch and the range switch of the ohmmeter to the "R × 100" scale.
2. Connect the test leads of the ohmmeter to the meter jacks as required.
3. Connect the negative lead of the ohmmeter to terminal 1 (T1).
4. Connect the positive lead of the ohmmeter to terminal 2 (T2). The ohmmeter should read infinity. The resistance should, in fact, be over 250,000 Ω.
5. Short-circuit the gate by closing switch S1. The ohmmeter should read approximately 0 Ω. (The resistance should be approximately 10 Ω to 50 Ω but should register zero on the "R × 100" scale.) When switch S1 is opened, the resistance reading of 0 Ω should remain.
6. Reverse the leads so that the positive test lead is on T1 and the negative test lead is on T2. The ohmmeter should read infinity. The resistance should, in fact, be over 250,000 Ω.

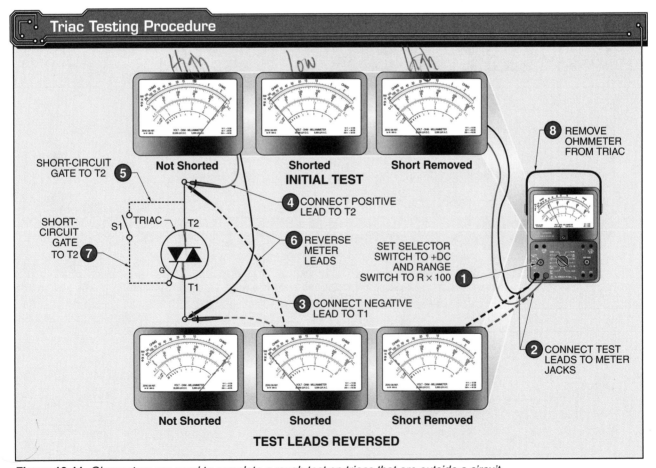

Figure 13-11. *Ohmmeters are used to complete a rough test on triacs that are outside a circuit.*

7. Short-circuit the gate by closing switch S1. The ohmmeter should read 0 Ω. (The resistance should be approximately 10 Ω to 50 Ω but should register zero on the "R × 100" scale.) When switch S1 is opened, the resistance reading of 0 Ω should remain.
8. Remove the ohmmeter from the triac.

Note: If the triac does not respond as indicated in each of these steps, it is probably defective and must be replaced.

DIACS

A *diac* is a three-layer, two-terminal bidirectional device that is typically used to control the gate current of a triac. **See Figure 13-12.** Unlike a transistor, the two junctions are heavily and equally doped. Each junction is almost identical to the other. A diac is a special diode that can be triggered into conduction in either direction.

Diac Operation

Electrically, a diac operates in a manner similar to two zener diodes that are connected in series in opposite directions. **See Figure 13-13.** The diac is used primarily as a triggering device. This operation is accomplished by the use of the negative resistance characteristic of the diac (current decreases with an increase of applied voltage).

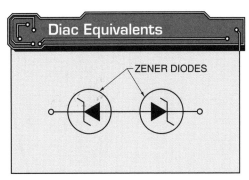

Figure 13-13. *Electrically, a diac operates in a manner similar to two zener diodes that are connected in series in opposite directions.*

A diac has negative resistance because it does not conduct current until the voltage across it reaches breakover voltage. **See Figure 13-14.** When a positive or negative voltage reaches the breakover voltage, the diac rapidly switches from a high-resistance state to a low-resistance state. Since the diac is a bidirectional device, it is ideal for controlling a triac, which is also bidirectional.

Diac Applications

The gate-control circuits of triacs can be improved by adding a breakover device in the gate lead, such as a diac. Using a diac in the gate-triggering circuit offers an important advantage over simple gate-control circuits. The advantage is that the diac delivers a pulse of gate current rather than a sinusoidal gate current. This results in a better-controlled firing sequence. Thus, diacs are used almost exclusively as triggering devices.

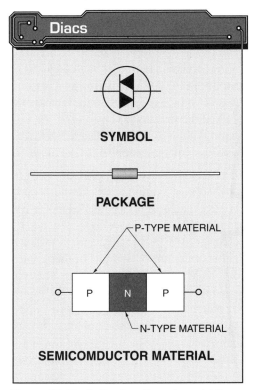

Figure 13-12. *A diac is a three-layer bidirectional device.*

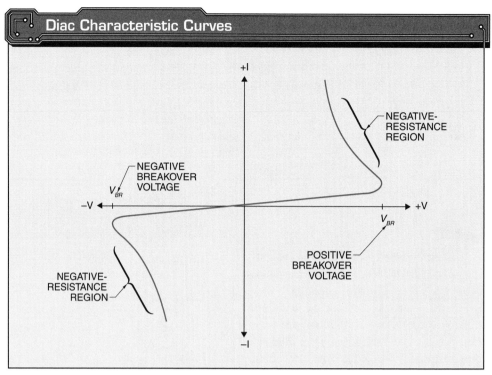

Figure 13-14. *A diac is bidirectional and is used primarily as a triggering device. When a positive or negative voltage reaches the breakover voltage, the diac rapidly switches from a high-resistance state to a low-resistance state.*

Lighting Control. A diac may be used to control the brightness of a lamp. **See Figure 13-15.** During the beginning of each half cycle of the applied AC voltage, the triac is in the OFF state and no voltage is applied to the lamp. However, the AC voltage is applied across the resistor/capacitor circuit connected in parallel with the triac. The applied voltage charges the capacitor (C1). The diac conducts when the voltage across the capacitor rises to the breakover voltage of the diac. The capacitor voltage is then discharged through the diac to the triac gate. The voltage at the triac gate triggers the triac into conduction, and voltage is applied to the lamp. This sequence of events is repeated for each half cycle of the AC sine wave.

The potentiometer (R2) may be adjusted to control the brightness of the lamp by determining at what point on the AC sine wave the triac is allowed to start conducting. A decrease in resistance allows the capacitor (C1) to charge faster and the breakover voltage of the diac to be reached earlier. This allows the triac to trigger into conductance sooner in the AC sine wave. As a result, a higher voltage is applied to the lamp. However, if the resistance is increased, the triggering of the triac occurs later in the cycle, and less voltage is applied to the lamp.

Universal Motor Speed Controllers. A diac/triac combination can be used to control the power to a universal motor. **See Figure 13-16.** In the circuit, capacitor C1 charges up to the firing voltage of the diac in either direction. Once fired, the diac applies a voltage to the gate of the triac. The triac conducts and supplies power to the motor.

Note: The triac will conduct in either direction. Since the universal motor is basically a series DC motor, current flowing in either direction will cause rotation in only one direction. The speed may be changed by varying the resistance of potentiometer R1, which in turn, varies the RC time constant.

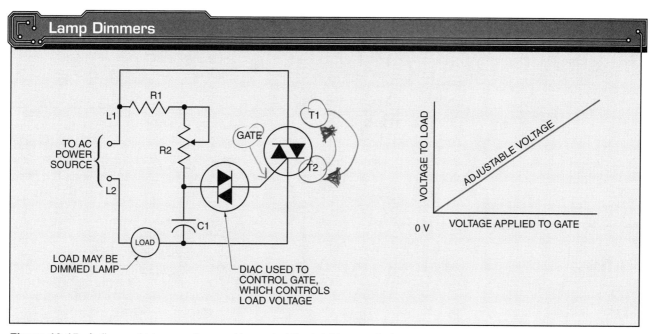

Figure 13-15. A diac and triac may be used to control the brightness of a lamp.

Diac Testing

An ohmmeter can be used to test a diac for a short circuit. **See Figure 13-17.** To test for a short circuit, apply the following procedure:

1. Place the ohmmeter on the "R × 100" scale.
2. Connect the ohmmeter leads to each terminal of the diac, and record the resistance reading.
3. Reverse the ohmmeter leads, connect to each terminal of the diac, and record the resistance reading.

Note: Since the diac is essentially two zener diodes connected in series, both readings should show high resistance. Also, testing a diac in this manner will only show that the component is shorted. If it is suspected that the diac is open, a second test using an oscilloscope should be performed.

An oscilloscope can be used to test for an open diac. **See Figure 13-18.** To test for an open diac using an oscilloscope, apply the following procedure:

1. Set up the test circuit.
2. Apply power to the circuit.
3. Adjust the oscilloscope.

Note: If the diac is good, a waveform similar to that shown on the oscilloscope should appear.

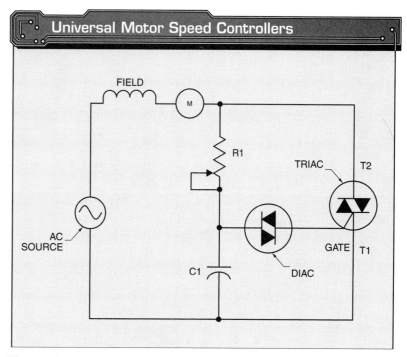

Figure 13-16. In this circuit, capacitor C1 charges up to the firing voltage of the diac in either direction. Once fired, the diac applies voltage to the gate of the triac. The triac conducts and applies power to the motor.

330 SOLID STATE DEVICES AND SYSTEMS

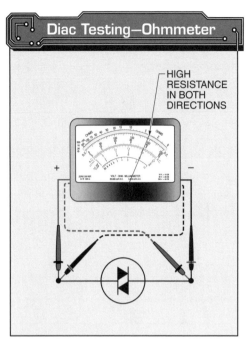

Figure 13-17. *An ohmmeter can be used to test a diac for a short circuit.*

UNIJUNCTION TRANSISTORS

A *unijunction transistor (UJT)* is a three-electrode device that contains one PN junction consisting of a bar of N-type material with a region of P-type material doped within the N-type material. **See Figure 13-19.** The N-type material functions as the base and has two leads, base 1 (B1) and base 2 (B2). The lead extending from the P-type material is the emitter (E).

> *Circuits or components that are tested with ohmmeters in normal operation are not required to be energized. Ohmmeters, and all test instruments that can measure resistance, have an internal battery that is used to supply voltage to test instrument leads and the component being tested.*

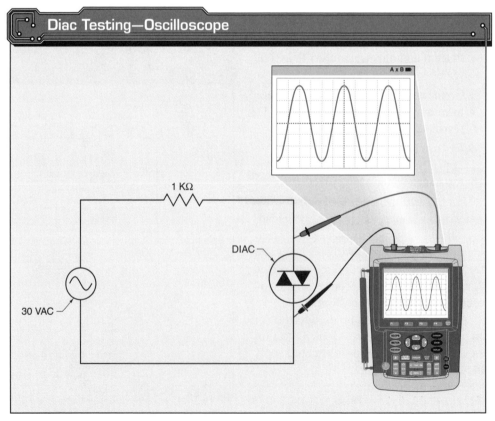

Figure 13-18. *An oscilloscope can be used to test for an open diac.*

Chapter 13—Triacs, Diacs, and Unijunction Transistors 331

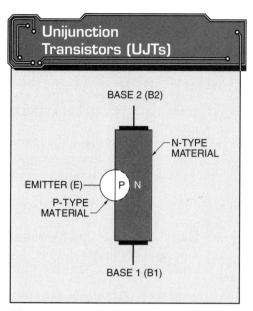

Figure 13-19. *A UJT consists of a bar of N-type material with a region of P-type material doped within the N-type material.*

The UJT is used primarily as a triggering device because it generates a pulse used to fire SCRs and triacs. Outputs from photocells, thermistors, and other transducers can be used to trigger UJTs, which in turn, fire SCRs and triacs. UJTs are also used in oscillators, timers, and voltage-current sensing applications.

UJT Equivalent Circuits

It is helpful to study the equivalent circuit of a UJT in order to understand how a UJT operates. A UJT can be considered a diode connected to a voltage divider network. **See Figure 13-21.** Potentiometer R1 and fixed resistor R2 form the voltage divider network in the UJT. The PN junction creates the diode, and the N-type material serves as the resistive voltage divider.

In the schematic symbol for a UJT, the arrowhead represents the emitter (E). Although the leads are usually not labeled, they can be easily identified because the arrowhead always points to B1. The case of a UJT may include a tab to identify the leads. **See Figure 13-20.**

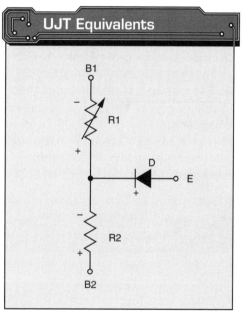

Figure 13-21. *The equivalent circuit of a UJT can be thought of as a diode connected to a voltage divider network.*

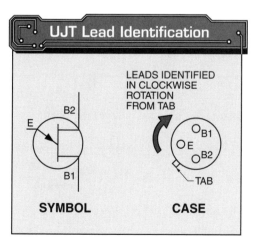

Figure 13-20. *In the schematic symbol for a UJT, the arrowhead represents the emitter and always points to base 1 (B1).*

UJTs have the ability to be used as relaxation oscillators. An oscillator is a circuit that produces a repetitive electronic signal, such as a sine wave, without AC input signals. Relaxation oscillators are characterized internally by short, sharp pulses of waveforms that can potentially trigger gates.

UJT Biasing

In normal operation, B1 is negative, and a positive voltage is applied to B2. The internal resistance between B1 and B2 is divided at E, with approximately 60% of the resistance between E and B1. The remaining 40% of the resistance is between E and B2. The net result is an internal voltage split. This split provides a positive voltage at the N-type material of the emitter junction, creating an emitter junction that is reverse-biased. As long as the emitter voltage remains less than the internal voltage, the emitter junction will remain reverse-biased, even at a very high voltage. However, if the emitter voltage rises above this internal value, a dramatic change will take place.

When the emitter voltage is greater than the internal value, the junction becomes forward-biased. Also, the resistance between E and B1 drops rapidly to a very low value. A UJT characteristic curve shows the dramatic change in voltage due to this change in resistance. **See Figure 13-22.**

The characteristic of the emitter-base 1 (E-B1) resistance is similar to that of potentiometer R1 of the equivalent circuit. Potentiometer R1 is normally higher than resistor R2 but collapses to a very low value when the emitter is forward-biased. **See Figure 13-23.**

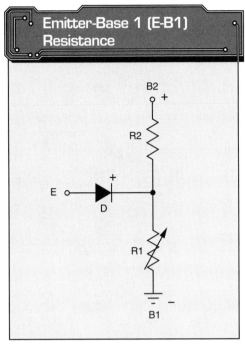

Figure 13-23. *The characteristic of the emitter-base 1 (E-B1) resistance is similar to that of potentiometer R1.*

In terms of real world application, the most significant region of the UJT characteristic curve falls between the peak current point and valley current point. It is at this point of the curve that the UJT is best applied as a triggering device.

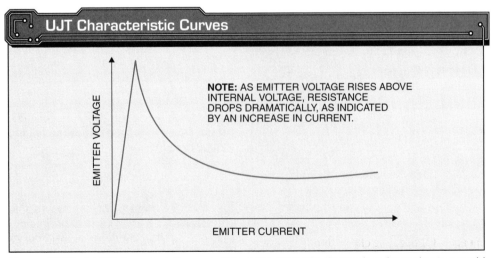

Figure 13-22. *A UJT characteristic curve shows the dramatic change in voltage due to a rapid change in resistance.*

Theory of Operation

As long as the E-B1 junction is reverse-biased and no current flows into the emitter, the current flow in the N-type material should be minimal. This is due to the small amount of doping that creates a high resistance. When the E-B1 junction is forward-biased, the junction turns on, causing carriers to be injected into the base region. These carriers create an excess of holes. Their presence in the N-type material increases conductivity, which lowers the resistance of the region. Once started, current flows easily between B1 and E. Therefore, the conductivity of this region is controlled by the flow of emitter current.

Saturation

The lowered resistance across the E-B1 junction, once it is forward-biased, means that less voltage drop should appear at the junction. With a lower voltage drop, the junction has an increased forward-bias voltage. The increased forward-bias voltage lowers the junction resistance further. Eventually, the emitter circuit saturates and no additional increase in current results. If the voltage at E is increased further, the UJT moves into its saturation region, and the current increases only gradually. The region between the peak current point and valley current point is the negative-resistance region. In the negative-resistance region, the UJT has its greatest application as a triggering device. **See Figure 13-24.**

Triggering Circuits

A UJT is typically used as a triggering circuit for a triac or similar device. **See Figure 13-25.** When switch S1 is closed, the voltage-divider action of the UJT produces a voltage between B1 and the N-type material of the emitter junction. At this same instant, the emitter voltage is zero since it is tied to capacitor C1. The emitter junction at that point is reverse-biased, and no current flows through the junction. As capacitor C1 begins to charge through resistor R1, the voltage across capacitor C1 should begin to increase.

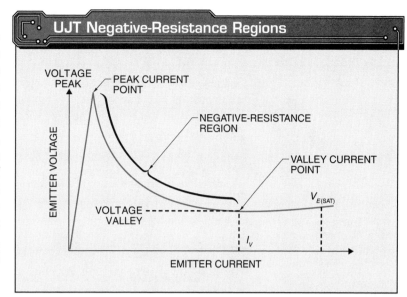

Figure 13-24. *In the negative-resistance region, a UJT has its greatest application as a triggering device.*

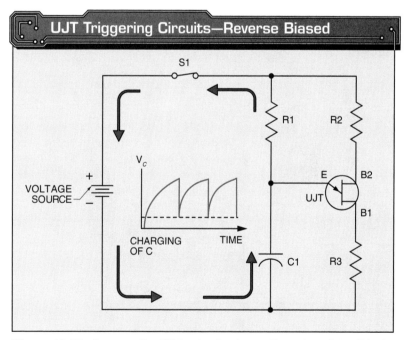

Figure 13-25. *As capacitor C1 begins to charge through resistor R1, the voltage across capacitor C1 begins to increase.*

For an emitter to be forward-biased, it must be more positive than the base (+0.6 V for silicon or +0.2 V for germanium). Assuming a silicon crystal is used in the UJT, the junction becomes forward-biased when the control voltage reaches 0.6 V beyond the junction voltage.

With the junction forward-biased, the internal resistance of the E-B1 region drops dramatically. This causes capacitor C1 to discharge its energy through base load resistor R3. **See Figure 13-26.** Once the capacitor has discharged enough to reduce the forward bias on the junction, the resistance of the junction returns to normal. The cycle of capacitor charging and discharging then repeats.

Each time the emitter becomes forward-biased, the total resistance between B1 and B2 drops, permitting an increase in current through the UJT. As a result, a positive pulse (V_{B1}) appears at B1, and a negative pulse (V_{B2}) appears at B2 at the time the capacitor discharges.

Note: The repetition rate, or frequency, of the discharge voltage is determined by the values of resistor R3 and capacitor C1. Increasing either one makes the device run more slowly. The pulses that appear across bases B1 and B2 are very useful in triggering SCRs and triacs.

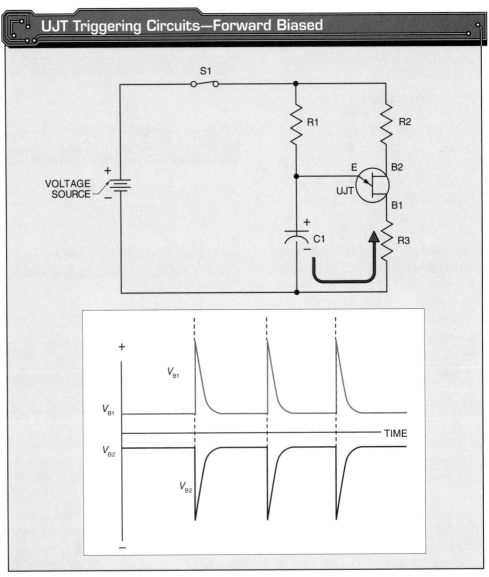

Figure 13-26. *With the emitter junction forward biased, the internal resistance of the E-B1 region drops dramatically and causes capacitor C1 to discharge its energy through base load resistor R3.*

UJT Applications

The UJT is used in switching and timing applications. A UJT often reduces the number of components needed to perform a given function. The number of components is often less than half of what is required when using bipolar transistors. Since UJTs are relatively inexpensive, it is economical to use them in many applications.

Emergency Flashers. A UJT can serve as a triggering circuit for an emergency flasher. As a triggering circuit, UJT Q1 provides base bias to drive transistors Q2 and Q3 through resistors R2 and R3. Transistors Q2 and Q3 are used to light an incandescent lamp load. **See Figure 13-27.**

The circuit repetition rate (frequency) is determined by the characteristics of the UJT, supply voltage, and emitter RC time constant of Q1. To change the flashing rate, the value of capacitor C1 must be changed. As capacitor C1 increases in value, the flashing rate decreases. As capacitor C1 decreases in value, the flashing rate increases.

Precision Temperature Controllers. A UJT can serve as a triggering circuit for a precision temperature controller. **See Figure 13-28.** Thermistor R4 is used as a temperature detector for feedback information. Unijunction transistor Q1 provides the triggering circuit for the triac. The triac turns the AC power applied to the heater element ON and OFF.

Carrier Corporation
HVAC equipment is regulated by temperature controllers such as electronic thermostats.

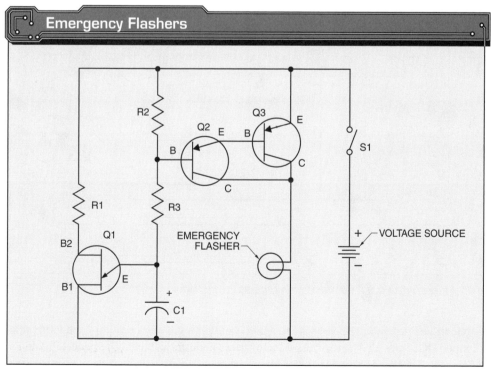

Figure 13-27. *A UJT can serve as a triggering circuit for an emergency flasher.*

Precision Temperature Controllers

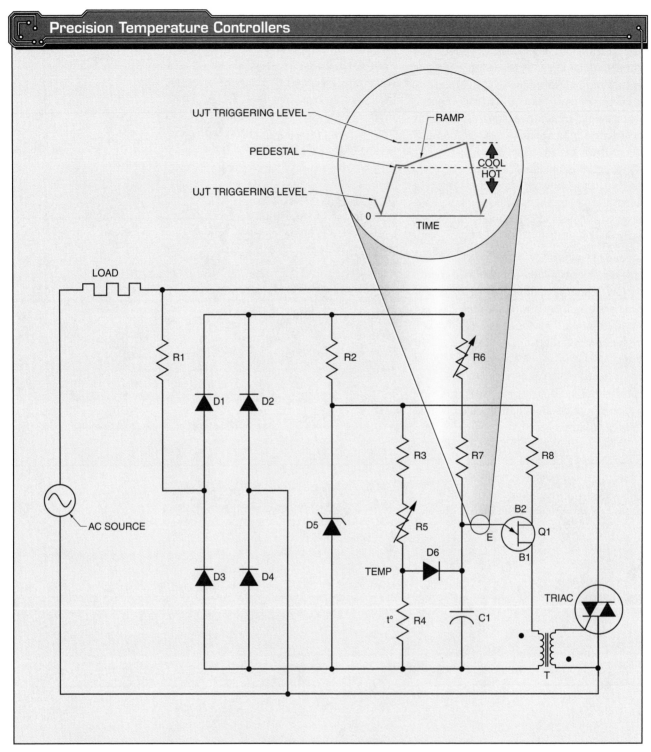

Figure 13-28. *A UJT can serve as a triggering circuit for a precision-temperature controller.*

More specifically, full-wave rectified and clamped DC is provided by resistors R1 and R2 and diodes D1 through D5 for driving the UJT. Emitter capacitor C1 is charged rapidly at the beginning of each half cycle to a pedestal voltage. The pedestal voltage is determined by the temperature-sensitive voltage divider resistors R4 and R5.

After reaching the pedestal voltage, capacitor C1 continues charging slowly. This charging process is shown graphically. The curve in the graph is called a ramp. The triac is triggered into conduction by a pulse from UJT Q1 whenever the UJT emitter voltage on capacitor C1 reaches the UJT triggering level. Raising the pedestal voltage (by cooling the thermistor) causes the ramp to reach the triggering level earlier in the cycle and provides more power to the heater. As more heat goes to the thermistor, the process is reversed. Therefore, the regulating action is provided.

Time-Dependent Lighting Control. A lamp dimmer automatically increases or decreases the brightness of a lamp over an adjustable time period and can have many functions. **See Figure 13-29.** Once the OFF function of the switch is activated on the dimmer, the bright lights fade away slowly over a period of 15 min to 20 min. The delay feature of the dimmer can also eliminate the blinding shock of turning on a light when a person's eyes are accustomed to darkness.

With a time-dependent lamp dimmer circuit, the DC voltage for the trigger circuit is derived from zener diodes D5 and D6. Zener diodes D5 and D6 clamp the pulsating DC voltage (from the full-wave rectifier bridge D1, D2, D3, and D4) to approximately 15 V. Unijunction transistor UJT Q3 delivers a trigger pulse to the triac. Depending on whether the trigger pulse is delivered late or early in the cycle, the output to the load (lamp) varies from completely OFF to completely ON.

The trigger circuit is designed so that a time-dependent output is obtained after initially energizing the circuit. Delay in turning the load ON or OFF is obtained after the position of switch S1 is changed. When switch S1 is placed in the ON position, capacitor C1 begins to charge through resistors R3 and R4. For time periods shortly after switching, capacitor C1 voltage is low. This holds the base current of transistor Q1 low. Therefore, the emitter of transistor Q2 is held at a low voltage (below the peak-point voltage on the UJT). Simultaneously, capacitor C2 is charged during each half cycle through resistor R7.

The R7-C2 time constant is long compared to a half cycle of the line voltage. The R7-C2 time constant is selected so that capacitor C2 voltage barely reaches the peak-point voltage at the end of the half cycle with zero voltage on capacitor C1. As the voltage on capacitor C1 increases, the voltage on capacitor C2 also increases. The R7-C2 charging curve then starts from a slightly higher voltage at each cycle. Therefore, the voltage on capacitor C2 reaches the peak-point voltage of the UJT slightly earlier during each cycle, and the output is slowly increased. The double emitter-follower configuration of transistors Q1 and Q2 provides extremely high impedance so that the charging and discharging currents to capacitor C1 are not shunted away from it.

When switch S1 is moved to the OFF position, capacitor C1 discharges through resistors R3 and R4. The operation proceeds as before but in reverse. Potentiometer R3 varies the speed that turns the light ON or OFF. The time duration from completely ON to completely OFF, or vice-versa, can be as long as 20 min. A bypass switch (S2) and resistor R2 provide a method for turning the light ON like a regular switch. Switch S3 is used to turn the device ON before it can perform the other functions described.

Refer to Chapter 13 Quick Quiz® on CD-ROM

Fluke Corporation

The proper test equipment must be used when bench testing solid state components to prevent damage to the component.

338 SOLID STATE DEVICES AND SYSTEMS

Time-Dependent Lamp Dimmers

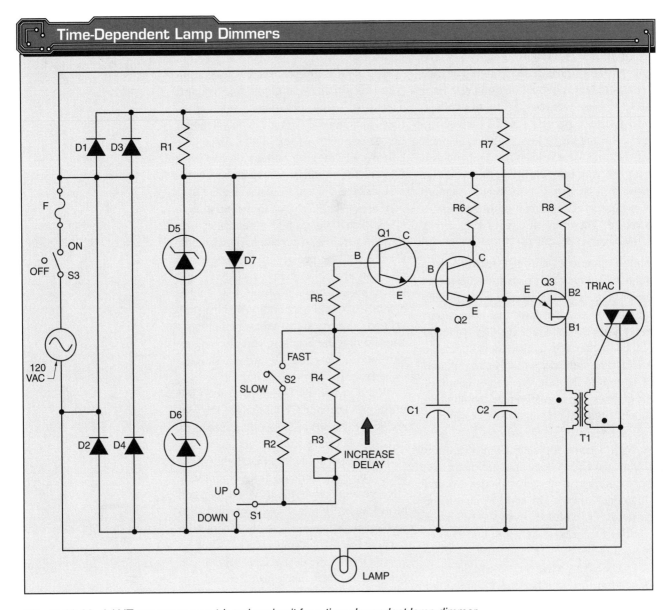

Figure 13-29. *A UJT can serve as a triggering circuit for a time-dependent lamp dimmer.*

KEY TERMS

- triac
- terminal 1 (T1)
- terminal 2 (T2)
- bidirectional
- reed switch
- diac
- unijunction transistor (UJT)
- base 1 (B1)
- base 2 (B2)
- saturation
- peak current point
- valley current point
- pedestal voltage

REVIEW QUESTIONS

1. What is a triac?

2. How are the terminals of a triac designated?

3. Which combined devices most closely resemble the triac?

4. What occurs when the voltage on a triac goes too high?

5. What have triacs replaced in control circuits?

6. Why would triacs be useful in explosive environments?

7. What is the procedure for testing a triac?

8. What is the primary use for a diac?

9. Define negative-resistance characteristic.

10. Describe the procedure for testing a diac with an ohmmeter.

11. How is a unijunction transistor (UJT) constructed?

12. How are the leads to a UJT labeled?

13. What is the equivalent circuit of a UJT?

14. In what region on a UJT curve does the UJT have its greatest application?

15. What is the primary use for a UJT?

OPERATIONAL AMPLIFIERS AND 555 TIMERS

14

Integrated circuits (ICs) are popular because they provide a complete circuit function in one small semiconductor package. Although many processes have been developed to create these devices, the result is always a totally enclosed system with specific inputs and specific outputs. Operational amplifiers (op amps) and 555 timers are ICs that may be included in devices with hundreds or thousands of components. Because of the nature of ICs, a technician must approach them in an entirely different manner.

OBJECTIVES

- Define integrated circuit (IC).
- Explain the operation of operational amplifiers (op amps).
- Describe IC timers.
- Explain the operation of 555 timers.
- List and describe IC timer applications.

INTEGRATED CIRCUITS

An *integrated circuit (IC)*, also known as a chip, is an electronic device in which all components (transistors, diodes, and resistors) are contained in a single package. An IC contains numerous circuits that are microscopically photoetched in layers on a silicon base. **See Figure 14-1.** ICs are used to process and store information in computers, calculators, programmable logic controllers (PLCs), digital watches, and almost all other electronic circuits.

A *linear IC* is an integrated circuit that produces an output signal proportional to the applied input signal and is used to provide amplification and regulatory functions. Linear ICs are used in radios, amplifiers, televisions, and power supplies. An operational amplifier (op amp) is the most common type of linear IC.

Most new equipment is designed with ICs for both analog (or linear) and digital circuits. These ICs are systems in themselves and may have thousands of transistors permanently interconnected inside them.

Integrated Circuit Construction

The circuitry contained inside an IC must be connected to a power supply and other electronic components. ICs are powered by and interconnected with other components through pins. Most ICs are mounted on a printed circuit (PC) board with the pins soldered to the copper traces of the PC board. ICs are available in different packages.

The more complex the internal circuitry of an IC, the greater the number of pins that are required to connect to the external circuitry. For example, a basic dual operational amplifier (op amp) may require only 8 pins. **See Figure 14-2.** The large-scale integration (LSI) circuit used in the microprocessor of a personal computer may require 64 pins.

Some equipment manufacturers use sockets to connect ICs to PC boards. Sockets make removing and replacing ICs easy compared to desoldering and soldering. The two major disadvantages of using sockets on a PC board are that they are additional components with the potential of failing, and they add to the cost of the assembly.

342 SOLID STATE DEVICES AND SYSTEMS

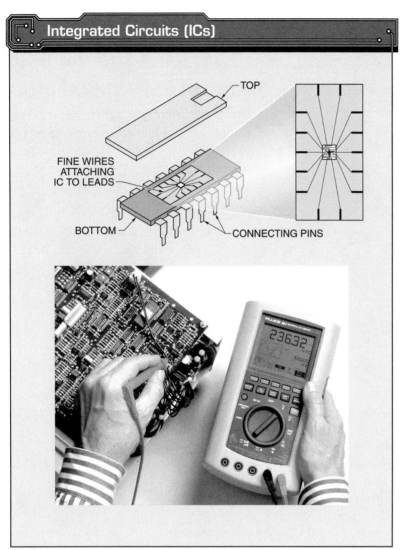

Figure 14-1. *An integrated circuit (IC) is composed of thousands of semiconductor devices providing a complete circuit function in one small semiconductor package.*

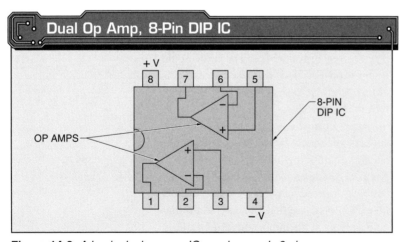

Figure 14-2. *A basic dual op amp IC may have only 8 pins.*

Troubleshooting Integrated Circuits

An IC that malfunctions cannot be repaired. The only duty of the troubleshooter is to test the IC to determine whether the IC is good or bad. If the IC is bad, the IC must be replaced.

An IC is tested in-circuit using the input/output method. If the input signal is normal, the power sources to the IC are tested. Any external components used with the IC, such as coils or capacitors (resistors are usually built in), are tested next. If the output is still abnormal, a new IC is substituted. Abnormal input usually means that the IC is not the problematic part, although an IC may sometimes distort the input.

When plug-in sockets are used, replacing ICs is easier and quicker than checking external components. However, the power source should always be tested before replacement to ensure that there is no excessive voltage. Excessive voltage may have destroyed the original IC and will most likely destroy the replacement. DC voltage sources for ICs are generally controlled by a regulator (often another IC). It is not unusual for a defective regulator to supply excess voltage.

Before replacing an IC, the power source must be turned off. Otherwise, the IC may be damaged by power surges if all terminals are not connected to the circuit at the same time.

Furthermore, an energized or "hot" circuit should not be soldered. Soldering may create problems that require a lot of time to identify, isolate, and correct. A small drop of solder can destroy a handful of parts instantly. Soldering also presents a safety hazard if there is high voltage.

Replacement dual in-line package (DIP) ICs are shipped with pins sprung slightly outward. **See Figure 14-3.** This allows them to press tightly against the socket wipers when they are installed. If the pins do not exactly match the socket holes, they must be bent slightly along both sides of the IC. If they are not bent evenly, the pins may fold outward or under.

Tips for Troubleshooting Integrated Circuits. Fortunately for the troubleshooter, ICs are usually either good or bad. Typically, they either work or do not work. They rarely develop marginal troubles, although intermittent faults are possible.

One problem with an IC that may occur is overheating. To test for overheating, each IC in the circuit should be touched by a bare hand. If an IC is hot to the touch, it must be determined whether it is good. If another IC of the same type elsewhere in the circuit is cool, then the hot IC is probably located in a problematic spot. If the supply voltages are normal, the best course of action is to replace the overheating IC.

Another sign of an internal IC problem is voltage on the IC output terminal that is too high or too low. If the power-supply voltages are normal, the terminal can be disconnected, and the input terminals can be shorted together. If the voltage is still incorrect, the IC must be replaced.

In most modern equipment, if an IC does not plug in on its own, usually an entire PC board, containing several other ICs and components, is designed to be plugged in. Plug-in circuit boards make troubleshooting a matter of deciding which PC board to try, trying the PC board, and then checking to see if the trouble has cleared.

Forcing Integrated Circuits to Work

A good IC can be forced to work if its operation in a particular circuit is known, its DC voltages are correct, and the sensing or signal voltages it receives are known. For example, when two inputs of a comparator reach the same voltage, an alarm should be tripped. A technician can short the two inputs together to see if the alarm trips. If the alarm trips, the IC is working properly, and the problem lies elsewhere. The problem may be in the sensing mechanism and caused by an absent or incorrect reference voltage.

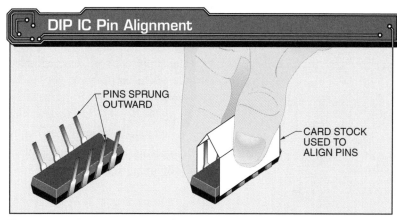

Figure 14-3. *Replacement DIP ICs are shipped with their pins sprung slightly outward and must be bent slightly along both sides of the IC to match the socket holes.*

Technicians should have a basic knowledge of electronic circuits and equipment. They should also have specific information about the particular piece of equipment they are troubleshooting. Technicians must also be able to develop a simple, systematic troubleshooting strategy to meet the demands of any assignment.

OPERATIONAL AMPLIFIERS

An operational amplifier (op amp) is one of the most widely used ICs. An *operational amplifier (op amp)* is a high-gain, directly coupled amplifier that uses external feedback to control response characteristics. **See Figure 14-4.** An example of this feedback control is gain.

The gain of an op amp can be controlled externally by connecting a feedback resistor between the output and input. A number of different amplifier applications can be achieved by selecting different feedback components and combinations. With the right component combinations, gains in the thousands are common.

An op amp is very versatile. Op amps can be made to perform math functions, such as addition, subtraction, multiplication, and division, by connecting it to a few components. Today, the main purpose of an op amp is to amplify small signals to levels that can be used for the control of another device.

344 SOLID STATE DEVICES AND SYSTEMS

The advantages of an op amp include:
- high input impedance (resistance)—an op amp does not draw much power from the input source; this feature is required when amplifying very weak input signals
- low output impedance—a high output impedance reduces the amplified output
- high gain—one op amp can replace many individual transistors

An op amp is often used in digital circuits that require the analog amplification of a weak signal. For example, weak audio or data signals must be amplified to be useful. In addition to being used for amplification, op amps may be used to perform other electronic functions.

Basic op amps have five leads attached to an amplifier. Two are input leads, one is an output lead, and two are power supply leads.

Schematic Symbols

The schematic symbol of an op amp may be shown as a triangle with the two inputs of the op amps inverting (−) and noninverting (+). **See Figure 14-5.** The two inputs are usually drawn with the inverting input at the top. The exception to the inverting input being at the top is when it would complicate the schematic. In either case, the two inputs should be clearly identified on the schematic symbol by polarity symbols.

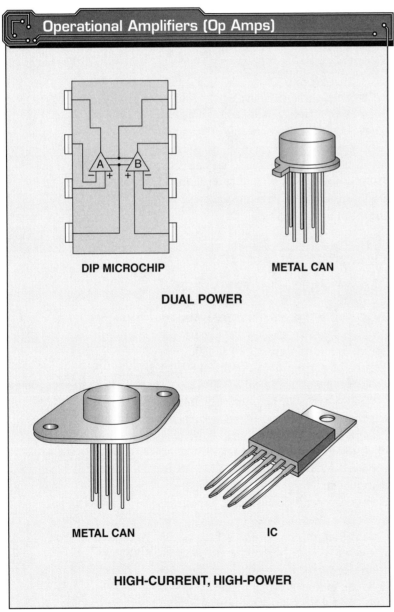

Figure 14-4. *Types of operational amplifiers (op amps) include dual power and high-current, high-power op amps.*

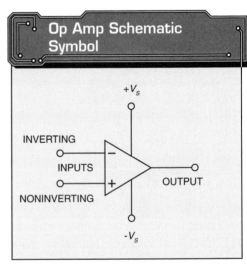

Figure 14-5. *The schematic for an op amp includes inverting and noninverting inputs.*

In 1941, Karl D. Swartzel Jr. of Bell Labs developed a vacuum tube op amp. The op amp was considered a general-purpose, DC coupled, high gain, inverting feedback amplifier.

Voltage Sources for Operational Amplifiers

Like other solid state devices, integrated circuits need DC operating voltages. The DC voltage pins on some ICs are labeled common-collector voltage (V_{CC}). Other pins are labeled input voltage (V_{in}) or supply voltage (V_S). DC voltage ratings in IC data books are labeled all three ways. The voltage source for op amps can be single supply. However, they are usually bipolar or dual supply. **See Figure 14-6.**

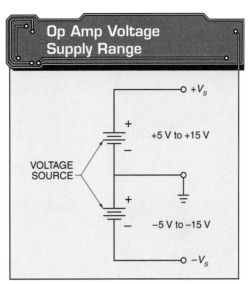

Figure 14-7. *Voltage supplies for op amps typically range from ±5 V to ±15 V.*

CAUTION: Never reverse the power supply polarity to an op amp. Applying a negative voltage to the positive pin or a positive voltage to the negative pin, even momentarily, will result in destructive current flow through the op amp.

Regardless of the power supply used, most op amp manufacturers suggest using bypass capacitors on the power supply leads. **See Figure 14-8.** The recommended capacitor size is about 0.1 µF (microfarads).

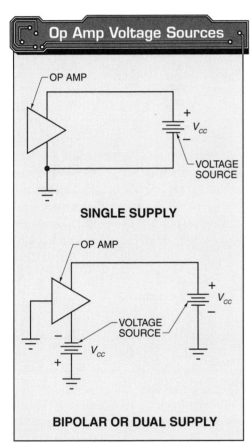

Figure 14-6. *Voltage sources for op amps can be single supply; however, they are usually bipolar or dual supply.*

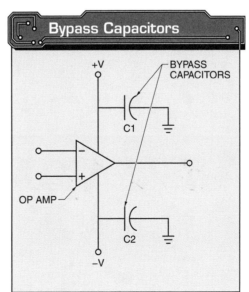

Figure 14-8. *Most op amp manufacturers suggest using bypass capacitors on power supply leads.*

The voltage supplies for op amps typically range from ±5 V to ±15 V. That is, one supply is +5 V to +15 V with respect to ground. The other supply voltage is –5 V to –15 V with respect to ground. **See Figure 14-7.**

Note: In some schematic circuits, the IC power supply connection is not shown on each symbol. In such cases, it is shown on the schematic, and each IC must be powered from this source, unless otherwise stated.

Electronic circuits are used in satellites and spacecrafts. *NASA*

Internal Operational Amplifier Operation

Internally, an op amp has three major sections. **See Figure 14-9.** It consists of a high-impedance differential amplifier, high-gain stage, and low-output impedance power-output stage. The differential amplifier provides the wide bandwidth and the high impedance. The high-gain stage boosts the signal. The power-output stage isolates the gain stage from the load and provides power.

The operation of the high-impedance differential amplifier is unique. **See Figure 14-10.** Current to the emitter-coupled transistors (Q1 and Q2) is supplied by the source (Q3). When manufactured, the characteristics of Q1 and Q2, along with their biasing resistors (R1, R2, and R3), are closely matched.

As long as the two input voltages (A and B) are either zero or equal in amplitude and polarity, the amplifier is balanced because the collector currents are equal. When balanced, zero voltage difference exists between the two collectors.

The sum of the emitter currents is always equal to the current supplied by Q3. Thus, if the input to one transistor causes it to draw more current, the current in the other decreases, and the voltage difference between the two collectors changes in a differential manner. The differential swing, or output signal, will be greater than the simple variation that can be obtained from only one transistor. Each transistor amplifies in the opposite direction so that the total output signal is twice that of one transistor. This swing will then be amplified through the high-gain stage and matched to the load through the power-output stage.

Operational Amplifier Characteristics

The noninverting input (+) of an op amp is used to create a voltage follower, or source follower. The inverting input (−) of an op amp is used to create a closed-loop (feedback inverting) amplifier capable of a 180° phase shift.

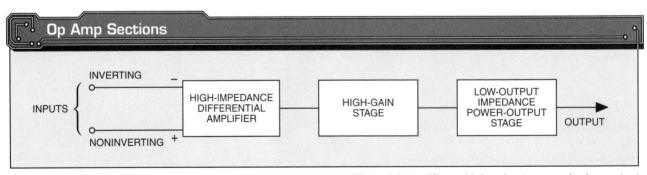

Figure 14-9. *Internally, an op amp consists of a high-impedance differential amplifier, a high-gain stage, and a low-output impedance power-output stage.*

High-Impedance Differential Amplifiers

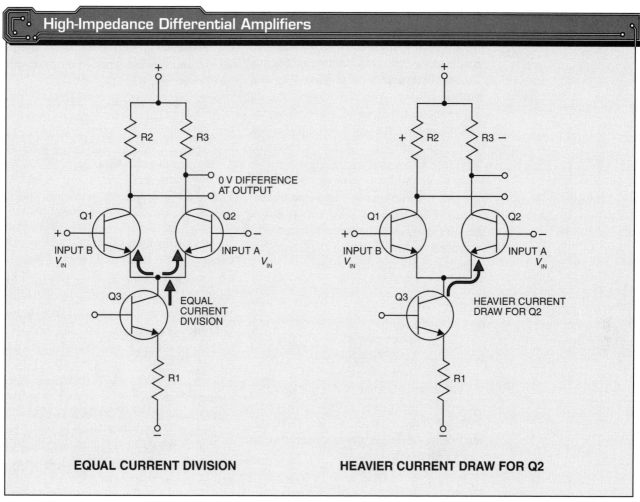

Figure 14-10. *In a high-impedance differential amplifier, current to the emitter-coupled transistors (Q1 and Q2) is supplied by the source (Q3).*

Gain. Gain is the ratio of the amplitude of the output signal to the amplitude of the input signal. Op amps may have gains of 500,000 or more, no gain (unity gain), or controlled gain. Gain in an op amp is controlled by external resistors that provide closed-loop feedback. **See Figure 14-11.**

In the high-gain mode, a very small change in voltage on either input results in a large change in the output voltage. These high-gain circuits can be far too sensitive and unstable for most applications. The gain is normally reduced to a much lower level due to this instability. The circuit is stabilized by feeding back some of the output signal to one of the inputs through a resistor.

Gain also plays a major role in the applied science of laser physics. In that context, gain is a medium with the capability of amplifying the power of light in the form of a beam.

Unity-Gain Condition

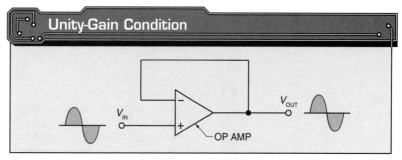

Figure 14-11. *The amplifier is operating in a unity-gain condition. The output voltage is an exact reproduction of the input voltage.*

In the no-gain mode, the output is connected to the input that does not have the input signal applied to it. In the no-gain mode, voltage out equals voltage in. For this reason, this circuit is often referred to as a voltage follower. Although there is no gain in this mode, the op amp is used as an isolation amplifier. An isolation amplifier has high input impedance and does not produce a loading effect on the input signal.

In the controlled-gain mode, the amount of gain is controlled by the value of two resistors ($R1$ and $R2$). The two resistors are the input and feedback resistors. To calculate the voltage gain of an inverting op amp, apply the following formula:

$$V = \frac{R2}{R1}$$

where
V = voltage gain
$R1$ = input resistor (in Ω)
$R2$ = feedback resistor (in Ω)

For example, what is the voltage gain of an inverting op amp circuit when the resistance of the input resistor equals 10 kΩ, and the resistance of the feedback resistor equals 90 kΩ?

$$V = \frac{R2}{R1}$$

$$V = \frac{90,000}{10,000}$$

$$V = 9$$

To calculate the voltage gain of a noninverting op amp, apply the following formula:

$$V = \frac{(R1 + R2)}{R1}$$

where
V = voltage gain
$R1$ = input resistor (in Ω)
$R2$ = feedback resistor (in Ω)

> Input, output, and supply connectors play an important role in overall signal stability. Proper connections and matches create phase-stable conditions that can directly result in better overall efficiency and performance.

For example, what is the voltage gain of a noninverting op amp circuit when the resistance of the input resistor equals 90 kΩ, and the resistance of the feedback resistor equals 100 kΩ?

$$V = \frac{(R1 + R2)}{R1}$$

$$V = \frac{(90,000 + 100,000)}{90,000}$$

$$V = \frac{190,000}{90,000}$$

$$V = 2.1$$

The gain of an op amp may be calculated if the input and output voltages are known. Op amp gain is calculated by dividing the output voltage by the input voltage. To calculate op amp gain, apply the following formula:

$$\text{gain} = \frac{V_{out}}{V_{in}}$$

where
V_{out} = output voltage (in V)
V_{in} = input voltage (in V)

For example, what is the gain of an op amp with a measured input voltage of 8 mV and an output voltage of 8 V?

$$\text{gain} = \frac{V_{out}}{V_{in}}$$

$$\text{gain} = \frac{8}{0.008}$$

$$\text{gain} = 1000$$

Voltage Followers (Unity-Gain). A voltage follower, or source follower, is a noninverting amplifier. The output voltage (V_{out}) is an exact reproduction of the input voltage (V_{in}). The function of the voltage follower is identical to that of the source follower created by a bipolar transistor and FET. The circuit is used to impedance-match an input signal to its load. With the voltage follower, the input impedance is high and the output impedance is low. It should be noted that the voltage follower has no input or feedback components. Because no feedback components are used in this type of circuit, the amplifier is operating in a unity-gain condition.

Open-Loop Control. Open-loop systems are used almost exclusively for manual control operations. The two variations of the open-loop system are full control and partial control. Full control operation simply turns a system on or off. For example, in an electrical circuit, current flow stops when the circuit path is opened. Switches, circuit breakers, fuses, and relays are used for full control. Partial control operation alters system operations rather than causing them to start or stop. Resistors, inductors, transformers, capacitors, semiconductor devices, and integrated circuits are commonly used to achieve partial control.

Closed-Loop Control. To achieve automatic control, interaction between the control unit and the controlled element must occur. In a closed-loop system this interaction is called feedback. Feedback can be activated by electrical, thermal, light, chemical, or mechanical energy. Both full and partial control can be achieved through a closed-loop system. **See Figure 14-12.**

Many of the automated systems used in industry today are of the closed-loop type. A closed-loop system may have automatic correction control. In the system, energy goes to the control unit and the controlled element. Feedback from the controlled element is directed to a comparator, which compares the feedback signal to a reference signal or standard.

A correction signal is developed by the comparator and sent to the control unit. This signal alters the system so that it conforms to the data from the reference source. Systems of this type maintain a specified operating level regardless of external variations or disturbances.

Automated control has gone through many changes in recent years with the addition of control devices that are not obvious to the casual observer. These include devices that change the amplitude, frequency, waveform, time, or phase of signals passing through the system.

Feedback Inverting Amplifiers (Closed-Loop). A feedback inverting amplifier produces a 180° phase inversion from input (V_{in}) to output (V_{out}). **See Figure 14-13.** When a positive-going voltage is applied to the input, a negative-going voltage will be produced at the output. The input signal is applied to the op amp inverting input through R1, while resistor R2 serves as the feedback element.

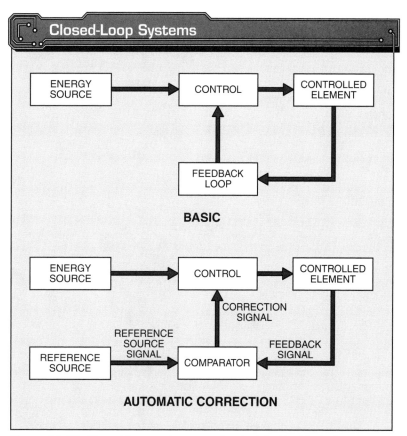

Figure 14-12. *Closed-loop systems may be depicted by block diagrams.*

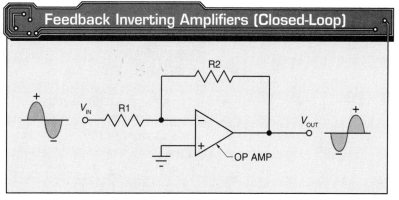

Figure 14-13. *A feedback inverting amplifier produces a 180° phase inversion from input to output.*

The voltage gain of the feedback inverting amplifier can be less than, equal to, or greater than 1.0. Its value depends on the values of resistors R1 and R2. Because a feedback component is used in this type of circuit, the amplifier is operating in a closed-loop condition. The closed-loop gain of the feedback inverting amplifier can be controlled by switching in different feedback resistors. **See Figure 14-14.** If resistors R1 and RF are equal, the op amp can be used as a simple unity (no-gain) signal inverter.

and phase-shift characteristics of the op amp must be compensated. Outside components, such as resistors and capacitors, are used for compensation. These components control the frequency response and phase-shift characteristics. **See Figure 14-15.** In this case, the amount of feedback increases as the frequency increases. This is because the reactance of the capacitor decreases. The upper limit for feedback is determined by the resistor value, which remains constant at high frequencies. If an op amp does not require compensation, it is identified as such in the manufacturer's specifications.

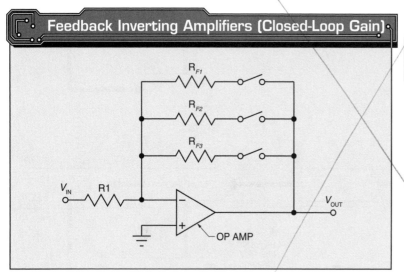

Figure 14-14. *The closed-loop gain of the feedback inverting amplifier can be controlled by switching in different feedback resistors.*

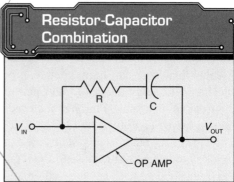

Figure 14-15. *Compensation can be accomplished through a series resistor-capacitor combination.*

Compensation. Because high-gain op amps usually use feedback, the feedback must be controlled to ensure that the op amp circuit is stable. If properly controlled, feedback should not affect changes in frequency nor cause oscillations when the input-output phase relationship changes.

If phase compensation is not furnished on an op amp, the gain of the feedback signal may be greater than the input when the phase angle approaches 180°. In this case, feedback that is negative at low frequencies becomes positive at high frequencies, and unwanted oscillations may result.

To overcome this tendency toward unwanted oscillation, the frequency response

Offset Error and Nulling. Although extreme care is taken in fabricating an op amp, a slight mismatch may still occur between the internal components. *Offset error* is a slight mismatch between internal components. It creates a problem when using the op amp in a DC circuit. The mismatch prevents the amplifier from having a zero output for a zero input.

Even with proper bias, the feedback inverting amplifier has an input bias current (I_B) through the input and feedback resistors (R1 and R2) with no signal applied. **See Figure 14-16.** The additional current flow through these resistors produces a voltage drop, which appears as DC input voltage. The op amp then amplifies this DC input voltage, compounding the offset error.

Feedback Inverting Amplifiers

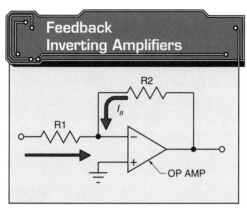

Figure 14-16. *Even with proper bias, the feedback inverting amplifier has an input bias current through the input and feedback resistors (R1 and R2) with no signal applied.*

To correct offset error, the nulling technique is often used through a nulling resistance network. **See Figure 14-17.** With this network, the nulling variable resistor is adjusted for zero output with zero input.

Bandwidth and Slew Rate. When a high-frequency signal is fed into an op amp, certain changes can occur with the output signal. Some elements within the op amp have capacitive characteristics. A certain time span is required to charge and discharge them. The capacitive effect prevents the output voltage from following the input signal immediately. Thus, these internal capacitances limit the rate at which the output voltage can change. *Slew rate* is the maximum time rate of change of output, measured in volts per microsecond (V/μsec).

Slew rate is the time an amplifier takes to reach 90% of the steady-state output level for a given input. The slew rate of an op amp is very fast. Slew rates in op amps may vary from 0.1 V/μsec to 100 V/μsec. The slew rate of a feedback inverting amplifier depends on a number of factors, including the value of the closed-loop gain components.

Slew rate can limit bandwidth. The bandwidth is usually expressed as the range of frequencies in which the amplifier develops a minimum of 70.7% of its rated output.

Nulling

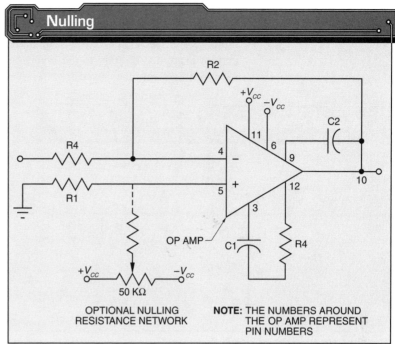

Figure 14-17. *Nulling is a technique often used to correct offset error.*

Operational Amplifier Applications

Op amps are used for a variety of amplification applications. For example, they are used in audio amplifiers and video amplifiers. They are also ideal for a variety of industrial and commercial control systems.

By adding various external components, op amp characteristics change to meet different circuit conditions. The number of circuit possibilities for op amps appears limitless.

Current-to-Voltage Converters. A current-to-voltage op amp uses the current sensitivity of the op amp to measure very small currents. The circuit can provide 1 V at the output for 1 μA at the input.

> *Low levels of harmonic distortion are ideal for operational amplifiers used in audio applications. The less total harmonic distortion in operational amplifier audio equipment results in a clearer, more accurate reproduction of frequencies.*

This basic current-to-voltage converter is essentially an inverting amplifier without an input resistor. **See Figure 14-18.** The input current is applied directly to the inverting input of the op amp.

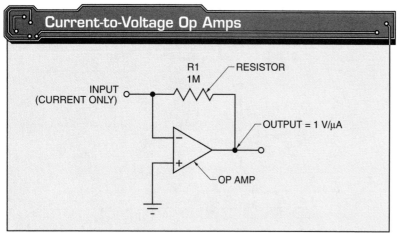

Figure 14-18. *A current-to-voltage op amp makes use of the current sensitivity of the op amp to measure very small currents. A current-to-voltage converter is essentially an inverting amplifier without an input resistor.*

A thermistor varies the amount of current (I_{in}) entering the inverting input of the op amp. **See Figure 14-19.** As the temperature increases, the resistance of the thermistor decreases, current to the input of the op amp increases, and voltage at the output increases. As the temperature decreases, the resistance increases, current decreases, and voltage output decreases.

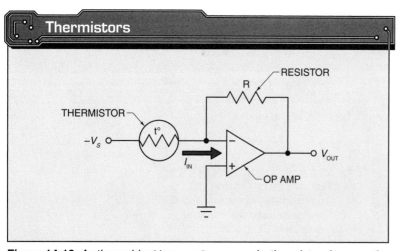

Figure 14-19. *As the ambient temperature around a thermistor changes, the resistance of the thermistor changes.*

Since the output of the circuit is now voltage, the op amp voltage can be used to drive a metal oxide semiconductor field effect transistor (MOSFET), which serves as the output switching element to the load. **See Figure 14-20.** Thus, an op amp and MOSFET combination can be used as an effective control device.

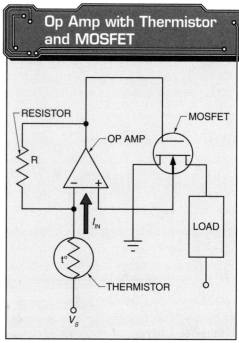

Figure 14-20. *Since the output of the current-to-voltage op amp is voltage, the thermistor and op amp can be used with a MOSFET to form an effective switching element to a load.*

The op amp can directly control the load if the output voltage of the op amp is large enough and the current draw of the load is within the op amp's limits. An interface is required if the load requires more power for operation than the op amp can safely deliver. The most common interface is a solid state relay (SSR). The input of the SSR is connected to the op amp. The output of the SSR is used to drive the load. In this circuit, an op amp can amplify the signal from a photoconductive cell (photocell), thermocouple, gas detector, or smoke detector. **See Figure 14-21.**

Op Amp Control

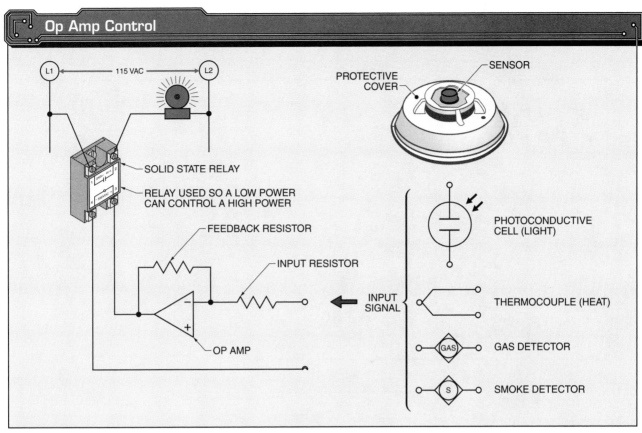

Figure 14-21. *An op amp can directly control a load if the output voltage of the op amp is large enough and the current draw of the load is within the op amp's limits.*

Voltage-to-Current Converters. In certain circuits, a change in voltage becomes the reference for the change in a circuit. When this is the case, a voltage-to-current op amp is used. **See Figure 14-22.** The output voltage of the bridge circuit is a function of the degree of imbalance present in the input bridge. For the bridge circuit, an imbalance can be created by changing the pressure on the pressure sensor (transducer). The potentiometer determines the pressure-set limits.

The unit can operate directly from the AC supply since it includes a step-down transformer and single-phase rectifier. When the pressure drops, the resistance of the pressure sensor decreases. Terminal 3 of the op amp becomes more positive than terminal 2. Under this condition, the output current at pin 6 causes the triac to conduct. With the triac conducting, power is applied to the coil, which turns on the air compressor. When the pressure in the tank is brought up to the preset limit, the value of the pressure sensor increases, balancing the bridge, and the air compressor shuts off.

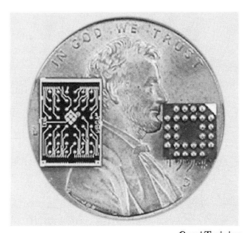

Omni Training
Electronic circuits can be made very small and take up little space, depending on the application.

354 SOLID STATE DEVICES AND SYSTEMS

Voltage-to-Current Op Amps

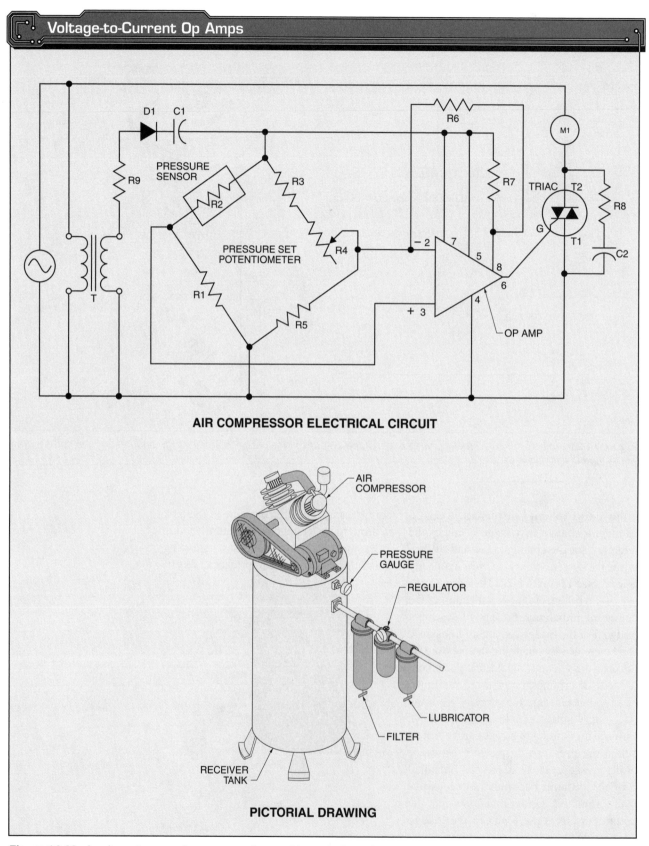

Figure 14-22. *A voltage-to-current op amp can be used to control an air compressor.*

Thermostat Control. An op amp can be used as a differential amplifier. A differential op amp is a type of op amp used in differential thermostat controls for solar heating units. Small signal changes can be detected easily in a circuit and are amplified for signaling or control purposes. **See Figure 14-23.**

When an op amp has signals of equal amplitude and polarity applied to each of its inputs simultaneously, the output will be zero. For the circuit, the bridge must be balanced for the op amp output to be zero. The bridge is formed by a tank sensor; collector sensor; and resistors R1, R2, and R3.

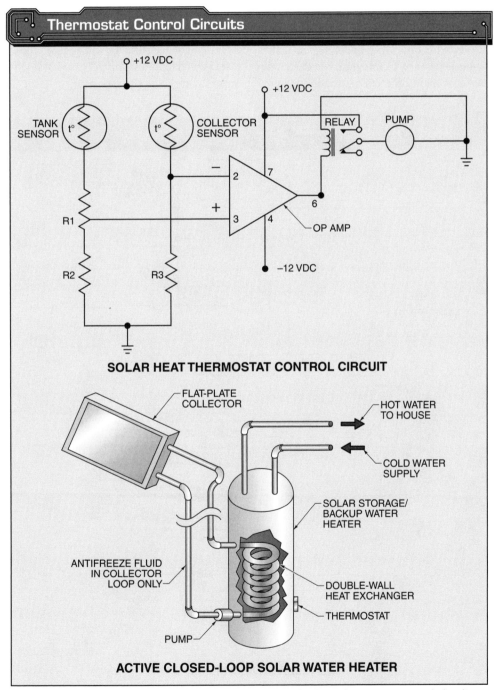

Figure 14-23. *A differential op amp can be used in a solar heat thermostat control circuit.*

If the tank senses a reduction in temperature, the bridge will become unbalanced, with the difference being presented to the input of the op amp. The signal is amplified to the output. The output is then connected to a relay that starts the pump. When enough hot water passes the collector and tank and heats the tank water, the bridge is again balanced and the relay is turned off.

Integrators. An op amp can be used as an integrator. It can integrate a voltage that varies with time. In control systems, it is often necessary to integrate, or determine the total energy content of, a series of signal pulses during some fixed period of time. **See Figure 14-24.** An electronic tachometer, for example, converts the rotational speed of a shaft into a series of pulses of fixed amplitude and duration. The faster the speed of rotation, the more pulses there will be. Each of these pulses contributes to the charge across capacitor C1 (connected to the op amp) during this time period. The output voltage of the op amp will be directly proportional to the number of pulses occurring during that time. At the end of each time period, the electronic reset switch in the op amp is closed momentarily to discharge the capacitor.

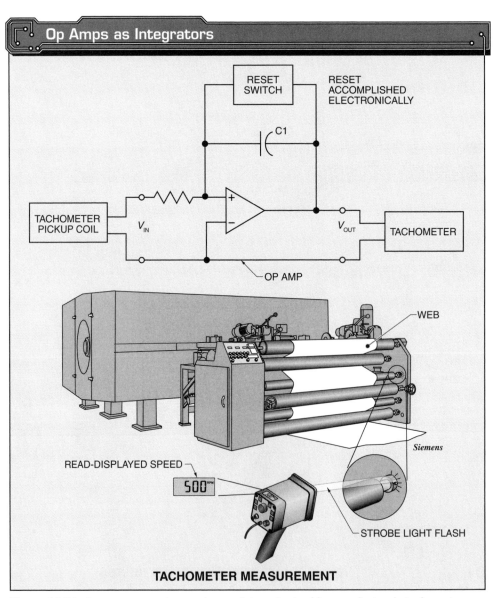

Figure 14-24. *An op amp can be used as an integrator to drive an electronic tachometer.*

Digital Multimeters. An op amp can be used in a solid state digital multimeter for DC voltage measurements. **See Figure 14-25.** If the 5 kΩ output resistor is changed to a different value, the sensitivity of the basic circuit can be changed. It is not unusual to have a full-scale range of 0.1 V to 100 V, with 100,000 Ω/V of input sensitivity.

The circuit has a null balance circuit since the typical DC offset multiplied by a gain of 100 would produce a significant zero offset on the meter. In this case, with the input shorted, the null balance potentiometer is adjusted to obtain a zero reading on the meter. This particular DC voltmeter is ideal for solid state testing since it has the necessary low-voltage scale readings and high impedance required for solid state devices.

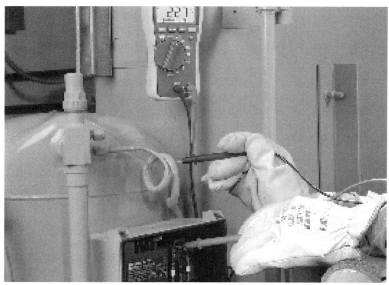

Fluke Corporation
Digital multimeters enable the testing of several different electrical characteristics, such as voltage, current, and resistance.

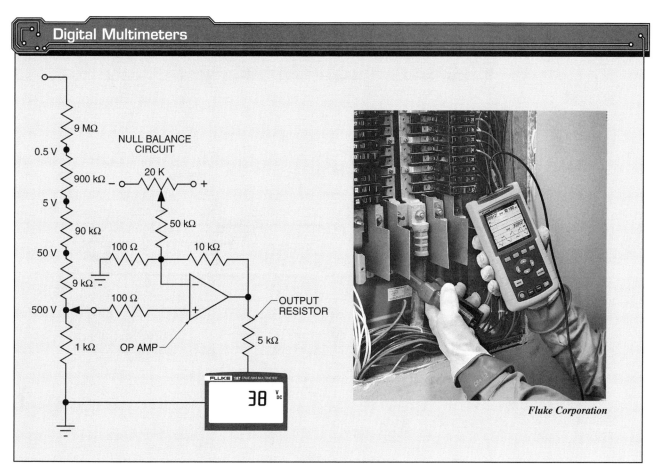

Figure 14-25. *An op amp can be used in a solid state digital multimeter.*

Comparators. A comparator circuit is used to compare a varied input voltage with a DC reference voltage. **See Figure 14-26.** The final output of the comparator will change depending upon whether the varied input signal is above or below the DC reference voltage to which it is compared.

A bar graph driver is one application of a comparator. Several op amps are connected as comparators, each with its own reference voltage connected to the inverting input. Each one monitors the same voltage signal on its noninverting inputs and builds a bar-graph-style meter. The meter is similar to the one commonly seen on the face of a stereo tuner and graphic equalizer. As the signal voltage representing radio signal strength or audio sound level increases, each comparator turns on in sequence and sends power to its respective LED. With each comparator switching on at a different sound level, the number of illuminated LEDs indicates the strength of the signal.

For example, LED 1 would illuminate first as the input voltage increases in a positive direction. As the input voltage continues to increase, the other LEDs would illuminate in succession until all are lit or the intended signal is reached.

INTEGRATED CIRCUIT TIMERS

An *integrated circuit (IC) timer* is a solid state timing device on a single IC. One type of IC timer is the astable multivibrator, also known as an astable timer, which is self-repeating. Another type of IC timer is the monostable multivibrator, which is externally triggered. The monostable multivibrator is also known as a monostable timer, one-shot timer, and one-shot multivibrator.

The 555 timer is one of the most popular ICs used for timing purposes. The 555 timer is a very stable IC that can operate as an astable or monostable timer.

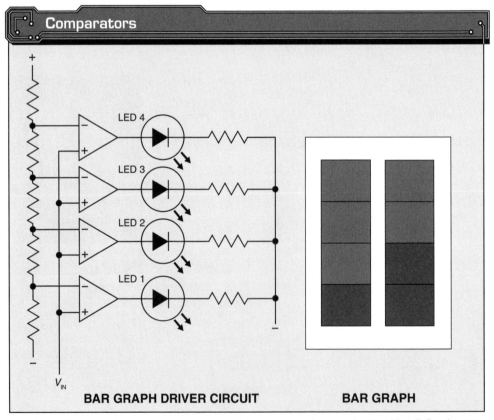

Figure 14-26. *A comparator circuit is used to compare an input voltage with a DC reference voltage.*

A 556 timer includes a 555 timer housed in a 14-pin package. With the 556 timer, the two timers share the same power supply pins. **See Figure 14-27.** A 556 timer could also be used as a single 555 timer by using half of the IC.

The circuit symbol for a 555 timer and 556 timer is a box with pins arranged the circuit diagram. For example, pin 8 is on the top right for the common-collector voltage (V_{CC}), and pin 4 is on the lower left. Typically, pins are represented by numbers and are not labeled by function. The 555 timer and 556 timer can be used with a V_{CC} of 4.5 V to 18 V.

The 555 timer can be connected to time on and off repeatedly by itself. It can also time out when the circuit receives an outside trigger signal. For the 555 timer to be fully functional, certain external components must be connected to the IC. The actual timing is accomplished with an RC time-delay network.

The accuracy of the timing sequence depends on the tolerance of the resistor and leakage in the capacitor. Timing circuits that are not critical may vary as much as ±20% of the stated delay. Other precision timers may vary ±1% or less of the stated delay. The time it takes for a capacitor to charge or discharge is determined by the value of the resistance and the size of the capacitor.

> The 555 timer was invented by Hans R. Camenzind in 1970. A typical 555 timer includes 23 transistors, 16 resistors, and several diodes. The timers can be found in an 8-pin package.

555 Timer Operation

A 555 timer consists of a voltage-divider network (R1, R2, and R3), two comparators (Comp 1 and Comp 2), a flip-flop, two control transistors (Q1 and Q2), and a power-output amplifier. **See Figure 14-28.** *Note:* A flip-flop is the electronic equivalent of a toggle switch. It has one high output and one low output. When one is high, the other is low.

The comparators compare the input voltages to the internal reference voltages created by the voltage divider, which consists of resistors R1, R2, and R3. Since the resistors are of equal value, the reference voltage provided by two of the resistors is two-thirds of the common-collector (supply) voltage (V_{CC}). The other resistor provides one-third of V_{CC}. The value of V_{CC} may change (e.g., from 9 V to 12 V to 15 V) from IC to IC. However, the ratios remain the same. When the input voltage to either one of the comparators is higher than the reference voltage, the comparator goes into saturation and produces a signal that will trigger the flip-flop. In this IC circuit, the flip-flop has two inputs: S and R.

Note: The two comparators feed signals into the flip-flop. Comp 1 is called the threshold comparator, and Comp 2 is called the trigger comparator. Comp 1 is connected to the S input of the flip-flop. Comp 2 is connected to the R input of the flip-flop.

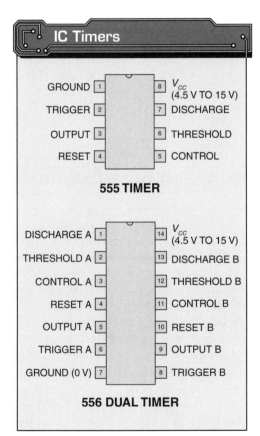

Figure 14-27. *IC timers are available with different pin arrangements.*

555 Timer Internal Circuits

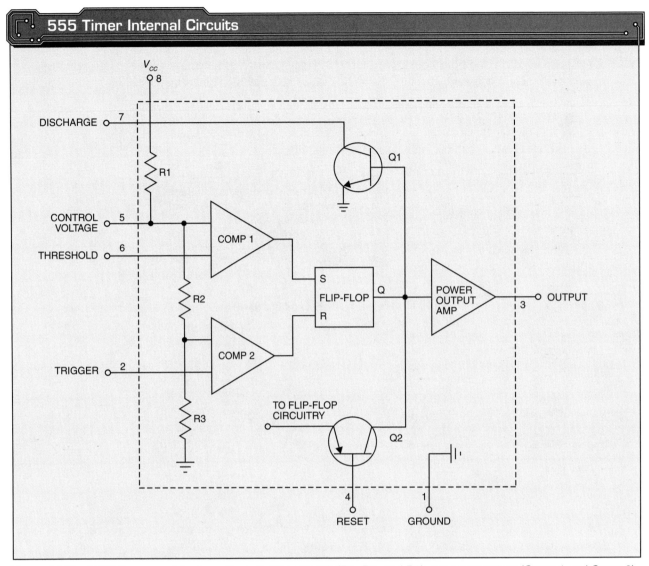

Figure 14-28. A 555 timer consists of a voltage-divider network (R1, R2, and R3), two comparators (Comp 1 and Comp 2), a flip-flop, two control transistors (Q1 and Q2), and a power-output amplifier.

Whenever the voltage at S is positive and the voltage at R is zero, the output of the flip-flop is high. Whenever the voltage at S is zero and the voltage at R is positive, the output of the flip-flop is low. The output from the flip-flop at point Q is then applied to transistors Q1 and Q2 and to the output amplifier simultaneously. If the signal is high, Q1 will turn on such that pin 7 (the discharge pin) will be grounded through the emitter-collector circuit. Then, Q2 will be in a position to turn on pin 7 to ground through the emitter-collector circuit.

The flip-flop signal is also applied to Q2. A signal to pin 4 can be used to reset the flip-flop. Pin 4 can be activated when a low-level voltage signal is applied. Once applied, this signal will override the output signal from the flip-flop. The reset pin (pin 4) will force the output of the flip-flop to go low, regardless of the state of the other inputs to the flip-flop.

The flip-flop signal is also applied to the power-output amplifier. The power-output amplifier boosts the signal, and the 555 timer delivers up to 200 mA of current when operated at 15 V. The output can be used to drive other transistor circuits and even a small audio speaker. The output will always be an inverted signal compared to the input. If the input to the power-output amplifier is high, the output will be low. If the input is low, the output will be high.

Monostable Multivibrators

Timing signals are needed to produce start/stop commands, synchronize pulses, and produce special sequences. These signals are often produced with time-delay circuits, which include 555 timers.

Some time-delay circuits produce an output until a certain time after the input appears. Monostable multivibrators can be used for this type of time-delay sequence. The monostable multivibrator has a stable state and an unstable state. An RC time constant determines the length of time that the circuit exists in the unstable state. The monostable multivibrator produces an output as soon as an input pulse appears. The duration of the output is determined by the time duration of the unstable state.

An external circuit is needed to operate the 555 timer as a monostable multivibrator. A variable resistor and a capacitor are used to determine the amount of time delay. **See Figure 14-29.** The capacitor is usually held in the discharge state by transistor Q1, which shorts the capacitor to ground. The timing cycle begins when a negative pulse is applied to the trigger input (pin 2). The negative pulse forces a flip-flop to go low on the output, which in turn removes the base bias to the discharge transistor (Q1). With transistor Q1 off, the short circuit (ground) is removed from the capacitor. Therefore, the capacitor is allowed to start charging with a time constant established by the values of the resistor and capacitor.

When the voltage across the capacitor reaches two-thirds of V_{CC}, as determined by the voltage divider, Comp 1 resets the flip-flop, returning it to its original high state. This change causes Q1 to turn on, which discharges the external capacitor. With the capacitor discharging, the charging cycle stops. The resetting of the flip-flop drives the power-output amplifier, causing the output to return to its normal low operating condition.

Henkel Corporation
Electronic circuits and components are often used in vehicle signaling applications.

The overall timing sequence can be seen by comparing input and output waveforms to the charging rate of the RC network. **See Figure 14-30.** A short input pulse at pin 2 produces a relatively long output pulse at pin 3.

Note: The leading edge of the output pulse starts with the leading edge of the input pulse. The width of the output pulse is determined by the voltage produced across the capacitor, which is determined by the RC time constant.

> *Monostable multivibrators have a stable state and an unstable state. They function as "one-shot" pulse generators for applications that require timing periods that last a certain amount of time. Triggers are needed to place the circuits into the unstable state.*

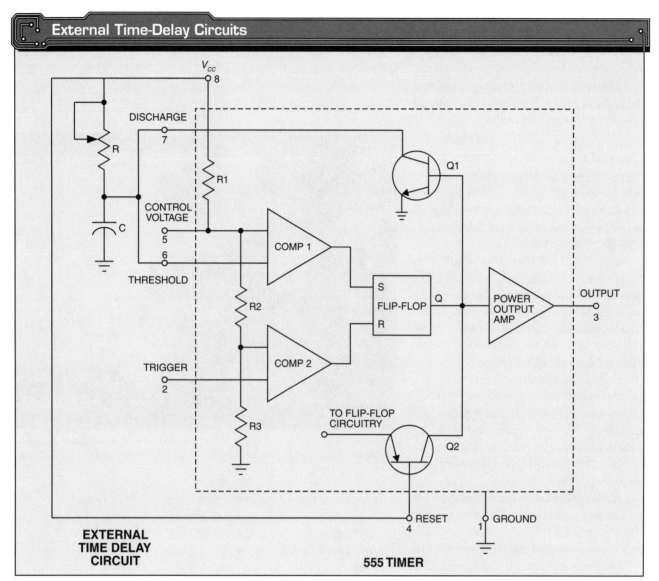

Figure 14-29. *An external time-delay circuit uses a resistor and capacitor to determine the amount of time delay for a 555 timer.*

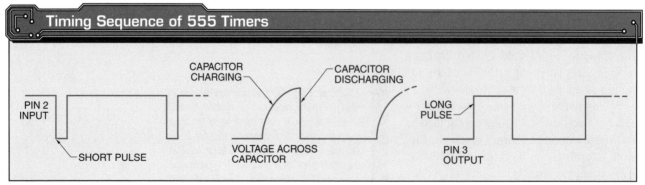

Figure 14-30. *The overall timing sequence of the 555 timer is shown by comparing input and output waveforms to the charging of the RC network.*

Security Alarm Circuits

Electronic alarms are among the most widely used electronic devices. Computers are equipped with fault alarm systems to prevent the loss of valuable data. Also, most businesses and automobiles have security alarms. **See Figure 14-31.**

In a security alarm circuit with a 555 timer, photocell R1 and resistor R2 form a voltage divider. The divider produces the biases for the base-to-emitter junction of Q1. With R1 in darkness, its resistance will be high. Therefore, when current flows through the voltage-divider network, a large voltage will be dropped across the resistance developed by R1. This voltage forward biases Q1 since its base is now positive.

With forward bias, Q1 conducts and drops the voltage at trigger pin 2 to a low value. When pin 2 is low, the output of the flip-flop will be low, and the output of the power amplifier will be high (approximately +10 V). This is because the power amplifier not only amplifies, but also inverts 180°, making the low input a high output. Since the voltage across pin 3 is approximately the same as V_{CC}, there is 0 V across the relay (K1). Thus, the relay is de-energized, and it prevents the security alarm signal from being connected to the speaker.

When light strikes R1, its resistance drops to a much lower value. The reduced resistance causes a reverse bias on Q1 by dropping the voltage at its base to a low value (less than 0.6 V). With reverse bias applied to the base-emitter junction, Q1 turns off. This raises the voltage at pin 2 to +10 V, the supply voltage. With pin 2 high, the 555 timer output at pin 3 will be low. With pin 3 low, +10 V appears across relay coil K1, and the relay will energize. This allows the security alarm signal to be applied to the speaker through the NO contacts of relay K1. Thus, the security alarm will be activated.

To shut off the alarm, the reset button must be pushed (closed). When the reset button is pushed, a pulse is applied to the reset of the 555 timer, placing the circuitry back to its original state.

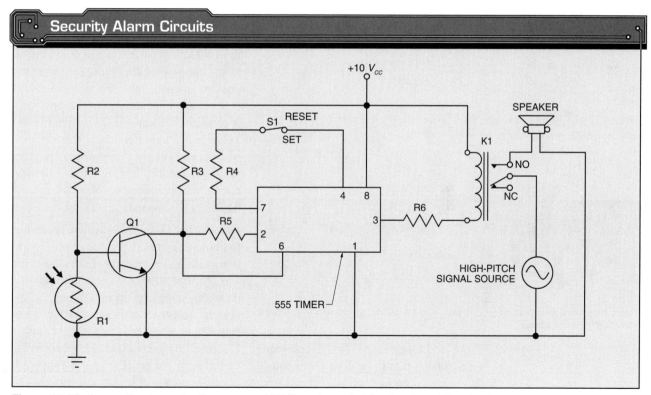

Figure 14-31. *A security alarm circuit may use a 555 timer to control the function of the circuit.*

Astable Multivibrators

Astable multivibrators are used extensively as waveform generators or oscillators. Multivibrators are basic electronic circuits found in many types of equipment. Astable multivibrator outputs are often used to control audible signals, such as sirens and other warning devices. Emergency vehicles often have sirens that are controlled electronically. Electronic circuits can change the pitch of the siren, cause the siren to warble fast or slow, or produce intermittent siren bursts.

In other applications, astable multivibrator square waves are used to control pulses of light. This is possible because square wave voltages act like electronic switches. The waves have straight sides that turn lights on and off very quickly. This effect can be seen with the barricade flashers used on highways. Other examples include automobile 4-way flashers, digital-display scoreboards, and strobe lights.

Refer to Chapter 14 Quick Quiz® on CD-ROM

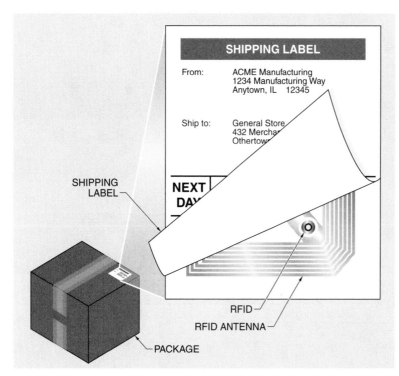

Radio frequency identification (RFID) tags may be placed on objects that require identification and tracking.

For a 555 timer to be used as an astable multivibrator, the IC must be continuously retriggered. The easiest way to retrigger the IC is to connect the trigger input (pin 2) to the threshold input (pin 6). **See Figure 14-32.** The timing resistors R1 and R2 are added with their junction point connected to the discharge terminal (pin 7).

Before power is applied to this circuit, the trigger and threshold inputs are both below one-third of V_{CC}, the timing capacitor is discharged, and the output is high. At the moment power is applied, the capacitor begins charging.

The capacitor continues charging until its voltage reaches two-thirds of V_{CC}. At that point, Comp 1 (in the 555 timer) triggers the internal flip-flop, causing the capacitor to discharge through resistor R2.

The amount of time it takes to charge and discharge the capacitor is determined by the value of internal resistors R1, R2, and R3 of the timer. The total time required to complete one charge/discharge cycle is found by using the following formula:

$$t = 0.693(R1 + 2R2)C1$$

The frequency of oscillation is found by using the following formula:

$$f = \frac{1.44}{(R1 + 2R2)C1}$$

Thickness Monitors. A 555 timer can use a photocell to vary its frequency. **See Figure 14-33.** Such circuits can be used to monitor the thickness of semitransparent materials on a production line. The material passes between the light source and the photocell. A change of thickness in the material causes a change in output frequency. This variation in frequency is then monitored by other electronic circuits. Finally, an indicator that notifies the operator of a problem in material thickness is activated.

Gas Alarms. Gas alarms are used, in addition to smoke detectors, to provide an early warning of toxic gases and fumes from a fire or gas leak. The operation of the gas alarm circuit is dependent upon a semiconductor sensor. The electrical resistance of the semiconductor sensor changes when the sensor's active surface is exposed to a gas, such as carbon monoxide, methane, or butane. The sensor element is normally enclosed in a small capsule and protected by a stainless steel mesh material. A low-power heater within the device activates the sensor element and purifies the element after exposure to gas.

A change in resistance is the key to triggering the alarm circuit using a 555 timer. **See Figure 14-34.** By connecting the resistive element of the gas sensor between the positive supply and pins 2 and 6 of 555 timer 1, the 555 timer can be used as a comparator.

When the resistance of the gas sensor resistive element drops due to the presence of gas, the resistance ratio of the gas sensor element to resistors R1 and R2 changes. Thus, the voltages at pin 2 and pin 6 are raised. The resistance of resistor R2 is adjusted so that the voltage increase exceeds two-thirds V_{CC}. When that occurs, the output of 555 timer 1 goes low, activating 555 timer 2. Timer 2 is designed to operate as an astable alarm. A speaker with a high-pitch signal is used to indicate the presence of gas.

When there is no gas present, the resistance of the gas sensor element will be high enough that the voltage applied by the diode will be below one-third V_{CC}. The output of 555 timer 1 is high, therefore, disabling the astable multivibrator.

Normally, the sensor heater element receives power from a stable voltage regulator at about 5 V. To clean the element, however, a higher voltage is required. A clean/run switch is installed to allow the voltage regulator to be bypassed and a higher voltage applied for cleaning.

Tilt Switches. A level switch with a 555 timer can be used as a tilt switch. **See Figure 14-35.** The switch is mounted in its normally open position, which allows the timer output to stay low, as established by capacitor C on startup. When switch S1 is disturbed, causing its contacts to be bridged by conductive liquid, the 555 latch is set to a high output level. The latch will stay at the high output level even if the switch is returned to its starting position. The high output can be used to trigger an alarm. Switch S2 silences the alarm and resets the latch. A tilt switch application may include crane operation.

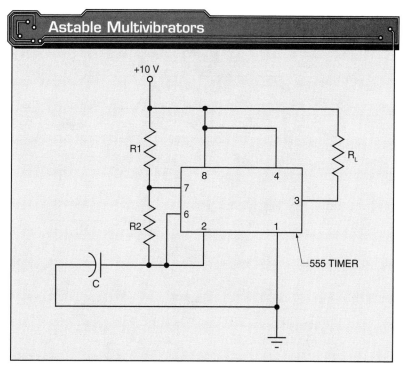

Figure 14-32. *A 555 timer can be used as an astable multivibrator by connecting the trigger input to the threshold input.*

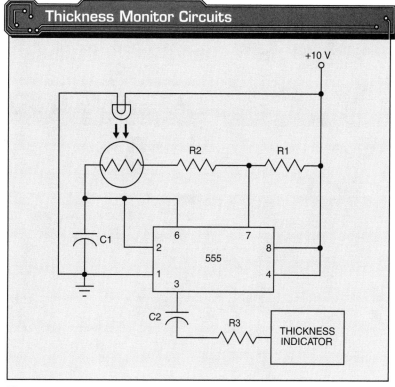

Figure 14-33. *A thickness monitor circuit may include an astable multivibrator.*

366 SOLID STATE DEVICES AND SYSTEMS

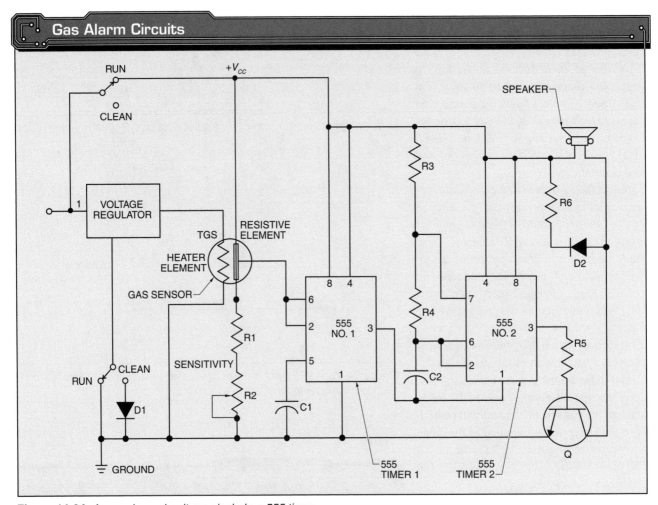

Figure 14-34. *A gas alarm circuit may include a 555 timer.*

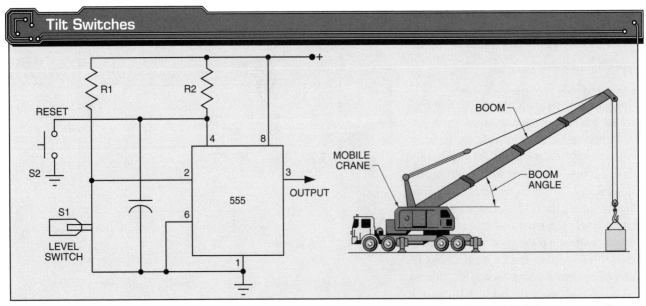

Figure 14-35. *A tilt switch circuit may include a 555 timer with a level switch. A tilt switch application may include crane operation.*

KEY TERMS

- integrated circuit (IC)
- linear integrated circuit (IC)
- operational amplifier (op amp)
- unity
- offset error
- nulling
- slew rate
- integrator
- comparator
- monostable multivibrator
- astable multivibrator
- 555 timer
- 556 timer
- oscillator
- differential op amp

REVIEW QUESTIONS

1. What is an integrated circuit (IC)?
2. What is a linear integrated circuit (linear IC)?
3. Explain how integrated circuits can be troubleshooted.
4. Describe how to force an IC to work.
5. What is an operational amplifier (op amp)?
6. Explain the operation of an op amp.
7. Describe the characteristics of an op amp.
8. Explain gain.
9. What is offset error?
10. What is nulling?
11. What is slew rate?
12. List and describe common op amp applications.
13. What is an integrated circuit timer?
14. Explain 555 timer operation.
15. What is a monostable multivibrator?
16. What is an oscillator?
17. Explain the operation of security alarm circuits.
18. Explain the operation of thickness monitors.
19. Explain the operation of gas alarms.
20. Explain the operation of tilt switches.

PHOTONICS 15

Photonics is a field that uses light to perform functions traditionally performed by electronics. The field of photonics became a reality with the invention of the laser and the development of fiber optics. Applications for photonics can now be found in almost all aspects of residential, commercial, and industrial equipment and processes. Photonics covers a wide range of applications, from sun lamps using ultraviolet light to communication devices using infrared light, with a majority of the systems operating in the range of visible to infrared light.

OBJECTIVES

- Explain photonics.
- List and describe the types of light sources used in photonics.
- List and describe the devices used to detect light.
- Explain fiber optics.

PHOTONICS

Photonics is a branch of science that deals with the detection, emission, generation, manipulation, and transmission of light. A *photon* is the basic unit of light or electromagnetic energy. Photons have no mass and no charge. They are carriers of electromagnetic energy and interact with electrons, atoms, and molecules. A beam of light is considered to be a stream of photons. **See Figure 15-1.**

Photonics plays a vital role in daily activities. For example, it is used in DVD players, in which lasers reflecting off DVDs transform the returning signals into movies. It is used in grocery store checkout lines, where laser beams read bar codes for product identification and prices. It is used by laser printers to record images on paper. It is used in digital cameras to capture images and allow these images to be displayed. It is the basis of the technology that allows computers and telephones to be connected to one another over fiber-optic cables. Photonics is also used in the medical field for the lasers used in surgery.

[Handwritten annotation: Quantity (a quant of light)]

Photonic Applications	
Consumer applications	Laser printing
	3D television
	Blue-ray Disc™ players
	Remote controls
	Laser shows
	Digital cameras
	Energy saving lighting
Commercial applications	Bar code scanners
	Fiber-optic communication
	Credit cards
	Holography
	Information processing
	Surveillance cameras
	Night vision goggles
Industrial applications	Laser welding
	Laser cutting
	Laser drilling
	Laser leveling
	Machine vision
	Precision measurements
	Optical data storage
	Optical scanners

Figure 15-1. *Photonics covers a range of applications that use light, from ultraviolet to infrared, with a majority of the systems operating in the range of visible to infrared light.*

Devices that use photonics have a number of advantages over those that use electricity. Light travels approximately 10 times the speed of electricity, which means that data transmitted through light can travel long distances in a fraction of the time. Furthermore, visible light and nonvisible infrared (IR) beams pass through each other without interference, unlike electric current. A single optical fiber has the capacity to carry three million telephone calls simultaneously.

Light

Light is electromagnetic radiation that may or may not be visible. The term "light" is often used to describe photonic operations instead of electromagnetic radiation. Light refers to not only visible electromagnetic radiation but also the entire nonvisible spectrum where semiconductor devices respond. The term "light" is not completely accurate because of the presence of nonvisible radiation. However, use of the term is accepted in some areas such as fiber optics.

Light consists of electromagnetic radiation, which is similar to the radiation of radio transmissions. The main difference is the frequency. Light waves are composed of frequencies that are much higher than those of radio transmissions. Therefore, the wavelength of visible light is much shorter than the wavelength generated by radio equipment. **See Figure 15-2.**

Propagation is the speed [Traveling] at which light travels through a substance. It is very common for substances such as fiber optics, glass, and plastics to be referred to as media through which light passes. Light energy may behave like a wave as it moves through space. Alternatively, it may behave like a particle with a certain amount of energy, which can be emitted or absorbed.

When light strikes the surface of certain materials, electrons are released. This is known as the photoelectric effect. The frequency of light (photons) determines the energy level necessary for an electron to escape from the surface of the material. Photons with frequencies below a certain level will not emit electrons, regardless of the intensity of light. Photons with frequencies above a certain level emit electrons even at low light intensity.

Light can be in the form of visible light or nonvisible light. Visible light is the portion of the electromagnetic spectrum to which the human eye responds. Visible light includes the part of the electromagnetic spectrum that ranges from the color violet to red. Nonvisible light is the portion of the electromagnetic spectrum on either side of the visible light spectrum. Nonvisible light includes ultraviolet and infrared light.

Light scatters as it travels farther from a light-producing source. The relationship between the amount of light produced at the source and the amount of illumination at different distances from the light source is expressed by the inverse square law. The inverse square law states that the amount of illumination on a surface varies inversely with the square of the distance from the light source.

> In 1962, Nick Holonyak, Jr. invented the first practical visible LED. He is known as the "father of the light-emitting diode."

Brightness is the perceived amount of light reflecting from an object. Brightness depends on the amount of light falling on an object and the reflecting ability of the object. Glare takes place when there is too much brightness. Glare is reduced when the brightness is reduced and/or the eye reduces the amount of light received. Proper lamp spacing, number of lamps, and watts determines the evenness of the brightness. *Contrast* is the difference of brightness between different objects. For example, when there are black letters on a white background, the contrast is high. However, when there are black letters on a gray background, the contrast is low. High contrast is helpful when distinguishing outlines and small objects. Contrast between surfaces is sharpened by greater brightness and/or contrasting surfaces. **See Figure 15-3.**

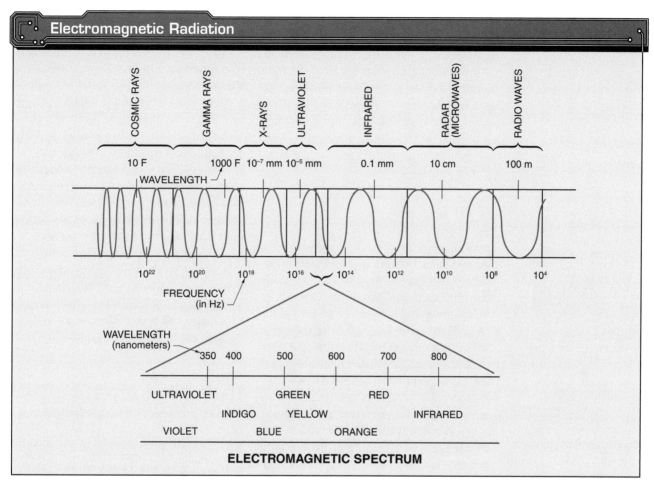

Figure 15-2. *Light consists of electromagnetic radiation, which is the same type of radiation used for radio transmission and radar.*

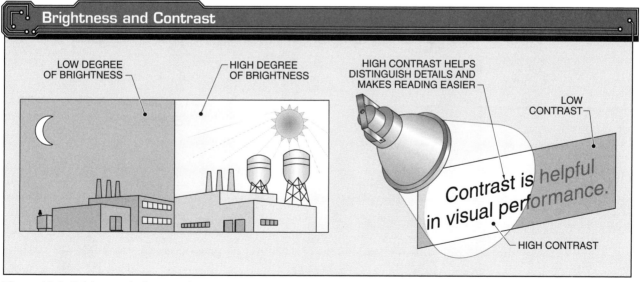

Figure 15-3. *Brightness is the perceived amount of light reflecting from an object. Contrast is the ratio of brightness between different objects.*

Technicians maintain pieces of equipment that use frequencies from one end of the electromagnetic spectrum to the other. An object can be seen because the light produced by or reflected from the surface reaches the eyes of the technician. If the object is the source of light, it is called luminous. If the object is not the source of light but reflects light, it is called an illuminated body.

Color

All objects absorb some of the light that strikes them. An object appears to be a certain color because it absorbs all of the light waves except those whose frequency corresponds to that particular color. Those waves are reflected from the surface of the object, strike the eyes, and cause the eye to see the particular color. Therefore, the color of an object depends on the frequency of the electromagnetic wave reflected.

The color of light is determined by its wavelength. The shortest wavelengths of visible light produce the color violet. The longest wavelengths of visible light produce red. Wavelengths of visible light between violet and red produce blue, green, yellow, and orange. The combination of the various colors of light produces white light. For example, the sun produces energy over the entire visible spectrum in approximately equal quantities. **See Figure 15-4.**

Both ultraviolet and infrared light are invisible to the human eye. However, ultraviolet and infrared light are used in many applications. The ultraviolet region of the spectrum is the region with wavelengths slightly shorter than the wavelengths that produce the color violet. The infrared region of the spectrum is the region with wavelengths slightly longer than the wavelengths that produce the color red. Depending on the wavelength, ultraviolet light is used in black lights, sun lamps, and sterilization lamps. Infrared light is used in heat lamps and in communication devices, such as television remote controls.

Speed of Light

Light travels through air at approximately 186,000 miles/sec. Light travels more slowly in other media, such as glass or water. A good example of the change in the speed of light is when light travels through air and into water. When light passes through air into water, it changes speed.

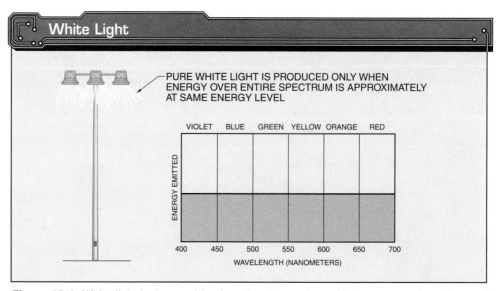

Figure 15-4. *White light is the combination of various colors of light at approximately equal quantities.*

Also, different wavelengths of light travel at different speeds in the same medium. A *medium* is any material through which light travels. This principle can be seen in a glass prism. White light entering a prism is composed of all colors representing different wavelengths. The prism refracts the light because each wavelength changes speed differently. Therefore, the light emerging from the prism is separated and it divides into colors of the visible spectrum. **See Figure 15-5.**

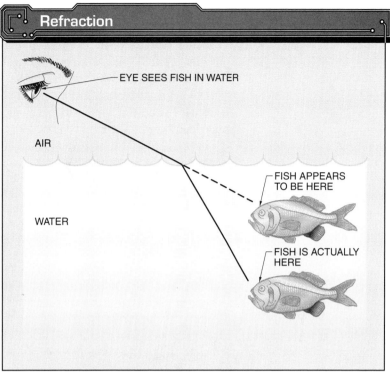

Figure 15-6. *Refraction is a deflection of light caused when light passes through air into water and changes speed.*

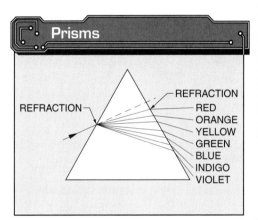

Figure 15-5. *A prism refracts white light into separate colors because the different wavelengths of light travel at different speeds through the prism.*

Refraction

Refraction is a deflection of light caused when light passes through air into water and changes speed. Knowledge of the refraction of light will help to understand how light moves through fiber-optic cables. **See Figure 15-6.**

Three important terms associated with refraction are normal, angle of incidence, and angle of refraction. A *normal* is an imaginary line perpendicular to the interface of two materials, such as glass and air. An *angle of incidence* is the angle formed by incoming light (an incident ray) and the normal. **See Figure 15-7.** An *angle of refraction* is the angle formed between a refracted ray and the normal.

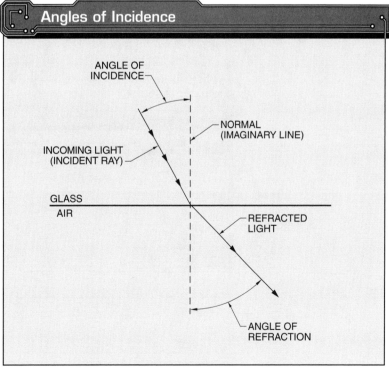

Figure 15-7. *The angle of incidence is the angle formed by the incoming ray (incident ray) and the imaginary line (normal).*

The angle of incidence determines what happens to an incident ray. As the angle of incidence increases, the chances that an incident ray will enter a second material decreases. A *critical angle* is the angle at which light no longer passes into a second material. The critical angle varies with different substances. **See Figure 15-8.**

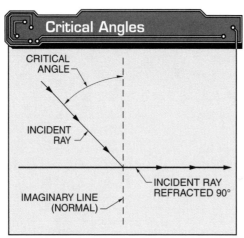

Figure 15-8. *The angle at which light no longer passes into a second material is the critical angle.*

When an incident ray reaches a 90° refraction, the ray does not enter the second material. If the angle of incidence increases past the critical angle, light is completely reflected back into the first material. **See Figure 15-9.**

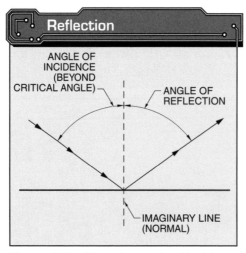

Figure 15-9. *When the angle of incidence passes beyond the critical angle, the light is totally reflected.*

Mediums include air, water, glass, and any transparent material through which light can pass. A medium to which light rays are directed can reflect, absorb, and transmit rays. **See Figure 15-10.**

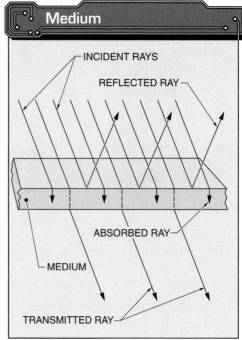

Figure 15-10. *A medium may reflect, absorb, or transmit rays of light directed at it.*

LIGHT SOURCES

A variety of light sources are used in photonics. A device that generates its own light is considered a light source. The most common light sources used in photonics include light-emitting diodes, liquid crystal displays, plasma display panels, and lasers.

Light-Emitting Diodes

A *light-emitting diode (LED)* is a diode that emits light when current flows through it. The light source for fiber optics is usually an LED or a laser diode. An LED provides less power and operates at slower speeds than a laser diode. However, an LED is suitable for most applications requiring transmission distances up to several thousand feet and speeds up to several hundred megabits per second. LEDs are also more reliable, less expensive, and easier to use than laser diodes in many applications. **See Figure 15-11.**

An LED produces light through the use of semiconductor materials. A diode junction can emit light when an electrical current is present. The presence of an electrical current produces light energy because electrons and holes are forced to recombine.

The energy level of an electron increases as it passes through the junction of a semiconductor diode. To get through the depletion region, the electron must acquire additional energy. **See Figure 15-12.** This additional energy comes from the positive field of the anode. If the field is not strong enough, the electron will not get through the depletion region and no light will be emitted. For a standard silicon diode, a minimum of 0.6 V must be present before the diode will begin to conduct. For a germanium diode, 0.3 V must be present before the diode will begin to conduct. Most LED manufacturers make a larger depletion region that requires 1.5 V for the electron to get across the depletion region. As the electron moves across the depletion region, it gives up its extra kinetic energy. The extra energy is converted to light.

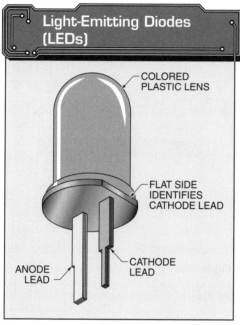

Figure 15-11. *An LED is a diode that produces light when current flows through it.*

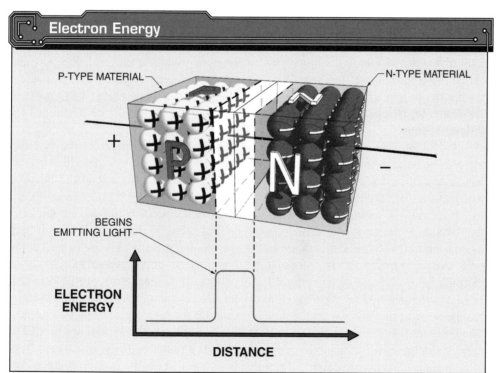

Figure 15-12. *Electron energy increases as it passes through the junction of the diode.*

LED Operation Media Clip

The light emitted can be either invisible (infrared) or visible light in the spectrum. LEDs for electronic applications, due to the spectral response of silicon and efficiency considerations, are usually infrared emitting diodes (IREDs).

An IRED is an LED that emits invisible light near the infrared region of the light spectrum. Generally, gallium arsenide and gallium arsenide phosphide are the materials used for the IRED. The materials are encased in either metal or plastic. The electrical characteristics of the IRED are similar to those of any other PN junction diode. The IRED has a slightly higher forward voltage drop than a silicon diode due to the higher energy necessary. It has a fairly low reverse breakdown voltage due to the doping levels required for efficient light production. The light output of the diode is dependent upon power supplies and is measured in milliwatts (mW).

LED Construction. Manufacturers of LEDs generally use a combination of gallium and arsenic with silicon or germanium to construct semiconductors. By adding impurities to the base semiconductor, and adjusting other impurities, different wavelengths of light can be produced. LEDs are capable of producing light that is not visible to the human eye, called infrared light, or they may emit a visible red or green light. If other colors are desirable, the plastic lenses may be of different colors.

Like standard semiconductor diodes, there must be a method for determining which end of an LED is the anode and which end is the cathode. The cathode lead in a through-hole mount is identified by the flat side of the device, or it may have a notch cut into its ridge. The cathode lead on a surface mount LED is colored for identification. The schematic symbol for an LED is the same as that of a photodiode, but the arrows point away from the diode. The LED must be forward-biased, and a current-limiting resistor is usually present to protect the LED from excessive current. **See Figure 15-13.**

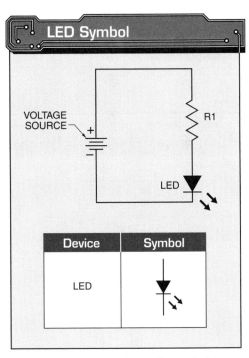

Figure 15-13. *The schematic symbol for an LED is similar to that of a photodiode except for the direction of the arrows.*

Colored plastic lenses cause different colors to be emitted. The colored plastic lens also focuses the light produced at the junction of the LED. Without the lens, the small amount of light produced at the junction would be diffused, and it would be virtually unusable as a light source. The size and shape of the LED package determines how it must be positioned for proper viewing.

LEDs can also be constructed as multicolor LEDs without a colored plastic lens. They are considered tricolored because they can produce red, green, and yellow, depending upon which leads on the LED are connected to the power source. If both the red and green LEDs are powered, the combination of red and green will create yellow. The three leads are of different lengths. The center lead (k) is the common cathode for both LEDs. The outer leads, A1 and A2, are the anodes to the LEDs, allowing each one to be lit separately or both together to give the third color. **See Figure 15-14.**

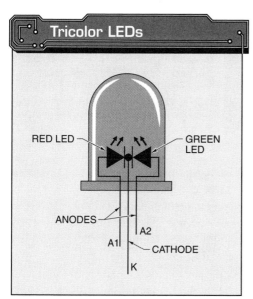

Figure 15-14. *A tricolor LED can produce red, green, and yellow, depending upon which leads on the LED are connected to a power source.*

Testing LEDs. The output of an LED or IRED is usually determined by using a test fixture. Electrically, the LED and IRED are powered and monitored by a test circuit. The object of the test is to determine the efficiency of the LED or IRED. If the fixture has been properly calibrated, it should measure the efficiency of the LED or IRED based on the amount of input current needed to drive the emitter. **See Figure 15-15.**

In an ideal emitter-receiver situation, such as the test fixture, every photon generated by the emitter should result in an electron being generated in the receiver. This ideal situation would result in a quantum efficiency ratio of 1. *Quantum efficiency* is a method of calculating the number of electrons produced for each photon striking a semiconductor. Mathematically, quantum efficiency is expressed as the following:

$$quantum\ efficiency = \frac{electrons}{photons}$$

Note: For all semiconductor devices, the quantum efficiency is less than 1.

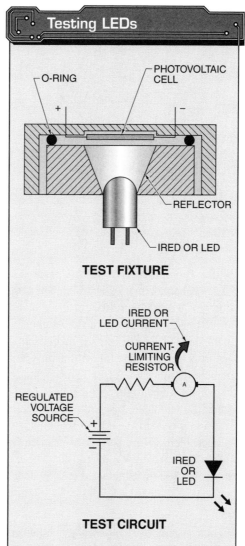

Figure 15-15. *The output of an LED or IRED is determined by using a test fixture and test circuit.*

LED Applications. As a light source, the LED can be used as a substitute for an incandescent light bulb. **See Figure 15-16.** Since the LED is inexpensive, has a nearly unlimited life span, and consumes little power, it is ideal for a variety of lighting applications.

LEDs have several advantages over conventional lamps. They do not have filaments that will burn out and can last as long as 10 to 15 years. In addition, their small plastic bulbs make them more durable.

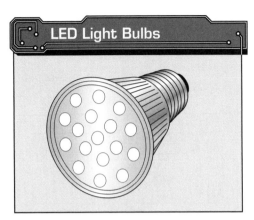

Figure 15-16. LEDs are being used for everyday applications once reserved for incandescent light bulbs.

> The word "photonics" comes from the Greek word "photos," which means light. Photonics first appeared in the 1960s to describe a research field involving the use of light to perform functions that traditionally required electronics.

Status Indicators. A *status indicator* is an LED application that indicates the operating status of a machine to the machine operator. It may be used to indicate whether the machine is ON, OFF, or malfunctioning. In a miniprocessor module with a programmable controller, the LEDs can be used for diagnostics. **See Figure 15-17.** In this case, the LEDs will indicate whether the machine is running properly (RUN) or whether there is a problem in the processor or with the memory. Selecting the diagnostic output is accomplished by using a key lock selector.

Seven-Segment Displays. The readout of test equipment is often presented on a seven-segment display. The display is made with LEDs and is called an LED display. Each segment of the display has several LEDs, but a colored plastic overlay makes the individual LEDs glow as a unit. The number 3 would be activated on the display by energizing segments a, b, c, d, and g. **See Figure 15-18.**

Dot-Matrix Displays. A 5 × 7 dot-matrix display is used to show symbols, letters, and numbers. This type of display uses one LED for each dot of the matrix. For example, when the proper sequence is energized, the letter E will appear. An *alphanumeric display* is a display that presents information in the form of symbols, letters, and numbers.

LED Nanotechnology. Technicians should become familiar with nanotechnology as it begins to affect electronics and photonics. Nanotechnology is embedded in photonics because in part of the impact it will have on LEDs. LEDs are very energy efficient when energy is converted to light. Unfortunately, not all of the light is actually emitted from the LED. However, with nanotechnology, thousands of microscopic holes can be created in the surface of the LED to increase the amount of light being emitted.

In the metric system, the prefix "nano" refers to one billionth. One nanometer (nm) is one billionth of a meter. To put this into perspective, 1 nm is approximately 1/50,000 of a strand of human hair. Manufacturing electronic devices of this size will not only change LED technology but also everything related to photonics/electronics.

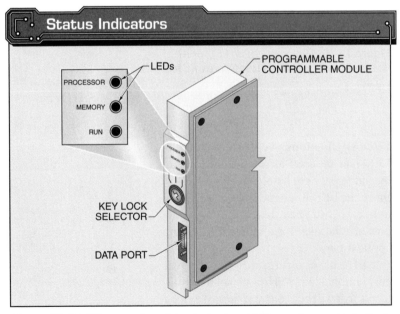

Figure 15-17. When used as status indicators, LEDs can be used to indicate the operating status of machines or systems.

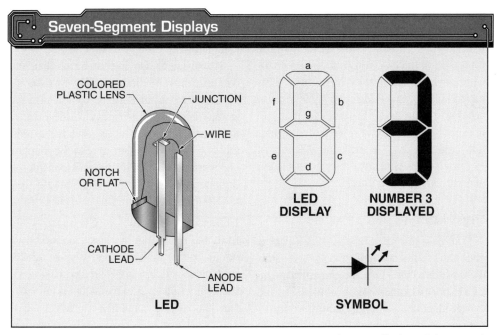

Figure 15-18. *LEDs can be arranged in seven-segment sections to display alphanumeric symbols.*

Organic Light-Emitting Diodes. An *organic light-emitting diode (OLED),* also known as a light-emitting polymer (LEP), is a solid state device composed of a thin film of organic molecules that create light with the application of voltage. They provide crisp, bright displays on electronic devices using less power than conventional LEDs. OLED displays are available as single color, multicolor, and full color displays. OLEDs are thinner than human hair. They can also be flexed and shaped easily.

OLEDs are used in television screens, computer monitors, mobile phones, and PDAs. OLED screens function without a backlight, and therefore are thinner and lighter than LCD displays. In low light, such as a dark room, OLED screens can achieve a higher contrast than LCD screens and LED-backlit screens.

A typical OLED is composed of an emissive layer, conductive layer, substrate, anode terminal, and cathode terminal. The layers are made of organic molecules that conduct electricity. When voltage is applied across an OLED, such that the anode is positive with respect to the cathode, electrons flow through the OLED from the cathode to the anode. Therefore, the cathode gives electrons to the emissive layer, and the anode withdraws electrons from the conductive layer. **See Figure 15-19.**

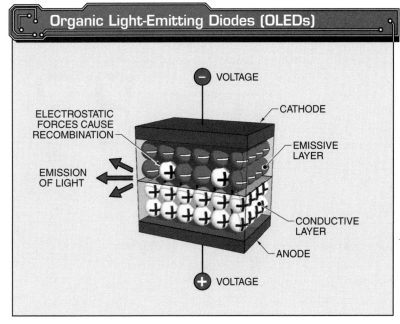

Figure 15-19. *OLEDs are composed of an emissive layer, conductive layer, substrate, anode terminal, and cathode terminal.*

The emissive layer becomes negatively charged, while the conductive layer fills with positively charged holes. Electrostatic forces bring the electrons and the holes toward each other and they recombine. This happens closer to the emissive layer because holes are more mobile than electrons in organic semiconductors. The recombination causes a drop in the energy levels of electrons. This is followed by an emission of light, which is visible.

Liquid Crystal Displays

A *liquid crystal display (LCD)* is a flat alphanumeric display that uses liquid crystals to modulate, but not directly emit, light. LCDs are packaged much like LEDs and can produce the same alphanumeric information. A significant difference between LCDs and LEDs is that LCDs control light, while LEDs generate light. LCDs require much less power to operate than LEDs. LCDs are less prone than LEDs to wash out under strong ambient light. The digits of an LCD actually become a deeper black, or they become clearer, as the intensity of the ambient light increases. Because of the advantages of the LCD, it is used extensively in newer test equipment, clock displays, and calculators.

However, LCDs cannot be read in the dark and must be illuminated by an external light source. A small light source may be installed in close proximity to the display to illuminate it. Users can switch the light ON and OFF as needed to read the display. Devices such as LCDs are not considered light sources in a strict sense because they require backlighting.

LCD Construction. An LCD consists of a front and rear piece of glass separated by a liquid crystal material. **See Figure 15-20.** Both pieces of glass are coated with a microscopically thin layer of metal that is transparent. The coating is applied to each piece of glass so that it faces the liquid crystal. The layer of metal applied to the front surface of the rear piece of glass covers the entire active area of the display. The layer of metal applied to the rear surface of the front piece of glass is broken into segments. The metal segments are then brought through a separator seal to the edges of the display to provide electrical connection points for the circuitry.

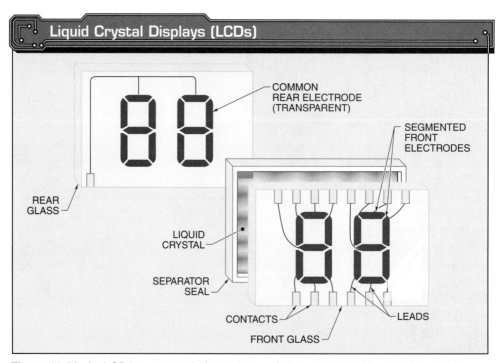

Figure 15-20. *An LCD is composed of two pieces of glass separated by a liquid crystal.*

The key to LCD operation is the liquid crystal that fills the space between the front and rear glass. The molecules of liquid crystal are normally parallel to the plane of the glass. However, when a voltage is applied to the liquid crystal, its molecules twist 90° to alter the light passing through it. **See Figure 15-21.**

A *polarizing filter* is a filter that allows light to pass through only at certain angles. Polarizing filters are used in LCDs to obtain black digits or clear digits. The only design change between the two types of displays is a 90° rotation of one of the polarizing filters for a vertical or horizontal axis. Depending on the application, LCDs can consist of black digits with a clear background or clear digits with a black background. Wristwatch manufacturers use displays with black digits because of the sharp contrast they have against the light-colored background. Clear digits are often preferred in digital-panel meters because the black background makes the display easier to read.

LCD Operating Voltages. LCDs require special operating voltages. Applying improper voltages can drastically reduce the life of a display. Improper voltages can also damage the display. The proper operating voltage for LCDs is AC voltage with no DC voltage present. Using DC voltage, or AC voltage with DC present, significantly reduces the life of the display. When the proper AC voltage is used, the life of the display may be several years. AC voltage applied to the display does not need to be sinusoidal. In fact, it is often in the form of a square wave.

Color LCDs. The light source in a color LCD monitor is a series of LEDs backlighting the screen. To achieve a full-color pallet on an LCD display, each pixel is divided into red, green, and blue subpixels that work in conjunction to determine the overall hue of the LCD pixel. These subpixels are created by subtracting certain wavelengths and the colors using special filters. By using a combination of red, green, and blue subpixels of various intensities, a single pixel triad can produce more than 16 million colors.

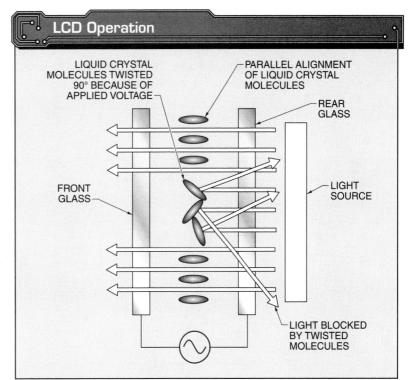

Figure 15-21. *Liquid crystal molecules are normally parallel to the plane of glass. However, when a voltage is applied to the liquid crystal, the molecules twist 90° to alter the light passing through it.*

Dot pitch is the distance between subpixels of the same color. Dot pitch affects the overall quality of a picture reproduced on LCDs. The closer these "dots" are to one another, the sharper the resolution will be. This is especially true when displaying computer images and graphs. Higher dot pitches also increase the viewing angles of LCD panels.

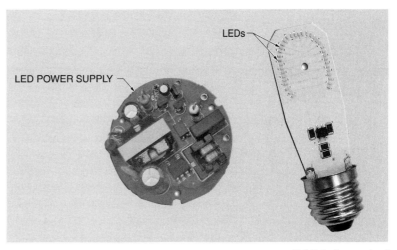

PolyBrite International
LEDs can be used to replace traditional incandescent light bulbs.

LED Backlighting. LED backlighting is very popular for small and medium LCD displays. The advantages of LED backlighting are its low cost, long life, and precise control over its intensity.

The two basic configurations of the LED backlight are array lighting and edge lighting. In both configurations, the LEDs are the light source that is focused into a diffuser that distributes the light evenly behind the viewing area. **See Figure 15-22.** In array lit configurations, many LEDs are mounted uniformly behind the displays. This arrangement offers more uniform and brighter lighting but consumes more power. In edge lit configurations, the LEDs are mounted to one side or on the top of the displays, with focused light on the diffuser. The edge light configuration offers a thinner package and consumes less power. However, the display is less uniform and not as bright.

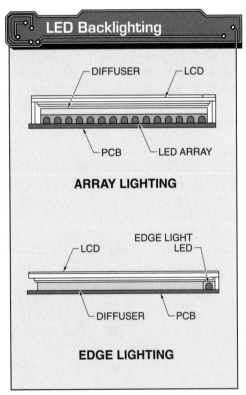

Figure 15-22. *LED backlighting consists of array lighting and edge lighting.*

In 1964, the monochrome plasma video display was co-invented at the University of Illinois at Urbana-Champaign by Donald Bitzer, H. Gene Slottow, and Robert Willson.

Plasma Display Panels

A *plasma display panel (PDP)* is a display panel that consists of small compartments or cells containing ionized gas, which forms plasma when a voltage is applied, and creates UV photons that strike phosphor pixels painted on the inside of the cells to produce light. **See Figure 15-23.** The designs of PDPs may vary but the basic operating principles are the same. A typical PDP has millions of pixels within the matrix behind the screen. The pixels are grouped in three segments of red, green, and blue. By activating any one or all three of the pixels, over 16 million different colors can be generated.

Phosphor pixels are sandwiched between two pieces of glass and two pieces of magnesium oxide dielectric layers. Gases are trapped between the front glass and the pixels. These gases are usually xenon, neon, and helium. The thin address electrodes are located behind the pixels along the back glass plate. Nearly transparent display electrodes are mounted in the insulating dielectric in front of the pixels along the front glass plate.

When a charge is placed in front of a pixel, the gases change to plasma and the phosphor in the pixel glows. Images are created by selectively charging and discharging millions of pixels.

Lasers

A *laser* is a device that emits and amplifies light. The term "laser" was originally an acronym for light amplification by stimulated emission of radiation. A laser typically consists of a power source, fully reflective mirror, partially reflective mirror, glass chamber, and gain medium. The gain medium is used to amplify light each time it reflects back and forth off the reflective mirrors.

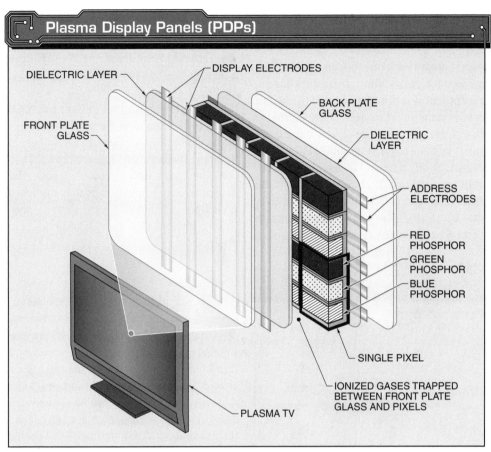

Figure 15-23. *PDPs create plasma from voltage applied to ionized gas.*

Gas Lasers. A *gas laser* is an electronic device in which electric current is used to ionize a gas to emit light, which is then amplified. Gas lasers are often used with sensing and control systems. A commonly used gas laser is the helium-neon laser. **See Figure 15-24.** Many other types of gas lasers are available.

High DC potential is applied to a plasma tube by means of a voltage multiplier circuit and a pulse transformer. The filament within the plasma tube is heated by an AC source. As electrons from the filament are accelerated, they strike helium-neon gas atoms in the tube and cause them to ionize. The ionized gas emits light. This is similar to a fluorescent light bulb. The light reflects from a flat, fully reflective mirror at the top. The plasma tube is cut at a precise angle to control reflection. The light is reflected back and forth several times to the partially reflective spherical mirror at the bottom. Here, it is concentrated into a laser beam that is emitted through the spherical mirror.

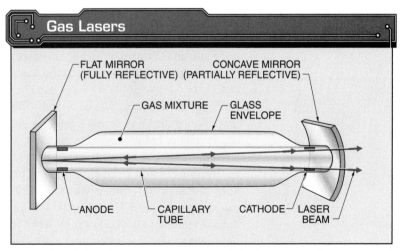

Figure 15-24. *Gas lasers are often used in industry with sensing and control systems.*

Semiconductor Lasers. A *semiconductor laser* is a solid state laser that emits and amplifies light through the use of a semiconductor material. Semiconductor lasers have a resonant cavity similar to other lasers, except that it is formed on a microchip of semiconductor material. Semiconductor lasers are very efficient and extremely small in size compared to other lasers. **See Figure 15-25.** The end faces of the microchip are made parallel and flat. Since the semiconductor material is reflective, no mirrors are needed. This is a distinct advantage in terms of complexity and cost.

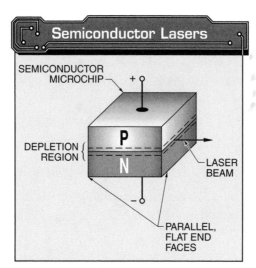

Figure 15-25. *Semiconductor lasers are small, efficient, and do not require mirrors to amplify light.*

As current flows through the semiconductor material, light is emitted. The atoms collide near the PN junction of the material and cause additional photons to be released. Due to the reflective properties of the semiconductor, a wave of photons develops between its flat surfaces. The back-and-forth movement of this wave creates the resonant action required to produce a laser beam.

Semiconductor lasers may be used in computers for a laser mouse. The laser mouse works similarly to the LED-based optical mouse. However, the LED is replaced with a laser. Since a laser beam is used, the mouse has better tracking capabilities, giving the user the ability to use the mouse on more surfaces.

LIGHT DETECTION

A *light detector* is an electronic device that detects the presence of light from a light source and converts the light into an electrical signal. The process by which a light source is converted into an electrical signal is the reverse of the process that produces the light. Once light rays have passed through a medium, they must be detected and converted back into electrical signals. The detection and conversion is accomplished with light-activated devices, such as photocells, photodiodes, PIN photodiodes, and photoelectric switches.

Photocells

A *photocell,* also known as a photoconductive cell, is an electronic device that varies resistance based on the intensity of the light striking it. A photocell is formed by a thin layer of semiconductor material, such as cadmium sulfide or cadmium selenide, deposited on a suitable insulator. Leads are attached to the semiconductor material, and the entire assembly is hermetically sealed with glass. The transparency of the glass allows light to reach the semiconductor material. **See Figure 15-26.**

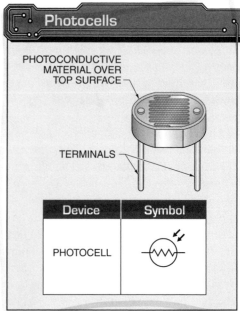

Figure 15-26. *Photocells vary resistance based on the intensity of light striking them.*

> An LCD TV panel contains a matrix of microthin transparent transistors across the whole screen. The transistor matrix sends information via electrical currents to each individual pixel. This process controls the color and temperature of the pixels.

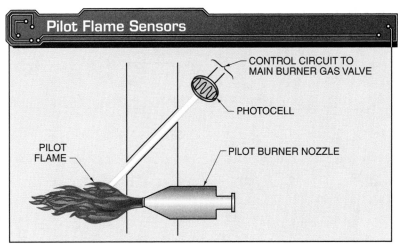

Figure 15-27. Photocells can be used to determine whether the pilot light of a gas furnace is ON or OFF.

For maximum current-carrying capacity, photocells are manufactured with a short conduction path having a large cross-sectional area. If a number of photocells are connected in parallel, they can control up to 0.5 A.

The resistance of a photocell changes as the light striking its surface changes. When light strikes a photocell, electrons are freed and the resistance of the material decreases. When light is removed, the electrons and holes recombine and the resistance increases. The resistance of a photocell ranges from several megohms (MΩ) in total darkness to less than one hundred ohms at full light intensity.

A certain amount of dark current flows even when it is dark. Dark current has no effect on a photocell used for ON/OFF operations. However, dark current may have to be taken into account in sensitive measuring circuits where small amounts of current distort the readings.

Photocell Applications. Photocells are used when time response is not critical. They would not be used where several thousand responses per second are needed to transmit accurate data. However, they can be used efficiently with slower responding electromechanical equipment such as pilot lights or street lights.

A photocell is used to determine whether the pilot light (flame) of a gas furnace is ON or OFF. When a pilot light is present, the light from the flame reduces the resistance of the photocell. Current is allowed to pass through the cell and activate a control relay. The control relay in turn allows the main gas valve to be energized when the thermostat calls for heat. **See Figure 15-27.**

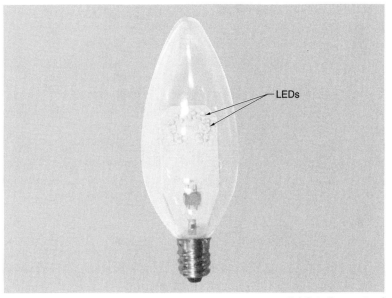

PolyBrite International
LEDs may be used for lighting applications such as energy-efficient candlelabra light bulbs.

A photocell is also used to determine whether it is day or night and to turn ON or OFF a street light accordingly. An increase of light at the photocell results in a decrease in resistance, and current is increased through the relay. The increased current in the relay causes the normally closed (NC) contacts to open, and the light turns OFF. With darkness, the resistance increases, causing the NC contacts to return to their original position and turn the light ON. **See Figure 15-28.**

386 SOLID STATE DEVICES AND SYSTEMS

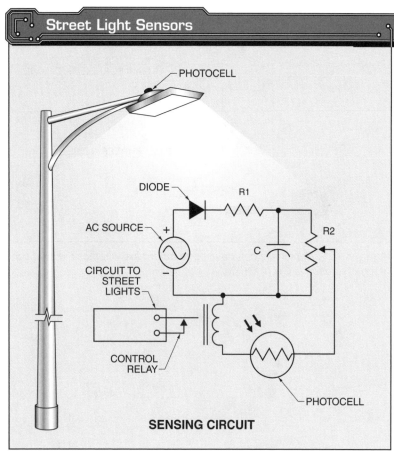

Figure 15-28. *Photocells can be used to control street lights.*

Advanced Assembly Automation, Inc.
Photoelectric sensors are used in manufacturing and assembly systems to control the positioning of products.

Testing Photocells. Humidity and contamination are the primary causes of photocell failure. The use of quality components that are hermetically sealed is essential for the long life and proper operation of photocells. Some plastic units are less rugged and more susceptible to temperature changes than glass units. The resistance of a photocell is tested with a digital multimeter (DMM) set to measure resistance, but the photocell must be disconnected from the circuitry. To test a photocell, apply the following procedure:

1. Connect the DMM leads to the photocell.
2. Cover the photocell and record its dark resistance.
3. Shine a light on the photocell and record its light resistance.
4. Compare these resistance readings with the manufacturer's specification sheets. If specification sheets are not available, use a similar photocell that is known to be good.

Note: As with other resistance-type sensors, the connection to the control circuit is very important. All connections should be tight and corrosion-free.

Bridge Circuits with Photocells. Resistive devices can be arranged to form bridge circuits. **See Figure 15-29.** More precise measurements and control can be obtained with a bridge circuit than a series or parallel circuit. A photocell is placed in one arm of the bridge so that it can detect changes in light. In this case, resistor R3 represents the sensing device. The bridge current is then balanced through the use of variable resistor R4. Resistors R1 and R2 are fixed-value precision resistors.

When the bridge is balanced, there is no current across the bridge circuit. Because there is no current, relay coil K1 is de-energized. Any changes in light or temperature changes the resistance of resistor R3 and the bridge becomes unbalanced. With the bridge unbalanced, current flows through relay coil K1 and the relay becomes energized. If a galvanometer is substituted for the relay coil of the bridge circuit, the circuit becomes an extremely sensitive light meter. **See Figure 15-30.**

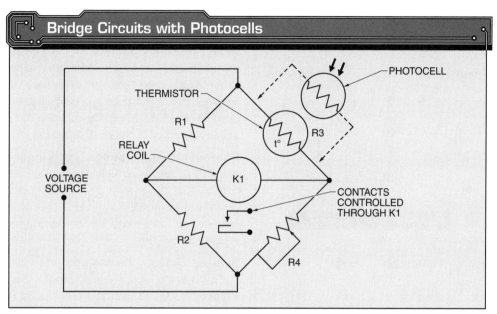

Figure 15-29. *Photocells can be used in a bridge circuit for precise light measurements.*

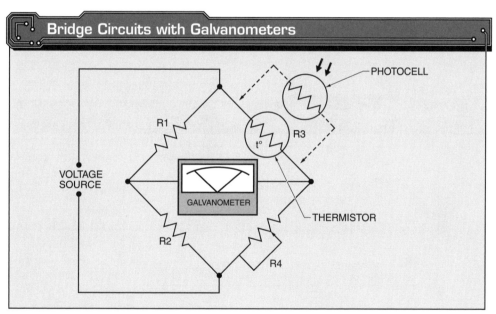

Figure 15-30. *A galvanometer can be substituted for the relay coil of a bridge, which creates an extremely sensitive light meter.*

In 1961, the helium-neon laser was the first gas laser invented at Bell Laboratories. Semiconductor diode lasers, which are by far the most common lasers, were invented around 1962. However, the diode lasers were unreliable until technical advances in the early 1980s dramatically lengthened their expected life.

Photodiodes

A *photodiode,* also known as a photoconductive diode, is a semiconductor diode that allows current to flow when exposed to light. Internally, a photoconductive diode is similar to a regular diode. The primary difference is the addition of a lens in the housing for focusing light on the PN junction area. **See Figure 15-31.**

Photodiodes

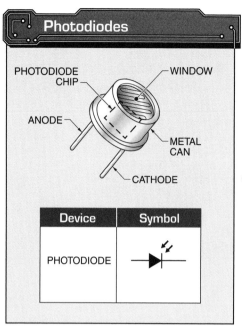

Figure 15-31. *A photodiode consists of a photodiode chip enclosed in a container with a window, which allows light to strike the PN junction and create current flow.*

With a photodiode, the conductive properties change when light strikes the surface of the PN junction. Without light, the resistance of the photodiode is high. When exposed to light, the resistance of the photodiode is reduced proportionately. **See Figure 15-32.** *Note:* The photodiode is usually operated in the reverse-bias mode.

Photodiode Applications. Photodiodes respond much faster than photocells, and they are usually more rugged. Photodiodes are found in movie equipment, conveyor systems, and other equipment requiring a rapid response time.

Sensing Devices. Photodiodes are used for a variety of sensing applications. **See Figure 15-33.** They are used to count, measure, maintain levels, and read coded marks for initiating operations in an automated process. Photodiodes can also be used to optically isolate circuits from noise and interference.

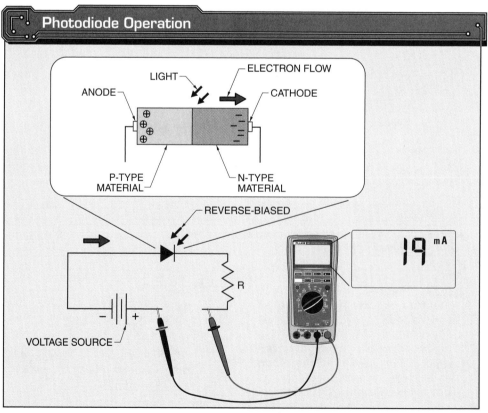

Figure 15-32. *Photodiodes are normally operated in the reverse-bias mode.*

Chapter 15—Photonics

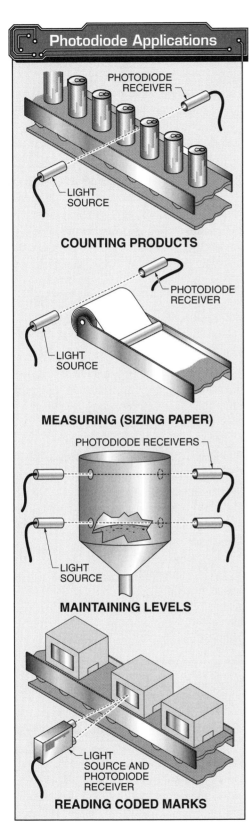

Figure 15-33. *Photodiodes can be used to count, measure, maintain levels, and read coded marks.*

PIN Photodiodes

A *PIN photodiode* is a photodiode with an intrinsic region between the P-type and N-type material. The intrinsic region is an undoped, semiconductor material. **See Figure 15-34.** The operation of a PIN photodiode is based on the principle that light, when exposed to a PIN junction, momentarily disturbs the structure of the PIN junction. The disturbance is due to a hole created when a high-energy photon strikes the PIN junction and causes an electron to be ejected from the junction. Therefore, light creates electron-hole pairs, which act as current carriers.

When connected in a circuit, the PIN photodiode is placed in series with the bias voltage and connected in reverse bias. A PIN photodiode is mounted to the base of a transistor-type header and sealed with a plane glass or plastic window. It is then capped to form a hermetically sealed package. **See Figure 15-35.**

PIN Photodiode Alignment. For the best operation of a PIN photodiode, light must be focused on the PIN junction. Current produced is the result of the collection, at the PIN junction, of the minority carriers freed near the junction by the photoelectric process. The electrical field in the crystal is negligible except near the PIN junction. Most of the carriers reach the junction by diffusion. The farther away from the junction that the minority carriers are freed, the less chance they have of being collected. Therefore, the response of the PIN photodiode decreases on either side of the junction. In some applications, a lens is provided to focus the light on the sensitive area of the PIN photodiode.

Response of PIN Photodiodes. The absorption of light in silicon decreases with increasing radiation wavelength. Therefore, as the radiation wavelength decreases, a larger percentage of the electron-hole pairs are created closer to the silicon surface. This results in the PIN photodiode exhibiting a peak-response point at some radiation wavelength. At this wavelength, a maximum number of electron-hole pairs are created near the collector-base junction.

390 SOLID STATE DEVICES AND SYSTEMS

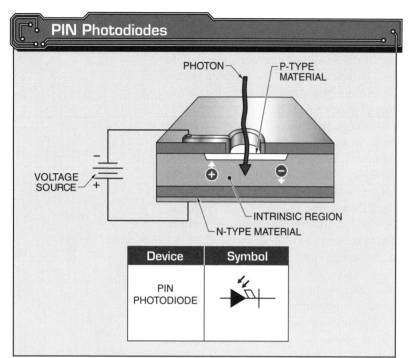

Figure 15-34. A PIN photodiode consists of a P-type and N-type material with an intrinsic region between them.

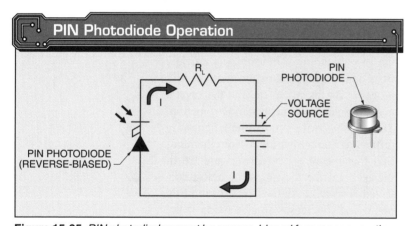

Figure 15-35. PIN photodiodes must be reverse-biased for proper operation.

The distance from the earth to the moon (about 384,000 km) can be measured to an accuracy of a few millimeters by timing the round trip of a laser pulse. The light is reflected off mirrors left by Apollo astronauts. The return signal is so weak that only a few photons are detected in each pulse. These experiments are used to test Einstein's theory of relativity.

Advantages and Disadvantages of PIN Photodiodes. PIN photodiodes can respond quickly to changes in light intensity. They are very useful for applications in which light changes at a rapid rate. The major disadvantage is that the output photocurrent is relatively low in relation to other photoconductive devices.

Compared to an ordinary PN photodiode, a PIN photodiode has a thicker depletion region, which allows more efficient collection of the carriers, and thus greater quantum efficiency. PIN diodes also have a lower capacitance and higher detection bandwidth. PIN photodiodes are small and efficient. Because of their low operating voltages, low power consumption, low noise, and simple circuitry, PIN photodiodes have the broadest application range.

Photoelectric Switches

A *photoelectric switch* is a solid state transmitter/receiver that can detect the presence of an object without touching the object. A photoelectric switch detects the presence of an object by means of a beam of light. Photoelectric switches can detect most materials and have a longer sensing distance than proximity sensors. Depending on the model, photoelectric switches can detect objects from several millimeters to over 100′ away. **See Figure 15-36.**

The maximum sensing distance of any switch is determined by the size, shape, color, and character of the surface of the object to be detected. Many switches include an adjustable sensing distance, which makes it possible to exclude detection of the background of the object. Photoelectric switches are used in applications in which the object to be detected cannot be touched or is excessively light, heavy, or hot.

Phototransistors. A *phototransistor* is a bipolar transistor with a transparent case that allows light to strike its base-collector junction. A phototransistor combines the effect of the photodiode and the switching capability of a transistor. Electrically, the phototransistor, when connected in a circuit, is placed in series with the bias voltage so that it is forward-biased. **See Figure 15-37.**

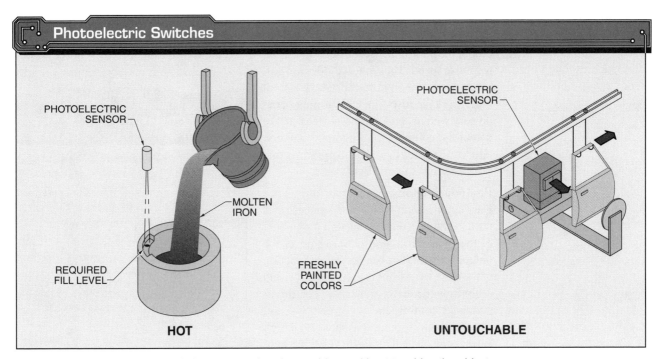

Figure 15-36. *Photoelectric switches are used to detect objects without touching the object.*

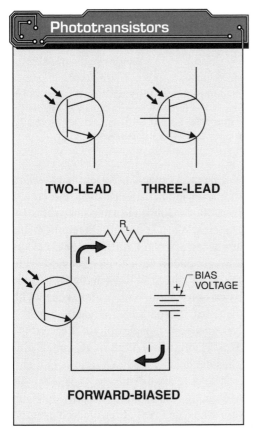

Figure 15-37. *A phototransistor combines the effect of the photodiode and the switching capability of a transistor.*

With a two-lead phototransistor, the base lead is replaced by a transparent case. This case allows light to fall on the base region. Light falling on the base region causes current to flow between the emitter and collector. The collector-base junction is enlarged and works as a reverse-biased photodiode controlling the phototransistor.

The phototransistor conducts more or less current depending on the light intensity. If light intensity increases, resistance decreases, and more emitter-to-base current is created. Although the base current is relatively small, the amplifying capability of the small base current is used to control the larger emitter-to-collector current. The collector current depends on the light intensity and the DC current gain of the phototransistor. In darkness, the phototransistor is switched OFF with the remaining leakage current. This remaining leakage current is called collector dark current.

Phototransistors have an advantage over photodiodes in that they have a sensitivity approximately 50 to 500 times greater. Two distinct disadvantages are that phototransistors are sensitive to temperature changes and protection against moisture is required.

Although the phototransistor can produce a higher output current than a photodiode, the phototransistor loses some of its response speed. Thus, photodiodes are still often used when phototransistors cannot meet the response time. Phototransistors do very well in applications, such as smoke detectors, counters, photographic meters, and mechanical positioning systems, where speed is not critical.

Photodarlingtons. A *photodarlington* is a Darlington circuit with a phototransistor as the driver transistor. The operating principle of the photodarlington is the same as the phototransistor. The base-collector junction of the driver transistor is light-sensitive, and it controls the driver transistor. The driver transistor controls the follower transistor. The Darlington configuration yields a high current gain, which results in very high light sensitivity. **See Figure 15-38.**

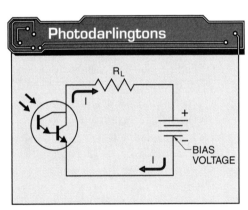

Figure 15-38. *Photodarlingtons consist of Darlington circuits with phototransistors as the driver transistors.*

Photodarlington circuits are usually formed simultaneously and packaged as single devices. Although a photodarlington is more sensitive than a photodiode or phototransistor, it has a slower response to changes in light intensity. Photodarlingtons are useful for detecting a low level of light.

Light-Activated SCRs. A *light-activated SCR (LASCR)* is an SCR that can be triggered by light. The schematic of an LASCR is identical to the schematic of a regular SCR. The only difference is that arrows are added in the LASCR schematic to indicate a light-sensitive device. **See Figure 15-39.**

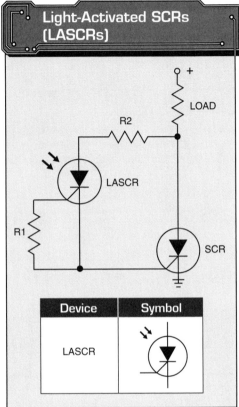

Figure 15-39. *An LASCR can be used as a trigger circuit for conventional SCRs.*

Like a photodiode, current is of a very low level in an LASCR. Even the largest LASCRs are limited to a maximum of a few amperes. When larger current requirements are necessary, the LASCR can be used as a trigger circuit for a conventional SCR. In the circuit, the SCR is normally OFF, since its gate circuit can be considered open with the LASCR in darkness. When the LASCR is triggered by a pulse of light, it turns ON. This in turn triggers the SCR, which supplies the heavier load current.

The primary advantage of an LASCR over an SCR is its ability to provide isolation. Since an LASCR is triggered by light, the LASCR provides complete isolation between the input signal and the output load current.

Phototriacs. A *phototriac* is a triac that can be triggered by light. The gate of the phototriac is light-sensitive. It triggers the triac at a specified light intensity. In darkness, the triac is not triggered. The remaining leakage current is called peak blocking current. Phototriacs are bilateral and designed to switch AC signals. **See Figure 15-40.**

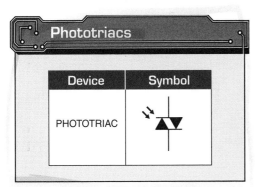

Figure 15-40. *Phototriacs are triacs that can be triggered by a specific light intensity.*

Optocouplers

An *optocoupler*, also known as an optoisolator, is an electronic device that electrically isolates switching circuits by coupling them with light. There are many situations in which information must be transmitted between switching circuits that are electrically isolated from each other. This isolation has been traditionally provided by relays, isolation transformers, and line drives and receivers. Optocouplers can be used effectively to solve the same problems that traditional devices have. The optocoupler is mostly used in applications where high voltage, noise isolation, and small size are considerations. By coupling two systems together with light, the need for a common ground is eliminated. **See Figure 15-41.**

Optocoupler Construction. Optocouplers are usually constructed as dual in-line packages. An optocoupler consists of an infrared-emitting diode (IRED) as the input stage and a silicon phototransistor as the output stage. Internally, the optocoupler uses a glass dielectric sandwich to separate the input from the output. This provides a one-way transfer of electrical signals from the IRED to the phototransistor without an electrical connection between the circuitry containing the devices.

Photons emitted from the IRED have wavelengths of about 900 nm. The phototransistor responds effectively to photons with this same wavelength. Therefore, input and output devices are always spectrally matched for maximum transfer characteristics. The signal cannot go back in the opposite direction because the emitters and detectors cannot reverse their operating functions.

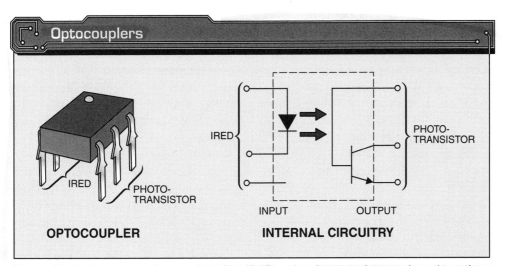

Figure 15-41. *An optocoupler consists of an IRED and a phototransistor packaged together.*

Optocoupler Isolation Voltage. The primary function of an optocoupler is to provide electrical separation between the input and output, especially in the presence of high voltages. An *isolation surge voltage rating* is a measure of the ability of an optocoupler to withstand a spike or surge in voltage and maintain electrical isolation. It is a measure of the integrity of the package and the dielectric strength of the insulating material. The amount of electrical isolation between the two devices is controlled by the material in the light path and the physical distance between the emitter and detector.

Although the DIP package is the most common for optocouplers, other packages are available to provide higher isolation surge voltage and other special requirements. For very high isolation surge voltage requirements (10 kV to 50 kV), an interrupter module can be modified at a very low cost. This modification is done by putting a suitable dielectric (glass, acrylic, or silicone) in the air gap and by insulating and encapsulating the lead wires.

Note: Isolation surge voltage may be expressed as a steady-state isolation voltage. Steady-state isolation voltage ratings are the ratings for working voltages that are continuously applied to the device. Isolation surge voltage is intended for short periods only and is usually higher than steady-state isolation voltage. Isolation surge voltage may be 1200 V peak, whereas steady-state isolation voltage may be 900 V peak.

Optical Interrupters. An optical interrupter is used to eliminate the mechanical positioning problem encountered when adjusting an emitter and detector for proper sensing. **See Figure 15-42.** These units are constructed so that the input and output are set as a coupled pair. All alignment and distance problems have already been taken into account. A technician only needs to apply the proper power source to the device. The schematic of an optical interrupter is similar to that of the optocoupler. These types of sensors are in the same category as precision mechanical limit switches. The activating mechanism blocks or reflects light instead of applying force.

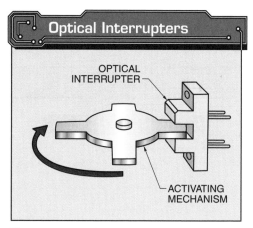

Figure 15-42. *Optical interrupters are activated by blocked or reflected light rather than applied force.*

Reflector Modules. Reflector modules are used when only one surface of an object is readily accessible. An example of such an object is an encoder wheel. An encoder wheel is a disk with alternating reflective and dark segments arranged in tracks. By bouncing a light source against the surface, light returns when a reflective segment is present. **See Figure 15-43.**

With a reflector module, the emitter and detector are mounted side by side and aimed so that the light sources converge at a point just beyond the surface of the module. When the reflective segment of the encoder wheel passes by the emitter, the detector receives a return light source. When the dark segment of the encoder wheel passes by the emitter, the detector will not receive a return light source.

Reflector modules are used in robotics to help control robot arms. Reflector modules are also used in printers, plotters, tape drives, positioning tables, and automatic handlers. A reflector module consists of a small plastic or metal case that can be mounted in a variety of positions.

Chapter 15—Photonics 395

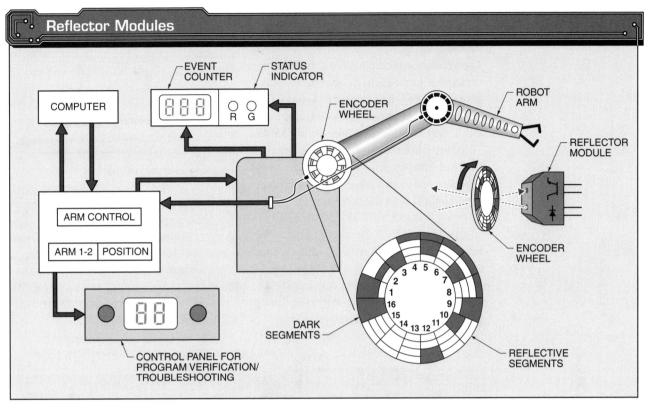

Figure 15-43. *Reflector modules can be used in robotics to help control robot arms.*

Bar Code Scanners

Most cash registers typically identify a product sold by scanning a bar code printed on the product label or packaging. A bar code is a series of alternating vertical black bars and white spaces. Data is encoded by varying the width of these bars and spaces. To retrieve data, a scanner is moved across the bar code.

A *bar code scanner* is a self-contained unit that has both a light source and a light detector for reading a bar code. **See Figure 15-44.** The light source is projected through an opening in the scanner. The beam of light strikes the bar code and is reflected back into the scanner and to the light detector.

Lasers were invented in 1960 and referred to as "a solution looking for a problem" in their early stage of development.

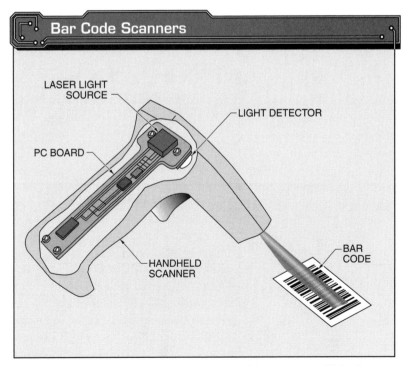

Figure 15-44. *A bar code scanner contains a light source and light detector and is used to read bar codes.*

Photo Tachometers

A *photo tachometer* is an electronic device that uses light to measure rotational speed. The reflective optical coupler contained in the tachometer probe consists of an IRED and a photodarlington. Resistor R1 is used to limit diode current, and resistor R2 is the load resistor for the phototransistor. **See Figure 15-45.**

When light emitted by the LED is reflected back to the phototransistor, the phototransistor conducts, and the collector voltage decreases. When light is not reflected back, the phototransistor does not conduct, and the collector voltage remains high. The output pulse of the tachometer probe frequency is proportional to the RPMs of a motor. The output can be viewed on an oscilloscope, measured on a frequency counter, or converted to an analog RPM reading with associated circuitry.

FIBER OPTICS

Fiber optics is the science of using light to transmit data from one location to another using optical fiber. Optical fiber is made of thin flexible glass or plastic that can easily transmit light. Fiber optics is most commonly used as a transmission link.

As a link, it connects two electronic circuits consisting of a transmitter and a receiver. The central part of the transmitter is the source. The source consists of an LED, IRED, or laser diode, which changes electrical signals into light signals. The receiver usually contains a photodiode that converts light back into electrical signals. The receiver output circuit also amplifies the signal and produces the desired results, such as voice transmission or video signals. **See Figure 15-46.**

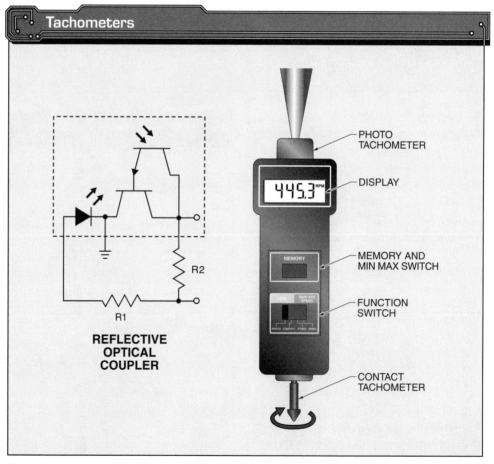

Figure 15-45. *A photo tachometer contains a reflective optical coupler that consists of an IRED and a photodarlington transistor.*

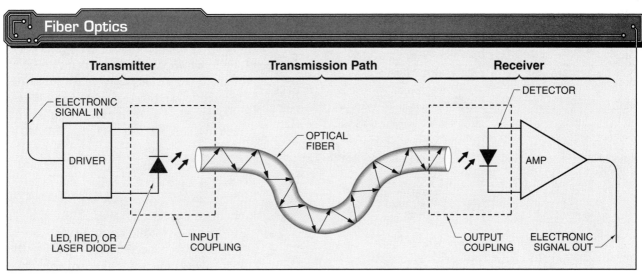

Figure 15-46. *Fiber-optic technology is used as a transmission link to connect two electronic circuits consisting of a transmitter and a receiver.*

> *Light is electromagnetic radiation that is visible to the human eye and is responsible for sight. In physics, light refers to the electromagnetic radiation of any wavelength, regardless of whether it is visible.*

Advantages of Fiber Optics

Fiber-optic technology offers many advantages for transmitting signals when compared to individual wire, twisted pairs of wires, and coaxial cable. Some advantages of fiber optics are the following:

- large bandwidth—Since the carrier wave for fiber-optic signals is light, fibers have bandwidths approaching 2 GHz/kM, which allows for the high-speed transfer of data. This is many times higher than the highest radio frequencies.
- low loss (attenuation)—Fibers can provide lower attenuation (loss) than copper wire. In addition, fibers do not react to changes in frequency. High frequency can attenuate signals in metal conductors.
- electromagnetic interference (EMI) immunity—EMI is the signal created in a conductor due to the presence of electromagnetic fields. Because fiber is an insulator, it is not affected by magnetic fields and the resulting EMI.
- small size—Fiber-optic cable is smaller than copper cable. A small fiber-optic cable has the same information-carrying capacity as a 900-pair copper cable. **See Figure 15-47.** Therefore, it is possible to use small conduits for optic fiber. It is also possible to increase the capacity of existing conduits. The latter is important when considering overcrowded conduits running under city streets.
- light weight—Glass and plastic weigh considerably less than copper.
- security—It is virtually impossible to tap fiber-optic cable without the tapping being detected. Any attempt to enter the cable will affect the quality of transmission. Also, fibers do not radiate energy, and conventional eavesdropping techniques do not work.
- safety and electrical isolation—The insulating property of fibers isolates the fibers electrically. In addition, there is no spark hazard, and it can be used in flammable environments and other hazardous locations where electrical codes might prohibit the use of other methods.

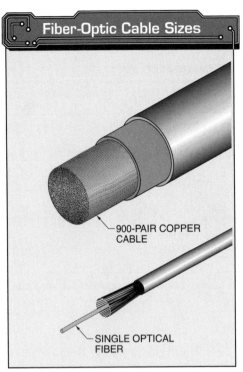

Figure 15-47. *A single optical fiber has the same information-carrying capacity as a 900-pair copper cable.*

Optical Fibers

Optical fiber consists of a core, cladding, and a protective jacket. The core is the actual path for light. Although the core occasionally is constructed of plastic, it is typically made of glass. A cladding layer, usually of glass or plastic, is bonded to the core. The cladding is enclosed in a jacket for protection. **See Figure 15-48.**

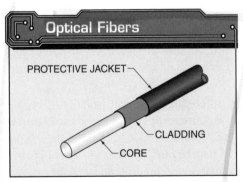

Figure 15-48. *Optical fibers consist of a core, cladding, and a protective jacket.*

A basic optical fiber forms two concentric layers with the core and cladding. The core has a higher refractive index than the cladding. *Refractive index* is the speed of light through a substance compared to that of free space. Since the core has a higher refractive index, light moves more slowly through the core than through the cladding.

Any light injected into a core that strikes the surface at an angle greater than the critical angle is reflected back into the core. Since the angles of incidence and reflection are equal, the ray of light continues to zigzag down the length of the core by total internal reflection. **See Figure 15-49.** The light is trapped in the core. Light that strikes the core surface at less than the critical angle passes into the cladding and is lost. The total amount of light moving through the fiber is determined by fiber size, fiber construction, and the nature of the light injected.

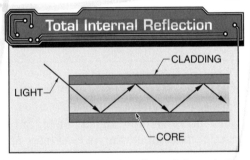

Figure 15-49. *Light is reflected through the fiber core by total internal reflection.*

Attenuation

Attenuation is the loss of signal strength during transmission. During transmission, light pulses lose some of their photons. The loss of photons reduces the amplitude of the light pulses. Attenuation for optical fiber is usually specified in decibels per kilometer (dB/km). For commercially available optical fibers, attenuation ranges from 1 dB/km for premium small-core glass fibers to over 2000 dB/km for large-core plastic fibers. Attenuation can be caused by impurities or imperfections in fiber-optic cables. **See Figure 15-50.**

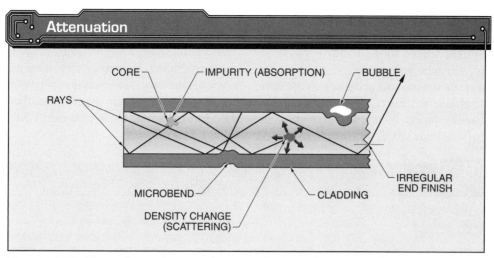

Figure 15-50. *Attenuation, or loss, can be caused by impurities or imperfections in fiber-optic cables.*

Absorption. Glass fibers are made of ultrapure glass that has purity exceeding that of semiconductors. If seawater were as pure as fiberglass, the bottom of the deepest part of the ocean could be seen from the surface. Some impurities still remain as residues in glass and other dopants that are added purposely to obtain certain optical qualities. Unfortunately, impurities in glass absorb light energy, turning photons into heat.

Scattering. Scattering results from imperfections in fibers and from the basic structure of fiber. Unintentional variations in the density and geometry of fiber occur during manufacturing and cabling. Small variations in core diameter, microbends, and irregularities (bubbles) in the core-to-cladding interface cause loss. The angle of incidence of rays striking such variations significantly changes some of the rays. They are refracted onto new paths and are not subject to total internal reflection.

Wavelength. Fiber attenuation is closely related to wavelength. Fiber attenuation requires a careful balance between light sources and fibers. Manufacturers specify the attenuation of fiber at a certain wavelength. Most manufacturers also provide a curve showing attenuation as a function of wavelength. Fiber attenuation data should be used when matching a fiber to a light source.

Numerical Aperture. The *numerical aperture* is a measure of the light gathering ability of a fiber. Unless light is placed into the fiber at the proper angle, light will be lost. Light must enter and strike the cladding at an angle less than the critical angle. **See Figure 15-51.** Any light striking within this area travels through the fiber. Light outside the area strikes the cladding at more than the critical angle and is lost. Loss created by misalignment is always a factor to consider.

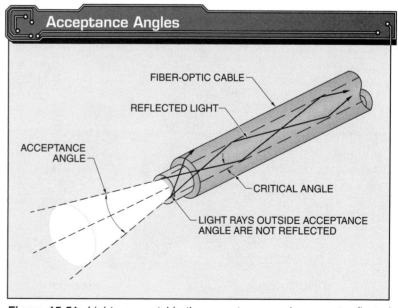

Figure 15-51. *Light rays outside the acceptance angle are not reflected and are lost.*

Fiber-Optic Cable Types

Fiber-optic cable consists of two or more optical fibers formed into a cable for convenience and protection. Coatings are used on the inside and outside of the cable to protect the fibers. The outer protective coating jacket is made of polyethylene, polyurethane, PVC, or Tefzel® to protect the cable from the environment. The inner jacket or jackets protect the individual fibers from dirt, moisture, abrasions, compression, and temperature variations. Classified by assembly method, the common types of fiber-optic cable include simplex, distribution, breakout, and loose tube cable. **See Figure 15-52.**

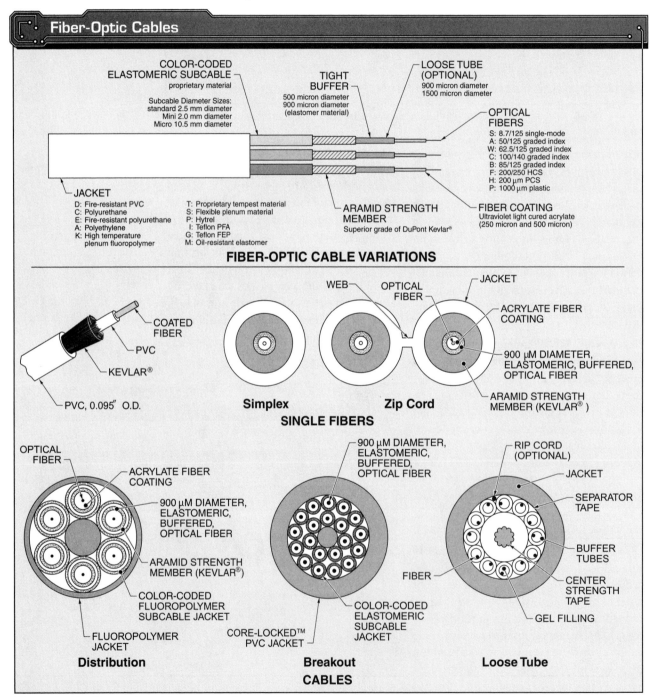

Figure 15-52. *Fiber-optic cables are manufactured in several types, depending on the operational requirements.*

All fiber-optic cables have a tension member that provides additional strength to the cable so that it may be safely pulled without damage. The tension member minimizes or eliminates any stretching force applied to the optical fiber. The tension member is commonly made of stranded steel wire, braided Kevlar®, aramid yarn, or a fiberglass compound. Simplex fiber-optic cable has one optic fiber that is coated and jacketed. When two simplex optical fibers are joined by a web, the resulting cable is called zip cord. The two optic fibers of zip cord are individually coated and jacketed and then joined.

Distribution cable contains more than one double-buffered (protected) fiber bundle under the same jacket. The jacket is made of a high-strength material and is often reinforced with fiberglass. Breakout cable is composed of several simplex fibers bundled together with reinforcement from a fiberglass rod. Loose tube cable is composed of several fibers brought together inside a small tube that is wound around a central tension member and jacketed. This cable is compact and has a high fiber count. All cables must carry identification markings and ratings specified by the National Electrical Code®. **See Figure 15-53.**

In 1876, Alexander Graham Bell invented the telephone. Four years later, he invented the photophone, a device that could transmit speech via a light beam. The light used in most fiber-optic communication systems is in the infrared portion of the spectrum.

Fiber-Optic Cable Couplers

There is little an electrician can do about the design of coupling materials. However, proper installation procedures should be understood and followed. Splices and fiber interconnections are often more of a negative factor than poor quality materials because of the alignment problems that can arise. **See Figure 15-54.** These problems can be eliminated by properly installing fiber splices, connectors, and couplers.

Fiber-optic splices and connectors are used for point-to-point applications that transport light from the source to the detector. Couplers allow light to be distributed among several fibers. Couplers are desirable in bus-structure applications or distributed networks. The two main types of couplers are the T-coupler and star coupler.

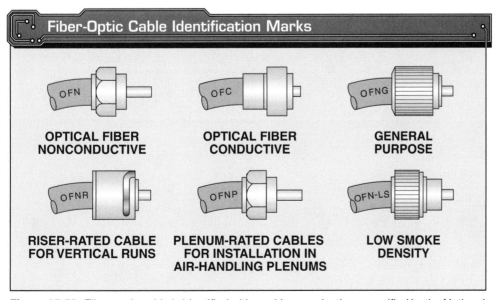

Figure 15-53. *Fiber-optic cable is identified with markings and ratings specified by the National Electrical Code®.*

Fiber-Optic Cable Alignment

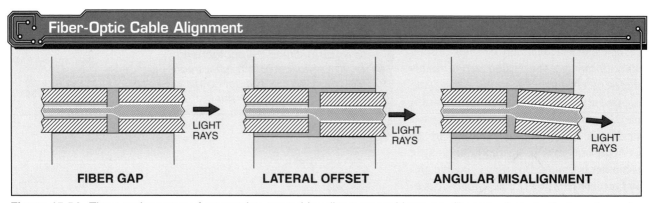

Figure 15-54. *Three major areas of attenuation caused by alignment problems are fiber gap, lateral offset, and angular misalignment.*

T-Couplers. A *T-coupler* is an in-line fiber-optic coupling device used for tapping a main fiber-optic bus. Lenses and a beam splitter can be used to tap the light. **See Figure 15-55.** A *beam splitter* is a coated plate that reflects a portion of light and transmits the rest. The structure of the beam splitter determines the reflection-to-transmission ratio. Loss of optical signal power increases linearly with the number of couplers in series with one another. There is not only an insertion loss associated with each coupler but also a division of light for each.

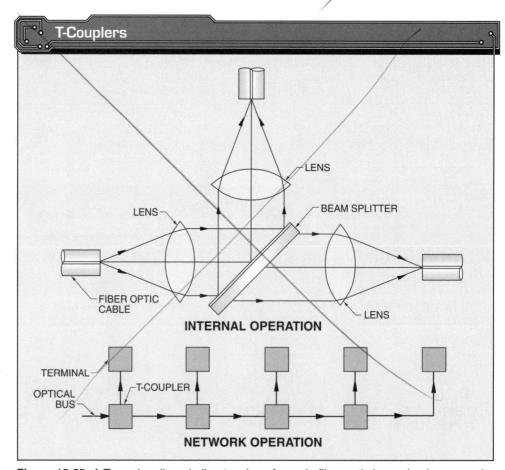

Figure 15-55. *A T-coupler allows in-line tapping of a main fiber-optic bus using lenses and a beam splitter.*

Star Couplers. A *star coupler* is a fiber-optic coupling device that uses a mirror to reflect light from one fiber into several fibers. Star couplers offer parallel arrangements. **See Figure 15-56.** As transmitted light enters a star coupler through its input ports, the light spreads out and strikes a mirror that reflects it into the output ports. The advantage of a star coupler over a T-coupler is that there is only one coupling loss associated with dividing light from one fiber into several fibers. In addition, as the number of output ports increases, the loss from beam splitting increases only slightly in relation to the number of ports.

Fiber-Optic Cable Installation

Fiber-optic cable and related components are becoming priced comparably to CAT V traditional copper wire installations. Fiber-optic cable is becoming increasingly easier to work with, and thus installation costs are becoming less. The pulling costs for fiber-optic cable and CAT V copper wire are now virtually the same. Termination and splicing techniques have also become easier and more reliable. With proper installation, fiber-optic cable is easier to maintain and more reliable. Network downtime and glitches can be expensive. The cable system that minimizes this cost can be less expensive in the long term.

Low-power lasers with only several milliwatts (mW) of output power can be hazardous to human eyesight when the laser beam hits the eyes directly or is reflected from a shiny surface.

To ensure that the fiber-optic cables are in proper condition for installation, a receiving inspection and a preinstallation test should be performed. During the receiving inspection, it is determined whether the manufacturer shipped what was ordered. The fiber-optic cables should be checked for breaks, correct cable length, and flaws. During the preinstallation check, the reels of the fiber-optic cables are checked for broken fibers and microbends, attenuation is measured, and the bandwidth characteristics are checked.

When installing fiber-optic cable, certain precautions should be taken to prevent cable damage. Working around heating ducts should be avoided since they carry airborne particles. Dust caps should always be kept on connectors and any associated equipment that is going to be used in the connection process. Lint-free pads and isopropyl alcohol should be used to clean connectors that may have been exposed to dust or foreign particles. **See Figure 15-57.**

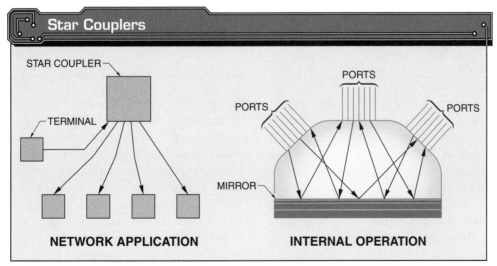

Figure 15-56. *A star coupler uses a mirror to reflect light from one fiber into several fibers.*

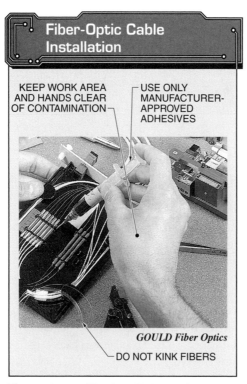

Figure 15-57. *The installation of fiber-optic cable requires cleanliness and taking precautions at all times.*

Pulling Fiber-Optic Cable. Fiber-optic cable can be pulled with much greater force than copper wire if manufacturer installation guidelines are followed. On long runs, lubricants should be used per the manufacturer's specifications. The use of an automated puller with tension control is recommended to ensure consistency on pulling tasks.

Fiber-optic cable is the strongest when it is pulled in a straight line. Breaks occur when the cable is forced into an excessively tight radius. Fiber-optic cable manufacturers normally specify the minimum bend radius for their product. Most manufacturers recommend a minimum bend radius that is 20 times the diameter of the cable. Allowing kinks or excessively tight radiuses to occur in cable can cause damage (immediate damage or damage becoming obvious only later) that will require cable replacement.

Fiber-optic cable should be rolled forward from a spool rather than being pulled off the end of the spool. Twisting cable can also put stress on the fibers. The use of a swivel-pulling eye reduces tension and twisting forces on the cable.

A fiber-optic cable has greater tensile strength than a copper or steel wire of the same diameter. Fiber-optic cables can withstand pulling forces of 600 lb, but pulling forces should not exceed 150 lb to 200 lb.

Splices. Splices are used primarily to extend a run of cable or restore a run interrupted by activities such as digging up buried cable, which results in separation. Fiber-optic cable is typically manufactured in lengths of up to 6.2 mi (10 km) without splices. Splices are intended to be permanent connections between two fiber-optic cables.

Regardless of the splicing method used, quality control measures must be followed during the cable-end preparation, protection, and storage of fiber-optic cable. Strict guidelines must be followed so that bare glass is protected from dirt and handling during the splicing process. If the bare fiber is scratched by dirt or dust particles, a small defect will remain in the glass.

When spliced fiber-optic cables are stored in splice trays (enclosures), the bare fiber of cables must not be exposed. For protection, the fiber must be completely coated with a soft compound or silicone sealant. Splice enclosures are environmentally sealed, stand-alone units that are mounted within the cable right-of-way.

The two splicing methods are fusion and mechanical. The splicing method required is determined by the cost and location.

A *fusion splice* is a splice made by welding two ends of optical fibers together. A fusion splicer is used to make a fusion splice. A fusion splicer precisely joins optical fiber by fusion using heat and pressure. The optical fiber ends to be joined are prepared and then secured using holders. The optical fiber cores are aligned into the proper position using a light source, microprocessor, and number of cameras. The light source directs light toward the cameras. The path of the light is obstructed by the optical fiber cores and is sensed by the cameras. A microprocessor analyzes location data from the cameras and then controls the movement of the holders. The alignment of the cores is also displayed on the camera screens for the operator to view.

> A single optical fiber the thickness of human hair can theoretically carry the equivalent of 300 million simultaneous telephone calls. Light channeled down optical fiber undergoes approximately 6000 "reflections" per meter.

After proper alignment, an electric arc formed between two electrodes heats the core ends to remove impurities. The aligned core ends are then joined with heat and pressure. Heat from another arc between the electrodes melts the cores together to complete the fusion process. A fusion splicer can join optical fiber automatically or manually. The fusion splicer is commonly located in a separate clean room or in a trailer brought into the facility.

A *mechanical splice* is a splice made by connecting two ends of optical fibers together using an adhesive without fusion from heat. Mechanical splices use alignment devices that hold the ends of two fibers together while the adhesive sets to bond the two ends. The most commonly used mechanical splice connectors include single and multiple connectors. The hardware to make mechanical splices is relatively inexpensive, but the splices are expensive because of the time involved. **See Figure 15-58.**

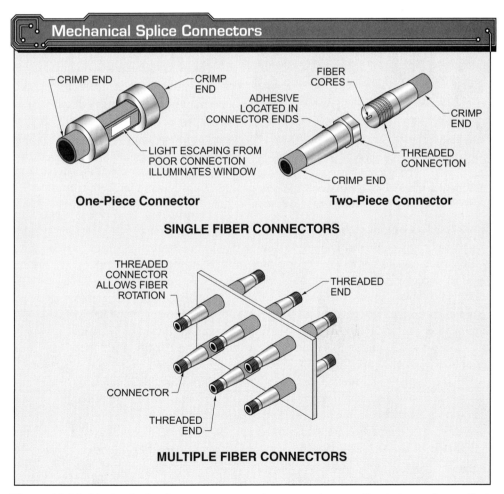

Figure 15-58. *Mechanical splices create a semipermanent connection between two optical fibers using adhesives and mechanical clamps.*

When attaching an optical connector, a technician or maintenance person strips the jacket from the optical fiber cable. The strength members are then trimmed and inserted into grommets on sleeves. Next, the prepared fiber is coated with an epoxy resin and inserted into a precision hole in a connector. The core of the fiber may protrude, depending on the type of connector. Some mechanical splicing connectors allow the fiber to be rotated to "tune" the fiber for minimum light loss.

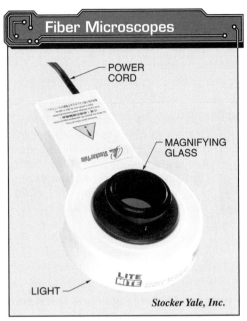

Figure 15-59. *Fiber microscopes are used to inspect fiber-optic cable and associated devices for flaws and foreign particles.*

Fluke Networks

Fiber-optic microscopes allow technicians to directly view the contaminants and faults that impair high-speed data transmissions.

Handling Fiber-Optic Cables. When working with fiber-optic systems, dirt must be carefully guarded against. Fiber microscopes are used to inspect connectors, terminators, and fiber-optic cables for any foreign particles. **See Figure 15-59.** Flaws in the manufacturing process and/or damage during installation can cause data transmission problems. Any fiber-optic cable may break in an area with a flaw while under stress. Even a small particle of dust or a random piece of Teflon® can scratch the surface of the fiber. Airborne particles are commonly silica-based and can scratch the surfaces of fiber-optic cable, causing a loss in data during data transfer. Even though fiber-optic cable has inherently low loss, light loss due to dust or scratches must still be minimized.

To prevent abrasions, manufacturers coat the fiber with a protective jacket immediately after the fiber is made. This jacket protects the surface of the fiber from moisture that can weaken the fiber-optic cable. The jacket also compensates for contraction and expansion caused by temperature variations.

Fiber-Optic Cable Safety. Fiber-optic systems can cause personal injury. Injury can occur from the lasers, LEDs, glass fibers, receiver power supplies, and materials and equipment used in the installation of fiber-optic cable. The electronic devices of fiber-optic systems commonly use less than 12 V. However, in some avalanche photodiode circuits, more than 300 V may be present. These voltages must be treated with extra care. Very high voltages are also present in fusion splicers and oscilloscopes. This equipment must be properly grounded.

In some field installation procedures, propanol is used as a lubricant. Adequate ventilation is required because the fumes from the propanol are flammable. Contact with acrylic adhesive, which is used in some cable splicing steps, should be avoided because it cures almost immediately upon contact with human skin. Prolonged skin contact with flame-retardant cables should also be avoided. The vapors of any solvents used to strip and clean fiber-optic cables, especially when in a confined space, should not be inhaled.

Glass optical fibers can puncture the skin and eyes. Safety glasses, gloves, and other personal protective equipment should be worn as required when handling exposed fibers and during fiber splicing, grinding, and polishing operations.

Fiber-optic cable can emit hazardous radiation as part of the fiber-optic system or part of the test equipment. Care should be taken to avoid viewing output flux (energy) using the magnification of a microscope or magnifier. It is quite dangerous to look at the end of a fiber-optic cable that is illuminated with visible or invisible light by a laser or an LED.

Troubleshooting (Testing)

A fiber-optic power meter or DMM with a power meter accessory can be used to test fiber-optic cable and fittings for any power loss. **See Figure 15-60.** With the DMM set to measure DC voltage (mV), light loss at a coupler can be detected by first taking a measurement at the cable before the coupler. With the DMM still set to measure DC voltage, a measurement is then taken at the coupler. Light loss is indicated by the difference between the two measurements.

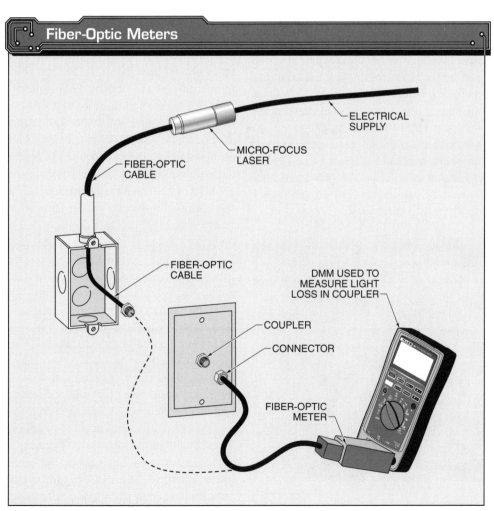

Figure 15-60. *The power loss at a fiber-optic coupler can be measured using a DMM and fiber-optic power meter accessory.*

When measuring power with a fiber-optic power meter, the meter must be set to the proper range (usually dB or 1 mW, but it can also be set for microwatts) and the proper wavelength of the signal. A fiber-optic cable has a metallic shield. This shield can be checked for shorts to ground by measuring resistance with an ohmmeter. If there is a short to ground, the cable is damaged and possibly the fiber as well.

Many fiber-optic power meters now use microprocessors and digital signal processing to enhance the performance of the instrument while simplifying its operation. When used with test software and a PC, the power meter can automatically set up tests, store data from the field, and provide powerful information for lab situations.

Fiber-Optic Cable Applications

The computer network is the backbone of data transmission in businesses, schools, and homes. Underestimating user requirements for bandwidth has been a common mistake in the computer and networking industry. High-speed, long-distance data transmission is the primary application of fiber-optic cable. Advances in video streaming, multimedia, teleconferencing, and other bandwidth applications require additional data transmission capacity.

Refer to Chapter 15 Quick Quiz® on CD-ROM

Voice-Data-Video Communications. Due to voice-data-video (VDV) communication, using means other than copper wire, the growth of fiber-optic cable and wireless communication has been sustained. **See Figure 15-61.** With modern communication, large amounts of data can be moved at one time. Entire volumes of printed material that once took years to manually copy may now be copied, moved over great distances, and reproduced in print or video all electronically. The ability to move large amounts of information at high speeds, without mistakes or file corruption, continues to be permitted by advances in technology. For example, using fiber optics to transmit VDV signals allows for much faster operating speeds than copper wire signal transmission.

The wireless transmission of VDV signals allows VDV communication from almost any location. Proper communication is ensured by testing newly installed cables and lines using VDV test instruments. After the cables and lines are operational, VDV test instruments are used to locate potential problems before data or communication is lost.

Also with modern communication, large amounts of data can be moved at faster speeds. Whereas one second was once considered a very short time period for almost any type of transmission, the transmission speeds of modern devices are in milliseconds (ms), microseconds (μs), and nanoseconds (ns). VDV transmission speed has advanced from kilohertz (kHz) to megahertz (MHz) and now to gigahertz (GHz).

Voice-data-video communication systems must be maintained to operate properly, prevent communication and storage mistakes, and prevent file corruption. To maintain these systems, most testing is performed using DMMs with fiber-optic test attachments. Other tests require fiber-optic microscopes, fiber-optic fault analyzers, or optical time domain reflectometers (OTDRs).

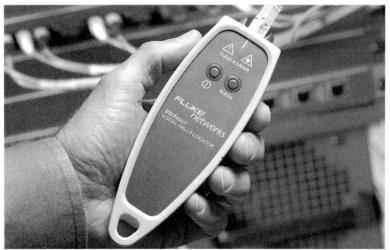

Fluke Networks

Simple fiber linking problems can be diagnosed with visual fault locators. Visual fault locators can locate fibers; verify continuity and polarity; and help find breaks in cables, connectors, and splices.

Chapter 15—Photonics 409

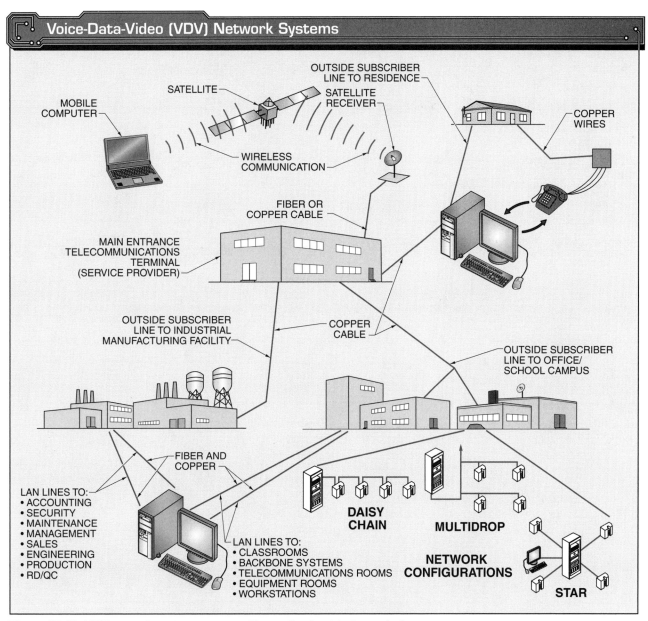

Figure 15-61. *VDV network systems may use fiber optics for data transmission.*

KEY TERMS

- photonics
- light
- refraction
- light-emitting diode (LED)
- organic light-emitting diode (OLED)
- liquid crystal display (LCD)
- plasma display panel (PDP)
- laser
- gas laser
- semiconductor laser
- light detector
- photocell
- photodiode
- PIN photodiode
- photoelectric switch
- phototransistor
- photodarlington
- light-activated SCR (LASCR)
- phototriac
- optocoupler
- bar code
- bar code scanner
- photo tachometer
- fiber optics
- attenuation
- T-coupler

REVIEW QUESTIONS

1. Describe photonics.
2. What is light?
3. Define refraction.
4. Explain light-emitting diodes (LEDs).
5. Describe organic light-emitting diodes (OLEDs).
6. What is a liquid crystal display (LCD)?
7. Explain lasers.
8. Define photocell.
9. Describe photodiodes.
10. What is a photoelectric switch?
11. Define phototransistor.
12. Describe light-activated SCRs.
13. Explain phototriacs.
14. What is an optocoupler?
15. Define fiber optics.
16. Explain voice-data-video (VDV) communication.

DIGITAL ELECTRONICS FUNDAMENTALS 16

Electronic devices use electronic components and circuits to produce sound, perform mathematical and logical functions, amplify signals, control operations, and increase operating speed. Most electronic systems use solid state components that have no moving parts. Solid state components, such as integrated circuits (ICs), led to higher quality and faster computers as well as higher quality electronic circuits in televisions, microwaves, telephones, and almost all other devices that use electronic circuitry.

OBJECTIVES

- Explain integrated circuits.
- Describe truth tables.
- List and describe logic gates.
- Describe logic families.
- Explain digital pulses.
- Describe the troubleshooting process for digital circuits.

DIGITAL ELECTRONICS

Electronic circuits that process quickly changing pulses are the basis for digital logic. Electronic systems are either analog or digital. An *analog signal* is the output of circuits with continuously changing quantities that may have any value between the defined limits. Quantities such as temperature, speed, voltage, current, and frequency are common analog signals. Many of the sensors required to monitor industrial processes are analog. For example, a heat sensor may change resistance gradually as the temperature rises and falls. A *digital signal* is the output of specific quantities that change in discrete increments, such as seconds, minutes, and hours. Analog signals are continually changing signals. Digital signals are either ON or OFF. **See Figure 16-1.**

The analog process can be compared to a light dimmer switch, whereas the digital process can be compared to a standard light switch. A light dimmer switch varies the intensity of light from fully ON to fully OFF over a range from dim to bright. However, a standard light switch has only two positions. Similar to a digital process, it is either fully ON or fully OFF. **See Figure 16-2.**

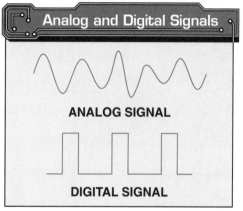

Figure 16-1. *Analog signals are continually changing signals. However, digital signals are abrupt and usually are either ON or OFF.*

INTEGRATED CIRCUITS

An *integrated circuit (IC)*, also known as a chip, is an electronic device in which all components (transistors, diodes, and resistors) are contained in a single package. An IC contains many circuits that are microscopically photoetched into layers on a silicon base. ICs are used to process and store information in computers, calculators, programmable logic controllers (PLCs), digital watches, and many other electronic devices. The two basic types of ICs are digital and linear. **See Figure 16-3.**

411

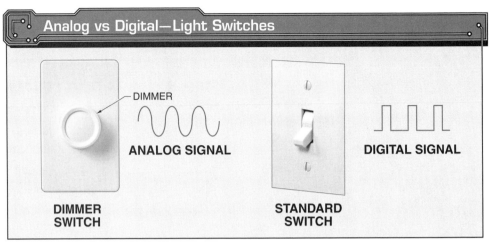

Figure 16-2. *The analog and digital processes can be exemplified with a simple comparison between a light dimmer switch and a standard single-pole light switch.*

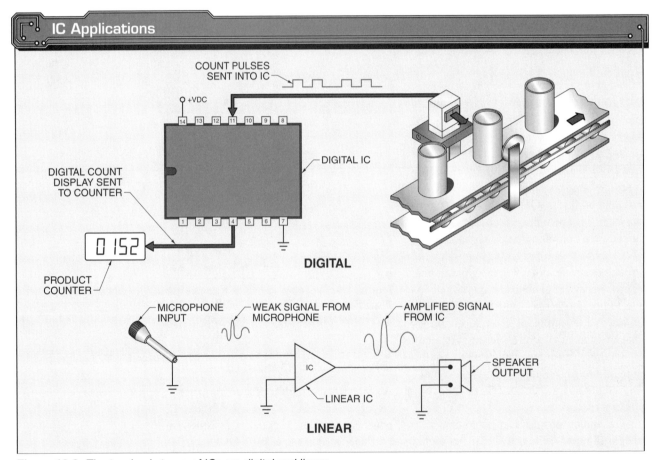

Figure 16-3. *The two basic types of ICs are digital and linear.*

A *digital IC* is an integrated circuit that processes finite (zero and maximum) signal levels. Digital ICs are used to perform mathematic, decision, and logic functions, such as AND, NAND, and NOT. They are used in the electronic circuits of devices such as computers and calculators. Digital ICs are the most widely used ICs.

A *linear IC* is an integrated circuit that produces an output signal proportional to the applied input signal and is used to provide amplification and regulation functions. They are used in radios, amplifiers, televisions, and power supplies. An operational amplifier (op-amp) is the most common type of linear IC.

When working with analog and digital signals, there may be a need to convert from one type of signal to the other. Converters are used to accomplish this task. An *analog-to-digital (A/D) converter* is an integrated circuit that converts an analog signal to a digital signal. A *digital-to-analog (D/A) converter* is an integrated circuit that converts a digital signal to an analog signal. **See Figure 16-4.**

IC Packages

ICs are powered and interconnected with other components through pins. Most ICs are mounted on printed circuit (PC) boards with pins soldered to the copper traces of the PC boards. ICs are available in various packages and pin configurations. **See Figure 16-5.** The more complex the internal circuitry of an IC, the greater the number of pins required. For example, an IC with basic logic functions, such as AND, OR, and NOT, may require only 8 or 14 pins. However, the microprocessor IC of a personal computer may require as many as 1000 pins to 2000 pins.

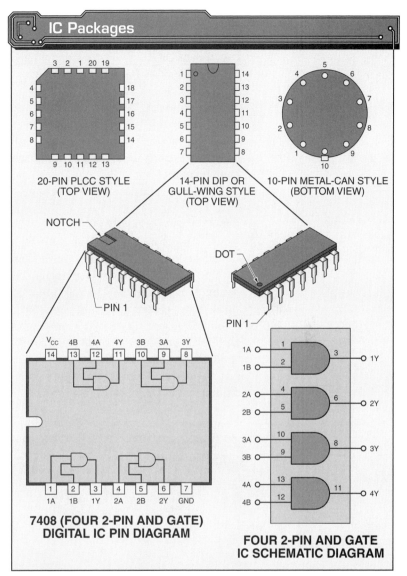

Figure 16-5. *ICs are available in various packages and pin configurations.*

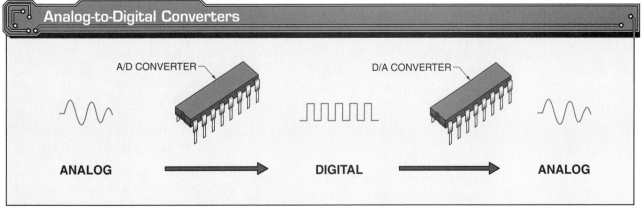

Figure 16-4. *ICs that convert analog signals to digital signals are called analog-to-digital (A/D) converters. ICs that convert digital signals to analog signals are called digital-to-analog (D/A) converters.*

DIGITAL LOGIC

ICs are manufactured in a variety of logic circuit configurations. These logic circuits may include basic logic functions, such as AND gates, or specialized circuits, such as counters. Most digital logic operations are performed by one of the basic logic gates or a combination of basic logic gates. Truth tables are used to show the output of each logic function.

Digital logic is designed to operate on binary (digital) signals. A *binary signal* is a signal that has only two states. A binary signal is either high (1) or low (0). A high signal is normally 5 V but may range from 2.4 V to 5 V. A low signal is normally 0 V but may range from 0 V to 0.8 V. A high signal may represent an ON, YES, or TRUE condition. A low signal may represent an OFF, NO, or FALSE condition. For example, a light may be ON or OFF, a valve may be open or closed, and a motor may be running or stopped. **See Figure 16-6.**

outputs as letters. Inputs are listed on the left side of the truth table and outputs are listed on the right side. Most digital truth tables use the output conditions of high (1) to indicate a signal and low (0) to indicate the absence of a signal. Letters may also be used, such as an "X" to represent a high signal and an "O" to represent a low signal. **See Figure 16-7.**

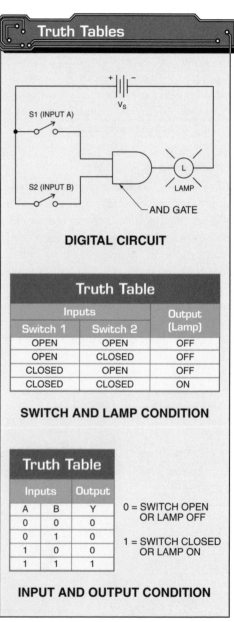

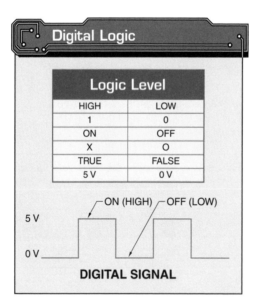

Figure 16-6. *Digital logic uses binary (digital) signals to determine logic level.*

Truth Tables

A *truth table* is a table used to describe the output condition of a logic gate or combination of gates for every possible input condition. A truth table normally lists inputs and

Figure 16-7. *A truth table is used to describe the output condition of a logic gate, or combination of gates, for every possible input condition.*

Logic Gates

The five basic logic gates are the AND, OR, NOT, NOR, and NAND gates. The basic AND, OR, NOR, and NAND gates have two or more inputs and one output. The NOT gate has one input and one output. In addition to the five basic gates, an exclusive OR gate is available that has two inputs and one output. **See Figure 16-8.**

AND Gates. An *AND gate* is a logic gate that provides logic level 1 only if all inputs are at logic level 1. AND gates are normally referred to by their number of inputs, such as two-input AND gate, three-input AND gate, etc. An example of AND logic is a control circuit that requires two or more switches to be closed before a load can be energized, such as a pressure pump circuit. **See Figure 16-9.**

Kone, Inc.

AND logic can be used in an elevator control circuit for the inner and outer doors of the elevator.

Basic Digital Logic Gates		
Device	**Symbol**	**Function/Notes**
AND GATE	INPUT A, B → Y OUTPUT	To provide logic level 1 only if all inputs are at logic level 1 HIGH = 1 LOW = 0
OR GATE	INPUT A, B → Y OUTPUT	To provide logic level 1 if at least one input is at logic level 1 HIGH = 1 LOW = 0
NOT (INVERTER) GATE	INPUT A → Y OUTPUT INDICATES INVERTED	To provide an output that is the opposite of the input HIGH = 1 LOW = 0
NOR GATE	INPUT A, B → Y OUTPUT	To provide logic level 1 only if all inputs are at logic level 0 HIGH = 1 LOW = 0
NAND GATE	INPUT A, B → Y OUTPUT	To provide logic level 0 only if all inputs are at logic level 1 HIGH = 1 LOW = 0
XOR (EXCLUSIVE OR GATE)	INPUT A, B → Y OUTPUT INDICATES EXCLUSIVE	To provide logic level 1 only if one input is at logic level 1 HIGH = 1 LOW = 0

Figure 16-8. *Logic gates include AND, OR, NOT, NOR, NAND, and XOR gates.*

AND Gates

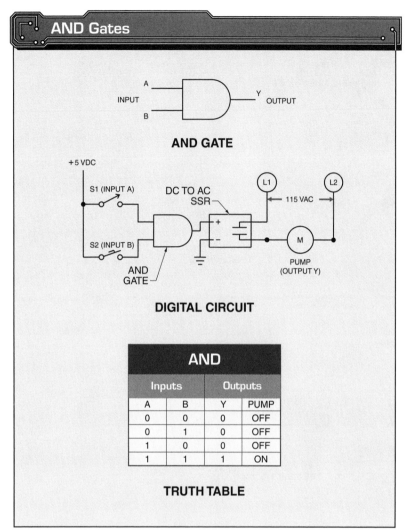

Figure 16-9. AND gates produce outputs with logic levels of 1 if all inputs have logic levels of 1.

The circuit operates in the same manner when a digital AND logic gate is used. The difference is that lower-rated control switches may be used because they only have to be rated for the voltage of the logic gate. This voltage rating is normally 5 VDC. This also allows the use of smaller wires connecting the control switches (18 AWG or smaller). A DC to AC solid state relay (SSR) is used as the interface between the DC control circuit and the 115 VAC power circuit containing the pump motor.

OR Gates. An *OR gate* is a logic gate that provides logic level 1 if at least one input is at logic level 1. Like AND gates, OR gates are normally referred to by their number of inputs, such as two-input OR gate, three-input OR gate, etc. An example of OR logic is a control circuit that requires one or more switches to be closed before a load can be energized, such as a doorbell circuit. **See Figure 16-10.**

> *The word "digital" is from the Latin word "digitus," meaning finger, which is used for discrete counting. The term "digital" is most commonly used in the realms of computing and electronics, especially when real-world information is converted to a different format such as in digital audio and digital photography.*

For example, an operator cannot turn on a 115 VAC pressure pump unless the ON/OFF and foot switches are both closed. Using a foot switch in this application allows the hands of the operator to be free for holding or moving an object. An electrical circuit that connects the two switches in series could be used to control the pressure pump. The disadvantage of this circuit is that the control switches must be at the same voltage level (115 VAC) as the load. In addition to being at the same voltage level, the control switches must be rated high enough to carry the current of the load.

In a doorbell circuit used for two entrances, pressing either pushbutton energizes the doorbell. An SSR is used as the interface between the digital circuit and the 24 VAC doorbell circuit because the doorbell draws more current and normally requires a higher voltage than the digital IC gate can deliver.

OR logic represents two or more inputs placed in parallel. In an OR switching circuit, the output is high if at least one input is high. Once the output is high, it will not be low until all inputs are low.

OR Gates

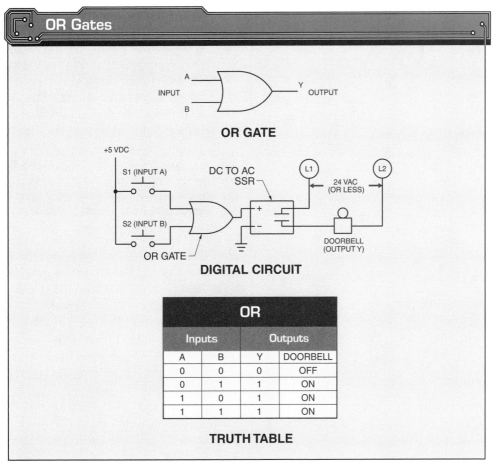

Figure 16-10. *OR gates produce outputs with logic levels of 1 when at least one input has a logic level of 1.*

NOT Gates. A *NOT (inverter) gate* is a logic gate that provides an output that is opposite of the input. A NOT gate is the simplest gate, with respect to inputs and outputs. An inverter is used in a circuit any time a signal must be changed to its opposite state. An inverter circuit may be used when a normally closed (NC) switch needs to operate as a normally open (NO) switch in the circuit. It may also be used when an NO switch needs to operate as an NC switch. The NOT gate has one input and one output. **See Figure 16-11.**

For example, an NC liquid level switch could be made to operate as an NO switch by adding an inverter into the digital circuit. Without inverters, NO and NC control switches are required to perform circuit operations. NO switches are commonly used to start or allow the flow of current in circuits. NC switches are commonly used to stop or remove the flow of current in a circuit. Any NO or NC switch may be changed to its opposite operating function by adding an inverter.

NOR Gates. A *NOR (negated OR) gate* is a logic gate that provides logic level 1 only if all inputs are at logic level 0. A NOR gate is represented by the OR gate symbol followed by a small circle, which indicates an inversion of the output. NOR gates are used in applications that require an output to turn off if any one of the input conditions is above a set limit, such as excessive temperature, pressure, or current. An example of NOR logic is when a heater must be turned OFF if the temperature or pressure in a system is above the set limit. **See Figure 16-12.** This function may be accomplished by using NC switches in the electrical circuit or NO switches and a NOR gate in the digital circuit.

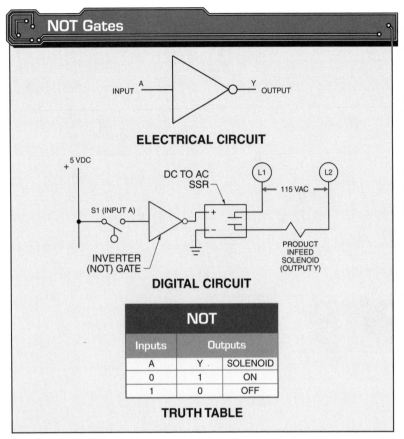

Figure 16-11. NOT gates produce outputs with logic levels that are opposite of the input logic levels.

In the electrical circuit of the heater, the temperature and pressure switches are used to control the heater contactor coil. The heater contactor coil is used to control the heating elements. In the digital circuit, the temperature and pressure switches are used to control the SSR. The SSR is used to control the heater contactor coil.

> An abacus, created sometime between 1000 BC and 500 BC, is a very basic digital calculator that uses beads on rows to represent numbers.

NAND Gates. A *NAND (negated AND) gate* is a logic gate designed to provide logic level 0 only if all inputs are at logic level 1. A NAND gate is represented by the AND gate symbol followed by a small circle, which indicates an inversion of the output. NAND gates are used in applications that require an output to turn off only if all the input conditions are met. An example of NAND logic is the control of a solenoid-operated valve that controls the flow of product critical to an operation. **See Figure 16-13.**

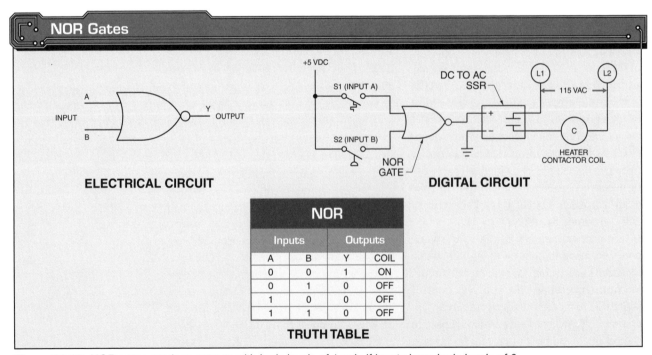

Figure 16-12. NOR gates produce outputs with logic levels of 1 only if inputs have logic levels of 0.

The solenoid-operated valve for the main water supply controls the flow of water through a pipe. The water may be flowing to a fire sprinkler system, cooling system, or other vital operation. Dangerous problems may develop if the flow of water is stopped. However, at times, the flow of water must be stopped. With NAND circuit logic, more than one switch must be activated to stop the flow. For example, if two key-operated switches are used, both switches must be activated before the water flow is stopped. An operator may have one key and a supervisor may have the other key. Both keys must be present to shut off the water.

XOR Gates. An *XOR (exclusive OR) gate* is a logic gate that provides logic level 1 only if one input is at logic level 1. An XOR gate is represented by the OR gate symbol with a curved line at the inputs. XOR gates are used in applications where the flow of two or more products into a common point must be controlled. An example of the need for XOR logic is when two products that should not be mixed must both use a common pipe. **See Figure 16-14.**

> The most useful inventions that aided in the development of digital electronics are microprocessors and integrated circuits.

For example, concentrated salt solution and dilute salt solution tanks use a common pipe at different periods of an operation. Concentrated salt solution and dilute salt solution are two different products that should not be mixed. XOR logic is used in the electrical circuit to ensure that the concentrated salt solution and dilute salt solution never mix. Switches with multiple contacts are required to develop XOR logic in the electrical circuit. For example, the tank and dilute salt solution control switches each require three contacts for proper circuit operation. An XOR gate with SSRs reduces the number of switch contacts required.

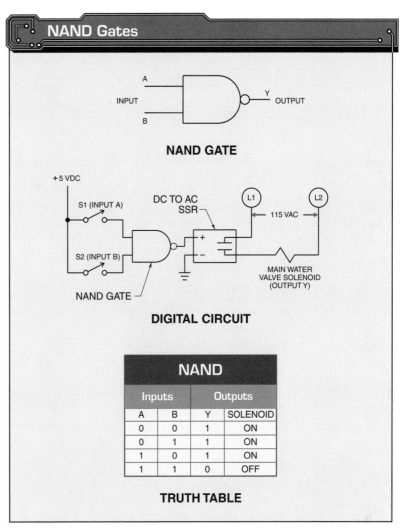

Figure 16-13. NAND gates produce outputs with logic levels of 0 only if inputs have logic levels of 1.

ASCO Valve, Inc.
NAND logic may be used with a solenoid-operated valve for water shutoff.

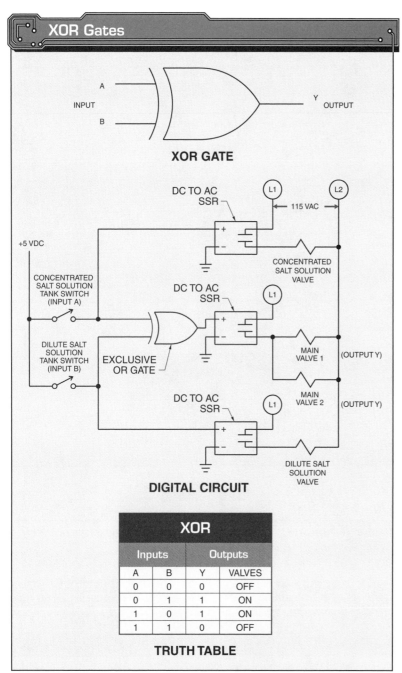

Figure 16-14. *XOR gates produce outputs with logic levels of 1 only if one input, but not both, has a logic level of 1.*

Buffer Gates. A *buffer gate* is a logic gate consisting of two NOT gates connected together so that the output of one is fed into the input of the other. The two NOT gates cancel each other out so that the input and final output have the same logic state. **See Figure 16-15.** Buffer gates are used when one inverter gate does not provide enough of a signal to drive a load. Each NOT gate acts as an amplifier and increases the output signal. A buffer gate is typically manufactured as a single gate. The buffer gate symbol is similar to that of the NOT gate but without the circle.

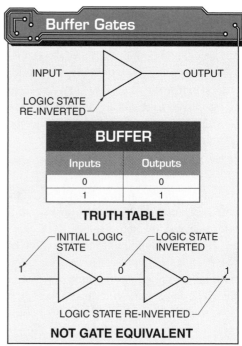

Figure 16-15. *Buffer gates are used to amplify weak signals.*

Pull-Up and Pull-Down Resistors

Digital circuits can last a relatively long time if they are properly installed and used. Basic considerations must be followed when designing a digital circuit, such as using pull-up and pull-down resistors to prevent problems that could occur. A *pull-up resistor* is a resistor that has one side connected to a power supply at all times and the other side connected to a gate input or output. A *pull-down resistor* is a resistor that has one side connected to circuit ground at all times and the other side connected to a gate input or output. Pull-up and pull-down resistors are used to force logic gate inputs or outputs either high or low when the logic circuit does not specify a state. Forced logic at gate inputs also ensure that the gate output is not overloaded. **See Figure 16-16.**

Digital inputs are designed to be either high (1) or low (0). High is equal to the input supply voltage and low is equal to ground potential. An input is floating when it is not connected to the supply voltage (high) or to ground (low). A *floating input* is a digital input that is not a true high or low at all times. A floating input should not occur in a digital circuit because it causes intermittent problems in circuit operation, such as false counts into a counter and false input signals.

Unused logic gate inputs may be tied together through a pull-up or pull-down resistor, or may be tied directly to the positive voltage source or ground. Unused inputs may also be tied to active inputs. **See Figure 16-17.**

> *Six digital states are used in Morse code to send messages via a variety of potential carriers such as telegraphs or flashing lights.*

Fan-Out

Fan-out is the number of loads that can be driven by the output of a logic gate. The output of a digital gate can supply a fixed amount of power that can be used to drive multiple logic gates of loads. The amount of power a gate can deliver is determined by the fan-out of the gate. **See Figure 16-18.** If fan-out is exceeded, a circuit may operate erratically. Interfaces, such as transistors and SSRs, must be used when a higher or different power output is required from the digital circuit.

Fan-out applies only within a given logic family. If there is a need to interface between two different logic families, the drive requirements and limitations of both families must be noted and met within the interface circuitry. Fan-out depends on the amount of current a gate can source or sink while driving other gates. A gate that exceeds its fan-out may have a weak signal source.

A weak signal source may be boosted with a buffer gate, which has high fan-out capability. The logic level is unchanged, but the full current-sourcing or current-sinking capability of the final output is available to drive loads as needed.

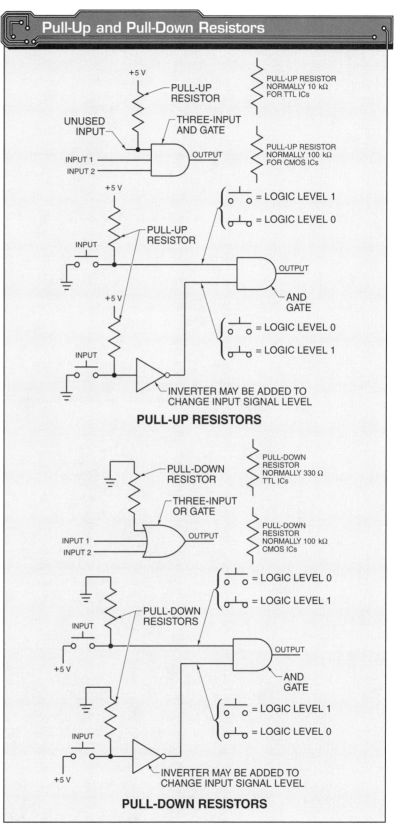

Figure 16-16. *Pull-up and pull-down resistors prevent unused inputs and switch contacts from floating.*

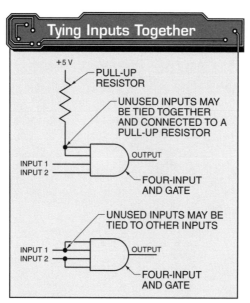

Figure 16-17. *Inputs that are not used may be tied together.*

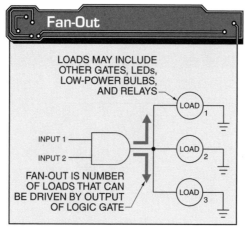

Figure 16-18. *Fan-out determines the amount of electric current a gate can source or sink while driving other gates.*

Switching

Logic gates are used to perform required logic functions. They can handle only small amounts of current, usually only a few milliamperes. In most cases, they must be aided by more powerful output devices that do the actual work. Logic gates are used to switch these devices.

For example, a gate may be used to send out a high signal with enough current to operate a small relay, which in turn triggers a larger motor relay capable of starting a motor. A gate may also be used to light a tiny lamp in an optical coupler, which in turn provides a pulse signal to turn on a light-operated SCR or triac control device.

In another application, a gate may drive a discrete transistor, which in turn supplies enough current to operate relays, lamps, and other electronic devices. A transistor output device may be connected to drive a small DC motor. The transistor, driven to saturation by high output from the gate, essentially acts as a switch for the motor, with less than a 0.2 V drop from the collector to the emitter. When the gate turns off the transistor, the transistor is considered open and draws no current, so the motor is OFF. **See Figure 16-19.** Even small silicon transistors can operate motors drawing 500 mA or more without overheating. These transistors must be operated only in the saturated or cutoff modes.

Logic Families

Manufacturers of digital chips have developed several families of digital ICs. Two major family groups are the transistor-transistor and complementary metal-oxide semiconductor families. Each family includes many of the same digital gates but have different operating specifications. The specifications of operating conditions, such as input voltage and current, output voltage and current, power consumption, noise margin level, and operating speed, vary from family to family. The prevailing characteristics of a logic family are operating speed and power consumption.

Transistor-Transistor Logic. Transistor-transistor logic (TTL) is a family of digital circuits built from bipolar junction transistors (BJTs) and resistors. It is called transistor-transistor logic because both the logic gate function and the amplifying function are performed by transistors. TTL ICs use transistors for inputs and outputs. The TTL family is commonly referred to as the 7400 series. The 74 identifies commercially available TTL ICs. ICs preceded with a 54 in the series number, such as 5400, are normally military TTL ICs that are functionally equivalent to the 7400 series. The supply voltage for TTL ICs is 5 VDC, ±0.25 V.

Complementary Metal-Oxide Semiconductor Logic. Complementary metal-oxide semiconductor (CMOS) ICs use complementary and symmetrical pairs of MOSFETs. The CMOS family of ICs uses less power than TTL ICs. The supply voltage for CMOS ICs ranges from 3 VDC to 18 VDC. The CMOS family is used in battery-operated electronic devices.

Very-large-scale integration (VLSI) circuit technology, which incorporates millions of basic logic operations on one chip, almost exclusively uses CMOS logic. This technology does not use regular bipolar transistors but instead employs field effect transistors (FETs). Because the technology used for the ICs is a complementary pair of FETs, it is called CMOS logic.

DIGITAL PULSES

Calculators, programmable controllers, and computers operate by simply speeding up the action of pulses. Semiconductor gates make this possible. Semiconductor gates have no moving parts and can switch from one state to another in a few nanoseconds. It is possible to have different types of pulses and combinations of pulses, such as square and rectangular waves. **See Figure 16-20.**

A pulse (clock) generator can be compared to an extremely fast traffic light. It controls every bit or byte of information traffic back and forth on the digital circuit. An individual pulse contains a single bit of information. A *bit* is a binary digit of either 0 or 1. A *byte* is a group of eight bits. A byte may also be referred to as a word or group of bits that are handled together by a computer system.

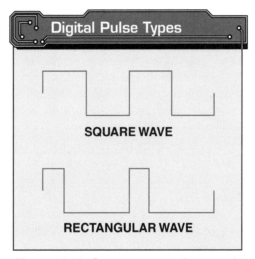

Figure 16-20. *Square waves and rectangular waves are the two main types of digital pulses.*

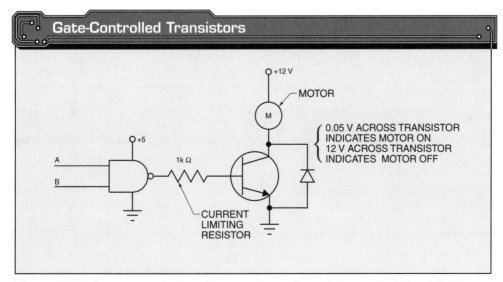

Figure 16-19. *A gate-controlled transistor can be used to switch a motor ON and OFF.*

Serial and Parallel Communication

Pulses, or digital information, may be transferred in a serial or parallel path. *Serial communication* is a method of communication that involves sending bits of data one bit at a time. **See Figure 16-21.** *Parallel communication* is a method of communication that involves sending several bits of data at the same time. Serial connections can be extremely slow. Bits are limited to being processed one after the other if there is only one line between the information and the computer.

Using an eight-lane parallel "highway" to send information to a computer is much faster than using a single-lane serial alternative. For example, using a car to represent a bit of information, and a highway to represent a pathway of parallel communication, it can be demonstrated how more cars could be handled on a multilane highway in the same amount of time as a single-lane road. The eight-bit (1 byte) signal is transferred to the computer as fast as a single bit. **See Figure 16-22.**

There are digital ICs designed to change serial bits to parallel bytes or parallel bytes into serial bits, as necessary. Computers are designed to handle a certain amount of information at a time. They are sophisticated, but the most critical characteristic is the speed at which they can process information when millions of bits of information must be processed.

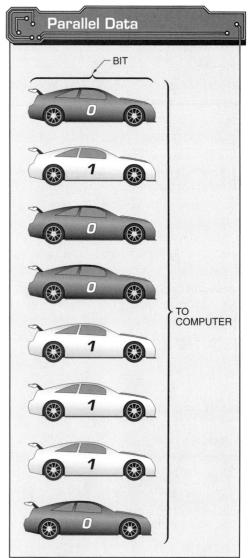

Figure 16-22. *Parallel communication is the process of sending several bits of data at the same time.*

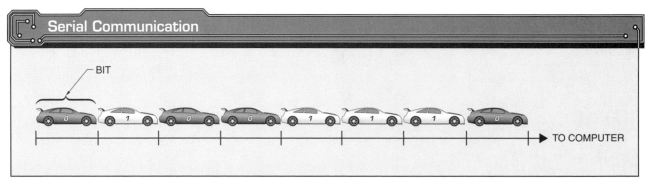

Figure 16-21. *Serial communication is the process of sending bits of data one bit at a time.*

Pulse Trains

A *pulse train* is a series of pulses that is typically plotted on a graph representing voltage versus time. Time is represented on the horizontal axis, and voltage is represented on the vertical axis. A *pulse interval* is the time elapsed between the beginning of one pulse and the beginning of the next pulse. *Pulse separation* is the time elapsed between the end of one pulse and the beginning of the next pulse. *Pulse width (pulse duration)* is the amount of time a pulse is present, or ON. **See Figure 16-23.**

Pulse amplitude is the peak-to-peak voltage in a pulse. A *pulse repetition rate* is the number of pulses during one second in a steady pulse train; measured in pulses per second (pps). For example, if 1000 pulses occur during one second, the pulse train has a repetition rate of 1000 pps.

Duty Cycles

A *duty cycle* is the amount of time a pulse is ON during a given period; expressed as a percentage. The longer a pulse is ON, the greater the amount of average current for which a power supply must be designed. In resistance welding, high-current pulses pass through the two metals to be welded. To ensure a proper weld, the number of pulses per second and the peak amplitude of the pulse must be long enough to fuse the two pieces of metal.

> A dual in-line package (DIP) switch is a group of tiny switches that are packaged together for placement on a PC board. DIP switches consist of slide switches that have two possible positions: OFF or ON (0 or 1). DIP switches may be used to control various electronic devices including remote controls and computer motherboards.

For example, in a certain series of identical pulses, the pulses move upward from zero to their positive maximum. After 25 μs, the pulses return to zero for a period of 125 μs between each pulse. **See Figure 16-24.** To find the duty cycle of a pulse train, apply the following formula:

$$\text{duty cycle} = \frac{W}{T} \times 100$$

where

W = pulse width (work time)
T = pulse period (ON time + OFF time)

Note: The units should be the same, expressed in either seconds or microseconds.

$$\text{duty cycle} = \frac{25}{25 + 125} \times 100$$

$$\text{duty cycle} = \frac{25}{150} \times 100$$

$$\text{duty cycle} = \mathbf{16.7\%}$$

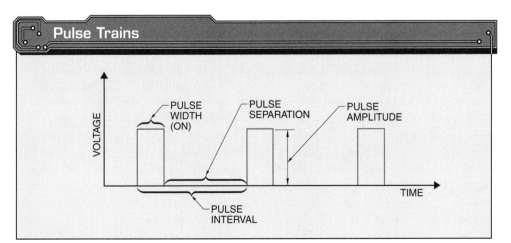

Figure 16-23. *A pulse train is a series of digital pulses.*

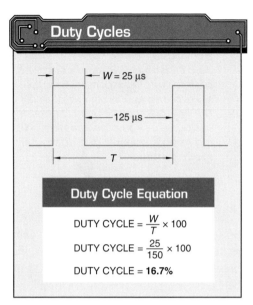

Figure 16-24. *Duty cycle is the percentage of time that a pulse is ON compared to the time it is OFF.*

Pulse Rise and Fall Times

For some pulse applications, it is necessary to specify how quickly a pulse rises from its zero, or minimum, value to its maximum value. Also, it is sometimes necessary to know how quickly the pulse falls back to zero. Digital pulses are not perfectly square and have a rise time and fall time. *Rise time* is the time required for a pulse to rise from 10% to 90% of its final steady value. Rise time is, therefore, a measure of the steepness of the leading edge of the wave. Usually the left-hand edge of a pulse is referred to as the leading edge and the right-hand edge as the trailing edge. *Fall time* is the time during which the voltage at the trailing edge of a pulse decreases from 90% to 10% of its steady maximum value.

The rise time and fall time of a pulse are measured with an oscilloscope that has a time base calibrated in microseconds. Digital pulses also contain voltage overshoot, voltage undershoot, and voltage droop. **See Figure 16-25.** In the past, it was common to measure pulse width between the 90% points (the 90% point on the leading edge and the 90% point on the trailing edge) of the pulse. Today, it is more common to measure pulse width between the 50% points of the pulse.

Clock Inputs

Clocks are pulse generators used to drive and synchronize digital circuits. Every calculator and computer and most digital devices include some type of digital clock. A digital clock is a stable source of digital square-wave pulses (signals). These pulses are fed into many different digital components in a circuit to start, stop, or synchronize some action necessary to make that circuit work. Without a clock, the digital calculator or computer could not operate properly. The advantage of a clock is its ability to provide a stable, virtually unchanging set of pulses.

Clock signals are used in digital circuits to synchronize the changing data in the various parts of the circuits. **See Figure 16-26.** In a circuit, five clock pulses are transmitted to a receiving unit when an enable signal is present for 5 μs. The enable signal allows the circuit to output data. A clock pulse may run rather slowly than typical digital clocks, only 50,000 pps. In a computer, the clock may run rather fast, at a rate of 50,000,000 pps.

Clock Operation. A clock is simply a device that outputs highs and lows, one after another, in a carefully timed sequence called a pulse. These highs and lows (0s and 1s) provide a stepping sequence to all the various gates. The clock also turns off whole sets of parallel lines, so that information can travel in one direction for an instant, and then in the opposite direction the next instant.

The clock allows a computer to check many different inputs, one after another. For example, a programmable controller may have ten sensors to check. It checks each sensor serially, one after the other. However, with such a high rate of speed, they appear to be in parallel. Checking a sensor more frequently than the measurement changes is, for practical purposes, the same as checking it continuously.

Clock waveforms should be checked against those listed on the schematic or elsewhere in the service information. On the schematic, the clock terminal of an IC or other test point should be marked "CLK" for clock or "OSC" for oscillator.

An oscilloscope shows the frequency of a clock and whether the clock is working properly. To avoid the possibility of overloading the clock oscillator, and to obtain the correct waveform display, measurements should be made using a 10x, low-capacity oscilloscope probe.

TROUBLESHOOTING DIGITAL ELECTRONIC CIRCUITS

With the PC board replacement method of troubleshooting, each PC board is viewed as a self-contained unit. Testing is not usually performed on individual PC board components. Furthermore, only voltage and current inputs into the board and voltage and current outputs from the board are measured. The input signal voltage and output voltage should be equal. For example, if the input signal at a PC board measures 5 VDC on a DMM when a pushbutton is closed, then the output from the PC board should also measure 5 VDC on the DMM.

The output signal from the PC board serves as the input to an SSR, which powers a motor. The SSR completes the circuit to a motor. The PC board should be replaced if the input to the PC board is correct but does not equal the output. For example, the PC board of a 120 VAC motor should be replaced if the input is 5 VDC but the output is not. Actual readings may vary from circuit to circuit, depending on the devices used. In this case, the voltage drop across the output of the SSR is 3 V because this amount of voltage is being used to force the electrons through the relay. This means that the motor may use the remaining 117 V.

This method of troubleshooting works only when the input and output voltage and current levels are known. This information is normally obtained from the manufacturer. Unless the technician has the proper training, the components on a PC board should not be soldered/desoldered. Manufacturers may want to repair damaged PC boards that are expensive, complex, or no longer available. If the manufacturer does not have the capability, the PC boards can be sent to a repair shop that specializes in electronic repair.

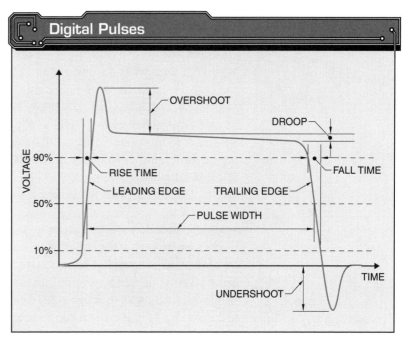

Figure 16-25. *Digital pulses are not perfectly square and have a rise time and a fall time.*

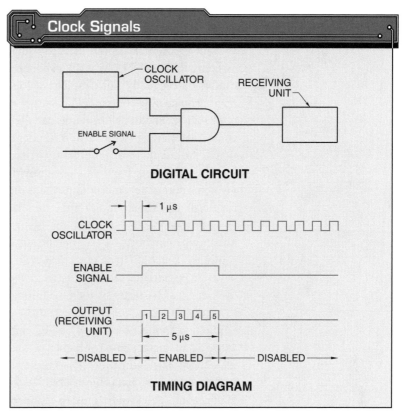

Figure 16-26. *Clock signals are used in digital circuits to synchronize the changing data in various parts of the circuits.*

Fault Isolation

Fault isolation is used to determine the cause and location of a fault. Troubleshooting digital circuitry is not as difficult as it appears. The thousands of interconnections are tucked inside carefully designed IC chips that either work exactly as they are designed or do not work at all. Most digital problems occur outside the IC chips. If a digital circuit is not working, it may be because the inputs to the system are defective.

For example, if a sensor is not working properly or the keyboard entry is incorrect, the chip reacts as though it has the correct input when it does not. Digital circuits should do exactly what their inputs tell them to do. If a chip fails, it is often due to external failures such as a runaway power supply, spikes on the power line, or incorrect troubleshooting. For example, accidental short circuits can occur while equipment is being adjusted or installed.

Built-In Troubleshooting Routines

In highly complex digital circuits, such as computers and PLCs, troubleshooting routines are usually built in. These routines evaluate how computer chips process information. The built-in routine can then compare its findings with the normal operation of the circuit. If an abnormality exists, it will trigger the proper response, providing some sort of readout noting where the troubleshooter should look for the trouble.

Less sophisticated digital circuits may not have such built-in troubleshooting routines. These control circuits usually have an instruction manual that lists the tests that can be made to localize trouble to a particular area. For example, it may be required to connect the pin of one IC chip to the pin of another chip to obtain a certain indication. The indication tells where to look for trouble or directs the troubleshooter to the next troubleshooting test to be made. The instructions are usually in the form of a troubleshooting chart.

Digital Logic Probes

Digital circuits fail because the digital signal is lost somewhere between the circuit input and output stages. Finding the point where the signal is missing and repairing that area typically corrects the problem. Repairs typically involve replacing a component or section of a PC board or the entire PC board.

A digital logic probe may be used to test a digital circuit. However, the power supply voltage must be tested with a DMM when a digital circuit or digital logic probe has intermittent problems. A digital logic probe may indicate a high signal when the supply voltage is too low for proper circuit or logic probe operation. **See Figure 16-27.**

Before taking any digital measurements using a digital logic probe, it should be ensured that the probe is designed to take measurements on the circuit being tested. The operation manual of the test instrument should be consulted for all measuring precautions, limitations, and procedures. The required personal protective equipment should always be worn and all safety rules followed when taking measurements. To take circuit measurements using a digital logic probe, apply the following procedure:

1. Connect the negative (black) lead to the ground side of the digital power supply.
2. Connect the positive (red) lead of the logic probe to the positive side of the digital power supply. The power supply is ±5 VDC for TTL circuits.
3. Set the selector switch of the digital logic probe to the logic family being tested (TTL or CMOS).
4. Touch the logic probe tip to the point on the digital circuit being tested. Start at the input side of the circuit and move to the output side.
5. Observe the condition of the LEDs on the logic probe. Single-shot pulses are stored indefinitely by placing the switch in the memory position.
6. Remove the logic probe from the circuit.

Refer to Chapter 16 Quick Quiz® on CD-ROM

Chapter 16—Digital Electronics Fundamentals 429

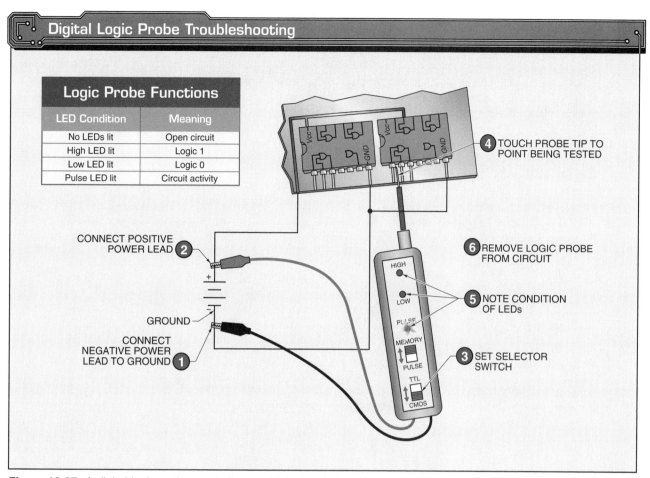

Figure 16-27. A digital logic probe can indicate a high signal when the supply voltage is too low for proper circuit or logic probe operation.

KEY TERMS

- digital logic
- integrated circuit (IC)
- AND gate
- OR gate
- NOT gate
- NOR gate
- NAND gate
- XOR gate
- buffer gate
- bit
- byte
- serial communication
- parallel communication
- pulse train
- pulse interval
- pulse separation
- pulse width (pulse duration)
- pulse amplitude
- pulse repetition rate
- duty cycle
- rise time
- fall time
- clock

REVIEW QUESTIONS

1. Explain the operation of digital and analog electronics.
2. What is the main feature of an integrated circuit (IC) that makes it more popular than individual components?
3. Which type of ICs are most widely used?
4. In the area of amplifiers, what are linear ICs used for?
5. What is the most common type of linear IC?
6. What is a truth table?
7. What are logic gates?
8. What is an AND gate?
9. What is an OR gate?
10. What is a NOT gate?
11. What is a NOR gate?
12. What is a NAND gate?
13. What is an XOR gate?
14. What is a buffer gate?
15. What is a pulse train?
16. What is a pulse interval?
17. What is pulse separation?
18. What is pulse width?
19. Explain serial communication.
20. Explain parallel communication.

SOLID STATE RELAYS

Solid state relays are prevalent in the industrial control market. Solid state relays are used as electronic switches and contain no moving parts. Due to their declining cost, high reliability, and vast capabilities, solid state devices are replacing many devices that operate on mechanical and electromechanical principles.

OBJECTIVES

- List and describe solid state relay switching methods.
- Explain solid state relay circuits.
- Describe solid state relay problems.
- Discuss two- and three-wire solid state switches.
- Explain troubleshooting solid state relays.
- List and describe solid state relay applications.

SOLID STATE RELAY SWITCHING METHODS

A *relay* is a device that controls one electrical circuit by opening and closing another circuit. A small voltage applied to a relay results in a larger voltage being switched. A *solid state relay (SSR)* is a switching device that has no contacts and switches entirely by electronic means. Solid state relays use an SCR, TRIAC, or transistor output instead of mechanical contacts to switch the controlled power. The output is optically coupled to an LED light source inside the relay. The relay is turned ON by energizing the LED, usually with low-voltage DC power. **See Figure 17-1.**

One disadvantage of SSRs is their tendency to "fail short" on their outputs, while electromechanical relay contacts tend to "fail open." Because a fail-open state is generally considered safer than a fail-closed state, electromechanical relays are still favored over SSRs in some applications. In a fail-open situation, the device will generally shut down. In a fail-short situation, the device may continue to run in an unsafe manner.

The SSR used in an application depends on the type of load to be controlled. Each SSR is designed to properly control certain loads. The four basic SSRs are the zero switching (ZS), instant-on (IO), peak switching (PS), and analog switching (AS) relays.

Zero Switching Relays

A *zero switching relay* is an SSR that turns ON a load when a control voltage is applied and the voltage at the load crosses zero (or within a few volts of zero). The relay turns the load OFF when the control voltage is removed and the current in the load crosses zero. **See Figure 17-2.**

> *Relays are used extensively in industrial assembly lines and commercial equipment. Relay applications include controlling machine tools, switching starting coils, and turning on small devices such as pilot lights and audible alarms. The two main types of relays are electromechanical relays (EMRs) and solid state relays (SSRs).*

432 SOLID STATE DEVICES AND SYSTEMS

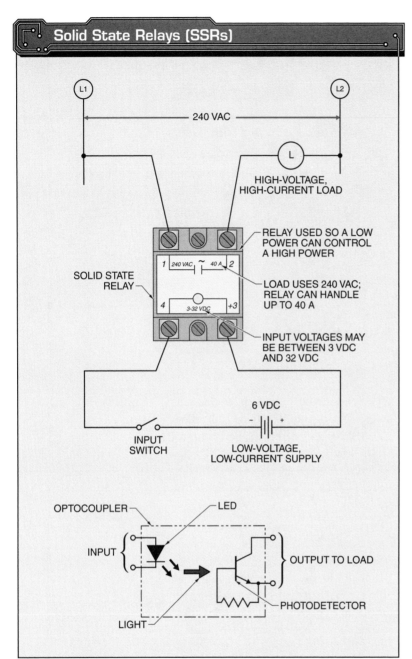

Figure 17-1. *A solid state relay (SSR) is an electronic switching device that has no moving parts.*

The zero switching relay is the most widely used relay. Zero switching relays are designed to control resistive loads. They control the temperature of heating elements, soldering irons, extruders for forming plastic, incubators, and ovens. They control the switching of incandescent lamps, tungsten lamps, flashing lamps, and programmable controller interfacing.

Instant-On Switching Relays

An *instant-on switching relay* is an SSR that turns ON a load immediately when a control voltage is present. This allows the load to be turned ON at any point on the AC sine wave.

Instant-on relays are designed to control inductive loads. In inductive loads, voltage and current are not in phase, and the loads turn ON at a point other than the zero voltage point that is preferred. Instant-on relays control the switching of contactors, magnetic valves and starters, valve positioning, magnetic brakes, small motors (used for position control), 1φ motors, small 3φ motors, lighting systems (fluorescent and HID), programmable controller interfaces, and phase control (by pulsing the input).

The relay turns OFF when the control voltage is removed and the current in the load crosses zero. Instant-on switching is exactly like electromechanical switching because both switching methods turn ON the load at any point on the AC sine wave. **See Figure 17-3.**

> *An SSR can be used to control many of the same circuits that an EMR is used to control. The difference between an SSR and an EMR is how the relays are connected to the control circuits. However, an SSR performs the same circuit functions as an EMR.*

Peak Switching Relays

A *peak switching relay* is an SSR that turns ON a load when a control voltage is present and the voltage at the load is at its peak. The relay turns OFF when the control voltage is removed and the current in the load crosses zero. Peak switching is preferred when the voltage and the current are about 90° out-of-phase because switching at peak voltage is switching at close to zero current. **See Figure 17-4.**

Chapter 17—Solid State Relays **433**

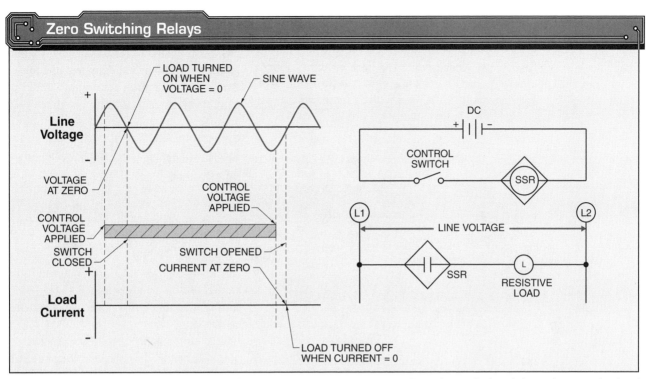

Figure 17-2. A zero switching relay turns ON the load when the control voltage is applied and the voltage at the load crosses zero.

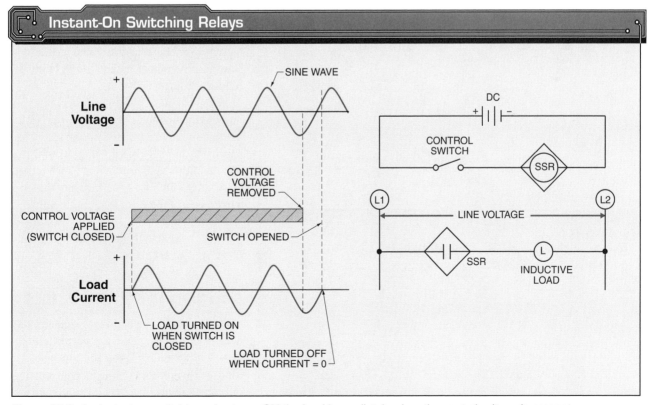

Figure 17-3. An instant-on switching relay turns ON the load immediately when the control voltage is present.

Peak switching relays control transformers and other heavy inductive loads and limit the current in the first half-period of the AC sine wave. Peak switching relays control the switching of transformers, large motors, DC loads, high inductive lamps, magnetic valves, and small DC motors.

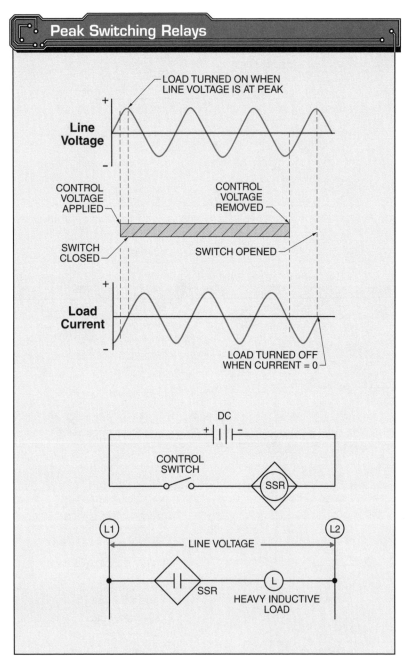

Figure 17-4. *A peak switching relay turns ON the load when the control voltage is present and the voltage at the load is at peak.*

Analog Switching Relays

An *analog switching relay* is an SSR that has a continuous range of possible output voltages within the rated range of the relay. An analog switching relay has a built-in synchronizing circuit that controls the amount of output voltage as a function of the input voltage. This allows for a ramp-up function of the load. In a ramp-up function, the voltage at the load starts at a low level and is increased over a period of time. The relay turns OFF when the control voltage is removed, and the current in the load crosses zero. **See Figure 17-5.**

Analog switching relays are designed for closed-loop applications. A closed-loop application is a temperature control with feedback from a temperature sensor to the controller.

In a closed-loop system, the amount of output is directly proportional to the amount of input. For example, if there is a small temperature difference between the actual temperature and the set temperature, the load (heating element) is given low power. However, if there is a large temperature difference between the actual temperature and the set temperature, the load is given high power. This relay may also be used for starting high-power incandescent lamps to reduce the inrush current.

A typical analog switching relay has an input control voltage of 0 VDC to 5 VDC. These low and high limits correspond respectively to no switching and full switching on the output load. For any voltage between 0 VDC and 5 VDC, the output is a percentage of the available output voltage. However, the output is normally nonlinear when compared to the input, and the manufacturer's data must be checked.

SOLID STATE RELAY CIRCUITS

An SSR circuit consists of an input circuit, a control circuit, and an output (load-switching) circuit. These circuits may be used in any combination, providing many different solid state switching applications. **See Figure 17-6.**

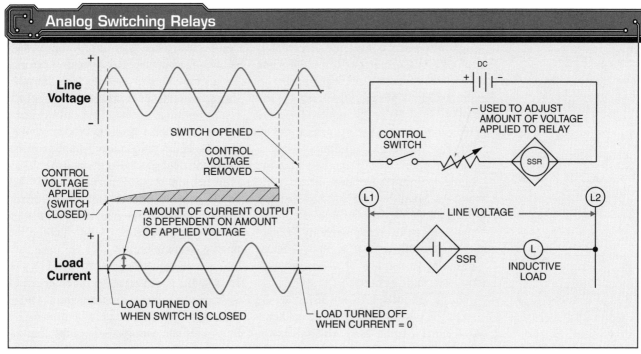

Figure 17-5. *An analog switching relay has an infinite number of possible output voltages within the rated range of the relay.*

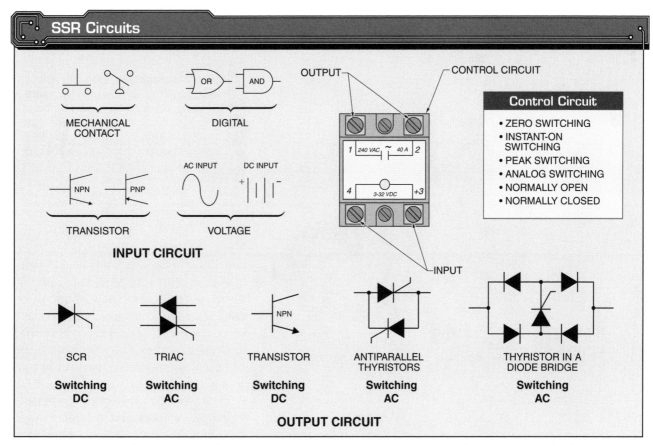

Figure 17-6. *An SSR circuit consists of an input circuit, a control circuit, and an output (load-switching) circuit.*

Input Circuits

An *input circuit* is the part of an SSR to which the control component is connected. The input circuit performs the same function as the coil of an electromechanical relay. The input circuit is activated by applying a voltage to the input of the relay that is higher than the specified pickup voltage of the relay. The input circuit is deactivated when a voltage less than the specified minimum dropout voltage of the relay is applied. Some SSRs have a fixed input voltage rating, such as 12 VDC. Most SSRs have an input voltage range, such as 3 VDC to 32 VDC. The voltage range allows a single SSR to be used with most electronic circuits.

The input voltage of an SSR may be controlled (switched) through mechanical contacts, transistors, digital gates, etc. Most SSRs may be switched directly by low-power devices, which include integrated circuits, without adding external buffers or current-limiting devices. Variable-input devices, such as thermistors, may also be used to switch the input voltage of an SSR.

Control Circuits

A *control circuit* is the part of an SSR that determines when the output component is energized or de-energized. The control circuit functions as the coupling between the input and output circuits. This coupling is accomplished by an electronic circuit inside the SSR. In an electromechanical relay (EMR), the coupling is accomplished by the magnetic field produced by the coil.

When the control circuit receives the input voltage, the circuit is switched, depending on whether the relay is a zero switching, instant-on, peak switching, or analog switching relay.

Each relay is designed to turn ON the load-switching circuit at a predetermined voltage point. For example, a zero switching relay allows the load to be turned ON only after the voltage across the load is at or near zero. The zero switching function provides a number of benefits, such as the elimination of high inrush currents on the load.

> In vehicle security systems, relays are used for alarm sensors and triggers, flashing lights, and starter interrupters. They are also used for door locks and courtesy lights that operate upon entering or exiting a vehicle.

Output (Load-Switching) Circuits

An *output (load-switching) circuit* is the load switched by an SSR that is part of the relay. The output circuit performs the same function as the mechanical contacts of an electromechanical relay. However, unlike the multiple output contacts of EMRs, SSRs normally have only one output contact.

Most SSRs use a thyristor as the output-switching component. Thyristors change from the OFF state (contacts open) to the ON state (contacts closed) very quickly

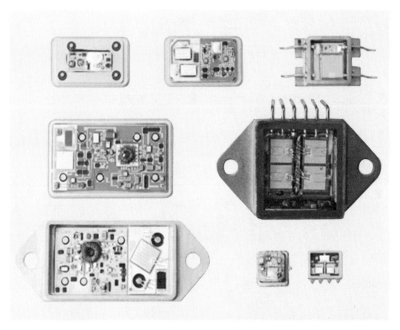

Micropac Industries, Inc.
SSRs have a faster switching time and longer life than EMRs.

when their gate switches ON. This fast switching action allows for high-speed switching of loads. The output-switching device used depends on the type of load to be controlled. Different outputs are required when switching DC circuits than are required when switching AC circuits. Common outputs used in SSRs include:

- SCRs–used to switch high-current DC loads
- triacs–used to switch low-current AC loads
- transistors–used to switch low-current DC loads
- antiparallel thyristors–used to switch high-current AC loads; able to dissipate more heat than triacs
- thyristors in diode bridges–used to switch low-current AC loads

Solid State Relay Circuit Capabilities

An SSR can be used to control most of the same circuits that an EMR is used to control. Because an SSR differs from an EMR in function, the control circuit for an SSR differs from that of an EMR. This difference is in how the relay is connected into the circuit. An SSR performs the same circuit functions as an EMR but with a slightly different control circuit.

Two-Wire Control. An SSR may be used to control a load using a momentary control, such as a pushbutton. **See Figure 17-7.** The pushbutton signals the SSR, which turns ON the load. To keep the load turned ON, the pushbutton must be held down. The load is turned OFF when the pushbutton is released. This circuit is identical in operation to the standard two-wire control circuit used with EMRs, magnetic motor starters, and contactors. For this reason, the pushbutton could be changed to any manual, mechanical, or automatic control device for simple ON/OFF operation. The same circuit may be used for liquid level control if the pushbutton is replaced with a float switch.

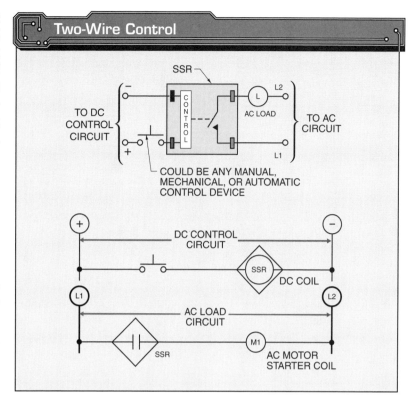

Figure 17-7. An SSR may be used to control a load using a momentary control, such as a pushbutton.

Three-Wire Memory Control. An SSR may be used with a silicon-controlled rectifier (SCR) for latching the load ON. **See Figure 17-8.** This is identical in operation to the standard three-wire memory control circuit. An SCR is used to add memory after the start pushbutton is pressed. An SCR acts as a current-operated OFF-to-ON switch.

The SCR does not allow the DC control current to pass through until a current is applied to its gate. There must be a flow of a definite minimum current to turn the SCR ON. This is accomplished by pressing the start pushbutton. Once the gate of the SCR has voltage applied, the SCR is latched in the ON position and allows the DC control voltage to pass through even after the start pushbutton is released. Resistor R1 is used as a current-limiting resistor for the gate and is determined by gate current and supply voltage.

Three-Wire Memory Control

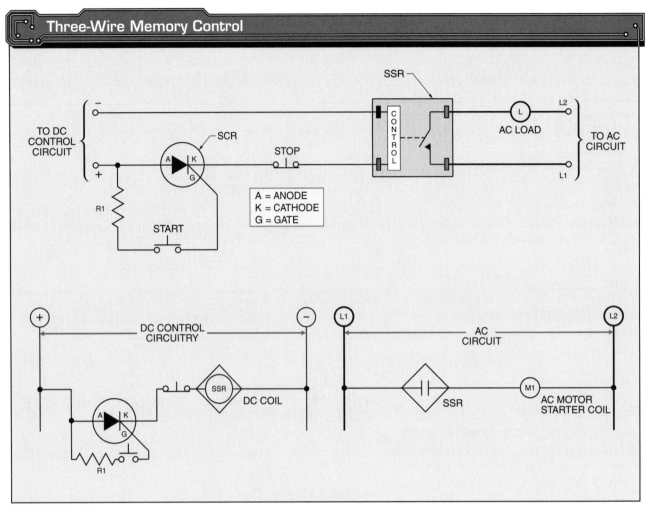

Figure 17-8. An SSR may be used with an SCR to latch a load ON.

The circuit must be opened to stop the anode-to-cathode flow of DC current to the SCR. This is accomplished by pressing the stop pushbutton. Additional start pushbuttons are added in parallel with the start pushbutton. Additional stops may be added to the circuit by placing them in series with the stop pushbutton. The additional start/stops may be any manual, mechanical, or automatic control.

Equivalent NC Contacts. An SSR may be used to simulate an equivalent NC contact condition. **See Figure 17-9.** An NC contact must be electrically made because most SSRs have the equivalent of an NO contact. This is accomplished by allowing the DC control voltage to be connected to the SSR through a current-limiting resistor (R). The load is held in the ON condition because the control voltage is present on the SSR. The pushbutton is pressed to turn the load OFF. This allows the DC control voltage to take the path of least resistance and electrically remove the control voltage from the relay. This also turns OFF the load until the pushbutton is released.

Transistor Control. All SSRs may be controlled by electronic control signals from integrated circuits and transistors. **See Figure 17-10.** In this circuit, the SSR is controlled through an NPN transistor, which receives its signal from IC logic gates, etc. The two resistors (R1 and R2) are used as current-limiting resistors.

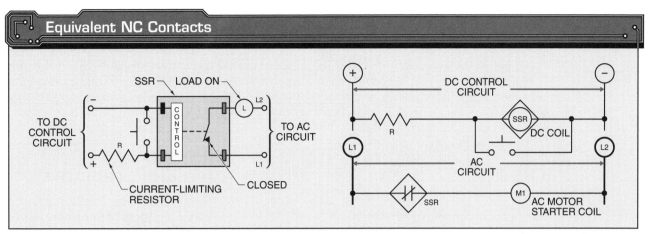

Figure 17-9. *An SSR with a load (current limiting) resistor may be used to simulate an equivalent NC contact condition.*

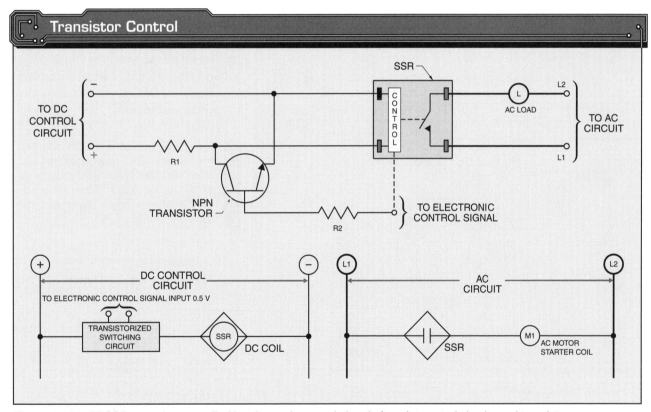

Figure 17-10. *All SSRs may be controlled by electronic control signals from integrated circuits and transistors.*

Series and Parallel Control of SSRs. An SSR can be connected in series or parallel to obtain multicontacts that are controlled by one input device. Multicontact SSRs may also be used. Three SSR control inputs may be connected in parallel so that, when the switch is closed, all three are actuated. **See Figure 17-11.** This controls the 3ϕ circuit.

In this application, the DC control voltage across each SSR is equal to the DC supply voltage because they are connected in parallel. When a multicontact SSR is used, there is only one input that controls all output switches.

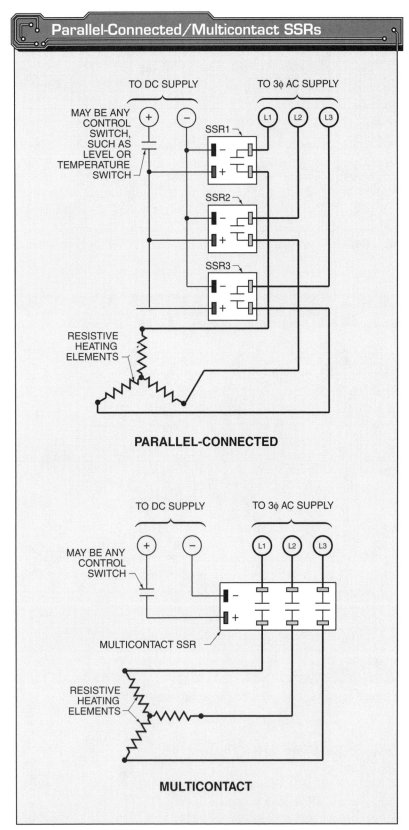

Figure 17-11. Three SSRs may be connected in parallel to control a 3φ circuit, or a multicontact SSR may be used.

To control a 3φ circuit, SSRs can be connected in series. **See Figure 17-12.** The DC supply voltage is divided across the three SSRs when the switch is closed. For this reason, the DC supply voltage must be at least three times greater than the minimum operating voltage of each relay.

SOLID STATE RELAY TEMPERATURE PROBLEMS

Temperature rise is the largest problem in applications using an SSR. As temperature increases, the failure rate of SSRs increases. As temperature increases, the number of operations of an SSR decreases. The higher the heat in an SSR, the more problems that can occur. **See Figure 17-13.**

The failure rate of most SSRs doubles for every 10°C temperature rise above an ambient temperature of 40°C. An ambient temperature of 40°C is considered standard by most manufacturers.

> *The failure rate of most SSRs doubles for every 10°C temperature rise above an ambient temperature of 40°C. An ambient temperature of 40°C is considered the standard by most manufacturers.*

Solid state relay manufacturers specify the maximum relay temperature permitted. The relay must be properly cooled to ensure that the temperature does not exceed the specified maximum safe value. Proper cooling is accomplished by installing the SSR to the correct heat sink. A heat sink is chosen based on the maximum amount of load current controlled.

Heat Sinks

The performance of an SSR is affected by ambient temperature. The ambient temperature of a relay is a combination of the temperature of the relay location and the type of enclosure used. The temperature inside an enclosure may be much higher than the ambient temperature of an enclosure that allows good airflow.

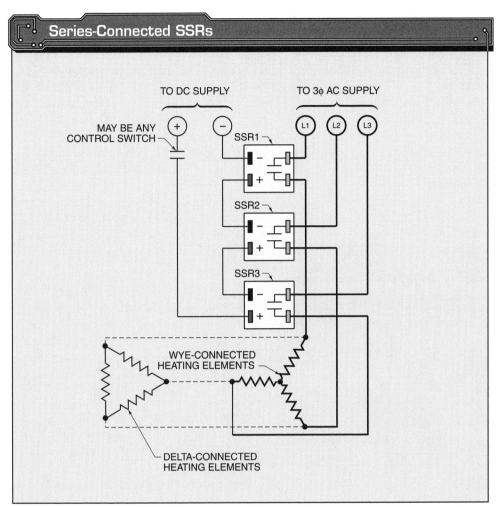

Figure 17-12. Three SSRs may be connected in series to control a 3ϕ circuit.

The temperature inside an enclosure increases if the enclosure is located next to a heat source or in the sun. The electronic circuit and SSR also produce heat. Forced cooling is required in some applications.

Selecting Heat Sinks. A low resistance to heat flow is required to remove the heat produced by an SSR. The opposition to heat flow is thermal resistance. *Thermal resistance (R_{TH})* is the ability of a device to impede the flow of heat. Thermal resistance is a function of the surface area of a heat sink and the conduction coefficient of the heat sink material. Thermal resistance is expressed in degrees Celsius per watt (°C/W).

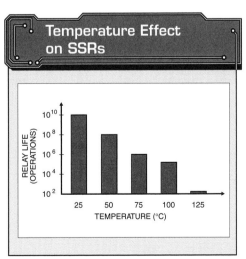

Figure 17-13. As temperature increases, the number of operations of an SSR decreases.

Heat sink manufacturers list the thermal resistance of heat sinks. The lower the thermal resistance number, the more easily the heat sink dissipates heat. The larger the thermal resistance number, the less effectively the heat sink dissipates heat. The thermal resistance value of a heat sink is used with an SSR load current/ambient temperature chart to determine the size of the heat sink required. **See Figure 17-14.**

A relay can control a large amount of current when a heat sink with a low thermal resistance number is used. A relay can control the least amount of current when no heat sink (free air mounting) is used. To maximize heat conduction through a relay and into a heat sink, the following considerations are applied:

- Use heat sinks made of a material that has a high thermal conductivity. Silver has the highest thermal conductivity rating. Copper has the highest practical thermal conductivity rating. Aluminum has a good thermal conductivity rating, and is a cost-effective and widely used heat sink.
- Keep the thermal path as short as possible.
- Use the largest cross-sectional surface area in the smallest space.
- Always use thermal grease or pads between the relay housing and the heat sink to eliminate air gaps and aid in thermal conductivity.

Mounting Heat Sinks. A heat sink must be correctly mounted to ensure proper heat transfer. The following procedure is used to properly mount a heat sink:

1. Choose a smooth mounting surface. The surfaces between a heat sink and a solid state device should be as flat and smooth as possible. Ensure that the mounting bolts and screws are securely tightened.
2. Locate heat-producing devices so that the temperature is spread over a large area. This helps prevent higher temperature areas.
3. Use heat sinks with fins to achieve as large a surface area as possible.
4. Ensure that the heat from one heat sink does not add to the heat from another heat sink.
5. Apply thermal grease between the heat sink and the solid state device to ensure maximum heat transfer.

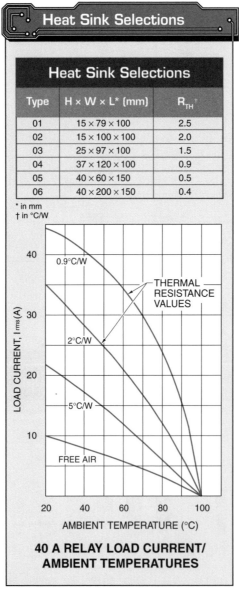

Figure 17-14. Thermal resistance (R_{TH}) is the ability of a device to impede the flow of heat.

Relays are used in circuits when a large amount of current or voltage must be controlled with a small electrical signal.

SOLID STATE RELAY CURRENT PROBLEMS

The current passing through an SSR must be kept below the maximum load current rating of the relay. An overload protection fuse is used to prevent overcurrents from damaging an SSR.

An overload protection fuse opens the circuit when the current is increased to a higher value than the nominal load current. The fuse should be an ultrafast fuse used for the protection of semiconductors. **See Figure 17-15.**

SOLID STATE RELAY VOLTAGE PROBLEMS

Most AC power lines contain transient voltage superimposed on the voltage sine wave. A *transient voltage* is temporary, unwanted voltage in an electrical circuit. Transient voltages are produced by switching motors, solenoids, transformers, motor starters, contactors, and other inductive loads. Large spikes are also produced by lightning striking the power distribution system.

The output element of a relay can exceed its breakdown voltage and turn ON for part of a half period if overvoltage protection is not provided. This short turn-on can cause problems in the circuit.

Varistors are added to the relay output terminals to prevent an overvoltage problem. A varistor should be rated 10% higher than the line voltage of the output circuit. The varistor bypasses the transient current. **See Figure 17-16.**

Voltage Drop

In all series circuits, the total circuit voltage is dropped across the circuit components. The higher the resistance of any component, the higher the voltage drop. The lower the resistance of any component, the lower the voltage drop. Thus, an open switch that has a meter connected across it shows a very high voltage drop because the meter and open switch have a very high resistance when compared to the load. Conversely, a closed switch that has a meter connected across it shows a very low voltage drop because the meter is closed, and closed

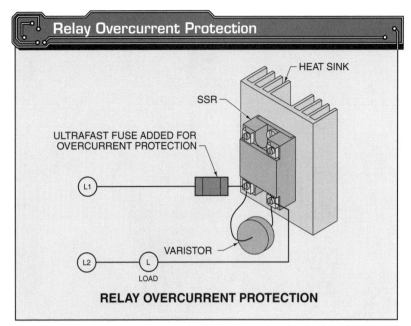

Figure 17-15. *An overload protection fuse opens the circuit when the current is increased to a higher value than the nominal load current.*

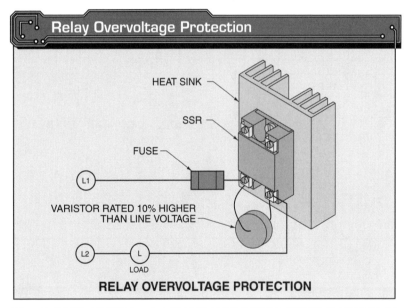

Figure 17-16. *Varistors are added to relay output terminals to prevent an overvoltage problem.*

switches have a very low resistance when compared to the load. **See Figure 17-17.**

A voltage drop in the switching component is unavoidable in an SSR. The voltage drop produces heat. The larger the current passing through the relay, the greater the amount of heat produced. The generated heat affects relay operation and can destroy the relay if not removed.

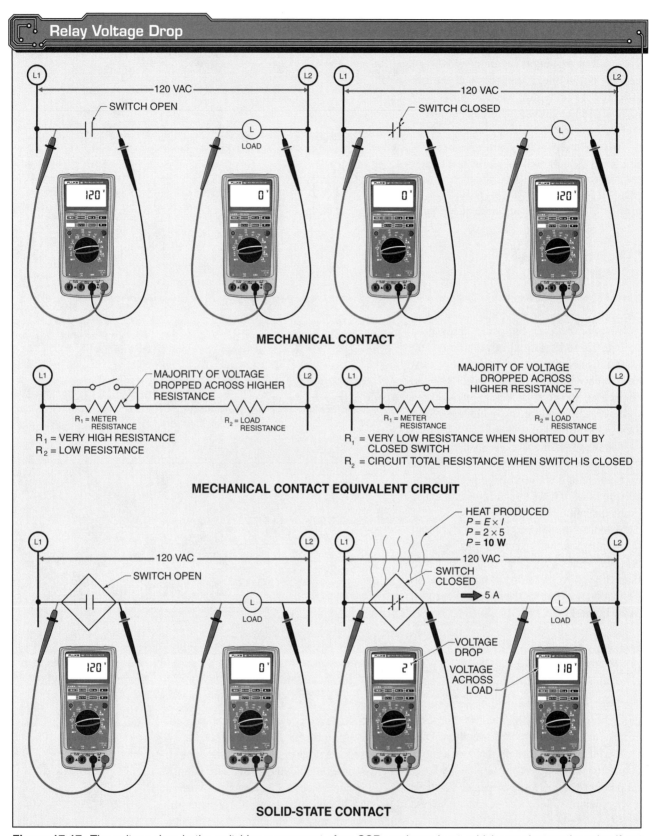

Figure 17-17. *The voltage drop in the switching component of an SSR produces heat, which can destroy the relay if not removed.*

The voltage drop in an SSR is usually 1 V to 1.6 V, depending on the load current. For small loads (less than 1 A), the heat produced is safely dissipated through the case of the relay. High-current loads require a heat sink to dissipate the extra heat. **See Figure 17-18.**

SSR Voltage Drop

Load Current (in A)	Voltage Drop (in V)	Power at Switch (in W)
1	2	2
2	2	4
5	2	10
10	2	20
20	2	40
50	2	100

Figure 17-18. *For small loads (less than 1 A), the heat produced in an SSR is safely dissipated through the case of the relay.*

For example, if the load current in a circuit is 1 A and the SSR switching device has a 2 V drop, the power generated is 2 W. The 2 W of power generates heat that can be dissipated through the relay's case.

If the load current in a circuit is 20 A and the SSR switching device has a 2 V drop, the power generated in the device is 40 W. The 40 W of power generates heat that requires a heat sink to safely dissipate the heat.

TWO-WIRE SOLID STATE SWITCHES

A *two-wire solid state switch* is a device that has two connecting terminals or wires (exclusive of ground). A two-wire switch is connected in series with the controlled load. A two-wire solid state switch is also referred to as a load-powered switch because it draws operating current through the load. The operating current flows through the load when the switch is not conducting (load OFF). This operating current is inadequate to energize most loads. Operating current is also referred to as residual current or leakage current. Operating current may be measured with a clamp-on ammeter set to measure amperes when the load is OFF. **See Figure 17-19.**

The current in a circuit is a combination of the operating current and load current when a switch is conducting (load ON). A solid state switching device must be rated high enough to carry the current of the load. Load current is measured with a clamp-on ammeter when the load is ON.

The current draw of a load must be sufficient to keep the solid state switch operating when the switch is conducting (load ON). Minimum holding current values range from 2 mA to 20 mA.

Operating current and minimum holding current values are normally not a problem when a solid state switch controls a low-impedance load, such as a motor starter, a relay, or a solenoid. Operating current and minimum holding current values may be a problem when a solid state switch controls a high-impedance load, such as a PLC or other solid state device.

The operating current may be high enough to affect the load when the switch is not conducting. For example, a programmable controller may see the operating current as an input signal. A load resistor must be added to the circuit to correct this problem. **See Figure 17-20.** A load resistor is connected in parallel with the load. The load resistor acts as an additional load, which increases the total current in the circuit. Load resistors range in value from 4.5 kΩ to 7 kΩ. A 5 kΩ, 5 W resistor is used in most applications.

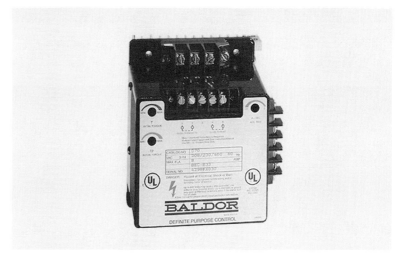

Baldor Electric Co.
An SSR with a timer allows the relay to be set so that the start or stop function occurs only after a set time.

446 SOLID STATE DEVICES AND SYSTEMS

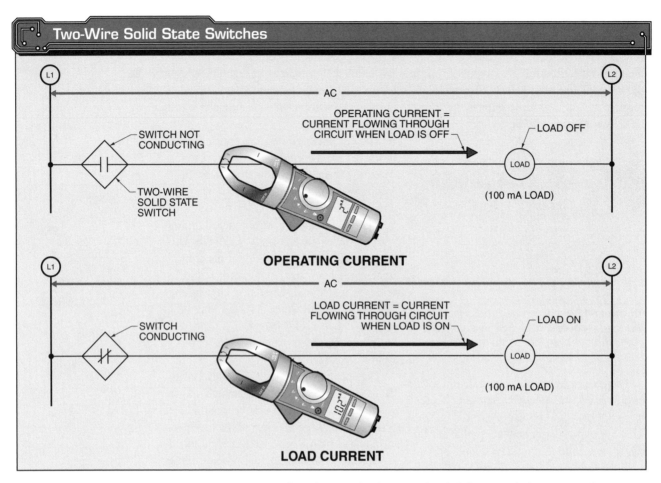

Figure 17-19. A two-wire solid state switch is also referred to as a load-powered switch because it draws operating current through the load.

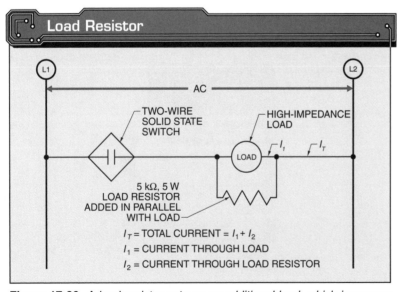

Figure 17-20. A load resistor acts as an additional load, which increases the total current in the circuit.

Testing Two-Wire Solid State Switches

Two-wire solid state switches connected in series affect the operation of a load because of the voltage drop across the switches. A two-wire solid state switch drops about 3 V to 8 V. The total voltage drop across the switches equals the sum of the voltage drops across each switch. No more than three solid state switches are allowed to be connected in series due to the voltage drop created by each switch. **See Figure 17-21.**

Temperature rise is the largest problem in applications involving SSRs. As temperature increases, the number of SSR operations decreases.

Chapter 17—Solid State Relays 447

Series and Parallel Two-Wire Solid State Switches

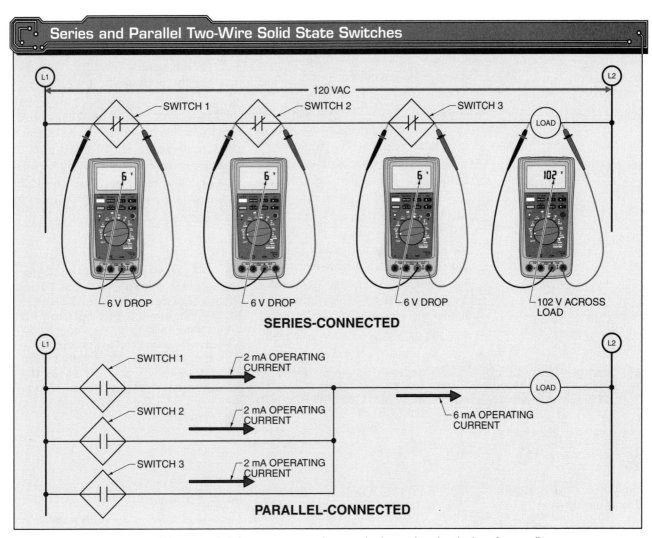

Figure 17-21. *A two-wire solid state switch has two connecting terminals or wires (exclusive of ground).*

Two-wire solid state switches connected in parallel affect the operation of a load because each switch has operating current that is flowing through the load. The load may turn ON when the operating current through the load becomes excessive. The total operating current equals the sum of the operating currents of each switch. No more than three solid state switches should be connected in parallel.

Caution should be used when troubleshooting SSRs because they may fail in the "short circuit" condition.

A suspected fault in a two-wire solid state switch is tested using a voltmeter. A voltmeter is used to test the voltage into the switch and out of the switch.

Before taking an AC voltage measurements using a digital multimeter, it must be ensured that the meter is designed to take measurements on the circuit being tested. The operations manual of the test instrument should be referred to for all measuring precautions, limitations, and procedures. The required personal protective equipment must be worn and all safety rules must be followed when taking a measurement. To test a two-wire solid state switch, apply the following procedure:

1. Set the function switch of the voltage meter to measure AC voltage. **See Figure 17-22.**
2. Connect the test leads of the voltage meter jacks as required.
3. Measure the supply voltage into the solid state switch. The problem is located upstream from the switch if there is no voltage present. The problem may be a blown fuse or open circuit. Voltage to the solid state switch must be reestablished before the switch is tested.
4. Measure the voltage out of the solid state switch. The voltage should equal the supply voltage minus the rated voltage drop (3 V to 8 V) of the switch when the switch is conducting (load ON). Replace the switch when the voltage output is not correct.
5. Remove the voltmeter from the circuit.

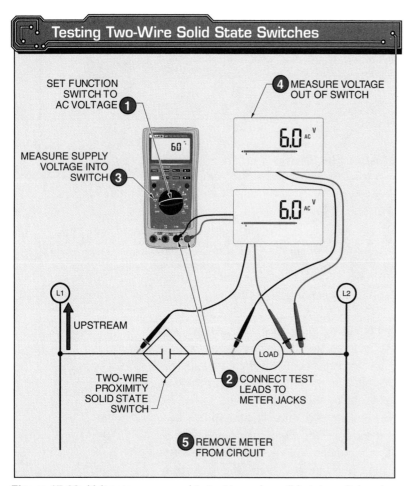

Figure 17-22. *Voltmeters are used to test two-wire solid state switches by testing the voltage into and out of the switches.*

THREE-WIRE SOLID STATE SWITCHES

A *three-wire solid state switch* is a device that has three connecting terminals or wires (exclusive of ground). A three-wire solid state switch draws operating current directly from a power source and does not allow the operating current to flow through the load. The two types of three-wire solid state switches are the current source (PNP) switch and the current sink (NPN) switch. **See Figure 17-23.**

An SSR may be used with an SCR for latching a load ON. The SCR acts as a current-operated OFF-to-ON switch. The SCR does not allow DC control voltage to pass through until current is applied to its gate. Once the gate of an SCR is latched in the ON condition, the SCR allows the DC control voltage to pass through even after the current is removed from its gate.

Testing Three-Wire Solid State Switches

Three-wire solid state switches connected in series affect the operation of the load. This is because each switch downstream from the previous switch must carry the load current and the operating current of each switch. A DMM set to measure current may be used to measure operating and load current values. The measured values must not exceed the manufacturer's maximum rating. **See Figure 17-24.**

Three-wire solid state switches connected in parallel affect the operation of the load because the nonconducting switch may be damaged due to reverse polarity. A blocking diode should be added to each switch output to prevent reverse polarity on the switch.

A suspected fault with a three-wire solid state switch is tested using a voltmeter. A voltmeter is used to test the voltage into the switch and out of the switch.

Chapter 17—Solid State Relays   449

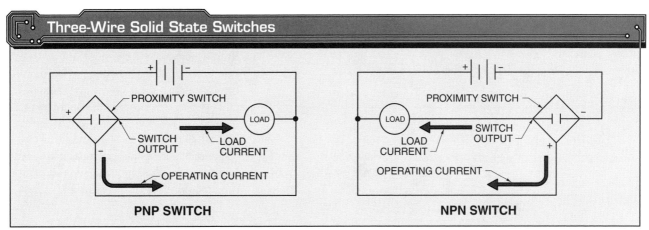

Figure 17-23. *Three-wire solid state switches (current source or current sink) are also called line-powered switches because the switch draws operating current from the power lines.*

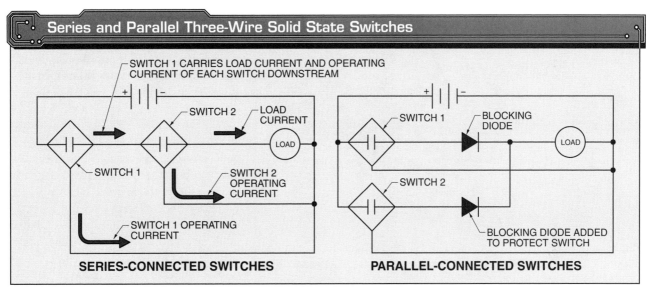

Figure 17-24. *A three-wire solid state switch draws its operating current directly from the power lines.*

Before taking any AC voltage measurements using a digital multimeter, it must be ensured that the meter is designed to take measurements on the circuit being tested. The operations manual of the test instrument should be referred to for all measuring precautions, limitations, and procedures. The required personal protective equipment must be worn and all safety rules must be followed when taking a measurement. To test a three-wire solid state switch, apply the following procedure:

1. Set the function switch of the voltmeter to DC voltage. **See Figure 17-25.**
2. Connect the test leads of the voltmeter to the proper jacks of the meter.
3. Measure the voltage into the solid state switch. The problem is located upstream from the switch when there is no voltage present or the voltage is not at the correct level. The problem may be a blown fuse or open circuit. Voltage to the solid state switch must be reestablished before the switch is tested.
4. Measure the voltage out of the solid state switch. The voltage should be equal to the supply voltage when the switch is conducting (load ON). Replace the solid state switch when the voltage out of the switch is not correct.
5. Remove the voltmeter from the circuit.

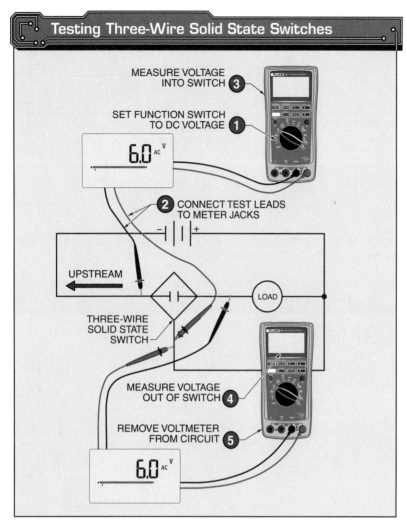

Figure 17-25. *Voltmeters are used to test three-wire solid state switches by testing the voltage into and out of the switches.*

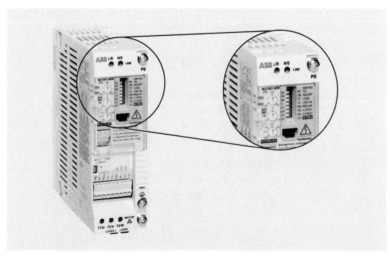

Motor drives may use solid state components for motor control.

TROUBLESHOOTING SOLID STATE RELAYS

Solid state relays require periodic inspection. Dirt, burning, or cracking should not be present on a solid state relay. Printed circuit (PC) boards should be properly seated. It must be ensured that the board locking tabs are in place, if used. Locking tabs may be added if a PC board without locking tabs loosens. Any cooling provisions should be working and free of obstructions.

Troubleshooting an SSR is accomplished by either the exact replacement method or the circuit analysis method. The *exact replacement method* is a method of SSR replacement in which a bad relay is replaced with a relay of the same type and size. The exact replacement method involves making a quick check of the input and output voltages of the relay. The relay is assumed to be the problem and is replaced when there is only an input voltage being switched.

The *circuit analysis method* is a method of SSR replacement in which a logical sequence is used to determine the reason for the failure of the SSR. Steps are taken to prevent the problem from recurring once the reason for a failure is known. The circuit analysis method of troubleshooting is based on three improper relay operations, which include the following:

- The relay fails to turn the load OFF.
- The relay fails to turn the load ON.
- The relay operates erratically.

Relay Fails to Turn Load OFF

A relay may not turn the load OFF to which it is connected when a relay fails. This condition occurs when either the load is drawing more current than the relay can withstand, the heat sink of the relay is too small, or transient voltages are causing a breakover of the relay's output. Overcurrent permanently shorts the switching device of the relay if the load draws more current than the rating of the relay. High temperature causes thermal runaway of the switching device if the heat sink does not remove the heat.

The relay is replaced with one of a higher voltage rating and/or a transient suppression device is added to the circuit if the power lines are likely to have transients (usually from inductive loads connected on the same line). To troubleshoot an SSR that fails to turn a load OFF, apply the following procedure:

1. Disconnect the input leads from the SSR. See Step 3 if the relay load turns OFF. The relay is the problem if the load remains ON and the relay is normally open. **See Figure 17-26.**

2. Measure the voltage of the circuit that the relay is controlling. The line voltage should not be higher than the rated voltage of the relay. Replace the relay with a relay that has a higher voltage rating if the line voltage is higher than the relay's rating. Check to ensure that the relay is rated for the type of line voltage (AC or DC) being used.

3. Measure the current drawn by the load. The current draw must not exceed the rating of the relay. For most applications, the current draw should not be more than 75% of the maximum rating.

4. Reconnect the input leads, and measure the input voltage to the relay at the time when the control circuit should turn the relay OFF. The control circuit is the problem and needs to be checked if the control voltage is present. The relay is the problem if the control voltage is removed and the load remains ON. Before changing the relay, ensure that the control voltage is not higher than the relay's rated limit when the control circuit delivers the supply voltage. Ensure that the control voltage is not higher than the rated drop-out voltage of the relay when the control circuit removes the supply voltage. This condition may occur in some control circuits using solid state switching.

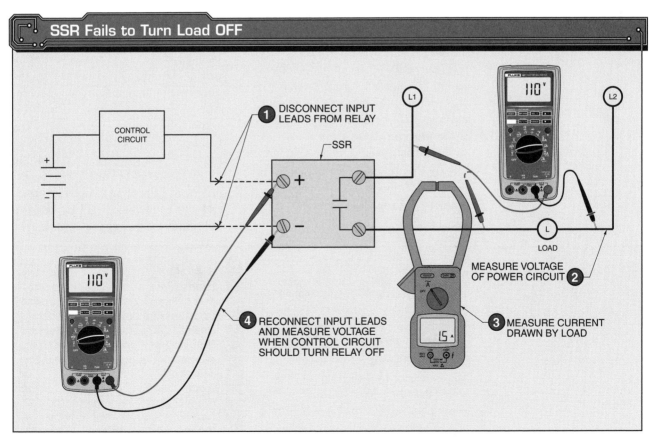

Figure 17-26. *A relay may not turn OFF the load to which it is connected when the relay fails.*

Relay Fails to Turn Load ON

A relay may fail to turn the load ON to which it is connected when the relay fails. This condition occurs when the switching device of the relay receives a very high voltage spike or the input of the relay is connected to a higher-than-rated voltage. A high voltage spike blows open the switching device, preventing the load from turning ON. Excessive voltage on the input side of the relay destroys the electronic circuit.

Replace the relay with one that has a higher voltage and current rating, and/or add a transient suppression device to the circuit if the power lines are likely to have high voltage spikes. **See Figure 17-27.** To troubleshoot an SSR that fails to turn ON a load, apply the following procedure:

1. Measure the input voltage when the relay should be ON. Troubleshoot the circuit ahead of the relay's input if the voltage is less than the rated pickup voltage. The circuit ahead of the relay is the problem if the voltage is greater than the rated pickup voltage of the relay. The higher voltage may have destroyed the relay. The relay may be a secondary problem caused by the primary problem of excessive applied voltage. Correct the high-voltage problem before replacing the relay. The relay or output circuit is the problem if the input voltage is within the pickup limits of the relay.

2. Measure the voltage at the output of the relay. The relay is probably the problem if it is not switching the voltage. See Step 3. The problem is in the output circuit if the relay is switching the voltage. Check for an open circuit in the load.

3. Use a clamp-on ammeter set to measure current with the input leads of the relay. Measure the current when the relay should be ON. The relay input is open if no current is flowing. Replace the relay. The relay is bad if the current flow is within the rating. Replace the relay. The control circuit is the problem if current is flowing but is less than that required to operate the relay.

Erratic Relay Operation

Erratic relay operation is the proper operation of a relay at times and the improper operation of the relay at other times. Erratic relay operation is caused by mechanical problems (loose connections), electrical problems (incorrect voltage), or environmental problems (high temperature). **See Figure 17-28.** To troubleshoot erratic relay operation, apply the following procedure:

1. Check all wiring and connections for proper wiring and tightness. Loose connections cause many erratic problems. No sign of burning should be present at any terminal. Burning at a terminal usually indicates a loose connection.

2. Ensure that the input control wires are not next to the output line or load wires. The noise carried on the output side may cause unwanted input signals.

3. The relay may be half-waving if the load is a chattering AC motor or solenoid. *Half-waving* is a phenomenon that occurs when a relay fails to turn OFF because the current and voltage in the circuit reach zero at different times. Half-waving is caused by the phase shift inherent in inductive loads. The phase shift makes it difficult for some solid state relays to turn OFF. Connecting an RC or another snubber circuit across the output load should allow the relay to turn OFF. An *RC circuit* is a circuit in which resistance (R) and capacitance (C) are used to help filter the power in a circuit.

Solid state devices are now widely used in the industrial control market. Due to their declining cost, reliability, and immense capabilities, solid state devices are replacing many devices that operate on mechanical and electromechanical principles. The selection of either a solid state or electromechanical relay is based on the electrical, mechanical, and cost characteristics of each device and the required application.

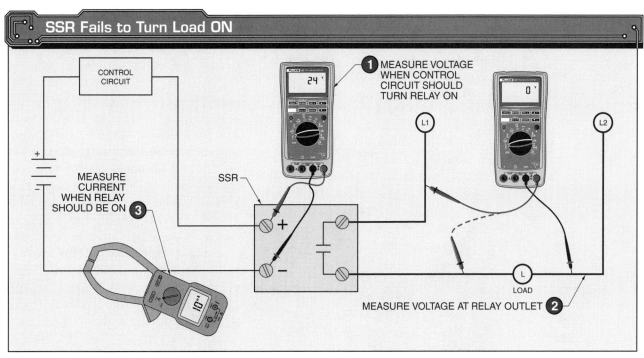

Figure 17-27. *A relay may fail to turn ON the load to which it is connected when the relay fails.*

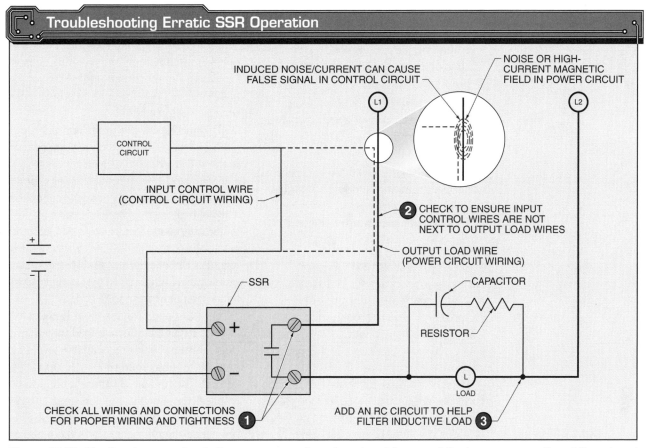

Figure 17-28. *Erratic relay operation is the proper operation of a relay at times and the improper operation of the relay at other times.*

PROXIMITY SENSOR APPLICATIONS

Proximity sensors use solid state outputs to control the flow of electric current. The solid state output of a proximity sensor may be a thyristor, NPN transistor, or PNP transistor. The thyristor output is used for switching AC circuits. The NPN and PNP transistor outputs are used for switching DC circuits. The output selected depends upon specific application needs. Considerations that affect the solid state output include the following:

- voltage type to be switched (AC or DC)
- amount of current to be switched (most proximity sensors can only switch a maximum of a few hundred milliamperes; an interface is needed if higher current switching is required; the SSR is the most common interface used with proximity sensors)
- electrical requirements of the device to which the output of the proximity sensor is to be connected (compatibility to a controller, such as a programmable controller, may require one type of solid state output or the other to be used as an input into that controller)
- required polarity of switched DC output (NPN outputs deliver a negative output and PNP outputs deliver a positive output)
- electrical characteristics (such as load current, operating current, and minimum holding current because solid state outputs are never completely open or closed)

AC Proximity Sensors

An AC proximity sensor is a switch used to switch alternating current circuits. An AC proximity sensor is connected in series with the load that it controls. The sensor is connected between line 1 and the load to be controlled. **See Figure 17-29.**

Because AC proximity sensors are connected in series with the load, special precautions must be taken. The three main factors considered when AC proximity sensors are connected include load current, operating (residual) current, and minimum holding current.

Load Current. *Load current* is the amount of current drawn by a load when energized. Since a solid state proximity sensor is wired in series with the load, the current drawn by the load must pass through the solid state sensor. For example, if a load draws 5 A, the proximity sensor must be able to safely switch 5 A. Five amperes burns most solid state proximity sensors because they are normally rated for a maximum of less than 0.5 A. An electromechanical or solid state relay must be used as an interface to control the load if a solid state proximity sensor must switch a load above its rated maximum current. A solid state relay is the preferred choice of interface with a proximity sensor. **See Figure 17-30.**

Operating Current. *Operating (residual or leakage) current* is the amount of current a sensor draws from the power lines to develop a field that can detect a target. When a proximity sensor is in the OFF condition (target not detected), a small amount of current passes through both the proximity sensor and the load. This operating current is required for the solid state detection circuitry housed within the proximity sensor. Operating currents are normally in the range of 1.5 mA to 7 mA for most proximity sensors. **See Figure 17-31.**

The small operating current normally does not have a negative effect on low impedance loads or circuits, such as mechanical relays, solenoids, and magnetic motor starters. However, the operating current may be enough to activate high impedance loads, such as programmable controllers, electronic timers, and other solid state devices. The load is activated regardless of whether a target is present. This problem may be corrected by placing a resistor in parallel with the load.

The resistance value should be selected to ensure that the effective load impedance (load plus resistor) is reduced to a level that prevents false triggering due to the operating current. The resistance value should also be selected to ensure that the minimum current required to operate the load is provided. This resistance value is normally in the range of 4.5 kΩ to 7.5 kΩ. A general rule is to use a 5 kΩ, 5 W resistor for most conditions.

Chapter 17—Solid State Relays 455

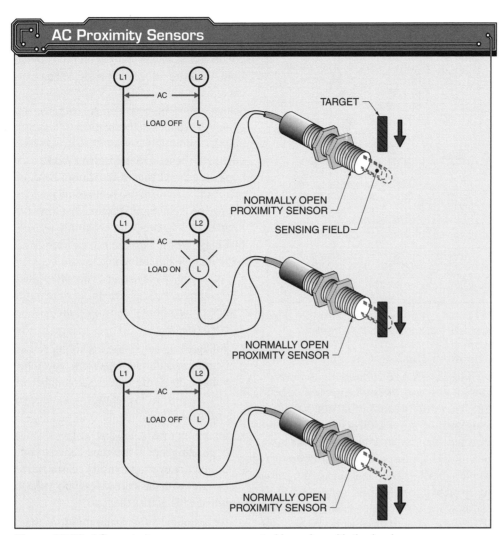

Figure 17-29. *AC proximity sensors are connected in series with the load.*

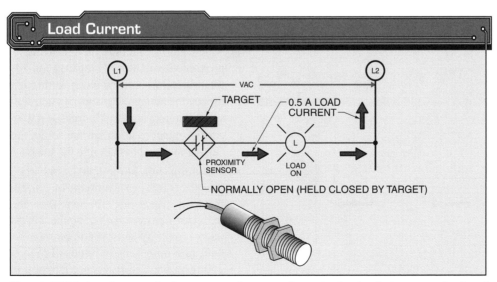

Figure 17-30. *Load current is the amount of current drawn by the load when energized and flows through the proximity sensor.*

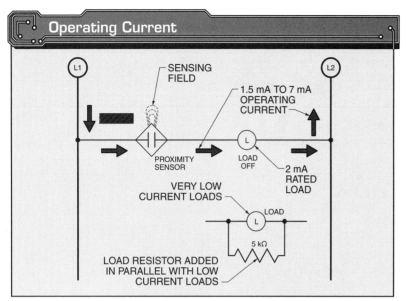

Figure 17-31. Operating current is the amount of current a sensor draws from the power lines to develop a field that can detect the target.

Minimum Holding Current. *Minimum holding current* is the minimum amount of current required to keep a sensor operating. When the proximity sensor has been triggered and is in the ON condition (target detected), the current drawn by the load must be sufficient to keep the sensor operating. Minimum holding currents range from 3 mA to 20 mA for most solid state proximity sensors. The amount of current a load draws is important for the proper operation of a proximity sensor. Excessive current (operating current) burns up the sensor. Low current (minimum holding current) prevents proper operation of the sensor. **See Figure 17-32.**

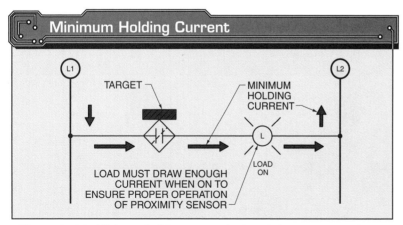

Figure 17-32. Minimum holding current is the minimum amount of current required to keep a sensor operating.

Series/Parallel Connections

All AC, two-wire proximity sensors may be connected in series and parallel to provide both AND and OR control logic. When connected in series (AND logic), all proximity sensors must be activated to energize the load. When connected in parallel (OR logic), any one proximity sensor that is activated energizes the load. **See Figure 17-33.**

As a general rule, a maximum of three proximity sensors may be connected in series to provide AND logic. Factors that limit the number of AC, two-wire proximity sensors that may be wired in series to provide AND logic include the following:

- AC supply voltage (generally, the higher the supply voltage, the higher the number of proximity sensors that may be wired in series)
- voltage drop across the proximity sensor (varies for different sensors; the lower the voltage drop, the higher the number of proximity sensors that may be connected in series)
- minimum operating load voltage (varies depending upon the load that is controlled; for every proximity sensor added in series with the load, less supply voltage is available across the load)

As a general rule, a maximum of three proximity sensors may be connected in parallel to provide OR logic. Factors that limit the number of AC, two-wire proximity sensors that may be wired in parallel to provide OR logic include the following:

- proximity switch operating current (the total operating current flowing through the load is equal to the sum of each sensor's operating current; the total operating current must be less than the minimum current required to energize the load)
- amount of current the load draws when energized (the total amount of current the load draws must be less than the maximum current rating of the lowest rated proximity sensor; for example, if three proximity sensors rated at 125 mA, 250 mA, and 275 mA are connected in parallel, the maximum rating of the load cannot exceed 125 mA)

Chapter 17—Solid State Relays   457

Series/Parallel Connections

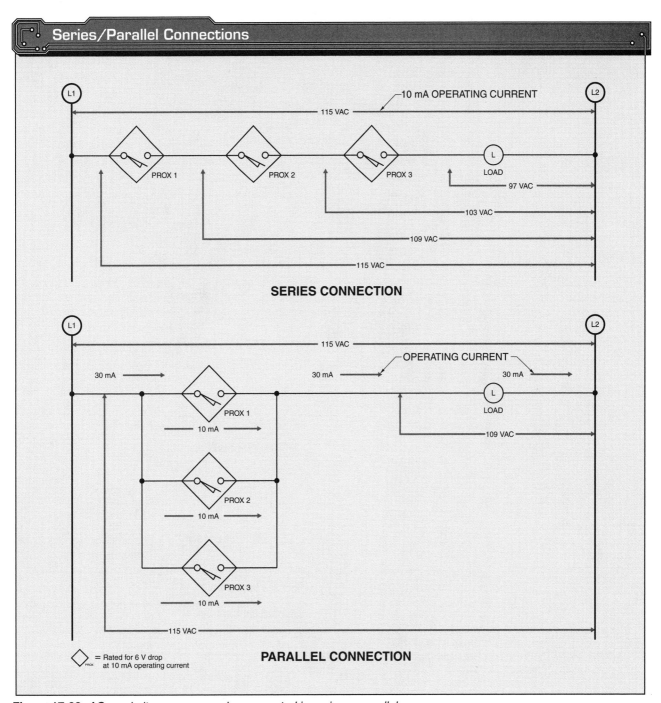

Figure 17-33. *AC proximity sensors may be connected in series or parallel.*

MOTOR STARTER APPLICATIONS

A solid state relay may be used to interface a computer that controls a motor. Solid state relays are used to control high-power loads with low-power electronic circuits. **See Figure 17-34.** The computer could never be directly connected to control the motor without damaging the electronic circuits of the computer. The solid state relay provides an interface for controlling the motor through the computer. In this application, the solid state relay also provides a high electrical isolation between the computer and the motor.

458 SOLID STATE DEVICES AND SYSTEMS

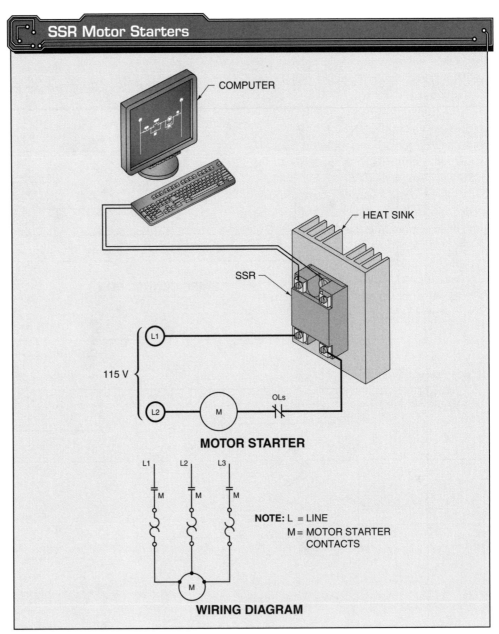

Figure 17-34. *Solid state relays are used to control high-power loads with low-power electronic circuits.*

Refer to Chapter 17 Quick Quiz® on CD-ROM

High electrical isolation prevents high-voltage surges and transients produced in the motor circuit from being transmitted to the electronic circuits of the computer. High-voltage surges and transients can destroy electronic circuits. The solid state relay controls the load and isolates the two circuits from each other.

Solid state relays, if properly applied, can switch billions of times because they have no moving parts. Solid state relays also provide fast-acting, arc-free switching. Because solid state relays have no exposed electronic parts, their circuits are not affected by outside environments, such as dusty conditions or volatile gases. However, heat affects the electronic circuits inside the relay and special care must be taken to prevent exposure to excessive heat.

SOFT START APPLICATIONS

Solid state controls are used in many applications that previously used mechanical switches. For example, SCRs and triacs are now used to control motors. The use of solid state controls to operate motors eliminates electrical contacts that may arc and cause a potential explosion hazard. Solid state controls allow motors to be started and stopped in a controlled manner. Solid state controls can soft start and soft stop a motor. Soft starting and soft stopping is the application and removal of voltage to and from the motor in a controlled manner. Soft starts and soft stops reduce mechanical stress on motors, shafts, gear boxes, belts, and products being moved. **See Figure 17-35.**

Rockwell Automation, Allen-Bradley Company, Inc.
Solid state motor starters use solid state components to turn motors ON and OFF, therefore electromechanical components are not required.

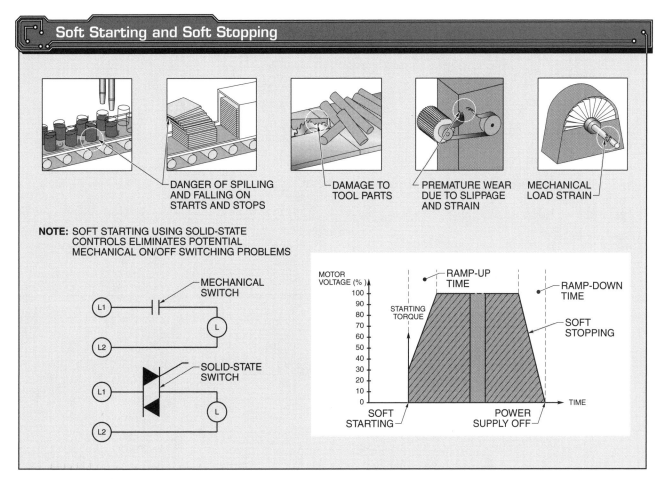

Figure 17-35. *Soft starting and soft stopping is the application and removal of voltage to and from the motor in a controlled manner.*

KEY TERMS

- relay
- solid state relay
- zero switching relay
- instant-on switching relay
- peak switching relay
- analog switching relay
- input circuit
- control circuit
- output (load-switching) circuit
- thermal resistance (R_{TH})
- two-wire solid state switch
- three-wire solid state switch
- exact replacement method
- circuit analysis method
- transient voltage
- half-waving
- RC circuit
- proximity sensors
- load current
- operating (residual or leakage) current
- minimum holding current
- soft start
- soft stop

REVIEW QUESTIONS

1. What is a relay?
2. Describe solid state relays (SSRs).
3. What is a zero switching relay?
4. What is an instant-on switching relay?
5. What is a peak switching relay?
6. What is an analog switching relay?
7. Explain an input circuit.
8. Explain a control circuit.
9. Explain an output (load-switching) circuit.
10. What is thermal resistance (R_{TH})?
11. What is a two-wire solid state switch?
12. What is a three-wire solid state switch?
13. What is the exact replacement method?
14. What is the circuit analysis method?
15. What is transient voltage?
16. What is half-waving?
17. What is an RC circuit?
18. Explain load current.
19. Explain operating current.
20. Explain minimum holding current.

SOLID STATE TECHNOLOGY IN PROGRAMMABLE CONTROLLERS 18

The programmable logic controller (PLC) typifies one of the most common and somewhat complex electronic systems. A PLC contains many solid state components such as microchips, sensors, and switches and thus can be used for many applications. Proper troubleshooting procedures must be followed when troubleshooting a PLC system or subsystem.

OBJECTIVES

- Explain programmable logic controllers (PLCs).
- Describe areas of PLC applications.
- List and describe the PLC sections.
- Describe interfacing circuits.
- Explain the troubleshooting of PLC systems.

PROGRAMMABLE LOGIC CONTROLLERS

A *programmable logic controller (PLC)* is a solid state control device that is programmed and reprogrammed to automatically control processes or machines. Programmable memory is used to store instructions by implementing functions, such as logic sequencing, timing, and counting, for process and machine control.

The PLC was first developed in response to the needs of the automotive industry. Annual model changes required constant modifications of production equipment controlled by relay circuitry. In some cases, entire control panels had to be scrapped and new ones designed and built with new components. This resulted in increased production costs.

The automotive industry was looking for equipment that could reduce the changeover costs associated with model changes. In addition, the equipment had to operate in harsh factory environments of dirty air, vibration, electrical noise, and wide temperature and humidity ranges.

To meet this need, the PLC, a rugged computer-like control, was constructed. The PLC could easily accommodate constant circuit changes using a keyboard to introduce new operation instructions. In 1968, the first PLC was delivered to General Motors (GM) in Detroit by Modicon.

> PLCs must be installed in an enclosure or cabinet. All cables connected to a PLC must remain in the enclosure or be protected by conduit or other means.

The first PLCs were large and costly. They were initially used in large systems with 100 or more relays. Today, PLCs are considerably smaller. PLCs are manufactured from small sizes that control a few input and output terminals (devices and components) to large sizes that control thousands of input and output terminals. PLC system programming can be viewed using software with integrated programming devices, handheld programming devices, operator interface panels, and personal computers. **See Figure 18-1.**

462 SOLID STATE DEVICES AND SYSTEMS

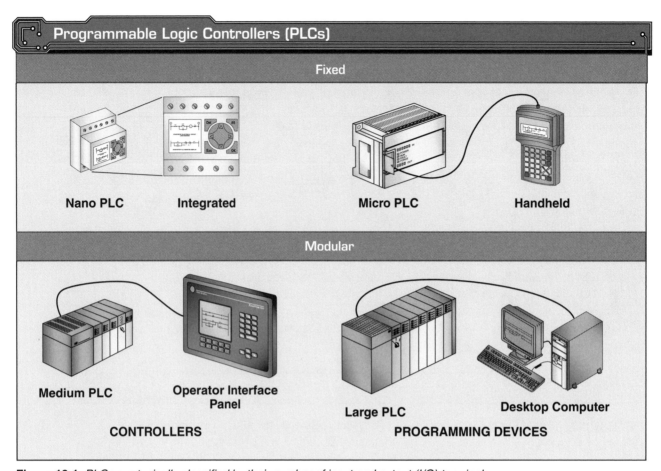

Figure 18-1. *PLCs are typically classified by their number of input and output (I/O) terminals.*

Integrated Programming Devices

An *integrated programming device* is a device that consists of a small LCD and a small keypad or set of buttons that are part of a PLC. Integrated programming devices are typically found on small-size PLCs. A PLC program is entered into the PLC using the keypad and viewed on the LCD display.

Using integrated programming devices is a convenient way to access PLC programs without additional hardware. However, small LCDs provide only limited viewing, and making programming changes with integrated keypads is typically time-consuming.

Handheld Programming Devices

A *handheld programming device* is a separate keypad device that is not an integral part of a PLC. A handheld programming device is about the same size as a personal digital assistant (PDA) but with a larger display and keypad than an integrated programming device. A handheld programming device connects to a PLC with a cable and connector.

As with integrated programming devices, handheld programming devices are typically used with small PLCs. Handheld programming devices are used to enter, copy, display, store, and transfer PLC programs. The ability to copy, store, and transfer a PLC program allows a handheld programming device to be used with multiple PLCs having the same type of programming. Although the display and keypad are larger than that of an integrated programming device, viewing and editing are still limited with handheld programming devices.

Operator Interface Panels

An *operator interface panel,* also known as a human machine interface (HMI), is a display panel that is connected to a PLC by a communication cable. An operator interface panel uses text, graphics, or a combination of text and graphics to provide real-time representation of a process or application. Buttons on an operator interface panel provide access to different screens and/or control functionality of the process depicted.

Operator interface panels come in a variety of sizes, with either monochrome or color displays. Although PLCs cannot be directly programmed from operator interface panels, the parameters of certain instructions can be adjusted, such as counter and timer preset values.

Personal Computers

Personal computers (PCs) are the most common programming device used with PLCs. Desktop and laptop PCs are used to program PLCs through an interface cable and connectors. Software and interface cables are specific to a PLC family and manufacturer and cannot be used on other PLC families of the same manufacturer or with a PLC of a different manufacturer. PCs are used as programming devices on all sizes of PLCs.

PLC APPLICATIONS

Four typical areas in which PLCs are used include medical, manufacturing, mining, and building automation. **See Figure 18-2.** Industries requiring high-speed monitoring and control of processes depend on the use of PLCs.

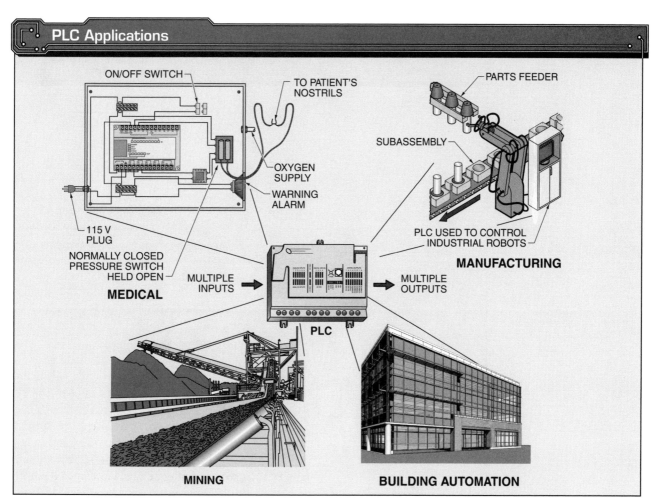

Figure 18-2. *PLC applications include medical, manufacturing, mining, and building automation applications.*

Beverage and food packaging industries use PLCs for conveyor belt control, container positioning, sensing, fill level detection, and counting products. PLCs are used in highway traffic signals and parking lots for timing and counting functions. PLCs are used in pipelines and water and wastewater facilities for flow detection and regulation. Airport baggage handling is also accomplished with PLC usage.

The future of programmable controllers relies not only on the continuation of new product developments but also on the integration of PLCs with other control and factory management equipment. PLCs are being incorporated, through networks, into computer-integrated manufacturing (CIM) systems. The power and resources of PLCs are being combined with numerical controls, robots, CAD/CAM systems, personal computers, management information systems, and computer-based systems. Programmable controllers will continue to play a substantial role in future control applications.

PLC SECTIONS

Although PLC sections are found in various shapes, sizes, and forms, they are physically constructed as individual fixed packages or modular designs, which allow greater design flexibility. **See Figure 18-3.** In many cases, programming is done off-line and the programming device is not shown.

All PLCs have five basic sections. A PLC consists of a processor, input, output, programming, and power supply section. The processor is used to control the systems. The input section is used to receive information from pushbuttons, limit switches, sensors, etc.

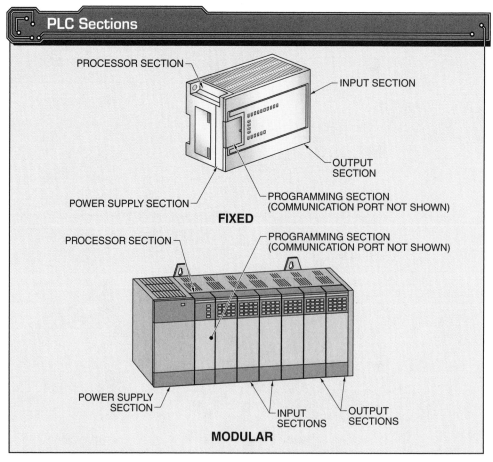

Figure 18-3. *A PLC consists of a processor, input, output, programming, and power supply section.*

The output section is used to control magnetic starter coils, lights, solenoids, etc. The programming section is used to allow technicians to view current programming and future edits. Using a schematic diagram of a modular programmable logic system is the easiest way to show how the basic sections of a PLC relate to one another. **See Figure 18-4.**

The programs used in manufacturing parts and equipment and in processing goods and other consumables are stored in and retrieved from memory as required. Sections of the PLC are interconnected and work together to allow the PLC to accept inputs from a variety of sensors, make logical decisions as programmed, and control outputs such as motor starters, solenoids, valves, and drives.

All PLC wiring should be installed in a neat and professional manner.

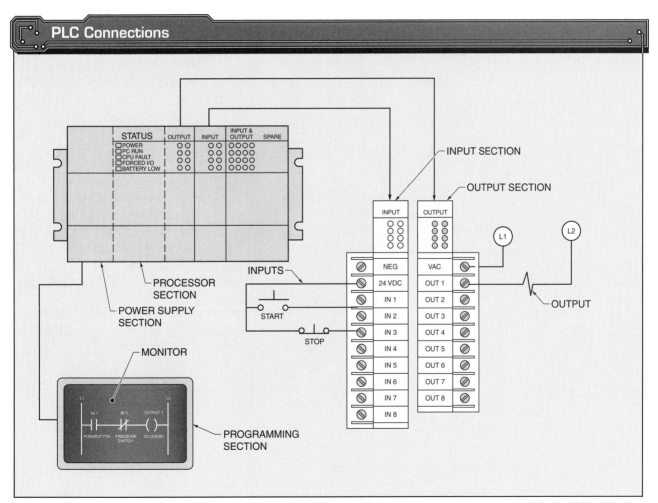

Figure 18-4. *A schematic diagram and pictorial are used to show how PLC sections are interconnected.*

Processor Section

A *processor section* is the section of a PLC that organizes all control activity by receiving inputs, performing logical decisions according to the program, and controlling outputs. **See Figure 18-5.** The processor section is the brain of the PLC.

The processor section evaluates all input signals and levels. This data is compared to the memory in the PLC, which contains the logic of how the inputs are interconnected in the circuit. The interconnections are programmed into the processor by the programming section. The processor section controls the outputs based on the input conditions and the program. The processor continuously examines the status of the inputs and outputs and updates them according to the program. **See Figure 18-6.**

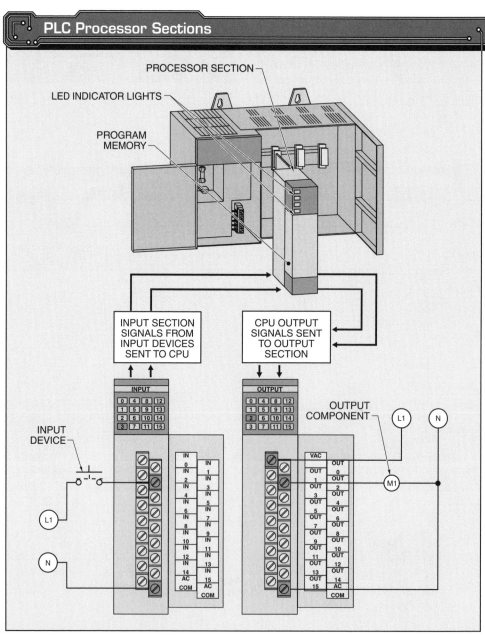

Figure 18-5. *The processor section organizes all control activity by receiving inputs, performing logical decisions according to the program, and controlling outputs.*

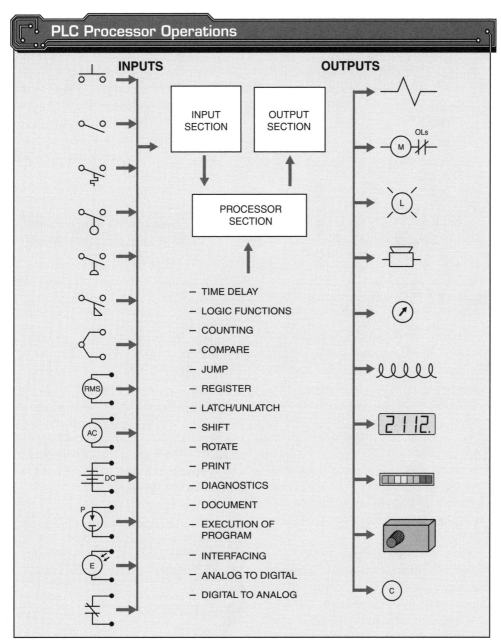

Figure 18-6. *The processor receives signals from input devices, compares the signals with information in the PLC program, and sends signals through the output section to turn output components ON and OFF.*

PLC Memory. *Memory* is an electronic storage device for digital data used in computers. Memory consists of a solid state microchip, onto which data is stored at specific addresses, inside the processor section. The processor of a PLC contains memory in which the PLC program and related data are stored. The two areas in the memory of a PLC are program files and data. Program files are files that contain the user-developed control program (PLC program). A data file is the section of PLC memory that contains the status of inputs and outputs, timer preset values, counter preset values, and other program instruction values. **See Figure 18-7.**

468 SOLID STATE DEVICES AND SYSTEMS

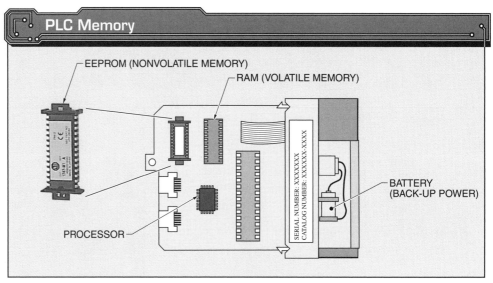

Figure 18-7. *PLC memory includes RAM and EEPROM.*

Memory can be volatile or nonvolatile. In the event of a power failure, volatile memory loses stored data while nonvolatile memory remains intact. Most PLCs contain both types of memory. *Random access memory (RAM)* is a type of computer memory that is lost when power is lost. RAM is used by PLCs to store program files and data. A PLC program is run while stored in RAM memory. A battery is used to provide back-up power so that data will not be lost when normal power is turned OFF or lost.

Erasable programmable read-only memory (EPROM) is a type of computer memory that can be programmed and erased, enabling it to be reused. Erasure is accomplished using ultraviolet (UV) light that shines through a quartz erasing window in the EPROM package. Electronically erasable programmable read-only memory (EEPROM) is similar to EPROM, but the erasure is accomplished using an electric field instead of a UV light source. This eliminates the need for a window.

EEPROM is a type of memory that is retained when power is lost. EEPROM is used in PLCs to provide a backup for the RAM memory. A copy of the PLC RAM program is stored in EEPROM memory. Typically, data is transferred from RAM to EEPROM upon power loss. At power up, data (including the PLC program) is transferred from EEPROM to RAM. The EEPROM can be an integral part of a PLC or a separate unit that plugs into a PLC socket. Most PLCs use a type of solid state EEPROM chip as a memory module. EEPROMs are configured using an EEPROM program that provides voltage at specified levels, depending on the type of EEPROM used.

> *Solid state motor starters produce minimal electronic noise. This is beneficial if PLCs, computers, microprocessor-based equipment, or other devices that may be affected by electronic noise are located nearby.*

PLC Input Section

An *input section* is the section of a PLC that receives signals from input devices and sends the signals to a CPU. Pushbuttons, limit switches, and proximity switches are common input devices. **See Figure 18-8.** Each input device is connected to a specific screw terminal that corresponds to a specific location (address) in the memory of the CPU. Typically, each PLC input (screw terminal) has an LED to indicate when an input device is closed (sending signal) and voltage is present at the input terminal.

Chapter 18—Solid State Technology in Programmable Controllers **469**

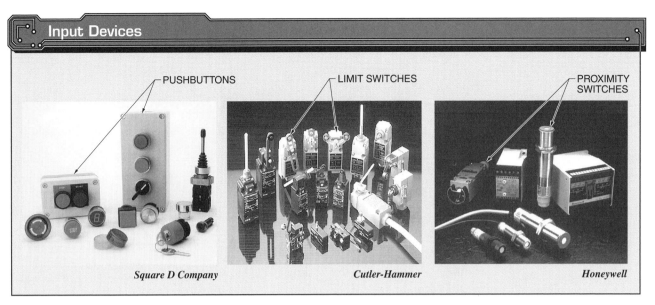

Figure 18-8. *Input devices, such as pushbuttons, limit switches, and proximity switches, are connected to the input modules of PLCs.*

PLC Input Section Internal Circuitry. Input devices typically operate at voltages higher than the voltage used by the CPU (5 V). Optical isolation is used by PLCs to convert high-voltage input signals to a voltage level the CPU can use. Inside AC-powered PLCs, AC input signals are converted to DC by a rectifier, and the DC signal turns on an LED. **See Figure 18-9.** The light from the LED is detected by a phototransistor, and the phototransistor sends a signal to the CPU. In most cases, the phototransistor and the CPU operate at the same voltage level.

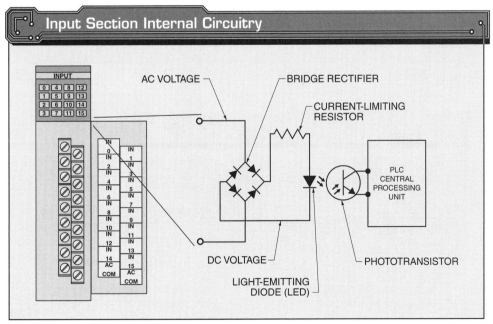

Figure 18-9. *The input section of a PLC is the section that receives signals from input devices and sends the signals through its internal circuitry to the CPU.*

The input section can be either located on the PLC (onboard) or part of an expansion module. An onboard input terminal is a permanent part of the PLC package. Expansion modules are removable units. Various input module types allow analog and digital input devices to send information to the PLC. An analog input device sends a continuously changing variable signal to the PLC. Temperature, pressure, flow, and level sensors are common analog input devices for a PLC-controlled system. ON/OFF switches such as pushbuttons, limit switches, and toggle switches are common digital input devices. A digital input device is a device that is either ON or OFF (open or closed).

PLC Input Voltage Ratings. All PLC input devices, such as pushbuttons, level switches, and proximity switches, are connected to input terminals. The input device circuit must be powered by a specified voltage or voltage range. PLC input terminals are available with DC, AC, or DC/AC ratings.

A low-voltage-rated input terminal (12 VDC or 24 VDC) is the most common and preferred type of PLC input terminal. Lower voltages are safer and provide for an easier installation because low voltages do not require input wires to be run in conduit. Many PLCs provide 12 VDC or 24 VDC output power terminals that can also be used to power input terminals.

See Figure 18-10. AC-rated PLC input terminals cannot be powered by DC power. Likewise, DC-rated PLC input terminals cannot be powered by AC power.

> *When inspecting a de-energized PLC, electrostatic discharge from the human body can damage integrated circuits and semiconductors of the PLC when internal components are touched.*

PLC Input Current Ratings. PLCs are designed for DC or AC input voltages. When using switches with mechanical contacts, such as standard pushbuttons or mechanical limit switches, the switches can be connected to either DC input terminals or AC input terminals. A DC-rated solid state input switch should only be connected to the input terminal of a DC-rated PLC. An AC-rated solid state input device should only be connected to the input terminal of an AC-rated PLC.

The amount of current in the input circuit of a PLC is low because the internal circuitry of the PLC input section is the load of the input circuit. The internal circuitry only requires a few milliamperes (mA) of electron flow to activate. A typical PLC input section only draws about 5 mA to 20 mA of current. **See Figure 18-11.** However, low-current inputs of a PLC are capable of controlling high-current outputs with the appropriate interfacing circuitry.

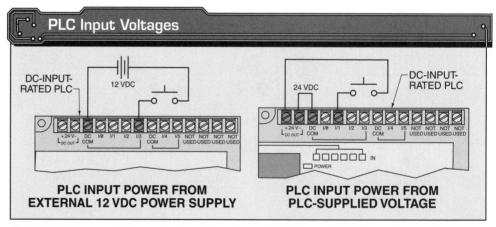

Figure 18-10. *Low-voltage-rated input terminals (12 VDC or 24 VDC) are the most common and preferred PLC input terminals.*

Chapter 18—Solid State Technology in Programmable Controllers 471

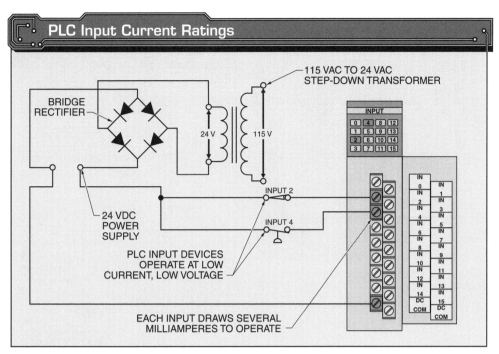

Figure 18-11. *The internal circuitry of a PLC only requires a few milliamperes of electron flow to be activated.*

DC Input Switching. DC power is commonly used with PLC input circuits. The advantage of using DC is that a DC supply is kept to a low level (typically 12 VDC or 24 VDC) for safety. Also, most photoelectric and proximity sensors typically use transistors as the switching element and are rated to operate in a range between 10 VDC and 30 VDC.

The two types of DC input switches are two-wire and three-wire input switches. A two-wire input switch circuit uses solid state (photoelectric and proximity) switches or mechanical (limit) switches for switching. A three-wire input switch circuit uses solid state switches for switching. **See Figure 18-12.**

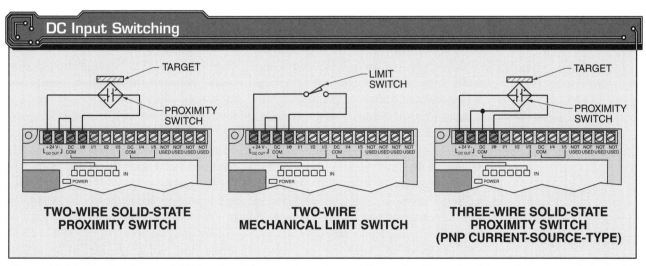

Figure 18-12. *The two types of DC input switches are two-wire and three-wire. The solid state switches use photoelectric and proximity sensors for switching.*

472 SOLID STATE DEVICES AND SYSTEMS

Most technicians wear Class E protective helmets for added protection against low-voltage and high-voltage electric shock and burns.

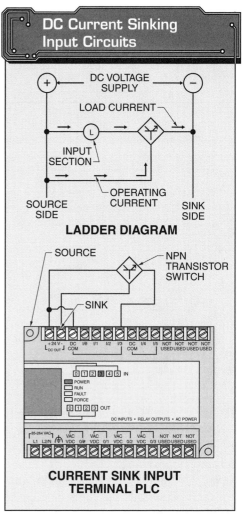

Figure 18-13. *Switching devices that use NPN transistors as switching elements are called current sinking, negative switching, or NPN devices.*

A three-wire input switch circuit typically uses an NPN or PNP transistor as the switching device. For most applications, the exact transistor used does not matter as long as the switch is properly connected into the input circuit of the PLC. However, NPN transistor switches are far more common than PNP transistor switches.

PLC input terminals are rated as current sink, current source, or current sink/source. PLC input terminals rated as current sink/source can be used with NPN or PNP solid state switching devices.

When using an NPN transistor-type input device, the load (PLC input section) is connected between the positive terminal of the supply voltage and the output terminal (collector) of the switch or sensor. When a switch, such as a photoelectric or proximity sensor, detects a target, electrons flow through the transistor switch, and the PLC input terminal is energized. **See Figure 18-13.**

Switching devices that use NPN transistors as the switching element are called current sink, negative switching, or NPN devices. A sink is the negative power supply terminal of a DC-powered PLC. Sinks use conventional flow through NPN transistors. A current-sinking switch "sinks" current from the load.

When using a PNP transistor-type input device, the load (PLC input section) is connected between the negative terminal of the supply voltage and the output terminal (collector) of the switch or sensor. When the switch detects a target, electrons flow through the transistor, and the PLC input terminal is energized. **See Figure 18-14.** PLC DC input devices can be configured as sinking or sourcing, depending on how the DC COM terminal is wired into the input circuit device.

Switching devices that use PNP transistors as the switching element are called current source, positive switching, or PNP devices. A source is the positive terminal of a DC-powered PLC. Sources use electron current flow through PNP transistors. A current-sourcing switch "sources" current to the load.

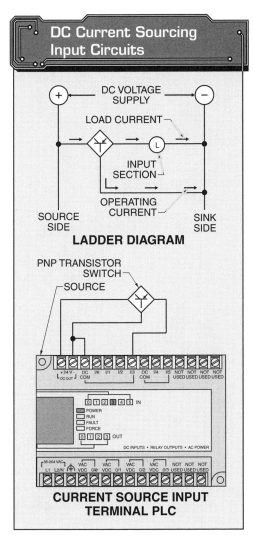

Figure 18-14. Switching devices that use PNP transistors as switching elements are called current sourcing, positive switching, or PNP devices.

Troubleshooting Input Modules. It is very important to choose the proper testing equipment when working on any electronic system. Some types of testing equipment can actually cause failures in the system they are testing. One example is a vibrating voltage tester (wiggie). It must never be used to measure the voltage level at any input to a PLC. A vibrating voltage tester contains a solenoid. When the test leads of a vibrating voltage tester are removed, the collapsing field of the solenoid can damage the solid state components of the input module.

Signals and information sent to a PLC using input devices are connected to the input module of the PLC. The controller cannot function if the proper information from the input device or input module is not communicated. **See Figure 18-15.** When troubleshooting, it is important to start with the input section and see if it is operating properly since there could be multiple input devices feeding the input module. To troubleshoot the input module of a PLC, apply the following procedure:

1. Measure the supply voltage at the input module or card to ensure that the required power is supplied to the input device(s) (voltage can vary as required). Test the main power supply of the controller when there is no power.

2. Measure the voltage from the control switch. Connect the DMM directly to the same terminal screw to which the input device is connected. The DMM should read the supply voltage when the control switch is closed. The DMM should read the full supply voltage when the control device uses mechanical contacts. The DMM should read nearly the full supply voltage when the control device is solid state. Full supply voltage is not read because 0.5 V to 6 V is dropped across the solid state control device. The DMM should read zero or little voltage when the control switch is open.

3. Monitor the status indicators on the input module. The status indicators should illuminate when the DMM indicates the presence of supply voltage.

4. Monitor the input device symbol on the programming terminal monitor. The symbol should be highlighted when the DMM indicates the presence of supply voltage. Replace the control device if the control device does not deliver the proper voltage. Replace the input module if the control device delivers the correct voltage but the status indicator does not illuminate.

474 SOLID STATE DEVICES AND SYSTEMS

Troubleshooting Input Modules

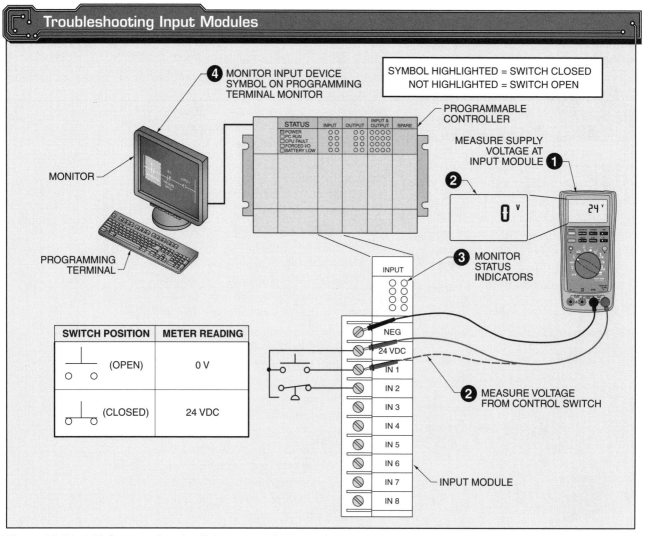

Figure 18-15. *A PLC cannot function if the proper information from the input device or input module is not communicated.*

Troubleshooting Input Devices. Input devices, such as pushbuttons, limit switches, pressure switches, and temperature switches, are connected to the input modules of a PLC. Input devices send information and data concerning circuit and process conditions to the controller. The processor receives the information from the input devices and executes the program. All input devices must operate correctly for the circuit to operate properly. **See Figure 18-16.** To troubleshoot the input device of a PLC, apply the following procedure:

1. Place the controller in the Test or Program mode. This step prevents the output devices from turning ON unexpectedly and creating a hazard. Output devices will be turned ON when the controller is placed in the Run mode.

2. Monitor the input devices. The input devices can be monitored using the input status indicators on each input module, the programming terminal monitor, or the data file. The data file consists of data values (inputs, timers, counters, and outputs) that are displayed on the monitor to give additional information regarding what should be taking place in the control process. This information often looks like truth tables.

3. Manually operate each input starting with the first input. Never reach into a machine when manually operating an input. Always use a wooden stick or other nonconductive device.

The input status indicator located on the input module will illuminate, and the input symbol will be highlighted in the control circuit on the monitor screen when a normally open input device is closed.

The input status indicator located on the input module should turn OFF. The input symbol should no longer be highlighted in the control circuit on the monitor screen when a normally closed input device is open. The next input device should be selected. Each input device should be tested until all inputs have been tested.

PLC Output Sections

An *output section* is the section of a PLC designed to deliver the output voltage required to control alarms, lights, solenoids, and other loads. The output section receives low-power digital signals from the processor and converts them into high-power signals. These high-power signals can drive loads that can light, move, grip, rotate, extend, release, heat, and perform a multitude of other functions. Onboard outputs usually contain a fixed number of outputs that will define the limits of the PLC. For example, a small PLC may include only eight outputs. Therefore, the PLC may only control eight devices unless an expansion module is added to the system.

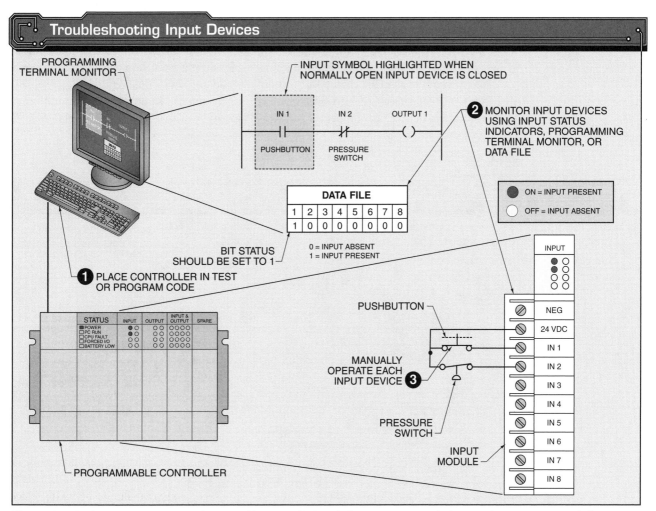

Figure 18-16. *All input devices, such as pushbuttons, limit switches, pressure switches, and temperature switches, must operate correctly for the circuit to operate properly.*

476 SOLID STATE DEVICES AND SYSTEMS

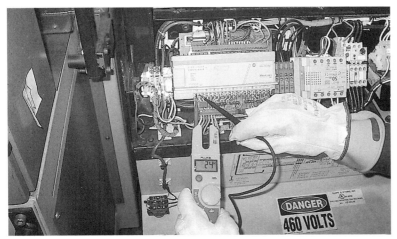

Controlling the coil of three-phase motor starters is a very common application for PLCs.

PLC Output Section Internal Circuitry.

Output components typically operate at a voltage that is higher than the operating voltage of the CPU (5 V). Optical isolation is used to convert low-level voltage signals from the CPU to a voltage that an output component can use. Inside the PLC, low-level voltage signals are used to turn ON an LED. The light from the LED is detected by a phototransistor, which turns on a triac. Triacs are used to turn output components ON and OFF. The CPU and the LED of the phototransistor operate at the same voltage level. **See Figure 18-18.**

Solid state outputs normally have fusing on the module to protect the triac or transistor from moderate overloads. If the output does not have internal fuses, then fuses should be installed externally (normally at the terminal block) during the initial installation. When adding fuses to an output circuit, manufacturer recommendations for the particular module should be followed. Only a properly rated fuse will ensure that the fuse will open quickly in an overload condition to avoid overheating of the output switching device.

Pilot lights, control relays, motor starters, and solenoids are typical components controlled by PLCs. **See Figure 18-17.** Each output component is connected to a specific screw terminal of the output section that corresponds to a specific location in the memory of the CPU. Each PLC output screw terminal has an LED to indicate when voltage to energize an output component is present at the terminal.

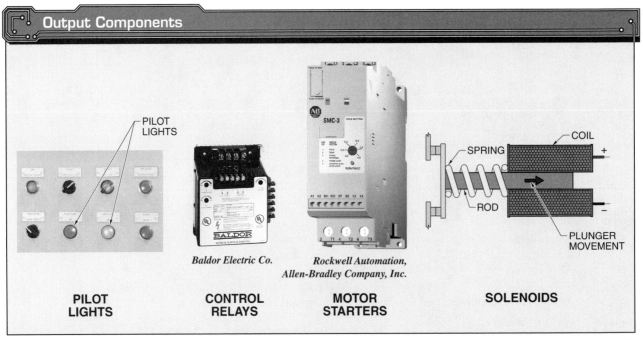

Figure 18-17. *Output components, such as pilot lights, control relays, motor starters, and solenoids, are connected to the output modules of a PLC.*

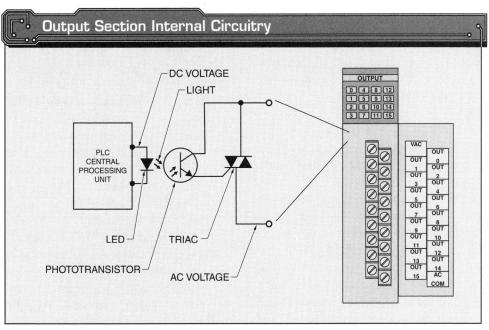

Figure 18-18. *The output section of a PLC is the section that receives signals from the CPU through internal circuitry and sends the signals through to the output components.*

PLC Output Voltage Ratings. All PLCs have output components such as lamps, solenoids, and motor starters connected to output terminals. The output terminals of a PLC are powered by a specific voltage or voltage range. The type of voltage (AC or DC) and voltage level (12 V, 24 V, or 115 V) used for a PLC are based on the loads controlled by the PLC. Loads that require more power typically have higher voltage ratings because power (*P*) is equal to voltage (*E*) multiplied by current (*I*). The higher the voltage rating for a load, the lower the current rating for any given power rating.

The output terminal switches of a PLC are either solid state switches or mechanical contact switches. **See Figure 18-19.** The advantage of a mechanical contact output terminal is that a mechanical contact can switch either AC or DC loads. An AC solid state output terminal of a PLC can only switch AC loads and a DC solid state output terminal can only switch DC loads. Solid state output switches of PLCs include triacs for switching AC circuits and transistors and SCRs for switching DC circuits.

PLC Output Terminal Switching

Output Type	Symbol
Transistor NPN or PNP DC switch; used to switch low-level DC	CATHODE, BASE, EMITTER — NPN; CATHODE, BASE, EMITTER — PNP
SCR DC switching device; one SCR used to switch high-level currents	ANODE, CATHODE, GATE — SCR
Triac AC switching device	TRIAC
Mechanical contact AC or DC switching device; used to switch high- or low-level currents	CONTACT; CONTACT

Figure 18-19. *The output sections of PLCs use transistors, triacs, and solid state switches or mechanical contact switches.*

Overtravel, interlocks, and stop pushbutton safety circuits of a machine must be hardwired in series directly to the master control relay, which allows the relay to be de-energized.

PLC Output Current Ratings. PLC output components connected to the output section of a PLC determine the amount of current a PLC must carry. The amount of current drawn by an output component can vary from a few milliamps (small indicating lamps), to several amps (solenoids, higher wattage lamps, small motors, and large-motor motor starters), to hundreds of amps (large three-phase motors and three-phase heating elements).

The current rating of PLC output circuitry is listed for the amount of current the internal switches can safely handle when the load is first turned ON ("make" current rating), when the load is operating ("continuous" current rating), and when the load is turned OFF ("break" current rating).

When AC loads are running, the break current rating will always be much lower than the make or continuous current ratings. This is because when an AC circuit is turned OFF, a large damaging current will try to arc across any switch that is beginning to open. With AC circuits, the arc dissipates somewhat because AC current is always fluctuating between zero and peak current. However, in a DC circuit there is no fluctuation, making switching DC circuits more damaging. Because of the possible damage, a PLC output relay contact or output switch will have a much lower DC rating than AC rating for the same output switch.

Troubleshooting Output Modules. A PLC turns ON and OFF the output devices (loads) in the circuit according to the process or work it was designed to control. The output devices are connected to the output module of the PLC. No work is performed in the circuit when the output module or the output devices are not operating correctly. The problem may lie in the output module, output device, or controller when an output device does not operate. **See Figure 18-20.** To troubleshoot the output module of a PLC, apply the following procedure:

1. Measure the supply voltage at the output module to ensure that there is power supplied to the output devices. Test the main power supply of the controller for voltage supplied to the output devices. Testing for voltage can be done in the Run mode but would be easier when using the Forced mode.

2. Measure the voltage available from the output module. Connect the DMM directly to the same terminal screw to which the output device is connected. The DMM should read the supply voltage when the program energizes the output device. The DMM should read full supply voltage when the output module uses mechanical contacts. The DMM should read almost full supply voltage when the output module uses a solid state switch. Full voltage is not read across solid state switches because 0.5 V to 6 V is dropped across the solid state switch. The DMM should read zero or little voltage when the program de-energizes the output device.

3. Monitor the status indicators on the output module. The status indicators should be energized when the DMM indicates the presence of the supply voltage.

4. Monitor the output device symbol on the programming terminal monitor. The output device symbol should be highlighted when the DMM indicates the presence of supply voltage. Replace the output module when the output module does not deliver the specified voltage. When the output module does deliver the correct voltage but the output device does not operate, troubleshoot the output device.

Troubleshooting Output Devices. As noted earlier, output devices, such as motor starters, solenoids, contactors, and lights, are connected to the output modules of a PLC. An output device performs the work required for an application. The processor energizes and de-energizes the output devices according to the programming entered. All output devices must operate properly for the circuit to operate properly. **See Figure 18-21.** To troubleshoot the output-device of a PLC, apply the following procedure:

1. Place the controller in the Test or Program mode. Placing the controller in the Test or Program mode prevents the output devices from turning ON. Output devices will be turned ON when the controller is placed in the Run mode. In some cases, it is more accurate to take measurements in the Run mode when full-load conditions are present. All safety issues must be addressed when testing in the Run mode.

2. Monitor the output devices using the output status indicators (located on each output module), the programming terminal monitor, or the data file.

3. Activate the input that controls the first output device. Check the program displayed on the monitor screen to determine which input activates which output device. Never reach into a machine to activate an input.

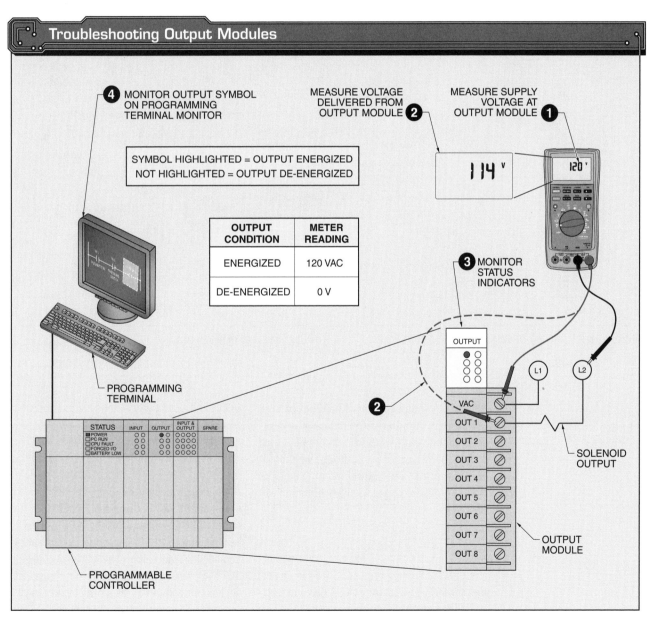

Figure 18-20. *When the output module is not operating properly, the output circuit does not operate properly.*

480 SOLID STATE DEVICES AND SYSTEMS

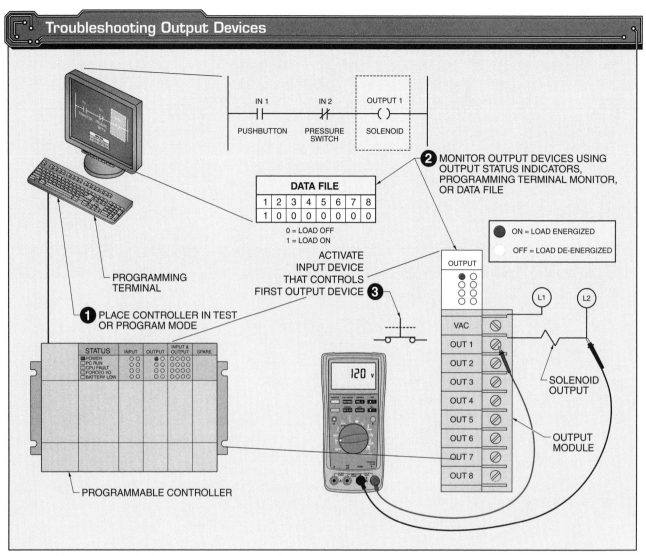

Figure 18-21. *All output devices must operate properly for the output circuit to operate properly.*

The next output device should be selected and tested when the status indicator and associated bit status match. Each output device should be tested. Troubleshooting of input and output devices should occur when the status indicator and associated bit status do not match.

PLC INTERFACING CIRCUITS

Once all loads (lamps, motors, or heating elements) are identified, the exact types of control devices and output components required to make the circuit operate as designed are selected. Low-power output components are connected directly to the output terminals of the PLC as long as the output component voltage, current, and power ratings are less than the maximum rating of the PLC output module. An interface device is required when the power requirements of the load are higher than the ratings of the output module. A high-power load is a load with a voltage and current rating greater than the voltage and current rating of the PLC output. The most common interface devices used to allow PLCs to control high-power loads are relays, contactors, and motor starters. **See Figure 18-22.**

Interfacing Devices

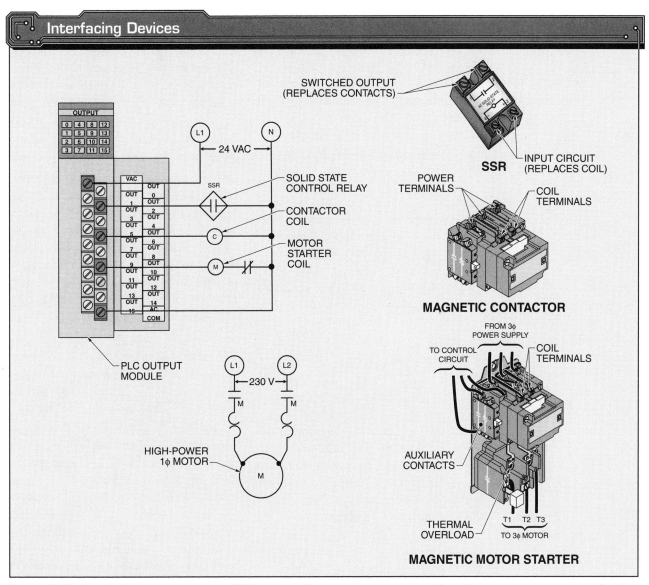

Figure 18-22. Interfacing devices, such as SSRs, magnetic contactors, and magnetic motor starters, are connected to the output modules of a PLC.

Relays

A *control relay* is a device that controls an electrical circuit by opening and closing contacts in another circuit. A *solid state relay (SSR)* is a switching device that has no contacts and switches entirely by electronic means. In an SSR, the coil is replaced by an input circuit and the contacts are replaced with a solid state switching component (transistor, SCR, or triac).

Relays typically have low-power-rated contacts and are used for switching low-power loads such as alarms, small solenoids, and very small (milliwatt) motors. Low-power loads can be connected directly to a PLC as long as the load is less than the PLC output section rating. However, by using a relay between the load and the PLC, the circuit and system is better protected.

The input circuit of an SSR is activated by applying a voltage to the circuit that is higher than the specified pick-up voltage of the relay. Most SSRs have an input voltage range of 3 VDC to 32 VDC. The voltage range allows a single SSR to be used with most electronic circuits and PLC output modules.

The switched circuit (output) of an SSR is controlled by the input signal. Most SSRs use thyristors (for AC switching) or SCRs (for DC switching) as the output switching components. Thyristors and SCRs (or gates) change from an OFF-state (contacts open) to an ON-state (contacts closed) very quickly when the input is energized. The gate of an SSR is switched ON when the relay receives the proper input signal.

The type of relay used for an application depends on the life expectancy, electrical requirements, and cost of the application. SSRs can be rated for billions of operations. In applications that require thousands of switching operations a day, such as with flashing lights and high-speed production lines, SSRs are essential.

Contactors

A *contactor* is a control device that uses a small control current to energize or de-energize the load connected to it. Contactors have high-power-rated contacts and are used for switching high-power loads such as lamps (fluorescent and HID), heating elements, transformers, and small motors (less than ¼ HP) that have built-in overload protection. Contactors are required between the PLC output section and high-power loads.

Motor Starters

A *motor starter* is an electrically operated switch that can turn a motor ON and OFF and includes motor overload protection. Motor starters are used for switching most motors over ¼ HP and all motors over 1 HP. Contactors and motor starters are available in sizes that switch loads ranging from a few amperes to several thousand amperes. Motor starters are required between the PLC output section and high-power loads.

TROUBLESHOOTING PLC SYSTEMS

PLC troubleshooting typically falls into one of two categories: software or hardware. A PLC software malfunction is a condition in which a PLC is properly installed and all inputs and outputs are working but there is a problem with the PLC program. The PLC does not execute the programmed instructions necessary for the completion of the tasks assigned.

PLC Software Problems

Technicians must be aware of the PLC software (program) for the type and model of PLC being serviced and how the electrical control system operates. PLC programs include symbols that represent series-, parallel-, and series-parallel-connected input devices and output components. An example of a PLC software error is two outputs, such as a motor and a light, which cannot be connected in series. **See Figure 18-23.**

PLC Hardware System Problems

As previously indicated, troubleshooting is the systematic elimination of various parts of a system or process to locate a malfunction. A system is a combination of components, units, or modules that are connected to perform work or meet a specific need. A process is a sequence of operations that accomplish desired results. A malfunction is the failure of a system, piece of equipment, or part to operate properly.

Since an electronic system controls the work performed by the electrical system of a machine or process, PLCs are one of the most logical places to start troubleshooting. **See Figure 18-24.** The advantages of starting at the PLC electrical system include the following:

- The input status lights (LEDs) of the PLC electronic input section or module provide a visual indication of which input devices are sending signals to the PLC (input status light ON) and which switches are not sending signals (input status light OFF).

- The PLC input section or module provides a central location for checking all input device signals for proper electrical levels using test instruments.

- The output status lights (LEDs) of the PLC output section or module indicate which output components are energized (output status light ON) and which output components are de-energized (output status light OFF).
- The PLC output section or module provides a central location for checking all output components (lamps and solenoids) for proper signals or system output interfaces (heating/lighting contactors or magnetic motor starters) for proper signal levels using test instruments.
- The operational status lights (LEDs) of a PLC provide a visual indication of when a PLC has power, when it is in the Run mode, which inputs and outputs are being forced, when a CPU fault occurs, and when the back-up battery is low.
- Laptops can be used to monitor a PLC by connecting the PLC to a laptop. In addition to monitoring the status of circuit input devices and output components, the timers, counters, and sequencers of a PLC can also be monitored.
- Once connected to a computer, input terminals can be forced ON or OFF and output terminals can be forced ON or OFF. The ability to force input devices and output components helps to isolate faulty equipment.

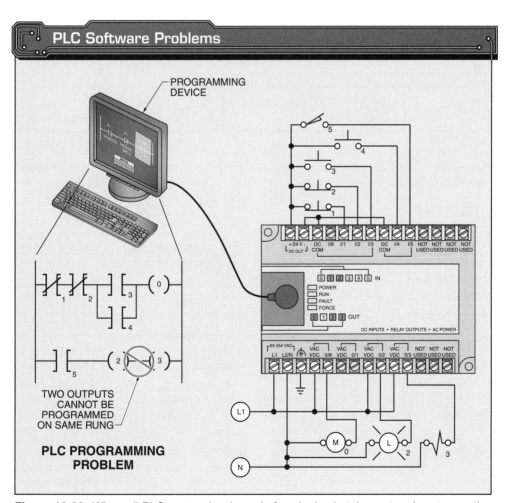

Figure 18-23. When all PLC system hardware is functioning but the system is not operating correctly, the problem may be with the software. For example, the program or symbols may be incorrect, or the program may not be keyed correctly into the PLC.

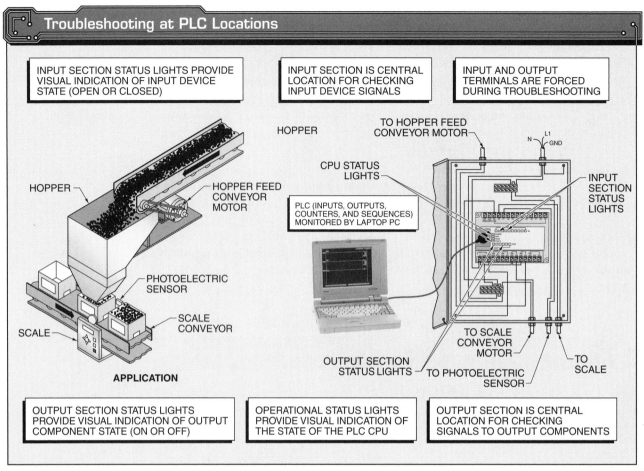

Figure 18-24. *Indicator lights that measure instruments and laptop computers can be used to determine whether a system, piece of equipment, or part is failing to operate properly in a PLC.*

When troubleshooting a PLC, heavy, electronic, noise-generating equipment should not be located too close to the PLC. Unnecessary items should be kept away from any equipment in an enclosure. Leaving items, such as drawings, installation manuals, or other materials, on top of the CPU rack or other rack enclosures can obstruct airflow and create hot spots, which can cause system malfunction. If the electronic system enclosure is in an environment that exhibits vibration, a vibration detector should be installed. This way, the electronic system (PLC) can monitor high levels of vibration, which can lead to loose connections.

Troubleshooting Procedures Based on Knowledge and Experience. Troubleshooting using knowledge and experience is a troubleshooting method used for finding a malfunction in a piece of equipment or process by applying information acquired from past malfunctions. In certain circumstances, troubleshooting using knowledge and experience is only partially effective because the primary malfunction is not corrected. For example, a fuse can blow or a circuit breaker can de-energize a part or all of a PLC-controlled system. Records may indicate that changing the fuse or resetting the breaker allows the system to continue operation, but the reason for the blown fuse or tripped circuit breaker may not be known. **See Figure 18-25.** Troubleshooting using knowledge and experience is improved when the following conditions exist:

- Information about system components is gathered, and the operation of various primary systems are studied and understood.
- Communication with other individuals (supervisors or operators) familiar with the system is used to gain information about the work or process of the equipment or process with which the technician is working. Supervisors or operators familiar with the system may not know the electrical, fluid power, or mechanical reasons that a machine or process works but may know about strange noises or any other unusual behavior that occurred before the malfunction.
- All service calls are documented for future reference. These documents include listings of all troubleshooting findings and repairs as well as all components checked and found to be good when a malfunction is found and corrected. All suggestions that may help prevent the malfunction from recurring are noted for preventive maintenance.
- The technician knows how input devices and output components are monitored and/or forced ON and OFF, how the program can be displayed and printed, and how system changes can be programmed into a PLC, including who is authorized to make any modifications.
- The test instruments used and the measurements taken are documented for future use. For example, voltage measurements taken over time can indicate a power quality problem, such as transients (high-voltage spikes) present on power lines.

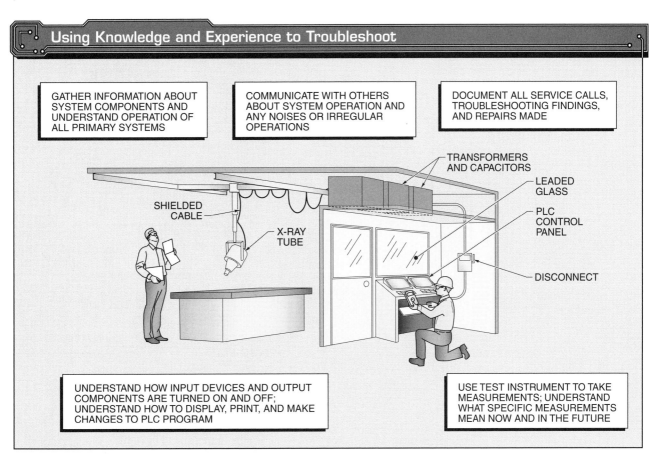

Figure 18-25. *Troubleshooting using knowledge and experience is a troubleshooting method used for finding a malfunction in a piece of equipment or process by applying information acquired from past malfunctions.*

Troubleshooting Using Manufacturer Flow Charts. Some PLC manufacturers include flow charts with the PLCs to aid in troubleshooting. A *flow chart* is a diagram that shows a logical sequence of steps for a given set of conditions. Flow charts help troubleshooters to follow logical paths when trying to solve problems. Symbols and interconnecting lines are used in flow charts to provide analytical direction to the troubleshooting process. **See Figure 18-26.**

The symbol shapes used in flow charts include ellipses, rectangles, diamonds, and arrows. An ellipse symbol is a symbol in a flow chart that indicates the beginning and end of a section of the chart. A rectangle symbol is a symbol in a flow chart that contains a set of instructions. A diamond symbol is a symbol in a flowchart that contains a question, worded so that the answer can be a "yes" or a "no." An arrow symbol is a symbol in a flow chart that indicates the direction to follow through the rest of the chart based on the answers to the questions.

Any change to the program of a PLC can dramatically change the operation of a process or machine.

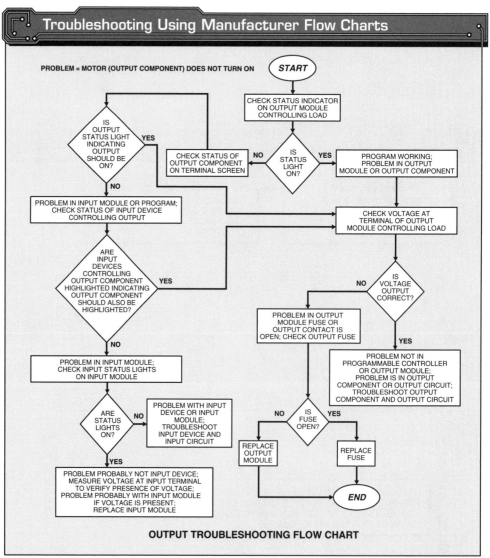

Refer to Chapter 18 Quick Quiz® on CD-ROM

Figure 18-26. *Flow charts help troubleshooters to follow logical paths when trying to solve problems.*

KEY TERMS

- programmable logic controller (PLC)
- integrated programming device
- handheld programming device
- operator interface panel
- personal computer (PC)
- PLC section
- processor section
- memory
- random access memory (RAM)
- erasable programmable read-only memory (EPROM)
- input section
- output section
- control relay
- solid state relay (SSR)
- contactor
- motor starter
- flow chart

REVIEW QUESTIONS

1. What is a programmable logic controller (PLC)?
2. What is an integrated programming device?
3. Define handheld programming device.
4. Explain operator interface panels.
5. Define personal computer (PC).
6. What are the main PLC sections?
7. Explain PLC memory.
8. What is random access memory (RAM)?
9. What is erasable programmable read-only memory (EPROM)?
10. Define control relay.
11. Define solid state relay (SSR).
12. What is a contactor?
13. What is a motor starter?
14. What is a flow chart?

APPENDIX

Ohm's Law 490
Power Formula 490
Voltage, Current, and Impedance Relationship 490
Parallel Circuit Calculations 491
Series Circuit Calculations 491
Power Formulas — 1ϕ, 3ϕ 491
Electrical/Electronic Abbreviations/Acronyms 492
Electrical Symbols 494
Periodic Table of the Elements 496

490 SOLID STATE DEVICES AND SYSTEMS

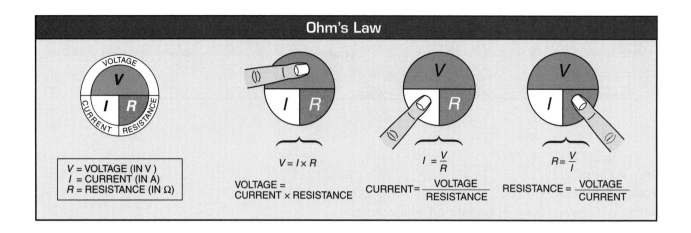

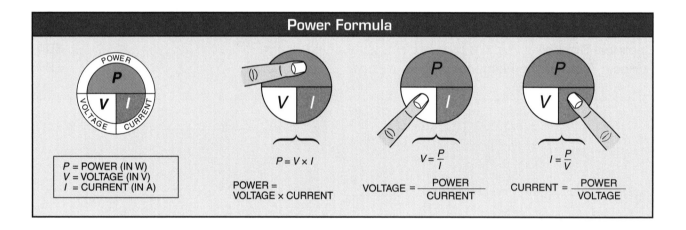

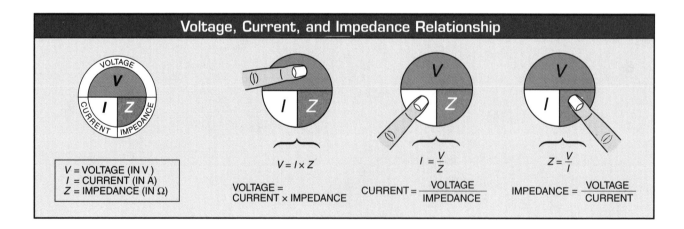

Parallel Circuit Calculations

RESISTANCE

$R_T = \dfrac{R_1 \times R_2}{R_1 + R_2}$

where
R_T = total resistance (in Ω)
R_1 = resistance 1 in (in Ω)
R_2 = resistance 2 in (in Ω)

VOLTAGE

$V_T = V_1 = V_2 = \ldots$

where
V_T = total applied voltage (in V)
V_1 = voltage drop across load 1 in (in V)
V_2 = voltage drop across load 2 in (in V) V)

CURRENT

$I_T = I_1 + I_2 + I_3 + \ldots$

where
I_T = total circuit current (in A)
I_1 = current through load 1 (in A)
I_2 = current through load 2 (in A)
I_3 = current through load 3 (in A)

Series Circuit Calculations

RESISTANCE

$R_T = R_1 + R_2 + R_3 + \ldots$

where
R_T = total resistance (in Ω)
R_1 = resistance 1 in (in Ω)
R_2 = resistance 2 in (in Ω)
R_3 = resistance 3 in (in Ω)

VOLTAGE

$V_T = V_1 + V_2 + V_3 + \ldots$

where
V_T = total applied voltage (in V)
V_1 = voltage drop across load 1 (in V)
V_2 = voltage drop across load 2 (in V)
V_3 = voltage drop across load 3 (in V)

CURRENT

$I_T = I_1 = I_2 = I_3 = \ldots$

where
I_T = total circuit current (in A)
I_1 = current through load 1 (in A)
I_2 = current through load 2 (in A)
I_3 = current through load 3 (in A)

Power Formulas – 1ϕ, 3ϕ

Phase	To Find	Use Formula	Example Given	Example Find	Example Solution
1ϕ	I	$I = \dfrac{VA}{V}$	32,000 VA, 240 V	I	$I = \dfrac{VA}{V}$ $I = \dfrac{32{,}000 \text{ VA}}{240 \text{ V}}$ **I = 133 A**
1ϕ	VA	$VA = I \times V$	100 A, 240 V	AV	$VA = I \times V$ $VA = 100 \text{ A} \times 240 \text{ V}$ **VA = 24,000 VA**
1ϕ	V	$V = \dfrac{VA}{I}$	42,000 VA, 350 A	V	$V = \dfrac{VA}{I}$ $V = \dfrac{42{,}000 \text{ VA}}{350 \text{ A}}$ **V = 120 V**
3ϕ	I	$I = \dfrac{VA}{V \times \sqrt{3}}$	72,000 VA, 208 V	I	$I = \dfrac{VA}{V \times \sqrt{3}}$ $I = \dfrac{72{,}000 \text{ VA}}{360 \text{ V}}$ **I = 200 A**
3ϕ	VA	$VA = I \times V \times \sqrt{3}$	2 A, 240 V	VA	$VA = I \times V \times \sqrt{3}$ $VA = 2 \times 416$ **VA = 831 VA**

Electrical/Electronic Abbreviations/Acronyms . . .

Abbr/Acronym	Meaning	Abbr/Acronym	Meaning
A	Ammeter; Ampere; Anode; Armature	DPST	Double-Pole, Single-Throw
AC	Alternating Current	DS	Drum Switch
AC/DC	Alternating Current; Direct Current	DT	Double-Throw
A/D	Analog to Digital	DVM	Digital Voltmeter
AF	Audio Frequency	EMF	Electromotive Force
AFC	Automatic Frequency Control	F	Fahrenheit; Fast; Forward; Fuse; Farad
Ag	Silver	FET	Field-Effect Transistor
ALM	Alarm	FF	Flip-Flop
AM	Ammeter; Amplitude Modulation	FLC	Full-Load Current
AM/FM	Amplitude Modulation; Frequency Modulation	FLS	Flow Switch
ARM.	Armature	FLT	Full-Load Torque
Au	Gold	FM	Frequency Modulation
AU	Automatic	FREQ	Frequency
AVC	Automatic Volume Control	FS	Float Switch
AWG	American Wire Gauge	FTS	Foot Switch
BAT.	Battery (electric)	FU	Fuse
BCD	Binary-Coded Decimal	FWD	Forward
BJT	Bipolar Junction Transistor	G	Gate; Giga; Green; Conductance
BK	Black	GEN	Generator
BL	Blue	GRD	Ground
BR	Brake Relay; Brown	GY	Gray
C	Celsius; Capacitance; Capacitor, Coulomb	H	Henry; High Side of Transformer; Magnetic Flux
CAP.	Capacitor	HF	High Frequency
CB	Circuit Breaker; Citizen's Band	HP	Horsepower
CC	Common-Collector; Configuration	Hz	Hertz
CCW	Counterclockwise	I	Current
CE	Common-Emitter Configuration	IC	Integrated Circuit
CEMF	Counter-Electromotive Force	INT	Intermediate; Interrupt
CKT	Circuit	IOL	Instantaneous Overload
cmil	Circular Mil	IR	Infrared
CONT	Continuous; Control	ITB	Inverse Time Breaker
CPS	Cycles Per Second	ITCB	Instantaneous Trip Circuit Breaker
CPU	Central Processing Unit	J	Joule
CR	Control Relay	JB	Junction Box
CRM	Control Relay Master	JFET	Junction Field-Effect Transistor
CT	Current Transformer	K	Kilo; Cathode
CW	Clockwise	kWh	kilowatt-hour
D	Diameter; Diode; Down	L	Line; Load; Coil; Inductance
D/A	Digital to Analog	LB-FT	Pounds Per Foot
DB	Dynamic Braking Contactor; Relay	LB-IN.	Pounds Per Inch
DC	Direct Current	LC	Inductance-Capacitance
DIO	Diode	LCD	Liquid Crystal Display
DISC.	Disconnect Switch	LCR	Inductance-Capacitance-Resistance
DMM	Digital Multimeter	LED	Light-Emitting Diode
DP	Double-Pole	LRC	Locked Rotor Current
DPDT	Double-Pole, Double-Throw	LS	Limit Switch

... Electrical/Electronic Abbreviations/Acronyms

Abbr/Acronym	Meaning	Abbr/Acronym	Meaning
LT	Lamp	RF	Radio Frequency
M	Motor; Motor Starter; Motor Starter Contacts	RH	Rheostat
MAX.	Maximum	rms	Root-Mean-Square
MB	Magnetic Brake	ROM	Read-Only Memory
MCS	Motor Circuit Switch	rpm	Revolutions Per Minute
MEM	Memory	RPS	Revolutions Per Second
MED	Medium	S	Series; Slow; South; Switch; Second; Siemen
MIN	Minimum	SCR	Silicon-Controlled Rectifier
MMF	Magnetomotive Force	SEC	Secondary
MN	Manual	SF	Service Factor
MOS	Metal-Oxide Semiconductor	1 PH; 1φ	Single-Phase
MOSFET	Metal-Oxide Semiconductor Field-Effect Transistor	SOC	Socket
MTR	Motor	SOL	Solenoid
N; NEG	North; Negative; Number of Turns	SP	Single-Pole
NC	Normally Closed	SPDT	Single-Pole, Double-Throw
NEUT	Neutral	SPST	Single-Pole, Single-Throw
NO	Normally Open	SS	Selector Switch
NPN	Negative-Positive-Negative	SSW	Safety Switch
NTDF	Nontime-Delay Fuse	SW	Switch
O	Orange	T	Tera; Terminal; Torque; Transformer
OCPD	Overcurrent Protection Device	TB	Terminal Board
OHM	Ohmmeter	3 PH; 3φ	Three-Phase
OL	Overload Relay	TD	Time-Delay
OZ/IN.	Ounces Per Inch	TDF	Time-Delay Fuse
P	Peak; Positive; Power; Power Consumed	TEMP	Temperature
PB	Pushbutton	THS	Thermostat Switch
PCB	Printed Circuit Board	TR	Time-Delay Relay
PH	Phase	TTL	Transistor-Transistor Logic
PLS	Plugging Switch	U	Up
PNP	Positive-Negative-Positive	UCL	Unclamp
POS	Positive	UHF	Ultrahigh Frequency
POT.	Potentiometer	UJT	Unijunction Transistor
P-P	Peak-to-Peak	UV	Ultraviolet; Undervoltage
PRI	Primary Switch	V	Violet; Volt
PS	Pressure Switch	VA	Voltampere
PSI	Pounds Per Square Inch	VAC	Volts Alternating Current
PUT	Pull-Up Torque	VDC	Volts Direct Current
Q	Transistor; Quality Factor	VHF	Very High Frequency
R	Radius; Red; Resistance; Reverse	VLF	Very Low Frequency
RAM	Random-Access Memory	VOM	Volt-Ohm-Milliammeter
RC	Resistance-Capacitance	W	Watt; White
RCL	Resistance-Inductance-Capacitance	w/	With
REC	Rectifier	X	Low Side of Transformer
RES	Resistor	Y	Yellow
REV	Reverse	Z	Impedance

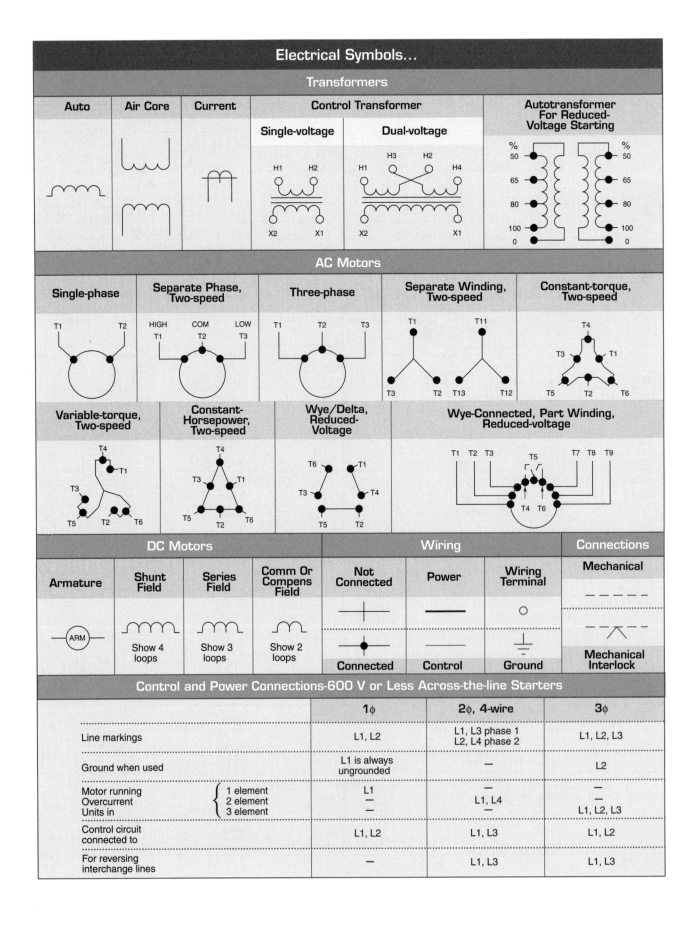

Appendix

...Electrical Symbols

Resistor	Capacitor	Inductor	Photocell	Photovoltaic Cell	Light-Emitting Diode (LED)
Thermistor	Battery	Diode	Zener Diode	Photodiode	Silicon-Controlled Rectifier

Diac | Triac | Bipolar Junction Transistors

Diac	Triac	PNP Transistor	NPN Transistor	PNP Phototransistor	NPN Phototransistor

Unijunction Transistor | Insulated Gate Bipolar Transistor | Junction Field-Effect Transistors | Metal-Oxide Semiconductor Field-Effect Transistors

Unijunction Transistor	Insulated Gate Bipolar Transistor	P-Channel JFET	N-Channel JFET	P-Channel MOSFET	N-Channel MOSFET

Logic Gates

AND Gate	OR Gate	NOT Gate	NOR Gate	NAND Gate	XOR Gate

Periodic Table of the Elements

Group	I	II		III	IV	V	VI	VII	0
	1 H Hydrogen 1.0079 1								2 He Helium 4.00260 2
	3 Li Lithium 6.941 2-1	4 Be Beryllium 9.01218 2-2		5 B Boron 10.81 2-3	6 C Carbon 12.01 2-4	7 N Nitrogen 14.0067 2-5	8 O Oxygen 15.9994 2-6	9 F Fluorine 18.99840 2-7	10 Ne Neon 20.179 2-8
	11 Na Sodium 22.98977 2-8-1	12 Mg Magnesium 24.305 2-8-2		13 Al Aluminum 26.98154 2-8-3	14 Si Silicon 28.086 2-8-4	15 P Phosphorus 30.97376 2-8-5	16 S Sulfur 32.06 2-8-6	17 Cl Chlorine 35.453 2-8-7	18 Ar Argon 39.948 2-8-8

Transition elements (periods 4–7):

19 K	20 Ca	21 Sc	22 Ti	23 V	24 Cr	25 Mn	26 Fe	27 Co	28 Ni	29 Cu	30 Zn	31 Ga	32 Ge	33 As	34 Se	35 Br	36 Kr
Potassium 39.098 -8-8-1	Calcium 40.08 -8-8-2	Scandium 44.9559 -8-9-2	Titanium 47.90 -8-10-2	Vanadium 50.9414 -8-11-2	Chromium 51.996 -8-13-1	Manganese 54.9380 -8-13-2	Iron 55.847 -8-14-2	Cobalt 58.9332 -8-15-2	Nickel 58.70 -8-16-2	Copper 63.546 -8-18-1	Zinc 65.38 -8-18-2	Gallium 69.72 -8-18-3	Germanium 72.59 -8-18-4	Arsenic 74.9216 -8-18-5	Selenium 78.96 -8-18-6	Bromine 79.904 -8-18-7	Krypton 83.80 -8-18-8

37 Rb	38 Sr	39 Y	40 Zr	41 Nb	42 Mo	43 Tc	44 Ru	45 Rh	46 Pd	47 Ag	48 Cd	49 In	50 Sn	51 Sb	52 Te	53 I	54 Xe
Rubidium 85.4678 -18-8-1	Strontium 87.62 -18-8-2	Yttrium 88.9059 -18-9-2	Zirconium 91.22 -18-10-2	Niobium 92.9064 -18-12-1	Molybdenum 95.94 -18-13-1	Technetium 97 -18-13-2	Ruthenium 101.07 -18-15-1	Rhodium 102.9055 -18-16-1	Palladium 106.4 -18-18-0	Silver 107.868 -18-18-1	Cadmium 112.40 -18-18-2	Indium 114.82 -18-18-3	Tin 118.69 -18-18-4	Antimony 121.75 -18-18-5	Tellurium 127.60 -18-18-6	Iodine 126.9045 -18-18-7	Xenon 131.30 -18-18-8

55 Cs	56 Ba	57–71	72 Hf	73 Ta	74 W	75 Re	76 Os	77 Ir	78 Pt	79 Au	80 Hg	81 Tl	82 Pb	83 Bi	84 Po	85 At	86 Rn
Cesium 132.9054 -18-8-1	Barium 137.34 -18-8-2		Hafnium 178.49 -32-10-2	Tantalum 180.9479 -32-11-2	Tungsten 183.85 -32-12-2	Rhenium 186.207 -32-13-2	Osmium 190.2 -32-14-2	Iridium 192.22 -32-15-2	Platinum 195.09 -32-17-1	Gold 196.9665 -32-18-1	Mercury 200.59 -32-18-2	Thallium 204.37 -32-18-3	Lead 207.2 -32-18-4	Bismuth 208.9804 -32-18-5	Polonium 209 -32-18-6	Astatine 210 -32-18-7	Radon 222 -32-18-8

87 Fr	88 Ra	89–103	104 Rf	105 Db	106 Sg	107 Bh	108 Hs	109 Mt	110 Ds	111 Uuu	112 Uub	113 Uut	114 Uuq	115 Uup	116 Uuh
Francium 223 -18-8-1	Radium 226.0254 -18-8-2		Rutherfordium 261 -32-10-2	Dubnium 262 -32-11-2	Seaborgium 263 -32-12-2	Bohrium 262 -32-13-2	Hassium 265 -32-14-2	Meitnerium 266 -32-15-2	Darmstadtium 269 -32-17-1	(Temporary name) 272 -32-18-1	(Temporary name) 277 -32-18-2	(Temporary name)	(Temporary name) 269	(Temporary name)	(Temporary name) 289

Lanthanide series:

57 La	58 Ce	59 Pr	60 Nd	61 Pm	62 Sm	63 Eu	64 Gd	65 Tb	66 Dy	67 Ho	68 Er	69 Tm	70 Yb	71 Lu
Lanthanum 138.9055 -18-9-2	Cerium 140.12 -19-9-2	Praseodymium 140.9077 -21-8-2	Neodymium 144.24 -22-8-2	Promethium 145 -23-8-2	Samarium 150.4 -24-8-2	Europium 151.96 -25-8-2	Gadolinium 157.25 -25-9-2	Terbium 158.9254 -26-9-2	Dysprosium 162.50 -28-8-2	Holmium 164.9304 -29-8-2	Erbium 167.26 -30-8-2	Thulium 168.9342 -31-8-2	Ytterbium 173.04 -32-8-2	Lutetium 174.97 -32-9-2

Actinide series:

89 Ac	90 Th	91 Pa	92 U	93 Np	94 Pu	95 Am	96 Cm	97 Bk	98 Cf	99 Es	100 Fm	101 Md	102 No	103 Lr
Actinium 227 -18-9-2	Thorium 323.0381 -18-10-2	Protactinium 231.0359 -20-9-2	Uranium 238.029 -21-9-2	Neptunium 237.0482 -22-9-2	Plutonium 244 -24-8-2	Americium 243 -25-8-2	Curium 247 -25-9-2	Berkelium 247 -27-8-2	Californium 251 -28-8-2	Einsteinium 254 -29-8-2	Fermium 257 -30-8-2	Mendelevium 258 -31-8-2	Nobelium 255 -32-8-2	Lawrencium 260 -32-9-2

Key:

14 Si Silicon 28.086 2-8-4
— Atomic Number
— Symbol
— Name
— Atomic Weight
— Electron Configuration

GLOSSARY

A

abbreviation: A letter or combination of letters that represents a word that is used for a device, connection, or process.

alphanumeric display: A display that presents information in the form of symbols, letters, and numbers.

amplifier coupling: The joining of two or more circuits so that power can be transferred from one amplifier stage to another.

analog multimeter: A meter that can measure two or more electrical properties and displays the measured properties along calibrated scales using a pointer.

analog signal: The output of circuits with continuously changing quantities that may have any value between the defined limits.

analog switching relay: An SSR that has a continuous range of possible output voltages within the rated range of the relay.

analog-to-digital (A/D) converter: An integrated circuit that converts an analog signal to a digital signal.

AND gate: A logic gate that provides logic level 1 only if all inputs are at logic level 1.

anemometer: An instrument that measures wind speed.

angle of incidence: The angle formed by incoming light (an incident ray) and the normal.

angle of refraction: The angle formed between a refracted ray and the normal.

array: A set of several connected modules.

attenuation: The loss of signal strength during transmission.

avalanche current: The flow of electrons when a diode breaks down and allows electrons to pass freely, which can damage the diode.

average voltage (V_{avg}): The average of all instantaneous voltages during a half cycle, either positive or negative.

B

bandwidth: The range of frequencies over which the gain of an amplifier is maximum and relatively constant.

bar code: A series of vertical lines and spaces of varying widths that are used to represent data.

bar code scanner: A self-contained unit that has both a light source and a light detector for reading a bar code.

base current (I_B): A critical factor in determining the amount of current flow in a transistor.

battery: A group of connected electrochemical cells that uses an electrochemical reaction to produce electrical energy.

beam splitter: A coated plate that reflects a portion of light and transmits the rest.

bench oscilloscope: A test instrument that displays the shape of a voltage waveform and is used mostly for bench testing electrical and electronic circuits.

biased clamping circuit: A diode clamping circuit that uses a battery to reverse-bias the diode.

binary signal: A signal that has only two states.

bipolar junction transistor (BJT): A three-terminal electronic device that switches or amplifies electrical signals.

bit: A binary digit of either 0 or 1.

blind via: A via that is exposed only on one outer layer of a PC board.

branch circuit identifier: A two-piece test instrument that includes a transmitter that is plugged into a receptacle and a receiver that provides an audible indication when located near the circuit to which the transmitter is connected.

brightness: The perceived amount of light reflecting from an object.

buffer gate: A logic gate consisting of two NOT gates connected together so that the output of one is fed into the input of the other.

buried via: A via that connects only conductors on the interior layers of a PC board and cannot be seen.

bus: A large trace or foil that extends around the edge of a PC board to provide a conducting path from several sources.

button cell: A button-shaped electrochemical cell manufactured with a separator layer between circular electrodes.

byte: A group of eight bits.

C

capacitive coupling: A method of amplifier coupling that uses an amplifier circuit consisting of two bipolar transistors connected by a capacitor.

capacitive filter: A circuit consisting of a capacitor and resistor connected in parallel.

carrier: *See* free electron.

cascade amplifier: A series of connected amplifiers in which the output of one amplifier is connected to the input of another amplifier.

chip: *See* integrated circuit (IC).

charging: The process of reversing the current of a cell or battery and converting the electrical energy back into chemical energy.

circuit analysis method: A method of SSR replacement in which a logical sequence is used to determine the reason for the failure of the SSR.

circuit breaker: An overcurrent protection device with a mechanical or electromechanical mechanism that manually or automatically opens a circuit when an overload condition or short circuit occurs.

clamping circuit: A circuit that holds voltage or current to a specific level.

clamping diode: A diode that prevents voltage in one part of a circuit from exceeding the voltage in another part.

clamp-on ammeter: A meter that determines the current in a circuit by measuring the strength of the magnetic field around a single conductor.

cogeneration: The use of a power source to generate electricity and useful heat at the same time. Also known as combined heat and power (CHP).

cold solder joint: A defective solder joint that results when the parts being joined do not exceed the liquidus temperature of the solder.

combined heat and power (CHP): *See* cogeneration.

common-base amplifier: A bipolar junction transistor with both the input and output signals connected to the base.

common-collector amplifier: A bipolar junction transistor with both the input and output signals connected to the collector.

common-emitter amplifier: A bipolar junction transistor with both the input and output signals connected to the emitter.

conductive tweezers: Tweezers that are heated in a manner similar to a soldering iron and used to solder and desolder small SMDs.

conductivity probe: A point level measuring system that uses two or more electrodes to determine the level of liquid in a vessel.

conductor: A material that has very little resistance to current and permits electrons to move through it easily.

contactor: A control device that uses a small control current to energize or de-energize the load connected to it.

contact temperature probe: A device used for taking temperature measurements at a single point by direct contact with the area being measured.

continuity tester: A test instrument that tests for a complete path for current to flow.

contrast: The ratio of brightness between different objects.

control circuit: The part of an SSR that determines when the output component is energized or de-energized.

control relay: A device that controls an electrical circuit by opening and closing contacts in another circuit.

control switch: A switch that controls the flow of current in a circuit.

converter: An electronic device that changes alternating current (AC) voltage to direct current (DC) voltage.

critical angle: The angle at which light no longer passes into a second material.

current gain: The ratio of output current to input current.

cutoff region: The region of a characteristic curve in which a transistor is turned OFF and no current flows through a collector.

cylindrical cell: A cylindrically shaped electrochemical cell with a positive electrode, negative electrode, synthetic separator, and electrolyte.

D

Darlington circuit: An amplifier circuit that consists of two bipolar transistors connected so that the emitter of the first transistor is connected to the base of the second transistor.

DC power supply: A device that converts alternating current (AC) to regulated direct current (DC) for use in electrical circuits.

decibel (dB): A unit of measure used to express the relative intensity of sound.

dendrite formation: The formation of crystals as a molten metal solidifies that results in a brittle solder joint.

depletion mode: The operation of a MOSFET with a negative gate voltage.

depletion region: A thin, neutral area created at the junction of P-type and N-type materials, where the two materials exchange carriers.

derating: The act of operating a diode at less than maximum operating current.

diac: A three-layer, two-terminal bidirectional device that is typically used to control the gate current of a triac.

diffused mode: A method of ultrasonic sensor operation in which the emitter and receiver are housed in the same enclosure.

digital display: An electronic device that displays readings as seven-segment numerical values.

digital IC: An integrated circuit that processes finite (zero and maximum) signal levels.

digital logic probe: A special DC voltmeter that detects the presence or absence of a signal.

digital multimeter (DMM): A meter that can measure two or more electrical properties and displays the measured properties as numerical values.

digital signal: The output of specific quantities that change in discrete increments, such as seconds, minutes, and hours.

digital-to-analog (D/A) converter: An integrated circuit that converts a digital signal to an analog signal.

diode: An electronic component that allows electrons to pass through it in only one direction.

diode characteristic curve: A curve on a graph that shows the relationship between voltage and current for a typical diode.

direct coupling: A method of amplifier coupling that uses an amplifier circuit consisting of two bipolar transistors connected so that the collector of the first transistor is directly connected to the base of the second transistor.

direct mode: A method of ultrasonic sensor operation in which the emitter and receiver are placed opposite each other so that the sound waves from the emitter are received directly by the receiver.

distortion: Any undesirable change in a signal.

doping: The addition of impurities to the crystal structure of a semiconductor to allow electron flow.

dot pitch: The distance between subpixels of the same color.

double-sided PC board: A PC board that has a layer of traces on each side of an insulating material.

dry cell: An electrochemical power source consisting of a cylindrical container made of zinc, a graphite rod in the center, and a paste electrolyte that allows electrons to flow.

dual-trace oscilloscope: An oscilloscope that displays two traces simultaneously.

duty cycle: The amount of time a pulse is ON during a given period; expressed as a percentage.

E

earplug: An ear protection device made of moldable rubber, foam, or plastic and inserted into the ear canal.

electrical circuit: An assembly of conductors and electrical and/or electronic devices through which electrons flow.

electrical shock: A shock that results any time a body becomes part of an electrical circuit.

electrochemical cell: A single unit that uses an electrochemical reaction to produce electrical energy.

electrochemical power source: A device that uses an electrochemical reaction to produce electrical energy.

electrostatic discharge (ESD): The movement of electrons from a source to an object across a gap.

enhancement mode: The operation of a MOSFET with a positive gate voltage.

equipment grounding conductor (EGC): An electrical conductor that provides a low-impedance ground path between electrical equipment and enclosures within the distribution system.

erasable programmable read-only memory (EPROM): A type of computer memory that can be programmed and erased, enabling it to be reused.

eutectic alloy: An alloy that has one specific melting temperature with no intermediate stage.

exact replacement method: A method of SSR replacement in which a bad relay is replaced with a relay of the same type and size.

exclusive OR gate: *See* XOR gate.

F

fall time: The time during which the voltage at the trailing edge of a pulse decreases from 90% to 10% of its steady maximum value.

fan-out: The number of loads that can be driven by the output of a logic gate.

fiber optics: The science of using light to transmit data from one location to another using optical fiber.

field-effect transistor (FET): A three- or four-terminal device in which output voltage is controlled by an input current.

filter: A circuit in a power supply section that smoothes the pulsating DC to make it more consistent.

flexible PC board: A PC board that bends or flexes.

floating input: A digital input that is not a true high or low at all times.

flow chart: A diagram that shows a logical sequence of steps for a given set of conditions.

flux-core solder: Solder that contains flux in its core.

focus: The sharpness of a displayed object.

foot protection: Protection consisting of safety shoes worn to prevent foot injuries that are typically caused by objects falling less than 4′ and having an average weight of less than 65 lb.

forward-bias voltage: Voltage applied with polarity that allows a diode to act as a conductor.

forward breakover voltage: The forward-bias voltage necessary for a semiconductor to act as a conductor.

forward operating current: The current range in which a semiconductor can safely operate once it reaches the forward breakover voltage.

free electron: An electron in a material that allows the conduction of electricity. Also known as a carrier.

frequency counter: *See* frequency meter.

frequency meter: A test instrument that is used to measure the frequency of an AC signal. Also known as a frequency counter.

frequency modulated continuous wave (FMCW) radar: A level-measuring sensor consisting of an oscillator that emits a continuous microwave signal that repeatedly varies its frequency between a minimum and maximum value, a receiver that detects the signal, and electronics that measure the frequency difference between the signal and the echo.

fuel cell: An energy source that transforms the chemical energy from fuel into electrical energy.

full-wave bridge rectifier: An electrical circuit that contains four diodes that allow both halves of a sine wave to be changed into pulsating DC.

full-wave doubler: An electrical circuit that uses both halves of the AC source to charge two capacitors on alternate half cycles so that both capacitor voltages can be combined to double the output.

full-wave rectifier: An electrical circuit containing two diodes and a center-tapped transformer used to produce pulsating DC.

function generator: *See* signal generator.

fuse: An overcurrent protection device with a fusible link that melts and opens the circuit when an overload condition or short circuit occurs.

fusion splice: A splice made by welding two ends of optical fibers together.

G

gain: The ratio of the amplitude of an output signal to the amplitude of an input signal.

gas detector: A transducer that detects gases or vapors.

gas laser: An electronic device in which electric current is used to ionize a gas to emit light, which is then amplified.

gas switch: A switch that detects a set amount of a specified gas and activates a set of electrical contacts.

grid-connected wind turbine: A wind energy system that is connected to an electric utility distribution system (grid).

grounded conductor: A conductor that has been intentionally grounded.

ground fault circuit interrupter (GFCI): A device that protects against electrical shock by detecting an imbalance of current in the normal conductor pathways and opening the circuit.

grounding: The connection of all exposed non-current-carrying metal parts to the earth.

grounding electrode conductor (GEC): A conductor that connects grounded parts of a power distribution system (equipment grounding conductors, grounded conductors, and all metal parts) to the NEC®-approved earth grounding system.

guided wave radar: A level-measuring detector consisting of a cable or rod as a wave carrier extending from an emitter down to the bottom of a vessel and electronics to measure the transit time.

H

half-power point: The point on a frequency-response curve where the power output is one-half the maximum value.

half-wave doubler: An electrical circuit containing two diodes, two capacitors, and a load resistor, where the output voltage is twice the peak input voltage.

half-wave phase control circuit: An SCR that has the ability to turn ON at different points of the conducting cycle of a half-wave rectifier.

half-wave rectifier: A circuit containing an AC source, a load resistor (R_L), and a diode that permits only the positive half cycles of the AC sine wave to pass, creating pulsating DC.

half-waving: A phenomenon that occurs when a relay fails to turn OFF because the current and voltage in the circuit reach zero at different times.

Hall effect sensor: A transducer that detects the proximity of a magnetic field.

handheld oscilloscope: A test instrument that displays the shape of a voltage waveform and is typically used for field testing.

handheld programming device: A separate keypad device that is not an integral part of a PLC.

heterogeneous material: A material that can be broken down by common mechanical processes.

high-rate discharge test: A battery test that measures battery terminal voltage after 15 sec of discharge when a heavy load is placed on the battery.

holes: The missing electrons in a crystal structure.

homogeneous material: A material that cannot be broken down by common mechanical processes.

horizontal positioning: The left or right shifting of a displayed voltage trace.

human machine interface (HMI): See operator interface panel.

hybrid circuit: A circuit that mixes mechanical devices, such as a relay, and electronic devices, such as an SCR, in the same circuit.

hybrid integrated circuit: A combination of two or more electronic components mounted and interconnected through a common PC board.

I

impedance matching: The process of setting the input impedance of a load equal to the output impedance of a signal source.

inductance: The property of a circuit to oppose a change in current due to energy stored in a magnetic field.

infrared temperature meter: A noncontact temperature meter that senses the infrared energy emitted by a material.

input circuit: The part of an SSR to which the control component is connected.

input impedance: The loading effect an amplifier presents to an incoming signal.

input section: The section of a PLC that receives signals from input devices and sends the signals to a CPU.

instant-on switching relay: An SSR that turns ON a load immediately when a control voltage is present.

insulated gate bipolar transistor (IGBT): A three-terminal switching device that combines an FET for control with a bipolar transistor for switching.

integrated circuit (IC): An electronic device in which all components (transistors, diodes, and resistors) are contained in a single package. Also known as a chip.

integrated circuit (IC) timer: A solid state timing device on a single IC.

integrated programming device: A device that consists of a small LCD and a small keypad or set of buttons that are part of a PLC.

intensity: The brightness level of a displayed object.

intermittent: An electronic device that works only part of the time.

International Electrotechnical Commission (IEC): An organization that develops international safety standards for electrical equipment.

inverter: An electronic device that changes direct current (DC) voltage to alternating current (AC) voltage.

inverter gate: *See* NOT gate.

isolation surge voltage rating: A measure of the ability of an optocoupler to withstand a spike or surge in voltage and maintain electrical isolation.

J

junction field-effect transistor (JFET): A simple field-effect transistor (FET) with a PN junction in which output current is controlled by an input voltage.

K

knee pad: A rubber, leather, or plastic pad strapped onto the knee for protection.

L

ladder diagram: *See* line diagram.

laser: A device that emits and amplifies light.

leakage current: *See* operating current.

light: Electromagnetic radiation that may or may not be visible.

light-activated SCR (LASCR): An SCR that can be triggered by light.

light detector: An electronic device that detects the presence of light from a light source and converts the light into an electrical signal.

light-emitting diode (LED): A diode that emits light when current flows through it.

light-emitting polymer (LEP): *See* organic light-emitting diode (OLED).

light-load test: A battery test that shows the ability of each cell to produce voltage under a light load.

linear amplifier: An amplifier that increases and maintains the exact duplicate of an input signal.

linear IC: An integrated circuit that produces an output signal proportional to the applied input signal and that is used to provide amplification and regulatory functions.

line diagram: A diagram that shows the logic of an electrical circuit or system using standard industry symbols. Also known as a ladder diagram.

liquid crystal display (LCD): A flat alphanumeric display that uses liquid crystals to modulate, but not directly emit, light.

lithium-ion battery: A rechargeable battery that consists of multiple secondary cells with a lithium compound as the electrolyte.

load: Any device that converts electrical energy to motion, heat, light, or sound.

load current: The amount of current drawn by a load when energized.

load line: A line drawn on a collector characteristic curve that shows the relationship between the collector current (I_C) and collector-emitter voltage (V_{CE}) in a circuit.

load-switching circuit: *See* output circuit.

lockout: The process of removing the source of electrical power, removing all sources of hazardous energy, and installing a lock that prevents the power from being turned ON.

L-section inductive filter: A filter that reduces surge currents by using a current-limiting inductor and a capacitor.

L-section resistive filter: A filter that reduces or eliminates the amount of DC ripple at the output of a circuit by using a resistor and capacitor as an RC time constant.

M

main bonding jumper (MBJ): A connection at the service equipment that connects the equipment grounding conductor, grounding electrode conductor, and grounded conductor (neutral conductor).

mechanical splice: A splice made by connecting two ends of optical fibers together using an adhesive without fusion from heat.

medium: Any material through which light travels.

memory: An electronic storage device for digital data used in computers.

metal-oxide semiconductor field-effect transistor (MOSFET): A three-terminal or four-terminal electronic switching device with metal-oxide or polysilicon insulating material that can be used for amplification.

minimum holding current: The minimum amount of current required to keep a sensor operating.

module: A photovoltaic device consisting of many electrically connected cells that form a panel.

momentary power interruption: A decrease to 0 V on one or more power lines lasting more than one half cycle to 3 sec.

motor starter: An electrically operated switch that can turn a motor ON and OFF and includes motor overload protection.

multilayered PC board: A PC board that has several layers of traces separated by layers of an insulating material.

multimeter: A meter that is capable of measuring two or more electrical quantities.

multistage amplifier: An amplifier circuit that allows several stages of amplification.

N

NAND gate: A logic gate designed to provide logic level 0 only if all inputs are at logic level 1. Also known as a negated AND gate.

nanotechnology: Technology that manipulates matter at the molecular level.

narrow bandwidth: A small range of frequencies over which an amplifier operates.

National Fire Protection Association (NFPA): A national organization that provides guidance in assessing the hazards of the products of combustion.

negated AND gate: *See* NAND gate.

negated OR gate: *See* NOR gate.

noncontact temperature probe: A device used for taking temperature measurements on energized circuits or on moving parts.

nonrechargeable cell: *See* primary cell.

NOR gate: A logic gate that provides logic level 1 only if all inputs are at logic level 0. Also known as a negated OR gate.

normal: An imaginary line perpendicular to the interface of two materials, such as glass and air.

NOT gate: A logic gate that provides an output that is opposite of the input. Also known as an inverter gate.

N-type material: Semiconductor material created by doping a region of a crystal with atoms of an element that has more electrons in its outer shell.

numerical aperture: A measure of the light gathering ability of a fiber.

O

off-grid wind turbine: *See* stand-alone wind turbine.

offset error: A slight mismatch between internal components.

one-line diagram: A diagram that uses single lines and graphic symbols to indicate the path and components of an electrical circuit.

operating current: The amount of current a sensor draws from the power lines to develop a field that can detect a target. Also known as residual or leakage current.

operational amplifier (op amp): A high-gain, directly coupled amplifier that uses external feedback to control response characteristics.

operator interface panel: A display panel that is connected to a PLC by a communication cable. Also known as a human machine interface (HMI).

optical time domain reflectometer (OTDR): A test instrument that is used to measure cable attenuation.

optocoupler: An electronic device that electrically isolates switching circuits by coupling them with light. Also known as an optoisolator.

optoisolator: *See* optocoupler.

OR gate: A logic gate that provides logic level 1 if at least one input is at logic level 1.

organic light-emitting diode (OLED): A solid state device composed of a thin film of organic molecules that create light with the application of voltage. Also known as a light-emitting polymer (LEP).

oscilloscope: A test instrument that provides a visual display of voltages.

oscilloscope trace: A reference point/line that is visually displayed on the face of the oscilloscope screen.

outage: *See* sustained power interruption.

output circuit: The load switched by an SSR that is part of the relay. Also known as a load-switching circuit.

output impedance: The loading effect an amplifier presents to another device.

output section: The section of a PLC designed to deliver the output voltage required to control alarms, lights, solenoids, and other loads.

overcurrent protection device (OCPD): A disconnect switch with circuit breakers or fuses added to provide overcurrent protection for the switched circuit.

P

parallel communication: A method of communication that involves sending several bits of data at the same time.

peak inverse voltage (PIV): The maximum reverse-bias voltage that a diode can withstand without breaking down and allowing electron flow.

peak switching relay: An SSR that turns ON a load when a control voltage is present and the voltage at the load is at its peak.

peak-to-peak voltage (V_{p-p}): The maximum value measured from the negative peak to the positive peak of a sine wave.

peak voltage (V_p): The maximum value of an alternating voltage that is reached on the positive or negative half cycle of a sine wave.

permissible exposure limit (PEL): A regulatory limit on the amount of workplace exposure to a hazardous chemical.

personal protective equipment (PPE): Clothing and/or equipment worn by a technician to reduce the possibility of injury in the work area.

phase control: The control of the time relationship between two events when dealing with voltage and current.

photocell: An electronic device that varies resistance based on the intensity of the light striking it. Also known as a photoconductive cell.

photoconductive cell: *See* photocell.

photoconductive diode: *See* photodiode.

photodarlington: A Darlington circuit with a phototransistor as the driver transistor.

photodiode: A semiconductor diode that allows current to flow when exposed to light. Also known as a photoconductive diode.

photoelectric switch: A solid state transmitter/receiver that can detect the presence of an object without touching the object.

photon: The basic unit of light or electromagnetic energy.

photonics: A branch of science that deals with the detection, emission, generation, manipulation, and transmission of light.

photo tachometer: An electronic device that uses light to measure rotational speed.

phototransistor: A bipolar transistor with a transparent case that allows light to strike its base-collector junction.

phototriac: A triac that can be triggered by light.

photovoltaic cell: A semiconductor device that directly converts solar energy into electrical energy.

photovoltaic effect: The production of electrical energy due to the absorption of light photons in a semiconductor material.

photovoltaic system: An electrical system consisting of photovoltaic cells, modules, and arrays used to convert light into electrical energy.

pictorial drawing: A drawing that shows the length, height, and depth of an object in one view.

piezoelectric effect: The electrical polarization of some materials when mechanically strained.

piezoelectricity: The voltage generated from the application of pressure.

piezoelectric sensor: A transducer that operates based on the interaction between the deformation of certain materials and an electric charge.

PIN photodiode: A photodiode with an intrinsic region between the P-type and N-type material.

pi-section filter: A filter made with two capacitors and an inductor or a resistor to smooth out the AC ripple in a rectified waveform.

plasma display panel (PDP): A display panel that consists of small compartments or cells containing ionized gas, which forms plasma when a voltage is applied, and creates UV photons that strike phosphor pixels painted on the inside of the cells to produce light.

plated hole: A hole drilled through a double-sided PC board and plated with a conductive material that interconnects two layers of traces.

polarizing filter: A filter that allows light to pass through only at certain angles.

pouch cell: An electrochemical cell enclosed in a flexible, heat-sealable foil with conductive foil tabs welded to the electrodes.

power gain: The ratio of output power to input power.

power source: A device that converts various forms of energy into electricity.

preamplifier: A circuit that provides gain for a weak signal before the signal goes through the normal stages of amplification.

primary cell: A cell that uses up its materials to the point where recharging is not practical. Also known as a single-use or nonrechargeable cell.

printed circuit (PC) board: A board of insulating material with electrical interconnections attached to its surface to create a circuit for components mounted on it.

prismatic cell: A rectangular-shaped electrochemical cell that contains a flat, rectangular anode and cathode with a separator in between enclosed in a case.

processor section: The section of a PLC that organizes all control activity by receiving inputs, performing logical decisions according to the program, and controlling outputs.

programmable logic controller (PLC): A solid state control device that is programmed and reprogrammed to automatically control processes or machines.

propagation: The speed at which light travels through a substance.

protective clothing: Clothing that provides protection from contact with sharp objects, hot equipment, and harmful materials.

protective helmet: A hard hat that is used in the workplace to prevent injury from the impact of falling and flying objects and from electrical shock.

proton-exchange-membrane fuel cell (PEMFC): A fuel cell that transforms chemical energy from a fuel into electrical energy and consists of a polymer electrolyte membrane that conducts protons (hydrogen ions) but not electrons.

P-type material: Semiconductor material created by doping a region of a crystal with atoms of an element that has fewer electrons in its outer shell, which creates empty spaces, or holes.

pull-down resistor: A resistor that has one side connected to ground at all times and the other side connected to unused gates.

pull-up resistor: A resistor that has one side connected to a power supply at all times and the other side connected to unused gates.

pulsating DC: A direct current that varies in amplitude but does not change polarity.

pulse amplitude: The peak-to-peak voltage in a pulse.

pulsed radar level sensor: A level-measuring sensor consisting of a radar generator that directs an intermittent pulse with a constant frequency toward the surface of the material in a vessel.

pulse duration: *See* pulse width.

pulse interval: The time elapsed between the beginning of one pulse and the beginning of the next pulse.

pulse repetition rate: The number of pulses during one second in a steady pulse train; measured in pulses per second (pps).

pulse separation: The time elapsed between the end of one pulse and the beginning of the next pulse.

pulse train: A series of pulses that is typically plotted on a graph representing voltage versus time.

pulse width: The amount of time a pulse is present, or ON. Also known as pulse duration.

Q

qualified person: A person who has special knowledge of the construction and operation of electrical equipment or a specific task and is trained to recognize and avoid electrical hazards that might be present.

quantum efficiency: A method of calculating the number of electrons produced for each photon striking a semiconductor.

R

radiation detector: A transducer that detects radioactive materials.

random access memory (RAM): A type of computer memory that is lost when power is lost.

RC circuit: A circuit in which resistance (R) and capacitance (C) are used to help filter the power in a circuit.

receptacle tester: A device that is plugged into a standard receptacle to determine if the receptacle is properly wired and energized.

rechargeable cell: *See* secondary cell.

rectification: The changing of AC to DC.

rectifier: A device consisting of diodes that convert AC power to DC power by allowing electrons to flow in only one direction.

refraction: A deflection of light caused when light passes through air into water and changes speed.

refractive index: The speed of light through a substance compared to that of free space.

regulated power supply: A power supply that maintains a constant voltage across an output even when loads vary.

relative mode test: A digital multimeter (DMM) test that records a voltage, current, or resistance measurement and displays the difference between the reading and subsequent measurements taken by a meter.

relay: A device that controls one electrical circuit by opening and closing another circuit.

residual current: *See* operating current.

reverse-bias voltage: Voltage applied with opposite polarity that allows a diode to act as an insulator.

rework station: A workstation with all of the tools required for soldering and desoldering plus many of the accessories required for rework.

ripple voltage: The amount of varying voltage present in a DC power supply.

rise time: The time required for a pulse to rise from 10% to 90% of its final steady value.

root-mean-square voltage (V_{rms}): The voltage equal to 0.707 multiplied by the peak voltage.

rubber insulating matting: A floor covering that provides technicians protection from electrical shock when working on live electrical circuits.

S

safety glasses: An eye protection device with special impact-resistant glass or plastic lenses, reinforced frames, and side shields.

safety label: A label that indicates areas or tasks that can pose a hazard to personnel and/or equipment.

saturation region: The region of a characteristic curve in which maximum current can flow in a transistor circuit.

schematic diagram: A diagram that shows the electrical connections and functions of a specific circuit arrangement with graphic symbols.

secondary cell: A cell that can be recharged. Also known as a rechargeable cell.

semiconductor: Material that has the electrical conductivity characteristics between that of a conductor (high conductivity) and an insulator (low conductivity).

semiconductor laser: A solid state laser that emits and amplifies light through the use of a semiconductor material.

separately derived system (SDS): A system that supplies electrical power derived (taken) from transformers, storage batteries, photovoltaic (solar) systems, and generators.

serial communication: A method of communication that involves sending bits of data one bit at a time.

signal generator: A test instrument that produces test signals or pulses for testing purposes. Also known as a function generator.

silicon-controlled rectifier (SCR): A four-layer (PNPN) semiconductor device that uses three electrodes for normal operation.

silk-screen layer: A layer of text that is normally applied to the outer surface of a PC board.

single-sided PC board: A PC board that has traces on only one side of an insulating material.

single-use cell: *See* primary cell.

sink circuit: A circuit in a transmitting device that takes a positive voltage generated by the receiving device and shunts it to a ground, lowering the voltage at the receiver circuit.

slew rate: The maximum time rate of change of output, measured in volts per microsecond (V/μs).

smoke detector: *See* smoke switch.

smoke switch: A switch that detects a set amount of smoke caused by smoldering or burning material and activates a set of electrical contacts. Also known as a smoke detector.

solder: An alloy consisting of specific percentages of two or more metals.

soldering: The process of making a sound electrical and mechanical joint between certain metals by joining them with solder.

soldering station: A workstation with all of the tools required for soldering.

solder mask: A permanent protective coating that protects circuitry from any unwanted solder.

solder pad: A small, flat conductor to which component leads are soldered.

solder wick: A fine copper braid impregnated with flux that is used to remove solder.

solid state relay (SSR): A switching device that has no contacts and switches entirely by electronic means.

source circuit: A circuit in a transmitting device that provides a positive voltage signal that can be detected in a receiver circuit.

spiral-wound cylindrical cell: A cylindrically shaped electrochemical cell that contains an anode and a cathode wound together with a microporous separator interspaced between thin electrodes.

stand-alone inverter: An inverter connected to a battery bank for its source of power and that operates independently.

stand-alone wind turbine: A wind energy system that is not connected to an electric utility distribution system (grid). Also known as an off-grid wind turbine.

star coupler: A fiber-optic coupling device that uses a mirror to reflect light from one fiber into several fibers.

static electricity: An electrical charge at rest.

status indicator: An LED application that indicates the operating status of a machine to the machine operator.

surface mount device (SMD): An electronic component that is soldered directly to the traces of a PC board.

surface mount technology (SMT): The technology used to directly mount components on the solder pads of a PC board.

surge current: A larger than normal current that is created when a device is first turned ON.

surge suppressor: An electrical device that provides protection from high-level transients by limiting the level of voltage allowed downstream.

sustained power interruption: A decrease to 0 V on all power lines lasting more than 1 min. Also known as an outage.

symbol: A graphic element that represents a quantity or unit.

T

tagout: The process of placing a danger tag on the source of electrical power and sources of hazardous energy, which indicates that the equipment may not be operated until the danger tag is removed.

T-coupler: An in-line fiber-optic coupling device used for tapping a main fiber-optic bus.

temperature coefficient: The percentage of change in voltage per degree change in temperature.

temperature probe: The part of a temperature test instrument that measures the temperature of liquids, gases, surfaces, and pipes.

temporary power interruption: A decrease to 0 V on one or more power lines lasting 3 sec to 1 min.

thermal instability: A change in bias due to heat.

thermal resistance (R_{TH}): The ability of a device to impede the flow of heat.

thermistor: A type of transducer that acts as a temperature-sensitive resistor.

thin-film battery: A rechargeable solid state battery that contains multiple secondary cells, is very thin, and may be flexible.

three-wire solid state switch: A device that has three connecting terminals or wires (exclusive of ground).

through-hole technology: The technology and methods used to mount components on a PC board that involves placing the leads (thin wires) of the components through small, conductive through-holes in the PC board and soldering the leads to the through-holes.

through-hole via: A via that passes completely through a PC board and connects all layers of conductors.

time/division (time per division): The length of time a cycle takes to sweep across a screen.

time-domain reflectometer (TDR): A cable tester used primarily for detecting faults in electrical conducting cables such as telephone cables.

toxic gas detector: A hazardous atmosphere detector used to measure toxic gases or vapors that are not combustible but are harmful to people.

trace: A conductive path found on various layers of a PC board.

transducer: A device that converts various forms of energy into electrical energy.

transfer current gain: The ratio of output current to input current.

transformer: A device with two windings (primary and secondary) used to step up or step down AC voltage in an AC circuit to the proper operating voltage.

transformer coupling: A method of amplifier coupling that uses an amplifier circuit consisting of two bipolar transistors connected by a transformer.

transient voltage: A temporary, unwanted voltage in an electrical circuit. Also known as a voltage spike.

transistor current gain (β): The ratio of collector current (I_C) to base current (I_B) of a transistor with a constant collector-emitter voltage (V_{CE}).

transistor outline (TO) number: A number determined by the manufacturer that represents the shape and configuration of a transistor.

triac: A three-electrode, bidirectional AC switch that allows electrons to flow in either direction.

truth table: A table used to describe the output condition of a logic gate or combination of gates for every possible input condition.

2D matrix bar code: A two-dimensional bar code that can represent more data than a standard, one-dimensional bar code.

two-wire solid state switch: A device that has two connecting terminals or wires (exclusive of ground).

U

ultrasonic sensor: A transducer that can detect the presence of an object by emitting and receiving high-frequency sound waves.

unijunction transistor (UJT): A three-electrode device that contains one PN junction consisting of a bar of N-type material with a region of P-type material doped within the N-type material.

uninterruptible power supply (UPS): A battery-based system that includes all of the additional power conditioning equipment to make a complete self-contained power source.

unregulated power supply: A power supply whose output varies, depending on changes of line voltage or load.

utility-interactive inverter : An inverter connected to, and operated in parallel with, an electric utility grid.

V

vapor: A gas that can be liquefied by compression without lowering the temperature.

vertical positioning: The up and down shifting of a displayed voltage trace.

via: A plated hole drilled through one or more layers of traces and insulating material that provides an electrical connection between the layers.

voltage divider: A circuit that provides several different voltages for several different loads.

voltage doubler: A circuit designed to produce a DC output level that is approximately twice that of the peak AC value.

voltage gain: The ratio of output voltage to input voltage.

voltage indicator: A test instrument that indicates the presence of voltage when the test tip touches, or is near, an energized hot conductor or energized metal part.

voltage limiter: An electronic circuit that consists of diodes used to control voltage where large voltage changes are expected.

voltage multiplier: An electrical circuit designed to supply higher voltages than an AC source voltage without using a transformer.

voltage regulator: An electronic circuit that maintains a relatively constant value of output voltage over a wide range of operating situations.

voltage spike: *See* transient voltage.

voltage tester: A device that indicates approximate voltage level and type (AC or DC) by the movement and vibration of a pointer on a scale.

volt/division (volts per division): The number that each horizontal division on a screen represents.

W

wet cell: An electrochemical power source consisting of two electrodes in a liquid electrolyte solution.

wide bandwidth: A large range of frequencies over which an amplifier operates.

wind turbine: A power generation system that converts the kinetic energy of wind into mechanical energy, which is used to rotate a generator that produces electrical energy.

wind vane: An instrument that measures wind direction.

wiring diagram: A diagram that shows the electrical connections of all components in a piece of equipment.

X

XOR gate: A logic gate that provides logic level 1 only if one input is at logic level 1. Also known as an exclusive OR gate.

Z

zener diode: A semiconductor device used as a voltage regulator.

zero switching relay: An SSR that turns ON a load when a control voltage is applied and the voltage at the load crosses zero (or within a few volts of zero).

INDEX

Page numbers is italic refer to figures.

A

abbreviations, 5, 35, *36*
acceptance angles, *399*
AC proximity sensors, 454, *455*
AC voltage measurements
  DMMs, *37*, 37
  oscilloscopes, 45–46, *46*
A/D converters, *413*, 413
additive (plate-up) circuitry creation, 58, *59*
adjustable temperature control (SCRs), 314, *315*
AFCs, 189, *190*
alarm circuits
  gas, 364–365, *366*
  security, *363*, 363
alkaline fuel cells (AFCs), 189, *190*
alphanumeric displays, 378
amplifier coupling, 259–262, *260–263*, 285–288, *286–289*
amplifier gain, *243*, 243–244, *347*, 347–348
amplifiers
  cascade amplifiers, 243, *244*
  classes of operation, 255–256, *256*
  common-base amplifiers, 249, *251*, 251–252, 257
  common-collector amplifiers, *252*, 252–253, 257
  common-emitter amplifiers, 246–249, *247*, *248*, *249*, *250*
  coupling of, 259–262, *260*, *261*, *262*, *263*
  determining gain, *243*, 243–244
  feedback inverting amplifiers, *349*, 349–350
  frequency-response curves, *245*, 245
  high-impedance differential amplifiers, 346, *347*
  impedances, *256*, 256–257, *257*
  linear amplifiers, 253, *254*
  multistage amplifiers, 259–262, *260–263*, 284–288, *286–289*
  op amps, 343–358, *344*

closed loop control, *349*, 349
  as comparators, *358*, 358
  and compensation, *350*, 350
  in current-to-voltage converters, 351–352, *352*, *353*
  definition, 343
  in DMMs, *357*, 357
  and gain, *347*, 347–348
  as integrators, *356*, 356
  nulling, *351*, 351
  open-loop control, 349
  operation of, *346*, 346
  schematic symbols, *344*, 344
  slew rate, 351
  in thermostat control circuits, *355*, 355–356
  voltage sources, *345*, 345–346
  in voltage-to-current converters, 353, *354*
  signal changes, *254*, 254–255, *255*
analog electronics, 411
analog multimeters, 26–27, *27*
analog signals, *411*, 411, *412*, *413*, 413
analog switching relays, 434, *435*
analog-to-digital (A/D) converters, *413*, 413
AND gates, *415*, 415–416, *416*
anemometers, 171, *172*
angles of incidence, *373*, 373–374, *374*
angles of reflection, *374*
angles of refraction, *373*, 373
antistatic brushes, 79
arrays, 163–164, *164*, *165*
astable vibrators, 364–365, *365*, *366*
attenuation, *33*, 33, 398–399, *399*
automatic security lights, *325*, 325
avalanche current, 113
avalanche diodes. *See* zener diodes
avalanche voltage, *123*, 123
average voltage (V_{avg}), 136–138, *137*

B

back protection, 18
bandwidth, 245

bar codes, *62*, 62, *395*, 395
bar code scanners, *395*, 395
base bias instability, 229, *230*
base current, *221*, 221
base (B) terminals, *217*, 217
base-emitter junctions, *219*, 219–221, *220*, *221*, 235, *236*
batteries
  battery cell designs, 180–182, *181*
  battery life, 183
  capacity, *175*, 175, *176*
  charging, 186–188
  connections, 176, *177*
  definition, 173
  disposal, 188
  lithium-ion batteries, 178–180, *179*
  maintenance, 182, *183*
  rechargeable batteries, 177–180, *179*
  single-use batteries, *178*
  storage, 182
  testing methods, 183–186, *184*, *185*, *186*
  thin-film batteries, 177–178, *179*
  troubleshooting, *184*
battery chargers, 187–188
beam splitters, *402*, 402
bench oscilloscopes, 31, *32*
bench testing, 40
beverage dispensing guns, *240*, 240
beverage gun sensors, *206*, 206
biased clamping circuits, *122*, 122
biasing, fixed, *221*, 221, 228, *229*
biasing diodes, *110*, 110–113, *111*, *112*, *113*
bias stabilization, *230*, 230–232, *231*, *232*, *233*
binary signals, *414*, 414
bipolar junction transistors (BJTs). *See* BJTs
bits, 423, 495
BJTs, *217*, 217–218, *218*
blind vias, *53*, 53
blocking capacitors, 260, *261*
blown fuses, 148, *149*
branch circuit identifiers, *25*, 25

509

breakover voltage, 327, *328*
brightness, 370, *371*
brushes, 78–79, *79*
buffer gates, *420*, 420
buried vias, *53*, 53
buses, *51*, 51
button cells, 180, *181*
bytes, 423

C

capacitive coupling, 259–260, *260, 261*, 285–286, *286, 287*
capacitive filters, 139–140, *140*
carriers, 107
cascade amplifiers, 243, *244*
CAT numbers, 40, *41*
cells. *See* electrochemical cells
CEMF, 141
characteristic curves
 diacs, *328*
 diodes, 112–113, *113*
 JFETs, *267*, 267
 NTC thermistors, *195*, 195
 PTC thermistors, *198*, 198
 SCRs, 292–293, *293*
 transistors (collector), *222*, 222–225, *223, 224, 225, 258*
 triacs, *322*, 322
 UJTs, *332*
charging (batteries), 186–188
Chip Quik®, *103*, 103
CHP, 190
circuit analysis troubleshooting method, 450
circuit boards. *See* printed circuit (PC) boards
circuit breakers, 2, 148, *149*
circuits. *See* electrical circuits
clamping circuits, 119–122, *120, 121, 122*
clamping diodes, 228
clamp-on ammeters, 27–28, *28*
classes of amplifier operation, 255, *256*
clipping circuits, 118, *119*, 126, *127*
clothing, protective, 17
cogeneration, 190
cold solder joints, 84
collector-base junctions, 219–221, *220, 221, 235, 236*
collector characteristic curves, *222*, 222–225, *223, 224, 225*
collector-emitter junctions, *236*
collector-feedback bias stabilization circuits, 230, *231*
collector (C) terminals, *217*, 217

color of light, *372*, 372
combination bias stabilization circuits, *231*, 231
combined heat and power (CHP), 190
common-base amplifiers, 249, *251*, 251–252, 257
common-collector amplifiers, *252*, 252–253, *257*, 257
common-emitter amplifiers, 246–249, *247, 248, 249, 250*
comparators, *358*, 358
complementary metal-oxide semiconductor (CMOS) logic, 423
conductive brushes, 79
conductive tweezers, *91, 100*, 100
conductors, 1–2, *2*
conformal coatings, 68–71, *69, 70, 71*
constant-current charging, 187
constant-voltage charging, 187
contactors, *481*, 482
contact temperature probes, 29, *30*
continuity testers, 23–24, *24*
continuous low-rate chargers, 188
contrast, 370, *371*
control circuits, 436
control relays, 481
controls, oscilloscopes, 42–45, *44, 45*
control switches, 1–2, *2*
converters
 current-to-voltage, 351–352, *352, 353*
 voltage-to-current, 353, *354*
 wind turbines, 169–170, *170, 171*
counter electromotive force (CEMF), 141
coupling, amplifiers, 259–262, *260–263, 284–288, 286–289*
critical angles, *374*, 374
crowbar circuits, 316, *318*
current flow, 1
current gain, 224, 244
current measurements
 with clamp-on ammeters, 28
 in-line, *39*, 39–40
cutoff region, *226*, 226
cylindrical cells, 180, *181*

D

D/A converters, *413*, 413
Darlington circuits, 262, *263*, 288, *289*
DC power supplies. *See* power supplies
DC voltage measurements
 DMMs, 37, *38*
 oscilloscopes, 46–47, *47*
decibels (dB), 18, 245–246, *246*
dendrite formation, *86*, 86

depletion-enhancement MOSFETs, *277*, 277–278, *278, 279*
depletion mode, *278*, 278
depletion regions, 110
derating, 117
desoldering, 97–103
 desoldering pump method, *99*, 99
 solder wick method, *98*, 98
 surface mount devices (SMDs), 99–103, *100, 101, 102, 103*
 through-hole components, *99*, 99
desoldering pumps, *99*, 99
diac characteristic curves, *328*
diacs, *327*, 327–330, *328, 329, 330*
diac/triac combination, 328, *329*
diagrams
 line (ladder), *10*, 10–11
 one-line, *10*, 10
 schematic, *9*, 9
 wiring, *9*, 9
diffused mode, ultrasonic sensors, 208, *209*
digital clock inputs, 426–427, *427*
digital displays, DMMs, 35, *36*
digital electronics, *411*, 411, *412*
digital electronics troubleshooting, 427–428, *429*
digital ICs, *412*, 412
digital logic, *414*, 414–423
 complementary metal-oxide semiconductor (CMOS) logic, 423
 fan-out, 421–422, *422*
 logic families, 422–423
 logic gates, *415*, 415–420
  AND gates, *415*, 415–416, *416*
  buffer gates, *420*, 420
  NAND gates, *415*, 415, 418–419, *419*
  NOR gates, *415*, 415, 417–418, *418*
  NOT gates, *415*, 415, 417, *418, 420*, 420
  OR gates, *415*, 415, 416, *417*
  XOR gates, *415*, 415, 419, *420*
 pull-down resistors, 420–421, *421*
 pull-up resistors, 420–421, *421*
 switching using logic gates, 422, *423*
 transistor-transistor logic (TTL), 422
 truth tables, *414*, 414
digital logic probes, 32, *33*
digital multimeters (DMMs), *27*, 27, 35
 abbreviations, 35, *36*
 AC voltage measurements, 37, *37*
 checking zener diodes, 152, *153*
 DC voltage measurement, 37, *38*
 digital display, 35, *36*

diode testing, 113–115, *114, 115*
frequency measurement, 30, *31*
in-line current measurement, *39,* 39–40
with op amps, *357,* 357
resistance measurement, 37–38, *38*
symbols, 35, *36*
transistor testing, 234–237
voltage protection, 40, *41*
digital pulses, *423,* 423–426, *425, 426, 427*
digital signals, *411,* 411, *412, 413,* 413
digital-to-analog (D/A) converters, *413,* 413
diode characteristic curves, 112–113, *113*
diodes, *107,* 107
   applications of, 117–122
   biasing, *110,* 110–113, *111, 112, 113*
   characteristic curves, 112–113, *113*
   clamping, 228
   in clamping circuits, 119–122, *120, 122*
   in clipping circuits, 118, *119*
   derating, 117
   installation of, *116,* 116–117
   materials in, 107–110, *108, 109, 110*
   mounting, *116,* 116
   power capacity of, 117
   in rectifiers, 117–118, *118, 119*
   servicing, *116,* 116–117, *117*
   for stabilization, 232, *233*
   testing, 113–115, *114, 115*
   zener. *See* zener diodes
direct coupling, 261–262, *263,* 287–288, *289*
direct mode, ultrasonic sensors, 208
dissipative brushes, 79
distortion, 253
DMM. *See* digital multimeters (DMMs)
doping, 108–109
dot-matrix displays, 378
dot pitch, 381
double-sided PC boards, 52, *53*
drawings, pictorial, *5,* 5
dry cells, *173,* 174–175, *175*
dual-gate MOSFETs, 278, *279*
dual-trace oscilloscopes, 42, *43*
duty cycles, 425, *426*

E

earplugs, *17,* 18
ear protection, 18
EGCs, *15,* 15
electrical abbreviations, 4

electrical circuits, 1
   abbreviations, 4
   components, 1–2, *2*
   diagrams, *9,* 9–11, *10, 11*
   symbols, 4
   terminology, 4
electrical energy sources, 159, *160*
electrical measurement safety, 34–35
electrical shock, 13–14, *14*
electrical symbols, 4
electrochemical cells
   button cells, 180, *181*
   capacity, *175,* 175, *176*
   connections, 176, *177*
   cylindrical cells, 180, *181*
   definition, 173
   designs, 180–182, *181*
   disposal, 188
   dry cells, *173,* 174–175, *175*
   pouch cells, 180, *181*
   primary cells, 176, *178*
   prismatic cells, *181,* 182
   secondary cells, 176–180, *179*
   spiral-wound cylindrical cells, 180–181, *181*
   wet cells, *173,* 173–174, *174*
electrochemical power sources, 173–188
   batteries
      battery cell designs, 180–182, *181*
      battery life, 183
      capacity, *175,* 175, *176*
      charging, 186–188
      connections, 176, *177*
      definition, 173
      disposal, 188
      lithium-ion batteries, 178–180, *179*
      maintenance, 182, *183*
      rechargeable batteries, 177–180, *179*
      single-use batteries, *178*
      storage, 182
      testing methods, 183–186, *184, 185, 186*
      thin-film batteries, 177–178, *179*
      troubleshooting, *184*
   electrochemical cells
      button cells, 180, *181*
      capacity, *175,* 175, *176*
      connections, 176, *177*
      cylindrical cells, 180, *181*
      definition, 173
      designs, 180–182, *181*
      disposal, 188
      dry cells, *173,* 174–175, *175*
      pouch cells, 180, *181*
      primary cells, 176, *178*

prismatic cells, *181,* 182
secondary cells, 176–180, *179*
spiral-wound cylindrical cells, 180–182, *181*
wet cells, *173,* 173–174, *174*
electromagnetic spectrum, *371*
electron flow, 1
electronic control devices (wind turbines), 169
electronic timers, 238, *239*
electrostatic discharge (ESD), 2–3, *3,* 63
electrostatic-discharge-safe soldering stations, 93
electrostatic discharge (ESD) workstations, 3–5, *3, 4, 5*
emergency flashers, *335,* 335
emitter arrow, *217,* 217
emitter-feedback bias stabilization circuits, *230,* 230
emitter (E) terminals, *217,* 217
enhancement mode, 278, *279*
enhancement MOSFETs, 275, *276,* 277
EPROM, 468
equipment grounding conductors (EGCs), *15,* 15
erasable programmable read-only memory (EPROM), 468
ESD, 2–3, *3,* 63
ESD-safe soldering stations, 93
ESD workstations, 3–5, *3, 4, 5*
eutectic alloy, 84
exact replacement troubleshooting method, 450
eye protection, *17,* 17

F

fall time, digital pulses, 426, *427*
fan-out, 421–422, *422*
feedback inverting amplifiers, *349,* 349–350
FETs, 265. *See also* IGBTs; JFETs; MOSFETs
fiber microscopes, *406,* 406
fiber-optic cable couplers, 401–403, *402, 403*
fiber-optic cables, *400,* 400–401, *401*
fiber-optic power meters, *407,* 407–408
fiber optics, 396–409, *397*
   advantages of, 397, *398*
   attenuation, 398–399, *399*
   fiber-optic cable applications, 408, *409*
   fiber-optic cable couplers, 401–403, *402, 403*

fiber-optic cable installation, 403–407, *404, 405, 406*
fiber-optic cables, *400,* 400–401, *401*
optical fibers, *398,* 398
troubleshooting, *407,* 407–408
field-effect transistors (FETs), 265. *See also* IGBTs; JFETs; MOSFETs
field repairs (PC boards), 73–77, *74, 75, 76, 77, 78*
field-service transistor testers, 258
filters, power supply, 139–142, *140, 141, 143*
fire safety, 19, *20*
555 timers, 358–366, *359*
    as astable multivibrators, 364–365, *365, 366*
    as monostable multivibrators, 361, *362*
    operation of, 359–361, *360*
    in security alarm circuits, *363,* 363
556 timers, *359,* 359
5S organization systems, 103–104, *104*
fixed biasing, *221,* 221, 228, *229*
flat-pack SCRs, 306, *307*
flexible PC boards, 54, *55*
flip-flops, 359
floating inputs, 421
flow charts, *486,* 486
flux-core solders, *87,* 87
FMCW radars, 209, *210*
focus control, oscilloscopes, 42
foot protection, 18
forward-bias voltage, 110, *111*
forward breakover voltage, 112
forward operating current, 113
free electrons (carriers), 107
frequency counters, 30, *31*
frequency measurements
    DMMs, 30, *31*
    oscilloscopes, 47–48, *48*
frequency meters, 30, *31*
frequency modulated continuous wave (FMCW) radars, 209, *210*
frequency-response curve, amplifiers, *245,* 245
fuel cells, 188–190, *189, 190*
full-wave bridge rectifiers, 134–135, *135,* 151, *152*
full-wave doublers, 147, *148*
full-wave rectifiers, 133, *134,* 150, *151*
function generators, 32, *33*
fuses, 2, 148, *149*
fusion splices, 404–405

G

gain, 243–244
gas alarm circuits, 364–365, *366*
gas detectors, 211–214, *212, 213, 214*
gas lasers, *383,* 383
gas switches (gas detectors), *214,* 214
GECs, *15,* 15
GFCIs, 15–16, *16*
graphical displays, oscilloscopes, 40, *42,* 42–43, *43*–45, 45
grid-connected wind turbines, 167, *168*
grounded conductors, *15,* 15
ground fault circuit interrupters (GFCIs), 15–16, *16*
grounding, 14–15, *15*
grounding electrode conductors (GECs), *15,* 15
guided wave radars, 210, *211*

H

half-power points, 245
half-wave doublers, *147,* 147
half-wave phase control circuits, 302, *303*
half-wave rectifiers, 131–133, *132*
    diode applications, *118,* 118, *119*
    troubleshooting, 148–149, *149–150*
half-waving, 452
Hall effect sensors, *203,* 203–207, *204, 205, 206, 207*
handheld oscilloscopes, 31, *32*
handheld programming devices, 462
hand protection, 18
hazardous locations, 19, *20*
head protection, 17
heater control using triacs, 323, *324*
heat guns, *101,* 101–102, *102*
heat sinking, 258
heat sinks
    for diodes, *116,* 116–117, *117*
    for soldering, 79, *80*
    for solid state relays (SSRs), 440–442
    thermal resistance of, 441–442, *442*
    for transistors, 233–234, *234*
helmets, protective, 17
heterogeneous materials, 68
high-impedance differential amplifiers, 346, *347*
high-rate discharge tests, *185,* 185, 499
HMIs, 463
holes, 108
homogeneous materials, 68

horizontal control, oscilloscopes, 43, *44*
hot air pencils, *102,* 102
hot air tools, *101,* 101–102, *102*
human machine interfaces (HMIs), 463
hybrid circuits, 312, *313*
hybrid integrated circuits, 54, *55*

I

ICs, 341–343, *342, 343,* 411–413, *412, 413*
IC timers, 358–359, *359. See also* 555 timers
IEC, 16
IGBTs, *280,* 280–284, *281, 282, 283, 284*
impedances, amplifiers, *256,* 256–257, *257*
incident rays, *373,* 373–374, *374*
in-circuit transistor testers, 258
inductance, 228
infrared emitting diodes (IREDs), 376
infrared temperature meters, 29, *30*
in-line current measurements, *39,* 39–40
input circuits, 436
input impedance, *256,* 256
input sections (PLCs), 468–475, *469*
    current ratings, *470,* 471
    DC input switching, *471,* 471–473, *472, 473*
    internal circuitry, *469,* 469–470
    troubleshooting, 473–475, *474, 475*
    voltage ratings, *470,* 470
instant-on switching relays, 432, *433*
insulated gate bipolar transistors (IGBTs). *See* IGBTs
insulative brushes, 79
integrated circuits (ICs), 341–343, *342, 343,* 411–413, *412, 413*
integrated circuit (IC) timers, 358–359, *359. See also* 555 timers
integrated programming devices, 462
integrators, *356,* 356
intensity control, oscilloscopes, 42
intermittents, 125
International Electrotechnical Commission (IEC), 16
inverters
    photovoltaic systems, 164–165, *165, 166*
    wind turbines, 169, *170, 171*
inverters (logic gates). *See* NOT gates
IREDs, 376
isolation amplifiers. *See* common-collector amplifiers
isolation surge voltage ratings, 394

J

JFETs, *265,* 265–273
   applications of, *272,* 272–273, *273, 274*
   characteristic curves, *267,* 267
   circuit configurations, *268,* 268–271, *269, 270, 271*
   definition, 265
   operation, *266,* 266–267
   power dissipation, 267–268
joystick sensors, 206, *207*
junction field-effect transistors (JFETs). *See* JFETs

K

knee pads, 18
knee protection, 18

L

labels, electrostatic discharge, *5*
laboratory-standard transistor testers, 258
ladder diagrams, *10,* 10–11
lamp dimmers, 328, *329,* 337, *338*
LASCRs, *392,* 392
lasers, 382–384, *383, 384*
LCDs, *380,* 380–382, *381, 382*
lead-free solders, 87–89
lead-mounted SCRs, 304, *305*
lead solders, *84,* 84–85
leakage current, 454, *456*
leak detectors, *213,* 213
LEDs. *See* light-emitting diodes (LEDs)
light, 370–372, *371*
light-activated SCRs (LASCRs), *392,* 392
light detectors, 384–396
   bar code scanners, *395,* 395
   optocouplers, *393,* 393–394, *394, 395*
   photocells, *384,* 384–386, *385, 386, 387*
   photodiodes, 387–388, *388, 389*
   photoelectric switches, 390–393, *391, 392, 393*
   photo tachometers, *396,* 396
   PIN photodiodes, 389–390, *390*
light-emitting diodes (LEDs), 374–380, *375*
   applications, 377–378, *378*
   construction of, *376,* 376, *377*
   displays, 378, *379*
   IREDs, 376
   and nanotechnology, 378
   OLEDs, *379,* 379–380
   testing, *377,* 377
lighting control, 328, *329,* 337, *338*

light-intensity meters, *166,* 166
light-load tests, 184
light sources, 374–384
   lasers, 382–384, *383, 384*
   LCDs, *380,* 380–382, *381, 382*
   LEDs, 374–380, *375*
      applications, 377–378, *378*
      construction of, *376,* 376, *377*
      displays, 378, *379*
      IREDs, 376
      and nanotechnology, 378
      OLEDs, *379,* 379–380
      testing, *377,* 377
   PDPs, 382, *383*
limiting circuits, 118, *119,* 126, *127*
limit switches, *11,* 11
linear amplifiers, 253, *254*
linear integrated circuits (ICs), 341, *412,* 413
line diagrams, *10,* 10–11
liquid crystal displays (LCDs), *380,* 380–382, *381, 382*
liquid level sensors, 204, *205*
lithium-ion batteries, 178–180, *179*
load current, 454, *455*
load lines, 226, *227*
loads, 1–2, *2*
load-switching circuits, 436–437
lockout, *19,* 19
logic families, 422–423
logic gates, *415,* 415–420
   AND gates, *415,* 415–416, *416*
   buffer gates, *420,* 420
   NAND gates, *415,* 415, 418–419, *419*
   NOR gates, *415,* 415, 417–418, *418*
   NOT gates, *415,* 415, 417, *418, 420,* 420
   OR gates, *415,* 415, 416, *417*
   XOR gates, *415,* 415, 419, *420*
logic probes, 32, *33*
low power supply voltage, 149
L-section inductive filters, *141,* 141
L-section resistive filters, 140, *141*

M

magnetic card reader sensors, *206,* 206
main bonding jumpers (MBJs), *15,* 15
materials, diode, 107–110, *108, 109, 110*
maximum power dissipation curve, *233,* 233
maximum ratings, transistors, 258
MBJs, *15,* 15

mechanical data, transistors, 258
mechanical splices, *405,* 405
mediums, *373,* 373, *374,* 374
memory, 467–468, *468*
metal-oxide semiconductor field-effect transistors (MOSFETs). *See* MOSFETs
minimum holding current, *456,* 456
modules, 163–164, *164*
momentary power interruptions, *153,* 153
monostable multivibrators, 361, *362*
MOSFETs, 274–279, *275, 276, 277, 278, 279*
motor starters, 457–458, *458, 481,* 482
multilayered PC boards, 52–53, *53*
multimeters, 26–27, *27. See also* digital multimeters (DMMs)
multistage amplifiers, 259–262, *260–263,* 284–288, *286–289*
multivibrators, 361, *362,* 364–365, *365, 366*

N

NAND (negated AND) gates, *415,* 415, 418–419, *419*
nanotechnology, 163
narrow bandwidth, 245
National Electrical Code® (NEC®), 12
National Fire Protection Association (NFPA), 11
NEC®, 12
negative-resistance regions, *333,* 333
negative temperature coefficient (NTC) thermistors, *195,* 195–197, *196, 197, 198*
NFPA, 11
NFPA 70E®, 16
noncontact temperature probes, 29, *30*
no output voltage, 149, *150*
NOR (negated OR) gates, *415,* 415, 417–418, *418*
normals, *373,* 373, *374*
NOT gates, *415,* 415, 417, *418, 420,* 420
NPN transistors, *217,* 217
NTC thermistors, *195,* 195–197, *196, 197, 198*
N-type material, *108,* 108
nulling, *351,* 351
numerical aperture, 399

O

OCPDs, 1–2, *2*
off-grid wind turbines, *168,* 168

offset error, 350
ohmmeters
    diac testing, 329, *330*
    triac testing, *326*
OLEDs, *379*, 379–380
180° phase shift, 248–249, *249, 250*
one-line diagrams, *10*, 10
operating characteristic curves.
        See transistors: collector characteristic curves
operating classes, amplifiers, 255, *256*
operating current, 454, *456*
operating points, *253*, 253, *254*
operational amplifiers (op amps), 343–358, *344*
    closed-loop control, *349*, 349
    as comparators, *358*, 358
    and compensation, *350*, 350
    in current-to-voltage converters, 351–352, *352, 353*
    definition, 343
    in DMMs, *357*, 357
    feedback inverting amplifiers, *349*, 349–350, *350, 351*
    and gain, *347*, 347–348
    high-impedance differential amplifiers, 346, *347*
    as integrators, *356*, 356
    nulling, *351*, 351
    offset error, 350–351, *351*
    open-loop control, 349
    operation of, *346*, 346
    schematic symbols, *344*, 344
    slew rate, 351
    in thermostat control circuits, *355*, 355–356
    voltage sources, *345*, 345–346
    in voltage-to-current converters, 353, *354*
operator interface panels, 463
optical fibers, *398*, 398
optical interrupters, *394*, 394
optical time domain reflectometers (OTDRs), *33*, 33
optocouplers, *393*, 393–394, *394, 395*
organic light-emitting diodes (OLEDs), *379*, 379–380
OR gates, *415*, 415, 416, *417*
oscilloscopes, 30–31, *32*, 40, *41*
    AC voltage measurement, 45–46, *46*
    bench, 31, *32*
    calibration, *127*, 127
    DC voltage measurement, 46–47, *47*
    diac testing, 329, *330*
    dual-trace, 42, *43*

frequency measurement, 47–48, *48*
graphical displays, 40, *42*, 42–43, 45
handheld, 31, *32*
triac testing, 326
zener diode testing, *125*, 125
OTDRs, *33*, 33
outages, *153*, 154
output circuits, 436–437
output impedances, *256*, 256
output sections (PLCs), 475–480, *476*
    current ratings, 478
    internal circuitry, 476, *477*
    troubleshooting, 478–480, *479, 480*
    voltage ratings, *477*, 477
overcurrent protection devices (OCPDs), 1–2, *2*

P

PAFCs, 189, *190*
paper sensors, *207*, 207
parallel communication, *424*, 424–425
PC boards. *See* printed circuit (PC) boards
PCs, 463
PDPs, 382, *383*
peak current points, *333*, 333
peak inverse voltage (PIV), 113, 138–139, 142, *143*
peak switching relays, 432, *434*, 434
peak-to-peak voltage (V_{p-p}), 136–137, *137*
peak voltage (V_p), 136–137, *137*
pedestal voltage, 336–337
PEL, 212
PEMFCs, *190*, 190
permissible exposure limit (PEL), 212
personal computers (PCs), 463
personal protective equipment (PPE), 16–19, *17*
phase control, 302
phase inversion, common-emitter, 248–249, *249, 250*
phosphoric acid fuel cells (PAFCs), 189, *190*
photocells, *384*, 384–386, *385, 386, 387*
photodarlingtons, *392*, 392
photodiodes, 387–388, *388, 389*
photoelectric switches, 390–393, *391, 392, 393*
photonics, *369*, 369–370
photons, 369
photo tachometers, *396*, 396
phototransistors, 390–392, *391*
phototriacs, *393*, 393

photovoltaic cells, 159–166, *160*
    applications of, 165–166, *166*
    and nanotechnology, 163
    operation of, 160–161, *161, 163*
    output of, 161–163, *162, 163, 164*
    photovoltaic effect, 160–161, *161, 164*
    photovoltaic systems, 163–165, *164, 165, 166*
photovoltaic effect, 160–161, *161*
photovoltaic systems, 163–165, *164, 165, 166*
pictorial drawings, *5*, 5
piezoelectric effect, 201–202, *202*
piezoelectricity, 201–202
piezoelectric sensors, 201–202, *202, 203*
PIN photodiodes, 389–390, *390*
pi-section filters, *141*, 141
PIV, 113, 138–139, 142, *143*
plasma display panels (PDPs), 382, *383*
plated-holes, 52
PLC applications, *463*, 463–464
PLC interfacing circuits, 480–482, *481*
PLCs, 461, *462*
PLC sections, *464*, 464–480, *465*
    input sections, 468–475, *469*
        current ratings, 470, *471*
        DC input switching, *471*, 471–473, *472, 473*
        internal circuitry, *469*, 469–470
        troubleshooting, 473–475, *474, 475*
        voltage ratings, *470*, 470
    output sections, 475–480, *476*
        current ratings, 478
        internal circuitry, 476, *477*
        troubleshooting, 478–480, *479, 480*
        voltage ratings, *477*, 477
    processor sections, *466*, 466–468, *467, 468*
PNP transistors, *217*, 217
polarizing filters, 381
positioning, horizontal and vertical, 43, *44*
positive temperature coefficient (PTC) thermistors, 195, *198*, 198–199, *199, 200*
pouch cells, 180, *181*
power dissipation, *233*, 233–234, *234*
power gain, 244
power interruptions, 152–154, *153*, 156
power MOSFETs, 278–279, *279*
power rating, zener diodes, 124–125
power sources, 1–2, *2*
power supplies, *131*, 131–147
    filters, 139–142, *140, 141, 143*
    rectifiers, 131–139, *132*

full-wave, 133, *134*
full-wave bridge, 134–135, *135*
half-wave, 131, *132, 133*, 133
voltage measurements, 135–139, *136*
regulators, 142–144, *143, 144*
transformers, 131
troubleshooting, 148–152, *149, 150, 151, 152*
voltage dividers, 144–145, *145*
voltage multipliers, 145–148, *146, 147, 148*
power supply filters, 139–142, *140, 141, 143*
power supply protection circuits, 316, *318*
PPE, 16–19, *17*
preamplifiers, *272*, 272
precision heat control (SCRs), 309–310, *310, 311*, 312
precision temperature controllers, 335–337, *336*
press-fit SCRs, 305, *306*
press-pack SCRs, 306–307, *307*
primary cells, 176, *178*
printed circuit (PC) boards
bar codes on, *62*, 62
building, 56–58, *57, 58, 59*
and component substitution troubleshooting, 66–67, *67*
conformal coatings, 68–71, *69, 70, 71*
defined, *51*, 51
designing, 54–56, *55*
double-sided, 52, *53, 56, 57, 58*
field repairs, 73–77, *74, 75, 76, 77, 78*
flexible, 54, *55*
handling, *63*, 63
hybrid integrated circuits, 54, *55*
information on, 62
manufacturing of, 54–59
materials, *52*, 52
mounting components on, *60*, 60–61, *61*
multilayered, 52–53, *53*
quality assurance tests, 58–59, *59*
recycling, *67*, 67–68
repair stations, 92, *93*
and signal tracing, 64
single-sided, 52, *53*
soldering procedure, 94
swapping of, *66*, 66
testing, 58–59, *59*
troubleshooting, *63*, 63–67, *64, 65, 66, 67*
visual inspection of, 64–65, *65*
prismatic cells, *181*, 182
prisms, *373*

processor sections (PLCs), *466*, 466–468, *467, 468*
programmable logic controllers (PLCs), 461, *462*
propagation, 370
protection
against electrical shock, 14–16, *15, 16*
with PPE, 16–19, *17*
with rubber insulating matting, 18–19
protective clothing, 17
protective helmets, 17
proton-exchange-membrane fuel cells (PEMFCs), *190*, 190
proximity sensors, 454, *455, 456*, 456, *457*
PTC thermistors, 195, *198*, 198–199, *199, 200*
P-type material, 108–109, *109*
pull-down resistors, 420–421, *421*
pull-up resistors, 420–421, *421*
pulse amplitude, *425*, 425
pulsed radar level sensors, 208–209, *209*
pulse duration, *425*, 425
pulse intervals, *425*, 425
pulse repetition rates, 425
pulse separation, *425*, 425
pulse trains, *425*, 425
pulse width, *425*, 425
pulse width modulation (PWM), 169, *171*

Q

qualified persons, 12–13
quantum efficiency, 377
quick charging, 187
quiescent (Q) points, *253*, 253, *254*

R

radar level sensors, 208–211, *209, 210, 211*
radiation detectors, 214, *215*
random access memory (RAM), 468, 503
receptacle testers, *24*, 24
rechargeable batteries, 177–180, *179*
rectification, 117–118
rectifiers, 117–118, 131–139, *132*
full-wave, 133, *134*
full-wave bridge, 134–135, *135*
half-wave, 131, *132, 133*, 133
voltage measurements, 135–139, *136*
reflection, *374*, 374
reflector modules, 394, *395*

refraction, *373*, 373–374, *374*
refractive index, 398
regulated power supply, 142
regulated soldering stations, 92
relative mode tests, 185–186, *186*
relays, 431. *See also* solid state relays (SSRs)
residual current, 454, *456*
resistance measurements, 37–38, *38*
Restriction of Hazardous Substances (RoHS) Directive, 68
reverse-bias voltage, *112*, 112
reversing contactors, replacement, 323, *324*
rework stations, 90, *91*
ripple voltage, 139–140, *140*
rise time, digital pulses, 426, *427*
RoHS Directive, 68
root mean square voltage (V_{rms}), 136–138, *137*
RPM sensors, 204, *205*
rubber insulating matting, 18–19

S

safety
electrical measurement, 34–35
fire, 19
test instrument, 34–35
voltage protection, 40, *41*
safety glasses, 18
safety labels, *13*, 13
satellite power sources, *166*, 166
saturation, UJT, *333*, 333
saturation region, 225–226, *226*
schematic diagrams, *9*, 9
SCRs. *See* silicon-controlled rectifiers (SCRs)
SDSs, 14
secondary cells, 176–180, *179*
security alarm circuits, *363*, 363
security alarm system circuits, 312–314, *314*
security lights, *325*, 325
semiconductor lasers, *384*, 384
semiconductors, 107, 128
separately derived systems (SDSs), 14
serial communication, *424*, 424–425
series/parallel connections, proximity sensors, 456, *457*
series regulators, 126, 142–143, *143*
seven-segment displays, *238*, 238, 378, *379*
shock, 13–14, *14*
shunt regulators, *126*, 126, 143–144, *144*

516 SOLID STATE DEVICES AND SYSTEMS

signal changes, *254,* 254–255, *255*
signal (function) generators, 32, *33*
signal tracing, 32
silicon-controlled rectifiers (SCRs), *291,* 291–318
  as AC switches, 301, *302*
  applications, 292, 309–318
    adjustable temperature control, 314, *315*
    hybrid circuits, 312, *313*
    power supply protection, 316, *318*
    precision heat control, 309–310, *310, 311,* 312
    security alarm systems, 312–314, *314*
    solid state starting, 314, *315, 316,* 316, *317*
    time-delayed relays, 312, *313*
  characteristic curves, 292–293, *293*
  construction of, *296,* 296–298, *297*
  cooling, 304–307, *305, 306, 307*
  as DC switches, 298–299, *299, 300, 301,* 301
  definition, 291
  equivalent circuits, *298,* 298, *299*
  flat-pack SCRs, 306, *307*
  gate control, 292
  lead-mounted SCRs, 304, *305*
  mounting, 304–307, *305, 306, 307*
  operating states, 292
  phase control, 301–302, *303,* 304
  press-fit SCRs, 305, *306*
  press-pack SCRs, 306–307, *307*
  SCR packages, *292*
  stud-mounted SCRs, 305, *306*
  triggering methods, 293–296, *294, 295*
  troubleshooting, 307–309, *308, 310*
silk-screen layers, *62,* 62
single-phase motor starters, 322–323, *323*
single-sided PC boards, 52, *53*
single-use batteries, *178*
smart chargers, 188
SMDs. *See* surface mount devices (SMDs)
smoke detectors, *214,* 214
smoke switches, *214,* 214
SMT, *60,* 60–61, 96. *See also* surface mount devices (SMDs)
SOFCs, 189, *190*
soft starting and stopping, *459,* 459
solar cells. *See* photovoltaic cells
solder fluxes, *85,* 85–87, *86, 87*
soldering
  aids to, 77–79, *78, 79, 80*
  avoiding bad solder joints, 90
  definition, 80

and safety, 90
  technician effects on, 93–94
soldering irons, *91*
soldering iron stands, *91*
soldering iron tips, 80–84, *81, 82, 83*
soldering process, 80, *81*
soldering stations, 90–93, *91, 93*
soldering temperature, 91
solder mask, *51,* 51
solder pads, *51,* 51
solders, 80, *84,* 84–85, *87,* 87–89
solder wicks, *98,* 98
solid oxide fuel cells (SOFCs), 189, *190*
solid state relay (SSR) circuits, 434–440, *435*
  control circuits, 436
  input circuits, 436
  load-switching circuits, 436–437
  NC contact simulation, 438, *439*
  output circuits, 436–437
  parallel control, 439–440, *440*
  series control, 439–440, *441*
  three-wire memory control, 437–438, *438*
  transistor control, 438, *439*
  two-wire control, *437,* 437
solid state relays (SSRs), 431–434, *432–435, 481,* 481–482
  current problems, *443,* 443
  and heat sinks, 440–442, *442*
  and motor starters, 457–458, *458*
  and proximity sensors, 454, *455, 456,* 456, 457
  temperature problems, 440, *441*
  troubleshooting, 450–452, *451, 453*
  voltage problems, *443,* 443–445, *444, 445*
solid state starting (SCRs), 314, *315, 316,* 316, *317*
solid state switches, 445–450, *446, 447, 448, 449, 450*
specification sheets, transistors, 258
specific gravity tests, 183–184
speed of light, 372–373, *373*
spiral-wound cylindrical cells, 180–182, *181*
SSRs, *481,* 481–482
stand-alone inverters, 164, *165*
stand-alone wind turbines, *168,* 168
star couplers, *403,* 403
static electricity, 2
status indicators, *378,* 378
stud-mounted SCRs, 305, *306*
subtractive (etch-down) circuitry creation, 56

surface mount devices (SMDs), *61, 61,* 94–97, 99–103
  definition, 94
  desoldering, 99–103, *100, 101, 102, 103*
  soldering, 94–97, *95, 96, 97*
surface mount technology (SMT), *60,* 60–61, 96. *See also* surface mount devices (SMDs)
surge currents, 140
surge suppressors, *155,* 155
sustained power interruptions, *153,* 154
switches, 445–450, *446, 447, 448, 449, 450*
switching applications, transistors, 238–240, *238, 239, 240*
switching using logic gates, 422, *423*
symbols, DMM, 35, *36*
symbols, electrical, 5

T

tachometer generators, 323
tagout, *19,* 19
T-couplers, *402,* 402
TDRs, 32, 210, *211*
temperature, soldering stations, 91
temperature coefficient, 125
temperature-controlled soldering stations, 92
temperature controllers, 335–337, *336*
temperature probes, 29, *30*
temporary power interruptions, *153,* 154, 505
testing diodes, 113–115, *114, 115*
test instruments, 23–48
  analog multimeters, 26–27, *27*
  branch circuit identifiers, *25,* 25
  clamp-on ammeters, 27–28, *28*
  continuity testers, 23–24, *24*
  digital logic probes, 32, *33*
  digital multimeters (DMMs), *27,* 27, 35
  frequency meters, 30, *31*
  multimeters, 26–27, *27*
  optical time domain reflectometers (OTDRs), *33,* 33
  oscilloscopes, 30–31, *32,* 40, *41*
  receptacle testers, *24,* 24
  and safety, 34–35
  signal (function) generators, 32, *33*
  temperature probes, 29, *30*
  visual inspection instruments, 28, *29*
  voltage indicators, *25,* 25–26
  voltage testers, *26,* 26

thermal instability, 229, *230*
thermal shunts, for soldering, 79
thermistors, 193, *194*, 195–201
　bias stabilization, *232*, 232
　NTC applications, *195*, 195–197, *196*, *197*, *198*
　PTC applications, 195, *198*, 198–199, *199*, *200*
　resistance values, 195
　schematic symbols, 195
　temperature resistance, 195
　troubleshooting, 200–201, *201*
　types of, 193, *194*
thermostat control circuits, *355*, 355–356
thickness monitors, 364, *365*
thin-film batteries, 177–178, *179*
three-wire solid state switches, 448–449, *449*, *450*
through-hole technology, *60*, 60, *99*, 99
through-hole vias, *53*, 53
tilt sensors, 206, *207*
tilt switches, 365, *366*
time-controlled chargers, 188
time-delayed relays, 312, *313*
time-dependent light dimmers, 337, *338*
time/division (time per division), *45*, 45
time/division control, oscilloscopes, *45*, 45
time-domain reflectometers (TDRs), 32, 210, *211*
tolerances, zener diodes, *124*, 124
TO numbers, *218*, 218
total internal reflection, *398*, 398
toxic gas detectors, 212–213, *213*
trace, oscilloscope, *42*, 42, *43*
traces, *51*, 51
transducers
　definition, 193
　gas detectors, 211–214, *212*, *213*, *214*
　Hall effect sensors, *203*, 203–207, *204*, *205*, *206*, *207*
　piezoelectric sensors, 201–202, *202*, *203*
　radar level sensors, 208–211, *209*, *210*, *211*
　radiation detectors, 214, *215*
　thermistors, 193, *194*, 195–201
　　NTC applications, *195*, 195–197, *196*, *197*, *198*
　　PTC applications, 195, *198*, 198–199, *199*, *200*
　　resistance values, 195
　　schematic symbols, 195
　　temperature resistance, 195

　　troubleshooting, 200–201, *201*
　　types of, 193, *194*
　ultrasonic sensors, 207–208, *208*, *209*
transformer coupling, 260, *262*, 287, *288*
transformers, 131, 171, *172*
transient voltages, 154, 443
transistor current gain, 224, 244
transistor operating characteristic curves. *See* transistors: collector characteristic curves
transistor outline (TO) numbers, *218*, 218
transistors. *See also* BJTs; IGBTs; JFETs; MOSFETs; UJTs
　as amplifiers. *See* amplifiers
　biasing, 226–232
　collector characteristic curves, *222*, 222–225, *223*, *224*, *225*
　and power dissipation, *233*, 233–234, *234*
　servicing of, 259
　specification sheets for, 258
　testers, 258
　testing of, 234–237
　　for opens and shorts, 234–235, *235*, *236*
　　switches, 236–237, *237*
　voltages in, *221*, 221
transistor switches, *225*, 225–226, 236–237, *237*. *See also* switching applications, transistors
transistor-transistor logic (TTL), 422
transistor voltages, *221*, 221
triac characteristic curves, *322*, 322
triacs, *319*, 319–327
　in automatic security lights, *325*, 325
　characteristic curves, *322*, 322
　construction of, 319–320, *320*
　for heater control, 323, *324*
　in single-phase motor starters, 322–323, *323*
　operation of, 320, *321*
　as reversing contactors, 323, *324*
　testing, *326*, 326–327
triboelectric effect, 78
trickle charging, 187
triggering circuits, *333*, 333–334, *334*
tripped circuit breakers, 148, *149*
troubleshooting
　batteries, *184*
　DC power supplies, 148–152, *149*, *150*, *151*, *152*
　digital electronics, 427–428, *429*
　fiber-optic cables, *407*, 407–408

　full-wave bridge rectifiers, 151, *152*
　full-wave rectifiers, 150, *151*
　half-wave rectifiers, 148–149, *149*, *150*
　IGBTs, 282–283, *283*
　input sections (PLCs), 473–475, *474*, *475*
　integrated circuits (ICs), 342–343, *343*
　light-sensing circuits, 307–309
　output sections (PLCs), 478–480, *479*, *480*
　PC boards, *63*, 63–67, *64*, *65*, *66*, *67*
　PLC systems, 482–486, *483*, *484*, *485*, *486*
　SCRs, 307–309, *308*, *310*
　SSRs, 450–452, *451*, *453*
　thermistors, 200–201, *201*
　zener diode regulators, 152, *153*
truth tables, *414*, 414
2D matrix bar codes, *62*, 62
two-wire solid state switches, 445–448, *446*, *447*, *448*

U

UJTs, 330–338, *331*
　biasing, *332*, 332
　characteristic curves, *332*
　equivalent circuits of, *331*, 331
　in emergency flashers, *335*, 335
　in lamp dimmers, 337, *338*
　and saturation, *333*, 333
　in temperature controllers, 335–337, *336*
　theory of operation, 333
　as triggering circuits, *333*, 333–334, *334*
ultrasonic sensors, 207–208, *208*, *209*
unijunction transistors (UJTs). *See* UJTs
uninterruptible power supply (UPS), *156*, 156
universal motor speed controllers, 328, *329*
unregulated power supplies, 142
unregulated soldering stations, 91–92
UPS, *156*, 156
utility-interactive inverters, 165, *166*

V

valley current points, *333*, 333
vapor, 214
VDV communications, 408, *409*
vertical control, oscilloscopes, 43, *44*

vias, *53*, 53
visual inspection test instruments, 28, *29*
voice-data-video (VDV) communications, 408, *409*
voltage dividers, 144–145, *145*, *231*, 231
voltage doublers, *147*, 147, *148*
voltage drop, SSRs, 443–445, *444*, *445*
voltage gain, 244
voltage indicators, *25*, 25–26
voltage limiters, 118, *119*, 126, *127*
voltage measurements
   AC, *37*, 37, 45–46, *46*
   DC, 37, *38*, 46–47, *47*
   with DMMs, *37*, 37, *38*
   with oscilloscopes, 45–47, *46*, *47*
   rectifiers, 135–139
voltage multipliers, 145–147, *146*, *147*, *148*
voltage protection, DMMs, 40
voltage ratings, zener diodes, *124*, 124
voltage regulators, *126*, 126, 142–144, *143*, *144*
voltages, transistors, *221*, 221
voltage spikes, 154
voltage testers, *26*, 26
volt/division control, oscilloscopes, 43, *44*

W

Waste Electrical and Electronic Equipment (WEEE) Directive, 68
water level detectors, *239*, 239
wattage, soldering stations, 91
WEEE Directive, 68
wet cells, *173*, 173–174, *174*
wide bandwidth, 245
wind turbines, *167*, 167–173
   converters, 169–170, *170*, *171*
   electronic control devices, 169
   grid-connected wind turbines, 167, *168*
   instruments, 171, *172*
   inverters, 169, *170*, *171*
   power output, 172–173
   stand-alone wind turbines, *168*, 168
   voltage controllers, 171, *172*
wind vanes, 171, *172*
wiring diagrams, *9*, 9
workstations. *See* soldering stations; rework stations; electrostatic discharge (ESD) workstations

X

XOR (exclusive OR) gates, *415*, 415, 419, *420*

Z

ZAFCs, 189, *190*
zener clipping, 126–127, *127*
zener diode regulators, *144*, 144, 152, *153*
zener diodes, *122*, 122–127
   applications, 125–127
   avalanche voltage, *123*, 123
   characteristic curves, *123*, 123
   in clipping circuits, 126–127, *127*
   for oscilloscope calibration, *127*, 127
   operation of, *123*, 123
   power rating, 124–125
   series connected, *124*, 124
   temperature coefficient, 125
   testing, *125*, 125
   tolerances, *124*, 124
   voltage rating, *124*, 124
   as voltage regulators, *126*, 126
zero switching relays, 432–433, *433*
zinc-air fuel cells (ZAFCs), 189, *190*

USING THE *SOLID STATE DEVICES AND SYSTEMS* INTERACTIVE CD-ROM

Before removing the Interactive CD-ROM from the protective sleeve, please note that the book cannot be returned for refund or credit if the CD-ROM sleeve seal is broken.

Windows System Requirements

To use this CD-ROM on a Windows® system, your computer must meet the following minimum system requirements:
- Microsoft® Windows® 7, Windows Vista®, or Windows® XP operating system
- Intel® 1.3 GHz processor (or equivalent)
- 128 MB of available RAM (256 MB recommended)
- 335 MB of available hard disk space
- 1024 × 768 monitor resolution
- CD-ROM drive (or equivalent optical drive)
- Sound output capability and speakers
- Microsoft® Internet Explorer® 6.0 or Firefox® 2.0 web browser
- Active Internet connection required for Internet links

Macintosh System Requirements

To use this CD-ROM on a Macintosh® system, your computer must meet the following minimum system requirements:
- Mac OS X 10.5 (Leopard) or 10.6 (Snow Leopard)
- PowerPC® G4, G5, or Intel® processor
- 128 MB of available RAM (256 MB recommended)
- 335 MB of available hard disk space
- 1024 × 768 monitor resolution
- CD-ROM drive (or equivalent optical drive)
- Sound output capability and speakers
- Apple® Safari® 2.0 web browser or later
- Active Internet connection required for Internet links

Opening Files

Insert the Interactive CD-ROM into the computer CD-ROM drive. Within a few seconds, the home screen will be displayed allowing access to all features of the CD-ROM. Information about the usage of the CD-ROM can be accessed by clicking on Using This Interactive CD-ROM. The Quick Quizzes®, Illustrated Glossary, Media Clips, Flash Cards, Virtual DMM, Interactive Logic Gates, Interactive Troubleshooting Simulations, Chapter Reviews, and ATPeResources.com can be accessed by clicking on the appropriate button on the home screen. Clicking on the ATP website button (www.go2atp.com) accesses information on related educational products. Unauthorized reproduction of the material on this CD-ROM is strictly prohibited.

Microsoft, Windows, Windows Vista, PowerPoint, and Internet Explorer are either registered trademarks or trademarks of Microsoft Corporation in the United States and/or other countries. Adobe, Acrobat, and Reader are either registered trademarks of Adobe Systems Incorporated in the United States and/or other countries. Intel is a registered trademark of Intel Corporation in the United States and/or other countries. Firefox is a registered trademark of Mozilla Corporation in the United States and other countries. Apple, Macintosh, and Safari are registered trademarks of Apple, Inc. PowerPC is a registered trademark of International Business Machines Corporation. Quick Quiz, Quick Quizzes, and Master Math are either registered trademarks of American Technical Publishers, Inc.